NEUROMETHODS

Series Editor
Wolfgang Walz
University of Saskatchewan
Saskatoon, SK, Canada

For further volumes:
http://www.springer.com/series/7657

Neuromethods publishes cutting-edge methods and protocols in all areas of neuroscience as well as translational neurological and mental research. Each volume in the series offers tested laboratory protocols, step-by-step methods for reproducible lab experiments and addresses methodological controversies and pitfalls in order to aid neuroscientists in experimentation. *Neuromethods* focuses on traditional and emerging topics with wide-ranging implications to brain function, such as electrophysiology, neuroimaging, behavioral analysis, genomics, neurodegeneration, translational research and clinical trials. *Neuromethods* provides investigators and trainees with highly useful compendiums of key strategies and approaches for successful research in animal and human brain function including translational "bench to bedside" approaches to mental and neurological diseases.

Clinical Trials In Parkinson's Disease

Edited by

Santiago Perez-Lloret

National Research Council (CAECHIS-UAI), Buenos Aires, Argentina

 Humana Press

Editor
Santiago Perez-Lloret
National Research Council (CAECHIS-UAI)
Buenos Aires, Argentina

ISSN 0893-2336   ISSN 1940-6045   (electronic)
Neuromethods
ISBN 978-1-0716-0914-9   ISBN 978-1-0716-0912-5   (eBook)
https://doi.org/10.1007/978-1-0716-0912-5

This Humana imprint is published by the registered company Springer Science+Business Media, LLC, part of Springer Nature.
The registered company address is: 1 New York Plaza, New York, NY 10004, U.S.A.

Dedication

To my wife and children, whose endless support was essential for the completion of this book.

Preface to the Series

Experimental life sciences have two basic foundations: concepts and tools. The *Neuromethods* series focuses on the tools and techniques unique to the investigation of the nervous system and excitable cells. It will not, however, shortchange the concept side of things as care has been taken to integrate these tools within the context of the concepts and questions under investigation. In this way, the series is unique in that it not only collects protocols but also includes theoretical background information and critiques which led to the methods and their development. Thus it gives the reader a better understanding of the origin of the techniques and their potential future development. The *Neuromethods* publishing program strikes a balance between recent and exciting developments like those concerning new animal models of disease, imaging, in vivo methods, and more established techniques, including, for example, immunocytochemistry and electrophysiological technologies. New trainees in neurosciences still need a sound footing in these older methods in order to apply a critical approach to their results.

Under the guidance of its founders, Alan Boulton and Glen Baker, the *Neuromethods* series has been a success since its first volume published through Humana Press in 1985. The series continues to flourish through many changes over the years. It is now published under the umbrella of Springer Protocols. While methods involving brain research have changed a lot since the series started, the publishing environment and technology have changed even more radically. Neuromethods has the distinct layout and style of the Springer Protocols program, designed specifically for readability and ease of reference in a laboratory setting.

The careful application of methods is potentially the most important step in the process of scientific inquiry. In the past, new methodologies led the way in developing new disciplines in the biological and medical sciences. For example, Physiology emerged out of Anatomy in the nineteenth century by harnessing new methods based on the newly discovered phenomenon of electricity. Nowadays, the relationships between disciplines and methods are more complex. Methods are now widely shared between disciplines and research areas. New developments in electronic publishing make it possible for scientists that encounter new methods to quickly find sources of information electronically. The design of individual volumes and chapters in this series takes this new access technology into account. Springer Protocols makes it possible to download single protocols separately. In addition, Springer makes its print-on-demand technology available globally. A print copy can therefore be acquired quickly and for a competitive price anywhere in the world.

Saskatoon, SK, Canada   *Wolfgang Walz*

Preface

Parkinson's disease is the second most frequent neurodegenerative disorder after Alzheimer's disease. Parkinson's disease is considered by many authors as a pandemic. Indeed, the prevalence of this disease, and thus disability and deaths owing to it, more than doubled from 1990 to 2015. The Global Burden of Disease Study estimated that 6.2 million individuals had Parkinson's disease in 2018. The number of affected persons will likely reach 14.2 million in 2040.

Parkinson's disease seriously affects quality of life. Indeed, the disease has a profound and progressive impact on various neurological functions. Motor symptoms such as bradykinesia, rigidity, and tremor are the cardinal symptoms of the disease. Notwithstanding, quality of life is more affected by non-motor symptoms. Patients with Parkinson's disease are frequently affected by cognitive impairment, mood disorders, autonomic disturbances, troubled sleep, and sensory symptoms. Therefore, both motor and non-motor symptoms need to be treated in order to achieve satisfactory quality of life in patients with Parkinson's disease.

Randomized, controlled, double-blinded clinical trials are the cornerstone of the assessment of drugs and medical devices efficacy and safety. Randomization reduces the possibility of selection bias, by assigning trial participants to treatments groups on a random basis. Double blind permits effective allocation concealment and reduces the risk of performance, attrition, and detection bias. Respectively, these biases occur when there are systematic differences between groups in the care that is provided or in exposure to factors other than the interventions of interest, or in withdrawals from a study, or in how outcomes are determined.

This book reviews major clinical trials for motor and non-motor symptoms in Parkinson's disease and discusses their most important aspects, including study designs, sample selection, and outcome selection. Therefore, it aids clinicians and researchers to better interpret results from clinical trials in Parkinson's disease and to design their own high-quality trials.

Buenos Aires, Argentina   *Santiago Perez-Lloret*

Contents

Contributors

AMY W. AMARA • *Division of Movement Disorders, Department of Neurology, University of Alabama at Birmingham, Birmingham, AL, USA*

AZMAN ARIS • *Maurice Wohl Clinical Neuroscience Institute and NIHR Biomedical Research Centre, Institute of Psychiatry, Psychology and Neuroscience, King's College, London, UK*

ORLANDO ARTAVIA • *Burrell College of Osteopathic Medicine, Eastern New Mexico Medical Center, Roswell, NM, USA*

MAREK BALAZ • *Department of Neurology, St. Anne's University Hospital, Masaryk University, Brno, Czech Republic*

BRANDON BARTON • *Parkinson Disease and Movement Disorder Section, Department of Neurological Sciences, Rush University Medical Center, Chicago, IL, USA; Neurology Service, Jesse Brown VA Medical Center, Chicago, IL, USA*

MANOLO CARTA • *Department of Biomedical Sciences, Section of Physiology, University of Cagliari, Cittadella Universitaria, Monserrato, Italy*

M. ANGELA CENCI NILSSON • *Basal Ganglia Pathophysiology Unit, Department of Experimental Medical Science, Lund University, Lund, Sweden*

LANA CHAHINE • *Department of Neurology, University of Pittsburgh, Pittsburgh, PA, USA*

K. RAY CHAUDHURI • *Maurice Wohl Clinical Neuroscience Institute and NIHR Biomedical Research Centre, Institute of Psychiatry, Psychology and Neuroscience, King's College, London, UK; National Parkinson Foundation, Centre of Excellence, Kings College Hospital, London, UK*

JAMES A. G. CRISPO • *Department of Neurology, University of Pennsylvania Perelman School of Medicine, Philadelphia, PA, USA; Division of Human Sciences, Northern Ontario School of Medicine, Sudbury, ON, Canada*

MARISSA N. DEAN • *Division of Movement Disorders, Department of Neurology, University of Alabama at Birmingham, Birmingham, AL, USA*

JOAQUIM J. FERREIRA • *Laboratory of Clinical Pharmacology and Therapeutics, Faculdade de Medicina, Universidade de Lisboa, Lisboa, Portugal; Instituto de Medicina Molecular, Lisboa, Portugal; Campus Neurológico Sénior (CNS), Torres Vedras, Portugal*

URBAN M. FIETZEK • *Department of Neurology, Ludwig-Maximilians-Universität München, Munich, Germany; Department of Neurology and Clinical Neurophysiology, Schön Klinik München Schwabing, Munich, Germany*

MATHIS GROSSMANN • *Department of Medicine, University of Melbourne, Melbourne, VIC, Australia; Department of Endocrinology, Austin Health, Melbourne, VIC, Australia*

PAOLO GUBELLINI • *Aix-Marseille Univ., CNRS (Centre National de la Recherche Scientifique), IBDM (Institut de Biologie du Développement de Marseille) UMR7288, Marseille, France*

PHILIPPE KACHIDIAN • *Aix-Marseille Univ., CNRS (Centre National de la Recherche Scientifique), IBDM (Institut de Biologie du Développement de Marseille) UMR7288, Marseille, France*

HORACIO KAUFMANN • *Department of Neurology, Dysautonomia Center, New York University School of Medicine, New York, NY, USA*

ZUZANA KOSUTZKA • *Second Department of Neurology, Faculty of Medicine, Comenius University, Bratislava, Slovakia*

DANIEL KREWSKI • *School of Epidemiology and Public Health, University of Ottawa, Ottawa, ON, Canada; Risk Sciences International, Ottawa, ON, Canada*

MÓNICA M. KURTIS • *Movement Disorders Unit, Neurology Department, Hospital Ruber Internacional, Madrid, Spain*

SHEN-YANG LIM • *Division of Neurology, Department of Medicine, Faculty of Medicine, University of Malaya, Kuala Lumpur, Malaysia; The Mah Pooi Soo and Tan Chin Nam Centre for Parkinson's and Related Disorders, University of Malaya, Kuala Lumpur, Malaysia; Neurology Laboratory, University of Malaya Medical Centre, Kuala Lumpur, Malaysia*

DONALD MATTISON • *School of Epidemiology and Public Health, University of Ottawa, Ottawa, ON, Canada; Risk Sciences International, Ottawa, ON, Canada*

TIAGO A. MESTRE • *Parkinson's Disease and Movement Disorders Center, Division of Neurology, Department of Medicine, The Ottawa Hospital Research Institute, University of Ottawa, Ottawa, ON, Canada*

JOSE-ALBERTO PALMA • *Department of Neurology, Dysautonomia Center, New York University School of Medicine, New York, NY, USA*

ISABEL PAREÉS • *Movement Disorders Unit, Neurology Department, Hospital Ruber Internacional, Madrid, Spain; Movement Disorders Unit, Neurology Department, Hospital Universitario Ramón y Cajal, Madrid, Spain*

SANTIAGO PEREZ-LLORET • *Center for Biomedical Research (CAECHIS), Universidad Abierta Interamericana, CONICET, Buenos Aires, Argentina; Department of Physiology, School of Medicine, University of Buenos Aires (UBA), Buenos Aires, Argentina; School of Medical Sciences, Pontifical Catholic University of Argentina (UCA), Buenos Aires, Argentina*

OLIVIER RASCOL • *Department of Clinical Pharmacology and Neurosciences, University Hospital and University of Toulouse 3, Toulouse, France; INSERM CIC1436 and UMR825, Toulouse, France*

MARIA VERONICA REY • *Center for Biomedical Research (CAECHIS), Universidad Abierta Interamericana, CONICET, Buenos Aires, Argentina*

ALEXANDRA RIZOS • *Parkinson's Foundation Centre of Excellence, King's College Hospital, London, UK; Maurice Wohl Clinical Neuroscience Institute and NIHR Biomedical Research Centre, Institute of Psychiatry, Psychology and Neuroscience, King's College Hospital, King's College, London, UK*

CARMEN RODRIGUEZ-BLAZQUEZ • *National Center of Epidemiology and CIBERNED, Institute of Health Carlos III, Madrid, Spain*

FRANCESCA ROSSI • *Department of Biomedical Sciences, Section of Physiology, University of Cagliari, Cittadella Universitaria, Monserrato, Italy*

KATARINA RUKAVINA • *Maurice Wohl Clinical Neuroscience Institute and NIHR Biomedical Research Centre, Institute of Psychiatry, Psychology and Neuroscience, King's College, London, UK; National Parkinson Foundation, Centre of Excellence, Kings College Hospital, London, UK*

ANNA SAUERBIER • *Maurice Wohl Clinical Neuroscience Institute and NIHR Biomedical Research Centre, Institute of Psychiatry, Psychology and Neuroscience, King's College Hospital, King's College, London, UK*

VÉRONIQUE SGAMBATO • *University of Lyon, CNRS UMR 5229, Marc Jeannerod Institut of Cognitive Sciences, Bron, France*

MATEJ SKORVANEK • *Department of Neurology, Faculty of Medicine, P. J. Safarik University, Kosice, Slovakia; Department of Neurology, University Hospital L. Pasteur, Kosice, Slovakia*

RAQUEL TADDEI • *Maurice Wohl Clinical Neuroscience Institute and NIHR Biomedical Research Centre, Institute of Psychiatry, Psychology and Neuroscience, King's College Hospital, King's College, London, UK*

AI HUEY TAN • *Division of Neurology, Department of Medicine, Faculty of Medicine, University of Malaya, Kuala Lumpur, Malaysia; The Mah Pooi Soo and Tan Chin Nam Centre for Parkinson's and Related Disorders, University of Malaya, Kuala Lumpur, Malaysia*

ELISABETTA TRONCI • *Department of Biomedical Sciences, Section of Physiology, University of Cagliari, Cittadella Universitaria, Monserrato, Italy*

PETER VALKOVIC • *Second Department of Neurology, Faculty of Medicine, Comenius University, Bratislava, Slovakia; Centre of Experimental Medicine, Slovak Academy of Sciences, Bratislava, Slovakia*

PATRICIO MILLAR VERNETTI • *Department of Neurology, Dysautonomia Center, New York University School of Medicine, New York, NY, USA*

Part I

Preclinical Models of Parkinson's Disease

Chapter 1

Toxin-Based Rodent Models of Parkinson's Disease

M. Angela Cenci and Véronique Sgambato

Abstract

A major pathological hallmark of Parkinson's disease (PD) is a severe degeneration of dopamine (DA)-producing neurons in the substantia nigra pars compacta (SNc) projecting to the motor part of the striatum. Therefore, there is a long-standing interest in using animal models with severe nigrostriatal degeneration for experimental research. Pathophysiological and behavioral features of PD are best studied in mammalian species endowed with well-developed corticobasal ganglia thalamocortical loops, such as rodents. Different toxins can be used to generate nigrostriatal damage, including 6-hydroxydopamine (6-OHDA), 1-methyl-4-phenyl-1,2,3,6-tetrahydropyridine (MPTP), paraquat, and rotenone. Models based on 6-OHDA lesions provide the main advantage of a severe and reproducible DA lesions. Models based on MPTP provide easy and versatile tools to rapidly evaluate potential neuroprotective treatments. Models based on paraquat and rotenone are appealing for their relevance to some well-known environmental risk factors of the human PD, although they yield only partial dopaminergic degeneration and entail a considerable risk of nonspecific toxicity. The main general limitation of neurotoxin-based models is that they do not replicate some characterizing features of PD pathology, such as the formation of Lewy body–like proteinaceous aggregates or the anatomical pattern of neurodegeneration, which also affects nondopaminergic brain regions.

   Key words Parkinson's disease, Animal models, Rodents, Toxin, 6-Hydroxydopamine, 1-Methyl-4-phenyl-1,2,3,6-tetrahydropyridine, Paraquat, Rotenone

1 Introduction

A major pathological hallmark of Parkinson's disease (PD) is a severe degeneration of dopamine (DA)-producing neurons in the substantia nigra pars compacta (SNc) projecting to the motor part of the striatum. It has been estimated that more than 50% of putaminal DA contents are already lost by the time when the disease become clinically manifest [1], and that the residual dopaminergic neurons degenerate very rapidly during the first 5 years after diagnosis, after which point the posterolateral putamen appears devoid of visible DA fibers [2]. The pronounced loss of dopaminergic input to striatal motor regions causes the typical motor features of PD, in particular, difficulty in movement initiation and

Santiago Perez-Lloret (ed.), *Clinical Trials In Parkinson's Disease*, Neuromethods, vol. 160,
https://doi.org/10.1007/978-1-0716-0912-5_1, © Springer Science+Business Media, LLC, part of Springer Nature 2021

poverty of movement (akinesia/hypokinesia), slow movements (bradykinesia), and muscle rigidity (for review see [3], whereas resting tremor (another cardinal sign of PD) appears to also depend on the degeneration of serotonergic raphe neurons [4]. All the above motor features are, however, improved by the DA precursor, L-DOPA, which continues to be the most effective symptomatic treatment for PD [5], although it causes long-term complications that limit its utility [6]. Approximately 6–8 years after starting L-DOPA treatment, it is estimated that 30% of the PD population at large develop abnormal involuntary movements (dyskinesia) and motor fluctuations, but the percentage is much larger in patients with young onset PD [7]. Like PD motor features, the development of motor complications is related to the pronounced nigrostriatal DA degeneration typical of this disease. Thus, studies in PD models have shown that L-DOPA–induced dyskinesia (LID) only can develop when the loss of DA innervation to striatal motor regions exceeds 80%, and postmortem investigations of PD patients affected by LID corroborate this concept (reviewed in [8]).

Given the central importance of the nigrostriatal DA system to the pathophysiology of PD and its treatment, there is a long-standing interest in using animal models with severe nigrostriatal degeneration for experimental research. Such models can be produced both in simple organisms (*C. elegans, Drosophila melanogaster*) and in mammalian species [9, 10]. Models in simple organisms are particularly useful for genetic and molecular studies, but they offer very limited possibilities to the investigation of circuit dysfunctions that cause motor and nonmotor symptoms in PD. Pathophysiological and behavioral features of PD are best studied in mammalian species endowed with well-developed corticobasal ganglia thalamocortical loops. In this regard, rodents have proven to be excellent models of PD-related circuit dysfunction and behavioral abnormalities. For example, detailed kinematic analyses have shown that rodent exhibit highly complex reaching-and-grasping movements quite homologous to those seen in humans [11], which are impaired after DA degeneration of the lateral caudate-putamen (the striatal region controlling forelimb function in rodents) [12]. In preclinical PD research, rodent species offer cost-effective models of both motor and cognitive-affective dysfunctions, in particular, executive deficits, apathy/anhedonia, and impulsive-compulsive behaviors [8, 13, 14]). Moreover, rodent models are being increasingly used to mimic autonomic deficits in PD, in particular, dysfunctions affecting the bladder, the gastrointestinal tract, and the cardiovascular system [15–19].

Although it is accepted that PD does not solely result from dopaminergic degeneration, a severe loss of nigrostriatal DA neurons is a prerequisite for animals to exhibit behavioral deficits having some resemblance to the motor features of PD [12, 20]. In this chapter we review parkinsonian rodent models obtained through systemic or intracerebral administration of toxins. All these toxins cause the demise of DA neurons primarily via oxidative stress and mitochondrial dysfunction, however, with a significant contribution of neuroinflammatory reactions. The behavioral PD-like phenotype accompanying nigrostriatal DA degeneration in rodents includes a generally reduced motor output (analogous to hypokinesia/akinesia and bradykinesia in PD patients), postural and gait abnormalities, and executive cognitive deficits [9, 12, 21, 22].

2 6-Hydroxydopamine

The first neurotoxin-based animal model of PD was based on the intracerebral injection of 6-hydroxydopamine (6-OHDA) [23, 24]. 6-OHDA cannot cross the blood–brain barrier (BBB) and is therefore delivered by intracerebral injection. As an analog of dopamine, 6-OHDA is taken up by neurons that express the dopamine transporter (DAT), in which it undergoes auto-oxidation producing highly reactive substances (such as hydrogen peroxide, superoxide, and hydroxyl radicals) [25, 26]. Neurons rapidly die because of oxidative stress and mitochondrial dysfunction, which are relevant pathogenic mechanisms of neurodegeneration in PD [27].

Rodent models of nigrostriatal DA degeneration can be obtained by injecting 6-OHDA anywhere along the course of the nigrostriatal dopaminergic pathway [12]. Intracerebroventricular administration of 6-OHDA has also been used in certain studies, this procedure causing degeneration of both dopaminergic and noradrenergic projections to the medial and ventral parts of the striatum and other periventricular brain structures [28, 29]. For the sake of producing PD models, the three most common types of 6-OHDA lesion are based on toxin injection into either the substantia nigra, the medial forebrain bundle (MFB), or the striatum. Intranigral 6-OHDA injection is associated with more variability than MFB injection [30], being used less frequently. Injection of 6-OHDA into the MFB leads to a virtually complete loss of dopaminergic innervation in the ipsilateral hemisphere, whereas intrastriatal injections lead to a less severe lesion and a regionally heterogeneous pattern of DA fiber loss in the striatum. With this procedure, varying degrees of DA denervation can be obtained by varying toxin dose and injection coordinates [30, 31].

For the sake of producing symptomatic models of PD, unilateral 6-OHDA lesions are most commonly used, because a complete bilateral 6-OHDA lesion would cause a dramatic akinetic state, requiring intense postoperative nursing protocols. In addition, a major practical advantage offered by unilateral models is that motor performance on the nonimpaired side of the body can serve as a control relative to the impaired side in all tests assessing lateralized behavior, for example, rotational locomotion, sensorimotor integration, and forelimb use. Several tests of forelimb use asymmetry are available and well characterized in unilateral 6-OHDA models; the most widely used of such tests evaluate spontaneous forelimb use during vertical exploration ("cylinder test") or forelimb placing movements during experimenter-imposed body translocations [30–35]. Rodents with unilateral lesions also show transient ipsilateral rotations when placed in a novel environment [30, 32] and a generally reduced performance in tests requiring motor coordination, such as the commonly used rotarod test [36, 37].

In experimental therapeutic research, 6-OHDA lesion models have proven useful to evaluate both symptomatic treatments and disease-modifying interventions. For the latter purpose, intrastriatal 6-OHDA lesion models are to be preferred because of the more protracted degenerative course and the partial loss of DA neurons, leaving a contingent of "rescuable" DA cells in the substantia nigra and a contingent of DA axons with regenerative potential in the relatively spared striatal regions [20]. On the other hand, MFB lesion models are to be preferred for evaluating symptomatic therapies, because they have more pronounced motor deficits and a more predictable and reproducible dose-response profile upon treatment with dopaminergic agents. When using MFB-lesioned rodents to test symptomatic therapies, some simple behavioral screening methods should be applied in order to exclude animals with unsuccessful lesions before initiating the actual drug testing study [38]. A particular application of 6-OHDA–based models in pharmacological PD research consists in the experimental study of L-DOPA–induced dyskinesia. When treated with L-DOPA, both rats and mice with unilateral 6-OHDA lesions develop abnormal involuntary movements (AIMs) on the side of the body contralateral to the lesion. These AIMs have been validated as a model of human LID using both pathophysiological and pharmacological criteria [39].

Like any other experimental model of human disease, 6-OHDA–based models have both pros and cons. Among the "cons" one should mention the rather acute time course of nigrostriatal DA degeneration, the lack of protein aggregation pathology, and the necessity of intracerebral delivery. The latter is usually associated with transient tissue reactions, including blood-brain barrier leakage and glial activation. Bolus injections of >7 microgram 6-OHDA into the same site have been shown to induce local, nonspecific tissue damage [40], a factor that should be controlled for when infusing large volumes of 6-OHDA solution. Nonspecific tissue damage is not an issue if the toxin is delivered in small volumes and using thin injection capillaries [31].

Because 6-OHDA also is toxic to noradrenergic neurons, its intracerebral delivery is often preceded by systemic administration of a selective noradrenaline reuptake inhibitor, such as desipramine, a procedure that has been found to significantly reduce the degree of noradrenergic injury [8, 41]. For many applications, however, the coexistence of noradrenergic and dopaminergic degeneration in the same animal model leads to a more overt parkinsonian phenotype and to more severe dyskinesia upon treatment with L-DOPA (reviewed in [8]). Furthermore, one may argue that the coexistence of noradrenergic and dopaminergic deficits improves the face validity of the animal model because it more closely resembles the neurochemical features of human PD [42].

3 1-Methyl-4-Phenyl-1,2,3,6-Tetrahydropyridine (MPTP)

1-Methyl-4-phenyl-1,2,3,6-tetrahydropyridine (MPTP) was accidentally discovered as a neurotoxin by William Langston and colleagues. This group described a parkinsonian syndrome developing in young drug addicts who had self-injected an illicit drug, and ultimately discovered that the drug was contaminated with MPTP [43]. This toxin can cross BBB, and it is then metabolized by MAO-B in glial cells to the potent dopaminergic neurotoxin 1-methyl-4-phenylpyridinium ion (MPP$^+$) [44], which is then transported into dopaminergic neurons by DAT because it is a structural analog of DA. Once transported into cells by the DAT, MPP$^+$ is concentrated in the mitochondria, binds to NADH dehydrogenase and inhibits an important step in the electron transport chain. More specifically, the neurotoxic effect of MPTP depends on inhibition of mitochondrial complex I, resulting in both reduction of ATP synthesis and accumulation of reactive oxygen species (ROS).

MPTP is administered to nonhuman primates and mice, since rats are resistant to MPTP toxicity [45], partly due to their increased capacity for vesicular sequestration of the toxin [46]. In general, rodent species are less sensitive than primate ones to the neurotoxic effects of MPTP [47]. Accordingly, MPTP-intoxicated monkeys exhibit robust motor impairments, whereas MPTP-treated mice may not do so unless the lesion is very severe. Nevertheless, MPTP-lesioned mice are a widely used model of PD, particularly within the area of neuroprotection research. There are both acute, subchronic, and chronic regimens of MPTP administration that can be used to induce nigrostriatal DA degeneration in mice. The acute model consists in four sequential injections of high-dose MPTP (20–30 mg/kg) given in one 24-h period, 2 h apart. This model provides a rapid screening model to obtain some indication on whether a certain treatment is neuroprotective or not, but it is advised to verify positive results using a more chronic model of nigrostriatal degeneration. Drawbacks of acute MPTP mouse models are both the partial degree of DA cell loss and the high mortality (which can reach a 50% rate or even higher [48]).

If the administration of MPTP is instead spread out over multiple days (subchronic or subacute regimen), the toxin appears to be better tolerated. In subchronic (subacute) MPTP models, the toxin is injected daily over 5–10 days at doses varying from 15 to 30 mg/kg (reviewed in [49]). In both acute and subchronic MPTP models, putative neuroprotective treatments are most often given during the phase of MPTP administration, which brings a necessity to control that the tested treatment does not interfere with toxin uptake or metabolism. More robust and severe models of nigrostriatal dopaminergic degeneration are achieved using chronic intoxication regimens. In this regard, three types of protocols have been developed, that is (1) 25 mg/kg MPTP injected every 3.5 days for 5 weeks; (2) continuous minipump delivery of MPTP, 23–46 mg/kg/day for 14 days or 46 mg/kg/day for 28 days, via subcutaneous or intraperitoneal routes; (3) escalating dose protocol with daily MPTP injections for 5 days (weekend off) for 4 weeks, where the injected doses are scaled up over time from 4 mg/kg to 32 mg/kg (a detailed literature review of these different models can be found in [49]). In chronic administration protocols, MPTP is often coadministered with probenecid, which retards the renal and CNS clearance of the toxic metabolites (although animal mortality may be a concern also with these slower intoxication methods). Chronic regimens of MPTP administration can lead to 70% SNc neuron loss and greater than 90% reduction in striatal DA levels [49]. Moreover, continuous MPTP administration with osmotic minipumps has been proposed as progressive PD model associated with the formation of nigral inclusions immunoreactive for ubiquitin and alpha-synuclein [50], but these findings have remained controversial [51].

MPTP-lesioned mice continue to provide an important tool for translational research. The interest in this model is furthermore supported by the observation that MPTP causes parkinsonism in humans, and that MPTP is technically easy to administer compared to toxins requiring stereotaxic surgery for intracerebral delivery [52]. MPTP lesions may, however, entail other disadvantages, in particular, a relatively large variability in behavioral and biochemical outcomes [9, 49, 51, 53].

4   Paraquat

The etiology of PD is multifactorial, also implying a probable environmental toxicity. Indeed, several epidemiological studies have demonstrated an association between rural residence and pesticide exposure with an increased risk of PD [54, 55].

The herbicide paraquat has a structure similar to MPP^+. Like MPTP, paraquat can cross the BBB, and the passage is presumably mediated by the neutral amino acid transporter [56]. Paraquat acts on the pentose phosphate pathway to increase NADPH reducing equivalents and stimulate paraquat redox cycling [57]. As a redox cycling compound, paraquat induces oxidative stress by impairing the redox recycling of glutathione and thioredoxin [58], which inhibits the function of intracellular antioxidant systems. The mitochondria-mediated apoptotic pathway is responsible for the proapoptotic activity of paraquat, leading to the upregulation of proapoptotic members of Bcl-2 family, followed by cytochrome c release and caspase-3 activation [59]. Paraquat exposure also activates microglia [60].

Paraquat is usually administrated orally or intraperitoneally. The parameters of its administration (dose and duration) have varied between studies, and so has the behavioral-histopathological phenotype of its corresponding animal models. One study in mice has reported locomotor deficits and decreased DA markers in both SNc and striatum after oral administration of 10 mg/kg paraquat for 4 months [61]. After intraperitoneal injection of 5 or 10 mg/kg paraquat once a week for 3 consecutive weeks, mice showed locomotor deficits and a dose-dependent loss of DA neurons in the SNc (36% versus 61%) and DA fibers in the striatum (87% versus 94%) [62]. In another study, mice were treated with paraquat (1–10 mg/kg) once a week for 3 weeks, exhibiting a dose- and age-dependent degree of nigral DA cell loss without any significant striatal DA depletion [63]. Other studies in mice have evidenced a higher susceptibility of SNc ventral tier to paraquat [64]. However, the robustness of paraquat-based PD models has been called into question by studies reporting that single-weekly or biweekly intraperitoneal injections of paraquat (20 or 10 mg/kg) for 3 weeks had no effect on the number of tyrosine hydroxylase positive cells, nor on markers of microglia activation, in the SNc [65]. Another study observed that one intraperitoneal injection of 15 mg/kg paraquat to prepubertal mice was sufficient to induce increases of oxidative stress and subsequent striatal DA loss, while a higher dose (20 mg/kg) induced 50% of lethality within 72 h [66]. Interestingly, one study has found that intraperitoneal injection of 10 mg/kg paraquat to mice resulted in locomotor deficits and triggered, besides a progressive DA degeneration, an increase of α-synuclein in the SNc [67]. The impact of paraquat on α-synuclein pathology has been investigated in mice overexpressing the human A53T mutated form of α-synuclein [68]. In these mice, it was found that 10 mg/kg oral paraquat administration for 6–8 weeks induced an earlier phosphorylation of α-synuclein (recognized as a key event in the formation of Lewy pathology) in the enteric nervous system, supporting the theory that ingested toxins (such as paraquat) may initiate idiopathic PD at the level of the gut. Similarly, the add-on administration of paraquat (7 mg/kg/day for 6 days, or once a week for 10 weeks) to mice receiving a nontoxic MPTP regimen (10 mg/kg/day for 5 days) aggravated both dopaminergic neurotoxicity and motor impairment [69]. Rats are also susceptible to paraquat toxicity. In rats, intraperitoneal injections of 10 mg/kg paraquat twice a week for 4 weeks [60] or once every 5 days [70] have been found to produce an approximately 50% loss of nigral DA neurons.

To produce more robust nigrostriatal dopaminergic degeneration, paraquat can be coadministered with maneb, which is a manganese polymer used as a fungicide in agriculture. It can cross the BBB and preferentially acts on mitochondrial complex III [71]. Rats intoxicated with paraquat (10 mg/kg) and maneb (30 mg/kg) twice a week for 6 weeks were not equally sensitive, some of them exhibiting motor or non-motor symptoms more or less rapidly, despite a significant (54%) reduction of striatal DA contents [72]. In a study using young adult rats, the administration of paraquat alone (10 mg/kg) or in combination with maneb (30 mg/kg), twice a week for 5 weeks, induced a significant loss of nigral DA neurons (15% for paraquat vs 21% for paraquat and maneb) but not of striatal DA fibers despite motor impairment (more severe in double pesticide-treated animals) [60]. In mice, a reduction of striatal DA contents by more than 20% was observed upon treatment with paraquat and maneb (10 mg/kg and 30 mg/kg, respectively) twice daily for 7 weeks [73]. In another study, mice exposed to both paraquat and maneb twice a week for 6 weeks exhibited a significant loss of DA nigrostriatal neurons, reactive gliosis and motor impairment compared to paraquat-only lesioned mice [74]. As seen for the paraquat, administration of maneb exacerbated MPTP neurotoxicity in mice [75]. Interestingly, there is also an enhanced neurotoxicity of these compounds with ageing, including reductions in locomotor activity, motor coordination and dopamine levels in response to the combined administration of paraquat and maneb [76]. These results suggest that environmental pesticides can act in conjunction with ageing to enhance the risk for PD. The hypothesis that exposure to toxins during development could have an impact on the vulnerability to subsequent insults during adulthood has also been investigated. Postnatal exposure to paraquat (0.3 mg/kg) and maneb (1 mg/kg) (daily for 15 days) reduced activity of mice and resulted into a mild DA lesion (37%) which could be exacerbated (62%) by reexposure to combined paraquat (10 mg/kg) and maneb (30 mg/kg) twice a week for 3 weeks at adult stage [77]. Prenatal exposure to maneb also potentiated the toxicity of paraquat at adult stages by inducing motor deficits and nigral DA loss [78]. Finally, genetic factors can also have an impact on the responses to pesticides. Indeed, it has been shown in mice that a reduced expression of the dopaminergic transporter (DAT) made a subgroup of nigral DA neurons resistant to paraquat/maneb toxicity [79, 80].

To conclude, most studies have shown that paraquat can induce a dose-dependent partial degeneration of nigrostriatal DA neurons although the effects on striatal DA levels have been quite variable, possibly due to compensatory mechanisms. If low-dose paraquat is administered over a long period, it can lead to upregulation and aggregation of alpha-synuclein [67]. When combined with other systemic toxins (MPTP, maneb), or with genetic overexpression of α-synuclein, paraquat exhibits an increased neurotoxicity towards the DA system leading to a more severe motor phenotype in the corresponding animal model. Moreover, the susceptibility to paraquat toxicity is enhanced during development or ageing, and appears to be influenced by genetic factors. Drawbacks of paraquat-based models are the only partial degree of nigrostriatal DA degeneration and the high systemic toxicity when high doses are used.

5   Rotenone

Rotenone (a natural extract from plants) is a broad-spectrum insecticide and pesticide, mainly used to kill agricultural pests. Because of its hydrophobicity, rotenone can easily cross the blood-brain barrier, independently of the DAT. Once in DA neurons, rotenone inhibits mitochondrial complex I and activates the production of ROS [81].

Greenamyre and collaborators were the first to develop a PD animal model based on the continuous administration of rotenone (with different doses and duration of exposure) by using osmotic minipumps. Rats were treated with rotenone at 3 mg/kg per day for 33 days, and a subgroup of them (only 12 of 25) exhibited a selective loss of nigrostriatal neurons, with nerve terminals being affected before cell bodies. The dopaminergic degeneration was sufficiently severe to produce hypokinesia and rigidity [82, 83]. Furthermore, nigral neurons from rotenone-treated rats accumulated α-synuclein, and noradrenergic neurons from the locus coeruleus were mildly to moderately affected [82]. This pioneering work was followed by other studies looking at the potential neurotoxicity of rotenone towards multiple neurotransmission systems [84]. Rats treated intraperitoneally with two doses of rotenone (1.5 and 2.5 mg/kg daily) for 2 months displayed catalepsy and DA depletion in striatal and prefrontal regions, without alterations of noradrenaline or serotonin levels [85]. However, in another study, rats treated with approximately the same rotenone regimen (2.5 mg/kg/day i.v. for 28 days), exhibited multiple DA and non-DA lesions [84]. Specifically, there was a loss of nigral DA neurons and striatal DA fibers, but also an injury of striatal serotonergic fibers, striatal DARPP-32–positive projection neurons and striatal cholinergic interneurons. A cell loss was also detected outside of the basal ganglia, in regions such as the pedunculopontine tegmental nucleus and the locus coeruleus [84]. These data indicated that rotenone intoxication affects multiple systems of neurons, being more suitable to model atypical parkinsonian syndromes rather than PD. In the same vein, using different rotenone doses and administration routes in rats, other studies have revealed a lack of correlation between loss of striatal DA innervation and motor deficits [86, 87], concluding that the motor phenotype induced by rotenone intoxication may depend on pervasive neurological effects of the toxin.

Despite this earlier controversy, several improved rat models of rotenone-based nigrostriatal degeneration have been published during the past 10 years. In one study [88], rats infused for 5 weeks with a low rotenone dose (2 mg/kg to avoid high mortality) developed motor deficits that were associated with selective oxidative damage in the striatum and strong nigral DA loss. Attempts to develop a highly reproducible rotenone PD model were also made using a more lipophilic injection vehicle and a more careful dose titration. In a study by the Greenamyre group, rats from different age groups (3, 7 or 12–14 months) received daily injections of rotenone (2.75 or 3.0 mg/kg i.p.) in a dimethylsulfoxide (DMSO)-based vehicle for 2 months, and animals finally exhibited bradykinesia, postural instability/rigidity that were reversed by apomorphine, consistent with a dependence on nigrostriatal DA degeneration [89]. Furthermore, the loss of DA neurons (45%) was associated with the presence of α-synuclein aggregates in the SNc [89]. As seen with paraquat, the behavioral phenotype was influenced by the animal's age, a higher neurotoxicity being seen in older animals [89].

Other studies in rats have focused on alternative administration routes. One such study reported that direct unilateral infusion of rotenone (5 μg) into the medial forebrain bundle induced a progressive nigrostriatal dopaminergic lesion, accompanied by an increased expression and aggregation of α-synuclein, whereas subcutaneous rotenone delivery (3 mg/kg/day for 28 days) resulted in extensive peripheral organ toxicity without any dopaminergic degeneration [90]. This study suggested that a direct intracerebral infusion of rotenone might be more efficient than systemic administration protocols. However, more variable results were reported by other studies using intracerebral administration routes. Thus, the stereotaxic injection of rotenone into medial forebrain bundle, ventral tegmental area, striatum, and/or SN at doses ranging from 0.4 to 12 μg was found to induce a significant dose-dependent loss of nigrostriatal DA neurons but an inconsistent α-synuclein pathology [91, 92]. Direct intragastric administration of rotenone has also been tested. In one study, low doses of intragastrically administered rotenone for 3 months induced changes in motor coordination and slight DA loss in the SNc (around 12%) [93]. The most interesting outcome of this administration procedure was the finding of α-synuclein phosphorylation and aggregation (associated with gliosis) in the enteric nervous system, the vagus nerve, and the substantia nigra [93]. These results supported the theory that environmental toxins may trigger the initiation and propagation of alpha-synuclein pathology acting at the level of the enteric nervous system. Other studies have applied regimens of rotenone intoxication (either by osmotic minipump [94] or by intranigral stereotaxic injection [95] to rat models of viral vector-mediated α-synuclein overexpression, reporting an exacerbation of motor impairments and nigrostriatal DA degeneration. A similar interaction between rotenone- and alpha-synuclein-driven degeneration was obtained using transgenic mice overexpressing the mutated A53T form of α-synuclein [96]. In the absence of alpha-synuclein pathology, mice appear to be more resistant than rats to the neurotoxic effects of rotenone. One study has reported that oral administration of high-dose rotenone to mice (50 mg/kg/day for 14 days) induces loss of DA neurons in the SNc, increases ROS accumulation, and activates mitochondrial apoptotic pathways [97]. However, other reports have been less positive about the neurotoxic effects of rotenone on the mouse dopaminergic system. In one study in mice, acute or subchronic administration of rotenone at high doses (5–15 mg/kg/day s.c.) did not induce nigrostriatal DA depletion but only an increase in DA turnover [98]. Moreover, rotenone failed to enhance the neurotoxicity induced by MPTP [98].

6   Concluding Remarks

Toxin-based rodent models of nigrostriatal dopaminergic degeneration continue to provide an essential tool in preclinical PD research. Models based on 6-OHDA lesions provide the main advantage of a severe and reproducible DA lesions which is required to induce L-DOPA-dependent motor features (including dyskinesia). Intrastriatal 6-OHDA lesion models are also used to assess neuroprotective treatments. Models based on MPTP provide easy and versatile tools to evaluate neuroprotective treatments. However, the above models do not replicate some characterizing features of PD pathology, in particular, the formation of proteinaceous aggregates resembling Lewy bodies, or the involvement of brain regions outside of the nigrostriatal DA system. Furthermore, animal mortality is a nonnegligible risk when producing severe dopaminergic lesions. Models based on herbicides and pesticides are appealing for their relevance to some well-known environmental risk factors of the human PD, although they yield only partial dopaminergic degeneration and entail a considerable risk of nonspecific toxicity. Herbicides and pesticides are, nevertheless, very interesting if applied at low-dose, as an "add-on" insult, to rodent models harboring genetic or nongenetic factors of DA neuron vulnerability. Undoubtedly, rodent models combining different types of lesions, or toxic and genetic factors relevant to PD, should continue to be explored for the evaluation of disease-modifying interventions.

References

1. Fearnley JM, Lees AJ (1991) Ageing and Parkinson's disease: substantia nigra regional selectivity. Brain 114(Pt 5):2283–2301. https://doi.org/10.1093/brain/114.5.2283

2. Kordower JH, Olanow CW, Dodiya HB, Chu Y, Beach TG, Adler CH, Halliday GM, Bartus RT (2013) Disease duration and the integrity of the nigrostriatal system in Parkinson's disease. Brain 136(Pt 8):2419–2431. https://doi.org/10.1093/brain/awt192

3. Schneider SA, Obeso JA (2015) Clinical and pathological features of Parkinson's disease. Curr Top Behav Neurosci 22:205–220. https://doi.org/10.1007/7854_2014_317

4. Pasquini J, Ceravolo R, Qamhawi Z, Lee JY, Deuschl G, Brooks DJ, Bonuccelli U, Pavese N (2018) Progression of tremor in early stages of Parkinson's disease: a clinical and neuroimaging study. Brain 141(3):811–821. https://doi.org/10.1093/brain/awx376

5. Olanow CW (2019) Levodopa is the best symptomatic therapy for PD: nothing more, nothing less. Mov Disord 34(6):812–815. https://doi.org/10.1002/mds.27690

6. Cenci MA, Ohlin KE, Odin P (2011) Current options and future possibilities for the treatment of dyskinesia and motor fluctuations in Parkinson's disease. CNS Neurol Disord Drug Targets 10(6):670–684

7. Manson A, Stirpe P, Schrag A (2012) Levodopa-induced-dyskinesias clinical features, incidence, risk factors, management and impact on quality of life. J Parkinsons Dis 2(3):189–198. https://doi.org/10.3233/JPD-2012-120103

8. Cenci MA (2014) Presynaptic mechanisms of l-DOPA-induced dyskinesia: the findings, the debate, and the therapeutic implications. Front Neurol 5:242. https://doi.org/10.3389/fneur.2014.00242

9. Francardo V (2018) Modeling Parkinson's disease and treatment complications in rodents: potentials and pitfalls of the current options. Behav Brain Res 352:142–150. https://doi.org/10.1016/j.bbr.2017.12.014

10. Masilamoni GJ, Smith Y (2018) Chronic MPTP administration regimen in monkeys: a model of dopaminergic and non-dopaminergic cell loss in Parkinson's disease. J Neural Transm (Vienna) 125(3):337–363. https://doi.org/10.1007/s00702-017-1774-z

11. Sacrey LA, Alaverdashvili M, Whishaw IQ (2009) Similar hand shaping in reaching-for-food (skilled reaching) in rats and humans provides evidence of homology in release, collection, and manipulation movements. Behav Brain Res 204(1):153–161. https://doi.org/10.1016/j.bbr.2009.05.035

12. Cenci MA, Whishaw IQ, Schallert T (2002) Animal models of neurological deficits: how relevant is the rat? Nat Rev Neurosci 3(7):574–579. https://doi.org/10.1038/nrn877

13. Espa E, Clemensson EKH, Luk KC, Heuer A, Bjorklund T, Cenci MA (2019) Seeding of protein aggregation causes cognitive impairment in rat model of cortical synucleinopathy. Mov Disord 34:1699. https://doi.org/10.1002/mds.27810

14. Magnard R, Vachez Y, Carcenac C, Krack P, David O, Savasta M, Boulet S, Carnicella S (2016) What can rodent models tell us about apathy and associated neuropsychiatric symptoms in Parkinson's disease? Transl Psychiatry 6:e753. https://doi.org/10.1038/tp.2016.17

15. Dodiya HB, Forsyth CB, Voigt RM, Engen PA, Patel J, Shaikh M, Green SJ, Naqib A, Roy A, Kordower JH, Pahan K, Shannon KM, Keshavarzian A (2018) Chronic stress-induced gut dysfunction exacerbates Parkinson's disease phenotype and pathology in a rotenone-induced mouse model of Parkinson's disease. Neurobiol Dis 135:104352. https://doi.org/10.1016/j.nbd.2018.12.012

16. Perez-Pardo P, Dodiya HB, Broersen LM, Douna H, van Wijk N, Lopes da Silva S, Garssen J, Keshavarzian A, Kraneveld AD (2018) Gut-brain and brain-gut axis in Parkinson's disease models: effects of a uridine and fish oil diet. Nutr Neurosci 21(6):391–402. https://doi.org/10.1080/1028415X.2017.1294555

17. Soler R, Fullhase C, Santos C, Andersson KE (2011) Development of bladder dysfunction in a rat model of dopaminergic brain lesion. Neurourol Urodyn 30(1):188–193. https://doi.org/10.1002/nau.20917

18. Taylor TN, Greene JG, Miller GW (2010) Behavioral phenotyping of mouse mod-

els of Parkinson's disease. Behav Brain Res 211(1):1–10. https://doi.org/10.1016/j.bbr.2010.03.004

19. Wang L, Fleming SM, Chesselet MF, Tache Y (2008) Abnormal colonic motility in mice over-expressing human wild-type alpha-synuclein. Neuroreport 19(8):873–876. https://doi.org/10.1097/WNR.0b013e3282ffda5e

20. Francardo V, Schmitz Y, Sulzer D, Cenci MA (2017) Neuroprotection and neurorestoration as experimental therapeutics for Parkinson's disease. Exp Neurol 298(Pt B):137–147. https://doi.org/10.1016/j.expneurol.2017.10.001

21. Lindgren HS, Dunnett SB (2012) Cognitive dysfunction and depression in Parkinson's disease: what can be learned from rodent models? Eur J Neurosci 35(12):1894–1907. https://doi.org/10.1111/j.1460-9568.2012.08162.x

22. Lindgren HS, Klein A, Dunnett SB (2014) Nigral 6-hydroxydopamine lesion impairs performance in a lateralised choice reaction time task--impact of training and task parameters. Behav Brain Res 266:207–215. https://doi.org/10.1016/j.bbr.2014.02.043

23. Ungerstedt U (1968) 6-Hydroxy-dopamine induced degeneration of central monoamine neurons. Eur J Pharmacol 5(1):107–110. https://doi.org/10.1016/0014-2999(68)90164-7

24. Ungerstedt U, Ljungberg T, Steg G (1974) Behavioral, physiological, and neurochemical changes after 6-hydroxydopamine-induced degeneration of the nigro-striatal dopamine neurons. Adv Neurol 5:421–426

25. Hernandez-Baltazar D, Zavala-Flores LM, Villanueva-Olivo A (2017) The 6-hydroxydopamine model and parkinsonian pathophysiology: novel findings in an older model. Neurologia 32(8):533–539. https://doi.org/10.1016/j.nrl.2015.06.011

26. Rotman A, Creveling CR (1976) A rationale for the design of cell-specific toxic agents: the mechanism of action of 6-hydroxydopamine. FEBS Lett 72(2):227–230. https://doi.org/10.1016/0014-5793(76)80974-x

27. Grunewald A, Kumar KR, Sue CM (2019) New insights into the complex role of mitochondria in Parkinson's disease. Prog Neurobiol 177:73–93. https://doi.org/10.1016/j.pneurobio.2018.09.003

28. Cenci MA, Kalen P, Duan WM, Bjorklund A (1994) Transmitter release from transplants of fetal ventral mesencephalon or locus coeruleus in the rat frontal cortex and nucleus accumbens: effects of pharmacological and behaviorally activating stimuli. Brain Res 641(2):225–248. https://doi.org/10.1016/0006-8993(94)90150-3

29. Leroux-Nicollet I, Panissaud C, Costentin J (1988) Involvement of norepinephrine neurons in the hypothermia induced by intracerebroventricular administration of 6-hydroxydopamine in mice, evidenced by antidepressants. J Neural Transm 74(1):17–27

30. Francardo V, Recchia A, Popovic N, Andersson D, Nissbrandt H, Cenci MA (2011) Impact of the lesion procedure on the profiles of motor impairment and molecular responsiveness to L-DOPA in the 6-hydroxydopamine mouse model of Parkinson's disease. Neurobiol Dis 42(3):327–340. https://doi.org/10.1016/j.nbd.2011.01.024

31. Winkler C, Kirik D, Bjorklund A, Cenci MA (2002) L-DOPA-induced dyskinesia in the intrastriatal 6-hydroxydopamine model of parkinson's disease: relation to motor and cellular parameters of nigrostriatal function. Neurobiol Dis 10(2):165–186

32. Francardo V, Bez F, Wieloch T, Nissbrandt H, Ruscher K, Cenci MA (2014) Pharmacological stimulation of sigma-1 receptors has neurorestorative effects in experimental parkinsonism. Brain 137(Pt 7):1998–2014. https://doi.org/10.1093/brain/awu107

33. Lundblad M, Andersson M, Winkler C, Kirik D, Wierup N, Cenci MA (2002) Pharmacological validation of behavioural measures of akinesia and dyskinesia in a rat model of Parkinson's disease. Eur J Neurosci 15(1):120–132. https://doi.org/10.1046/j.0953-816x.2001.01843.x

34. Olsson M, Nikkhah G, Bentlage C, Bjorklund A (1995) Forelimb akinesia in the rat Parkinson model: differential effects of dopamine agonists and nigral transplants as assessed by a new stepping test. J Neurosci 15(5 Pt 2):3863–3875

35. Schallert T, Fleming SM, Leasure JL, Tillerson JL, Bland ST (2000) CNS plasticity and assessment of forelimb sensorimotor outcome in unilateral rat models of stroke, cortical ablation, parkinsonism and spinal cord injury. Neuropharmacology 39(5):777–787. https://doi.org/10.1016/s0028-3908(00)00005-8

36. Lundblad M, Vaudano E, Cenci MA (2003) Cellular and behavioural effects of the adenosine A2a receptor antagonist KW-6002 in a rat model of l-DOPA-induced dyskinesia. J Neurochem 84(6):1398–1410. https://doi.org/10.1046/j.1471-4159.2003.01632.x

37. Rozas G, Guerra MJ, Labandeira-Garcia JL (1997) An automated rotarod method for quantitative drug-free evaluation of overall motor deficits in rat models of parkinsonism. Brain Res Brain Res Protoc 2(1):75–84

38. Cenci MA, Lundblad M (2007) Ratings of L-DOPA-induced dyskinesia in the unilateral 6-OHDA lesion model of Parkinson's

disease in rats and mice. Curr Protoc Neurosci Chapter 9:Unit 9.25. https://doi.org/10.1002/0471142301.ns0925s41

39. Cenci MA, Crossman AR (2018) Animal models of l-dopa-induced dyskinesia in Parkinson's disease. Mov Disord 33(6):889–899. https://doi.org/10.1002/mds.27337

40. Kirik D, Rosenblad C, Bjorklund A (1998) Characterization of behavioral and neurodegenerative changes following partial lesions of the nigrostriatal dopamine system induced by intrastriatal 6-hydroxydopamine in the rat. Exp Neurol 152(2):259–277. https://doi.org/10.1006/exnr.1998.6848

41. Sgambato-Faure V, Buggia V, Gilbert F, Levesque D, Benabid AL, Berger F (2005) Coordinated and spatial upregulation of arc in striatonigral neurons correlates with L-dopa-induced behavioral sensitization in dyskinetic rats. J Neuropathol Exp Neurol 64(11):936–947. https://doi.org/10.1097/01.jnen.0000186922.42592.b7

42. Weinshenker D (2018) Long road to ruin: noradrenergic dysfunction in neurodegenerative disease. Trends Neurosci 41(4):211–223. https://doi.org/10.1016/j.tins.2018.01.010

43. Langston JW (2017) The MPTP story. J Parkinsons Dis 7(s1):S11–S19. https://doi.org/10.3233/JPD-179006

44. Chiba K, Trevor A, Castagnoli N Jr (1984) Metabolism of the neurotoxic tertiary amine, MPTP, by brain monoamine oxidase. Biochem Biophys Res Commun 120(2):574–578. https://doi.org/10.1016/0006-291x(84)91293-2

45. Sundstrom E, Samuelsson EB (1997) Comparison of key steps in 1-methyl-4-phenyl-1,2,3,6-tetrahydropyridine (MPTP) neurotoxicity in rodents. Pharmacol Toxicol 81(5):226–231. https://doi.org/10.1111/j.1600-0773.1997.tb00051.x

46. Staal RG, Hogan KA, Liang CL, German DC, Sonsalla PK (2000) In vitro studies of striatal vesicles containing the vesicular monoamine transporter (VMAT2): rat versus mouse differences in sequestration of 1-methyl-4-phenylpyridinium. J Pharmacol Exp Ther 293(2):329–335

47. Johannessen JN, Chiueh CC, Burns RS, Markey SP (1985) Differences in the metabolism of MPTP in the rodent and primate parallel differences in sensitivity to its neurotoxic effects. Life Sci 36(3):219–224. https://doi.org/10.1016/0024-3205(85)90062-1

48. Hallman H, Lange J, Olson L, Stromberg I, Jonsson G (1985) Neurochemical and histochemical characterization of neurotoxic effects of 1-methyl-4-phenyl-1,2,3,6-tetrahydropyridine on brain catecholamine neurones in the mouse. J Neurochem 44(1):117–127. https://doi.org/10.1111/j.1471-4159.1985.tb07120.x

49. Meredith GE, Rademacher DJ (2011) MPTP mouse models of Parkinson's disease: an update. J Parkinsons Dis 1(1):19–33. https://doi.org/10.3233/JPD-2011-11023

50. Fornai F, Schluter OM, Lenzi P, Gesi M, Ruffoli R, Ferrucci M, Lazzeri G, Busceti CL, Pontarelli F, Battaglia G, Pellegrini A, Nicoletti F, Ruggieri S, Paparelli A, Sudhof TC (2005) Parkinson-like syndrome induced by continuous MPTP infusion: convergent roles of the ubiquitin-proteasome system and alpha-synuclein. Proc Natl Acad Sci U S A 102(9):3413–3418. https://doi.org/10.1073/pnas.0409713102

51. Alvarez-Fischer D, Guerreiro S, Hunot S, Saurini F, Marien M, Sokoloff P, Hirsch EC, Hartmann A, Michel PP (2008) Modelling Parkinson-like neurodegeneration via osmotic minipump delivery of MPTP and probenecid. J Neurochem 107(3):701–711. https://doi.org/10.1111/j.1471-4159.2008.05651.x

52. Jackson-Lewis V, Przedborski S (2007) Protocol for the MPTP mouse model of Parkinson's disease. Nat Protoc 2(1):141–151. https://doi.org/10.1038/nprot.2006.342

53. Rousselet E, Joubert C, Callebert J, Parain K, Tremblay L, Orieux G, Launay JM, Cohen-Salmon C, Hirsch EC (2003) Behavioral changes are not directly related to striatal monoamine levels, number of nigral neurons, or dose of parkinsonian toxin MPTP in mice. Neurobiol Dis 14(2):218–228

54. Ascherio A, Schwarzschild MA (2016) The epidemiology of Parkinson's disease: risk factors and prevention. Lancet Neurol 15(12):1257–1272. https://doi.org/10.1016/S1474-4422(16)30230-7

55. Chade AR, Kasten M, Tanner CM (2006) Nongenetic causes of Parkinson's disease. J Neural Transm Suppl (70):147–151

56. Shimizu K, Ohtaki K, Matsubara K, Aoyama K, Uezono T, Saito O, Suno M, Ogawa K, Hayase N, Kimura K, Shiono H (2001) Carrier-mediated processes in blood–brain barrier penetration and neural uptake of paraquat. Brain Res 906(1–2):135–142. https://doi.org/10.1016/s0006-8993(01)02577-x

57. Powers R, Lei S, Anandhan A, Marshall DD, Worley B, Cerny RL, Dodds ED, Huang Y, Panayiotidis MI, Pappa A, Franco R (2017) Metabolic investigations of the molecular mechanisms associated with Parkinson's disease. Meta 7(2). https://doi.org/10.3390/metabo7020022

58. Niso-Santano M, Gonzalez-Polo RA, Bravo-San Pedro JM, Gomez-Sanchez R, Lastres-Becker I, Ortiz-Ortiz MA, Soler G, Moran JM, Cuadrado A, Fuentes JM, Centro de Investigacion Biomedica en red sobre Enfermedades N (2010) Activation of apoptosis signal-regulating kinase 1 is a key factor in paraquat-induced cell death: modulation by the Nrf2/Trx axis. Free Radic Biol Med 48(10):1370–1381. https://doi.org/10.1016/j.freeradbiomed.2010.02.024

59. Fei Q, McCormack AL, Di Monte DA, Ethell DW (2008) Paraquat neurotoxicity is mediated by a Bak-dependent mechanism. J Biol Chem 283(6):3357–3364. https://doi.org/10.1074/jbc.M708451200

60. Cicchetti F, Lapointe N, Roberge-Tremblay A, Saint-Pierre M, Jimenez L, Ficke BW, Gross RE (2005) Systemic exposure to paraquat and maneb models early Parkinson's disease in young adult rats. Neurobiol Dis 20(2):360–371. https://doi.org/10.1016/j.nbd.2005.03.018

61. Ren JP, Zhao YW, Sun XJ (2009) Toxic influence of chronic oral administration of paraquat on nigrostriatal dopaminergic neurons in C57BL/6 mice. Chin Med J 122(19):2366–2371

62. Brooks AI, Chadwick CA, Gelbard HA, Cory-Slechta DA, Federoff HJ (1999) Paraquat elicited neurobehavioral syndrome caused by dopaminergic neuron loss. Brain Res 823(1–2):1–10. https://doi.org/10.1016/s0006-8993(98)01192-5

63. McCormack AL, Thiruchelvam M, Manning-Bog AB, Thiffault C, Langston JW, Cory-Slechta DA, Di Monte DA (2002) Environmental risk factors and Parkinson's disease: selective degeneration of nigral dopaminergic neurons caused by the herbicide paraquat. Neurobiol Dis 10(2):119–127

64. Fernagut PO, Hutson CB, Fleming SM, Tetreaut NA, Salcedo J, Masliah E, Chesselet MF (2007) Behavioral and histopathological consequences of paraquat intoxication in mice: effects of alpha-synuclein over-expression. Synapse 61(12):991–1001. https://doi.org/10.1002/syn.20456

65. Smeyne RJ, Breckenridge CB, Beck M, Jiao Y, Butt MT, Wolf JC, Zadory D, Minnema DJ, Sturgess NC, Travis KZ, Cook AR, Smith LL, Botham PA (2016) Assessment of the effects of MPTP and Paraquat on dopaminergic neurons and microglia in the substantia Nigra pars compacta of C57BL/6 mice. PLoS One 11(10):e0164094. https://doi.org/10.1371/journal.pone.0164094

66. Hosamani R, Krishna G, Muralidhara (2016) Standardized Bacopa monnieri extract ameliorates acute paraquat-induced oxidative stress, and neurotoxicity in prepubertal mice brain. Nutr Neurosci 19(10):434–446. https://doi.org/10.1179/1476830514Y.0000000149

67. Manning-Bog AB, McCormack AL, Li J, Uversky VN, Fink AL, Di Monte DA (2002) The herbicide paraquat causes up-regulation and aggregation of alpha-synuclein in mice: paraquat and alpha-synuclein. J Biol Chem 277(3):1641–1644. https://doi.org/10.1074/jbc.C100560200

68. Naudet N, Antier E, Gaillard D, Morignat E, Lakhdar L, Baron T, Bencsik A (2017) Oral exposure to Paraquat triggers earlier expression of phosphorylated alpha-synuclein in the enteric nervous system of A53T mutant human alpha-Synuclein transgenic mice. J Neuropathol Exp Neurol 76(12):1046–1057. https://doi.org/10.1093/jnen/nlx092

69. Shepherd KR, Lee ES, Schmued L, Jiao Y, Ali SF, Oriaku ET, Lamango NS, Soliman KF, Charlton CG (2006) The potentiating effects of 1-methyl-4-phenyl-1,2,3,6-tetrahydropyridine (MPTP) on paraquat-induced neurochemical and behavioral changes in mice. Pharmacol Biochem Behav 83(3):349–359. https://doi.org/10.1016/j.pbb.2006.02.013

70. Muthukumaran K, Leahy S, Harrison K, Sikorska M, Sandhu JK, Cohen J, Keshan C, Lopatin D, Miller H, Borowy-Borowski H, Lanthier P, Weinstock S, Pandey S (2014) Orally delivered water soluble coenzyme Q10 (Ubisol-Q10) blocks on-going neurodegeneration in rats exposed to paraquat: potential for therapeutic application in Parkinson's disease. BMC Neurosci 15:21. https://doi.org/10.1186/1471-2202-15-21

71. Zhang J, Fitsanakis VA, Gu G, Jing D, Ao M, Amarnath V, Montine TJ (2003) Manganese ethylene-bis-dithiocarbamate and selective dopaminergic neurodegeneration in rat: a link through mitochondrial dysfunction. J Neurochem 84(2):336–346. https://doi.org/10.1046/j.1471-4159.2003.01525.x

72. Tinakoua A, Bouabid S, Faggiani E, De Deurwaerdere P, Lakhdar-Ghazal N, Benazzouz A (2015) The impact of combined administration of paraquat and maneb on motor and non-motor functions in the rat. Neuroscience 311:118–129. https://doi.org/10.1016/j.neuroscience.2015.10.021

73. Kachroo A, Schwarzschild MA (2014) Allopurinol reduces levels of urate and dopamine but not dopaminergic neurons in a dual pesticide model of Parkinson's dis-

ease. Brain Res 1563:103–109. https://doi.org/10.1016/j.brainres.2014.03.031

74. Thiruchelvam M, Brockel BJ, Richfield EK, Baggs RB, Cory-Slechta DA (2000) Potentiated and preferential effects of combined paraquat and maneb on nigrostriatal dopamine systems: environmental risk factors for Parkinson's disease? Brain Res 873(2):225–234. https://doi.org/10.1016/s0006-8993(00)02496-3

75. Takahashi RN, Rogerio R, Zanin M (1989) Maneb enhances MPTP neurotoxicity in mice. Res Commun Chem Pathol Pharmacol 66(1):167–170

76. Thiruchelvam M, McCormack A, Richfield EK, Baggs RB, Tank AW, Di Monte DA, Cory-Slechta DA (2003) Age-related irreversible progressive nigrostriatal dopaminergic neurotoxicity in the paraquat and maneb model of the Parkinson's disease phenotype. Eur J Neurosci 18(3):589–600. https://doi.org/10.1046/j.1460-9568.2003.02781.x

77. Thiruchelvam M, Richfield EK, Goodman BM, Baggs RB, Cory-Slechta DA (2002) Developmental exposure to the pesticides paraquat and maneb and the Parkinson's disease phenotype. Neurotoxicology 23(4–5):621–633

78. Barlow BK, Richfield EK, Cory-Slechta DA, Thiruchelvam M (2004) A fetal risk factor for Parkinson's disease. Dev Neurosci 26(1):11–23. https://doi.org/10.1159/000080707

79. Rappold PM, Cui M, Chesser AS, Tibbett J, Grima JC, Duan L, Sen N, Javitch JA, Tieu K (2011) Paraquat neurotoxicity is mediated by the dopamine transporter and organic cation transporter-3. Proc Natl Acad Sci U S A 108(51):20766–20771. https://doi.org/10.1073/pnas.1115141108

80. Richter F, Gabby L, McDowell KA, Mulligan CK, De La Rosa K, Sioshansi PC, Mortazavi F, Cely I, Ackerson LC, Tsan L, Murphy NP, Maidment NT, Chesselet MF (2017) Effects of decreased dopamine transporter levels on nigrostriatal neurons and paraquat/maneb toxicity in mice. Neurobiol Aging 51:54–66. https://doi.org/10.1016/j.neurobiolaging.2016.11.015

81. Cannon JR, Greenamyre JT (2010) Neurotoxic in vivo models of Parkinson's disease recent advances. Prog Brain Res 184:17–33. https://doi.org/10.1016/S0079-6123(10)84002-6

82. Betarbet R, Sherer TB, MacKenzie G, Garcia-Osuna M, Panov AV, Greenamyre JT (2000) Chronic systemic pesticide exposure reproduces features of Parkinson's disease. Nat Neurosci 3(12):1301–1306. https://doi.org/10.1038/81834

83. Sherer TB, Kim JH, Betarbet R, Greenamyre JT (2003) Subcutaneous rotenone exposure causes highly selective dopaminergic degeneration and alpha-synuclein aggregation. Exp Neurol 179(1):9–16. https://doi.org/10.1006/exnr.2002.8072

84. Hoglinger GU, Feger J, Prigent A, Michel PP, Parain K, Champy P, Ruberg M, Oertel WH, Hirsch EC (2003) Chronic systemic complex I inhibition induces a hypokinetic multisystem degeneration in rats. J Neurochem 84(3):491–502. https://doi.org/10.1046/j.1471-4159.2003.01533.x

85. Alam M, Schmidt WJ (2002) Rotenone destroys dopaminergic neurons and induces parkinsonian symptoms in rats. Behav Brain Res 136(1):317–324. https://doi.org/10.1016/s0166-4328(02)00180-8

86. Fleming SM, Zhu C, Fernagut PO, Mehta A, DiCarlo CD, Seaman RL, Chesselet MF (2004) Behavioral and immunohistochemical effects of chronic intravenous and subcutaneous infusions of varying doses of rotenone. Exp Neurol 187(2):418–429. https://doi.org/10.1016/j.expneurol.2004.01.023

87. Lapointe N, St-Hilaire M, Martinoli MG, Blanchet J, Gould P, Rouillard C, Cicchetti F (2004) Rotenone induces non-specific central nervous system and systemic toxicity. FASEB J 18(6):717–719. https://doi.org/10.1096/fj.03-0677fje

88. Chen Y, Zhang DQ, Liao Z, Wang B, Gong S, Wang C, Zhang MZ, Wang GH, Cai H, Liao FF, Xu JP (2015) Anti-oxidant polydatin (piceid) protects against substantia nigral motor degeneration in multiple rodent models of Parkinson's disease. Mol Neurodegener 10:4. https://doi.org/10.1186/1750-1326-10-4

89. Cannon JR, Tapias V, Na HM, Honick AS, Drolet RE, Greenamyre JT (2009) A highly reproducible rotenone model of Parkinson's disease. Neurobiol Dis 34(2):279–290

90. Ravenstijn PG, Merlini M, Hameetman M, Murray TK, Ward MA, Lewis H, Ball G, Mottart C, de Ville de Goyet C, Lemarchand T, van Belle K, O'Neill MJ, Danhof M, de Lange EC (2008) The exploration of rotenone as a toxin for inducing Parkinson's disease in rats, for application in BBB transport and PK-PD experiments. J Pharmacol Toxicol Methods 57(2):114–130. https://doi.org/10.1016/j.vascn.2007.10.003

91. Mulcahy P, Walsh S, Paucard A, Rea K, Dowd E (2011) Characterisation of a novel model of Parkinson's disease by intra-striatal infusion of the pesticide rotenone. Neuroscience 181:234–242. https://doi.org/10.1016/j.neuroscience.2011.01.038

92. Xiong N, Huang J, Zhang Z, Zhang Z, Xiong J, Liu X, Jia M, Wang F, Chen C, Cao X, Liang Z, Sun S, Lin Z, Wang T (2009) Stereotaxical infusion of rotenone: a reliable rodent model for Parkinson's disease. PLoS One 4(11):e7878. https://doi.org/10.1371/journal.pone.0007878

93. Pan-Montojo F, Anichtchik O, Dening Y, Knels L, Pursche S, Jung R, Jackson S, Gille G, Spillantini MG, Reichmann H, Funk RH (2010) Progression of Parkinson's disease pathology is reproduced by intragastric administration of rotenone in mice. PLoS One 5(1):e8762. https://doi.org/10.1371/journal.pone.0008762

94. Mulcahy P, O'Doherty A, Paucard A, O'Brien T, Kirik D, Dowd E (2013) The behavioural and neuropathological impact of intranigral AAV-alpha-synuclein is exacerbated by systemic infusion of the Parkinson's disease-associated pesticide, rotenone, in rats. Behav Brain Res 243:6–15. https://doi.org/10.1016/j.bbr.2012.12.051

95. Mulcahy P, O'Doherty A, Paucard A, O'Brien T, Kirik D, Dowd E (2012) Development and characterisation of a novel rat model of Parkinson's disease induced by sequential intra-nigral administration of AAV-alpha-synuclein and the pesticide, rotenone. Neuroscience 203:170–179. https://doi.org/10.1016/j.neuroscience.2011.12.011

96. George S, Mok SS, Nurjono M, Ayton S, Finkelstein DI, Masters CL, Li QX, Culvenor JG (2010) Alpha-Synuclein transgenic mice reveal compensatory increases in Parkinson's disease-associated proteins DJ-1 and parkin and have enhanced alpha-synuclein and PINK1 levels after rotenone treatment. J Mol Neurosci 42(2):243–254. https://doi.org/10.1007/s12031-010-9378-1

97. Chiu CC, Yeh TH, Lai SC, Wu-Chou YH, Chen CH, Mochly-Rosen D, Huang YC, Chen YJ, Chen CL, Chang YM, Wang HL, Lu CS (2015) Neuroprotective effects of aldehyde dehydrogenase 2 activation in rotenone-induced cellular and animal models of parkinsonism. Exp Neurol 263:244–253. https://doi.org/10.1016/j.expneurol.2014.09.016

98. Thiffault C, Langston JW, Di Monte DA (2000) Increased striatal dopamine turnover following acute administration of rotenone to mice. Brain Res 885(2):283–288. https://doi.org/10.1016/s0006-8993(00)02960-7

Chapter 2

Assessment of Nonmotor Symptoms in Rodent Models of Parkinson's Disease

Francesca Rossi, Manolo Carta, and Elisabetta Tronci

Abstract

In the last decades, several experimental animal models have been developed in order to elucidate molecular mechanisms involved in Parkinson's disease (PD) neuropathology and to develop new therapeutic strategies. The toxin-based rodent models of PD offer a valuable tool as they allow to mimic many of the features of the disease, particularly motor symptoms such as akinesia and bradykinesia. Interestingly, PD nonmotor symptoms, such as neuropsychiatric, cognitive, and other autonomic dysfunctions, which often precede motor alterations by several years, can be also reproduced in experimental animal models, although with some limitation.

This chapter will provide an overview of the PD nonmotor symptoms in the most common toxin-based rodent models, highlighting their translational value for clinical investigations.

Key words PD, Rodent models, 6-OHDA, MPTP, Rotenone, Paraquat, Nonmotor symptoms

1 Introduction

Parkinson's disease (PD) is the second most common neurodegenerative disease, resulting from the progressive degeneration of the dopaminergic neurons in the substantia nigra pars compacta (SNc), and the subsequent reduction of dopamine (DA) content in the dorsal striatum [1]. However, it is now well accepted that other systems, such as the serotonergic, noradrenergic, glutamatergic, GABAergic and cholinergic ones are severely involved in the pathophysiology of the disease [2]. PD is classified as a motor disorder, being the principal motor symptoms akinesia and bradykinesia, consisting of poverty in the voluntary movements and slow movements, respectively. In addition to these motor manifestations, PD patients also experience rigidity and tremor [3], as well as gait abnormalities, which strongly affects the daily-life activities [4, 5]. Motor symptoms often emerge in one side of the body, affecting

Santiago Perez-Lloret (ed.), *Clinical Trials In Parkinson's Disease*, Neuromethods, vol. 160,
https://doi.org/10.1007/978-1-0716-0912-5_2, © Springer Science+Business Media, LLC, part of Springer Nature 2021

the body posture, with axial and limb rigidity and lack of arm movements during walking, but soon affect also the contralateral side.

Although PD diagnosis is based on the appearance of motor impairments, a number of nonmotor symptoms, unresponsive to the classical medical treatments, are now well recognized [6]. Thus, nonmotor symptoms, which include cognitive impairments, depression, anxiety, sleeping dysfunction, and psychosis-like behaviors, as well as gastrointestinal and sensory dysfunctions, are also becoming increasingly relevant in the diagnosis of the disease, as well as in the pharmacological management [3, 7].

Depressive disorders affect about 40% of PD patients, and may develop during the early phases of the pathology [8]. Similarly, a number of patients display anxiety [9]. Although, the molecular mechanisms involved in psychiatric symptoms still need to be fully elucidated, several neurotransmitter systems are involved [9]. Furthermore, a smaller but significant percentage of PD patients, also experiences cognitive impairments (particularly in advanced stages of disease), as dysfunctions in learning and memory, where both striatum and frontal cortex appear to be implicated [10].

In addition to neuropsychiatric symptoms, other prodromal symptoms may manifest years before the onset of motor impairments. Among them, olfactory and gastrointestinal dysfunctions affect over 70% of PD patients [11–13]. Furthermore, REM behavioral disorders (RBD) are observed in 60–80% of patients, and may appear up to 10 years before PD diagnosis [13].

To date, the most effective therapy for PD is represented by oral administration of the DA precursor L-DOPA, which is able to alleviate most of the motor symptoms of the disease, ameliorating patients' quality of life. Unfortunately, positive effects of L-DOPA therapy are not permanent and the vast majority of patients develop motor side effects including motor fluctuations, a decrease in the duration of L-DOPA effect, as well as L-DOPA-induced dyskinesia (LID) [14]. Moreover, nonmotor symptoms do not respond to L-DOPA even in early stages. Thus, the understanding of the mechanisms involved in motor and nonmotor symptoms of PD represents an important goal in preclinical and clinical research.

In this context, animal models are widely used in preclinical studies not only to understand the molecular mechanisms of PD, but also to develop new therapeutic strategies. Although the animal models currently available present many limitations due to their inability to recapitulate all the motor and nonmotor features of the human illness, over the years these models allowed to clarify important aspects of the disease. The first contribution came from the evidence that administration of reserpine in rabbits induced depletion of monoamines, including DA, and subsequent transient parkinsonian-like state that was reverted by the administration of L-DOPA [15]. These findings posed the basis for the use of animal models as a tool to investigate PD pathology. Thus, pharmacological, genetic, and toxin-based models have been developed, in the last decades, providing a valid tool for preclinical investigation of PD and development of new therapeutic strategies [16].

In this chapter, we will focus on the assessment of nonmotor symptoms in toxin-based rodent models of PD. Among all, the most widely used are the 6-hydroxydopamine (6-OHDA) and 1-methyl-1,2,3,6-tetrahydropyridine (MPTP); in addition, the rotenone- and paraquat-induced models of PD will also be described.

2 6-OHDA Rodent Models of PD

The neurotoxin 6-OHDA was first discovered in the 1950s [17], and nowadays represents the most used neurotoxin to model PD in rodents. 6-OHDA is the hydroxylated analog of DA and it is able to cause lesion of the DA nigrostriatal pathway both in rats [18] and mice [19, 20]. Due to its hydrophilic properties, 6-OHDA is unable to cross the brain-blood barrier; for this reason, it requires a direct injection into the brain in order to cause cell death [18]. It has been demonstrated that, once injected, 6-OHDA uptake takes place in DA neurons via the DA transporter (DAT) and accumulates in the cytosol, where it stimulates the production of reactive oxygen species, leading to oxidative stress and subsequent cell damage (Fig. 1). Particularly, it appears that 6-OHDA triggers the production of toxic species as quinones, hydrogen peroxide, superoxide radicals due to mechanisms of auto-oxidation [21]. In addition to this, the effect of 6-OHDA appears to be deleterious for the mitochondria, specifically for the mitochondrial complex I, causing the production of superoxide free radicals which, in turn, leads to inhibition of the respiratory chain and oxidative stress [22]. Although 6-OHDA can be administered into both hemispheres and ventricles, bilateral lesion of the striatum produces severe disabilities, such as adipsia, aphagia, and, in turn, mortality, unless injected at a very low dose. Thus, the unilateral lesion, which guaranties a very high survival rate of the animals, is usually preferred [21]. Depending on the toxin dose and injection site, 6-OHDA-lesioned rodent model allows to produce different degrees of DA denervation, and therefore mimic different stages of disease. In particular, when administered into the SNc, or the medial forebrain bundle (MFB), 6-OHDA induces a nearly complete DA neuron loss within 24 h [23, 24]. In contrast, a partial DA degeneration, which occurs over several weeks postinjection, is achievable following delivery of the toxin into the striatum [23, 25, 26]. The latter model, which induces mild motor symptoms, is useful to mimic early stages of disease, and it has been used for conducting studies on the mechanisms of cell death or

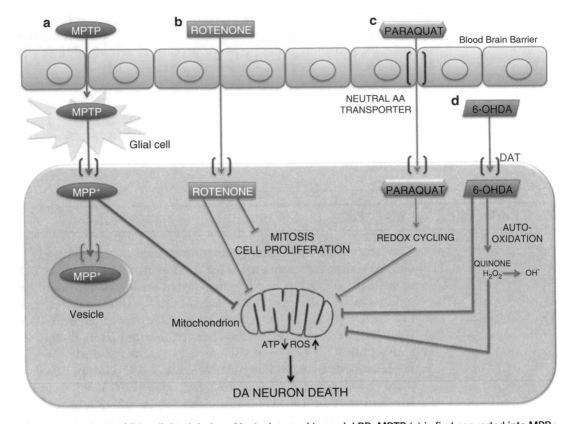

Fig. 1 Mechanisms of DA cell death induced by toxins used to model PD. MPTP (a) is first converted into MPP+ in glial cells, then it enters the DA neuron through the dopamine transporter (DAT) and inhibits the mitochondrial complex I, causing ROS production and ATP depletion and, in turn, DA cell death. MPP+ can be alternatively stored into the synaptic vesicles. Similarly, rotenone (b) and paraquat (c) induce DA cell death through inhibiting complex I. Paraquat also blocks mitosis and cell proliferation. 6-OHDA (d) triggers the production of toxic species as quinones, hydrogen peroxide, superoxide radicals through mechanisms of auto-oxidation and inhibits mitochondrial complex I, causing the production of superoxide free radicals, ATP depletion and cell death

neuroprotection. By contrast, MFB or SNc lesions are useful to test therapeutic strategies aimed at treating motor symptoms.

2.1 Nonmotor Symptoms in 6-OHDA Rodent Models

Whereas the 6-OHDA rodent model has mostly been used to mimic the motor symptoms that characterize this pathology, a number of evidences have demonstrated that this animal model can also be considered a useful tool to study the nonmotor symptoms, which might appear many years before the motor complications. The majority of the studies focused on modeling nonmotor symptoms use striatal injection of 6-OHDA, as it causes a slow retrograde DA neurodegeneration occurring over few weeks [9, 27]. Among nonmotor symptoms, neuropsychiatric symptoms, including cognitive impairment, depression, and anxiety, are extensively studied in rodents subjected to bilateral administration of 6-OHDA into the striatum [9, 28]. In support of this, it has been demonstrated that rats receiving 12 µg of 6-OHDA manifest impairment in several tests used to assess cognitive functions, depression and anxiety-like behaviors. In fact, lesioned rats show a decrease in sucrose consumption, a test used as a measure of anhedonia, and an increased immobility time in the forced swimming test, which is an indicator of behavioral despair [10]. In addition to this, the elevated plus-maze test revealed an increased anxiety-related behavior after 3 weeks of surgery, as 6-OHDA-lesioned rats tended to spend less time on open arms, compared to sham control rats [10]. Finally, cognitive performance assessed with water maze task revealed that lesioned rats spend significantly more time to individuate the platform [10]. Similarly, nonmotor symptoms can also be modeled in mice. In fact, bilateral injection of 4 µg of 6-OHDA (in 1 µl) into the mouse dorsal striatum, by producing a partial DA depletion, allows to reproduce an early stage of PD, where impairment in the long-term recognition memory is evident [29].

Although MFB lesion produces a rapid DA degeneration, mimicking a later stage of the pathology, evidence in the literature demonstrated that some of the nonmotor symptoms might be investigated in this model. In fact, rats subjected to unilateral injection of 6-OHDA in MFB show a reduced preference for a sweet solution, in the sucrose preference test, suggesting that 6-OHDA lesioned rats presented a depressive-like behavior. However, these rats do not show evidence for anxiety-related behavior, in the elevated plus maze test [30]. Moreover, it has been demonstrated that administration of reboxetine, an inhibitor of noradrenaline reuptake, reverses the anxiety and depressive-like behavior, in 6-OHDA-lesioned mice [29], suggesting that lesion of noradrenaline neurons in the locus coeruleus may play a role also in nonmotor symptoms.

In addition to neuropsychiatric symptoms, other nonmotor symptoms are also present in 6-OHDA-lesioned rodents. Among them, gastrointestinal dysfunctions have been assessed. Unilateral injection of 12 µg of 6-OHDA into the SNc produces a delay in stomach emptying and a decrease in colon motility, in rats [31]. Furthermore, bilateral 6-OHDA injection into the mouse striatum was shown to impair olfactory discrimination, which is one of the earliest symptoms of PD. The olfactory deficit was analyzed by measuring the decrease in exploratory behavior when animals were presented to the same odor, in the so-called habituation phase, and the time spent to explore a new odor, in the dishabituation phase [29].

Lastly, alterations in sleep architecture were also observed in lesioned rats, that closely replicate RBD seen in PD patients, demonstrating the validity of this model in studying the pathophysiological mechanisms of PD sleep disturbances [32].

An important aspect to be considered, when selecting a rodent model for nonmotor symptoms studies, is that the expression of these symptoms involves not only changes in the DA pathway but also in the noradrenergic, serotonergic, and cholinergic neurotransmitter systems [6]. On the other hand, the pathogenesis of nonmotor symptoms still needs to be fully elucidated. For this reason, it is difficult, at present, to clearly understand the role that each neurotransmitter system plays in the appearance of these symptoms.

3   MPTP-Treated Mouse Model of PD

MPTP was first discovered in 1982 when exposure of young drug users to illicit meperidine-containing MPTP induced both behavioral changes and DA neuronal damage, closely mimicking motor symptoms of idiopathic PD [33]. MPTP is a lipophilic molecule that crosses the brain blood barrier when administered systemically [34]; for this reason, very stringent safety precaution must be taken when using this neurotoxin [35]. After crossing the brain blood barrier, MPTP is converted to its oxidized metabolite 1-methyl-4-phenylpyridinium (MPP+) in the astrocytes by monoamine oxidase B (MAO-B) in a two-step reaction (Fig. 1). MPP+ enters the DA neurons via the DA transporter and can be accumulated into vesicles by the vesicular monoamine transporter (VMAT2). Once inside the neuron, MPP+ acts at the mitochondrial levels inhibiting the mitochondrial complex I, with subsequent arrest of the electron transport chain. This causes, in turn, production of ROS and reduction of ATP production [36]. The massive production of ROS is also induced by the auto-oxidation of DA released, which is from the vesicles due to accumulation of MPP+. These events trigger dopaminergic neurodegeneration, as a consequence of activation of apoptotic molecular pathways.

To date, the MPTP-treated nonhuman primate (including macaques and marmosets) remains the gold standard experimental model for preclinical evaluation of the efficacy of antiparkinsonian therapies [37]. Indeed, systemic administration of MPTP to nonhuman subjects induces a parkinsonian syndrome showing bradykinesia, postural deficit and reduction in movement [38]. A subjective rating scale has been developed in order to determine motor disability and evaluate the efficacy of symptomatic therapies [37, 39, 40]. However, most studies on PD preclinical research have been performed in mice, mostly for both ethical and financial issues concerning the use of primates.

Few injections of MPTP in mice reproduce the main hallmarks of the disease by inducing selective damage of the nigrostriatal dopaminergic pathway, as well as mitochondrial dysfunction with oxidative stress, which have been shown to be also implicated in the cell death that occurs in PD [41]. However, several studies have reported that the regimen of MPTP administration is a key factor for manifestation of symptomatic and histopathological hallmarks of PD [42]. The most popular regimens of intoxication are acute or sub-acute. Acute intoxication consists of four injections of MPTP at a concentration of 20 mg/kg every 2 h [43], leading to a striatal DA depletion of about 70% in 7 days, and rapid cell death that contrasts with the characteristic slow progression of the human disease. Sub-acute intoxication comprises daily injection of MPTP at a concentration of 30 mg/kg for 5 days, with about 40% of DA depletion in 21 days, and 30% of DA cell loss [35, 44]. In addition, MPTP can be chronically administered via osmotic minipumps or in association to the clearance inhibitor probenecid for several weeks. It has been reported that, in contrast with the acute or sub-chronic MPTP intoxication, the chronic regime over several weeks is able to induce a progressive and persistent DA neurodegeneration associated with motor deficits and chronic inflammatory responses, thus reproducing several aspects of PD [45–48].

The MPTP mouse model is mainly used to test mechanisms of cell death and allows to investigate molecular mechanisms observed in PD, through the use of genetically modified mouse models [49, 50]. A consideration that must be taken into account when using toxin based animal models of PD, and particularly the MPTP mouse model, is that different strains of animal might present a significant variability in the sensitivity to the various toxins; moreover, gender, age, and body weight can influence the reproducibility of the lesion. In support of this, it has been demonstrated that female mice are more sensitive to the MPTP lesion, therefore presenting a higher mortality rate, when compared to male mice [51].

The use of MPTP presents some advantages when compared to the 6-OHDA model. First of all, MPTP crosses the brain–blood barrier, for this reason it can be systemically administered, without stereotaxic surgery, as required for inducing the 6-OHDA lesion. In addition to this, MPTP produces a bilateral lesion, which is more representative to that occurring in PD. Finally, the MPTP model develops specific features of the pathology, such as increased levels of glutamate in the SNc. Importantly, inflammatory markers are also highly expressed in SNc and striatum of the MPTP-treated mice [52]. On the other hand, 6-OHDA model reproduces several behavioral features of PD and it has a higher reproducibility, compared to the MPTP model [51]. However, a limitation of both models is the absence of the Lewy-body formations characteristic of idiopathic PD [35].

3.1 Nonmotor Symptoms in the MPTP-Treated Mouse Model

A number of nonmotor symptoms can be observed in MPTP-treated mice although the transient nature of the symptomatology makes the analysis difficult. In particular, mice display impairments in associative memory and conditioned fear, measured with social transmission of food preference and auditory fear conditioning task, respectively, after 30 days of MPTP injected at a dose of 20 mg/kg for 4 days. In contrast, no signs of anxiety and depression were observed, when this regimen of intoxication was applied [53]. MPTP-treated mice might also develop alteration in the REM sleep, resulting in an increase during light and dark phase, when mice are subjected to a regimen dose of 25 mg/kg over 5 days, that caused a 30% decrease in DA neuron in the SNc. In addition, no gross GI dysfunctions are evident in the MPTP mouse model. It has been observed that mice developed a temporary increase in colon motility, which is in contrast to what occurs in PD patients [54]. Changes in olfactory function have been observed in mice subjected to intranasal administration of MPTP. These changes were accompanied by reduction of DA in the prefrontal cortex, olfactory bulb and striatum, and NE in the hippocampus, olfactory bulb, and striatum, both in mice and rats [55, 56]. Furthermore, it has been recently suggested that mice intranasally treated with MPTP suffer for cognitive, emotional and motor functions and that this condition was correlated with a time-dependent loss of tyrosine hydroxylase in the olfactory bulb and SNc, resulting in significant DA depletion in different brain areas [57]. However, it has been reported that DA was not altered in the olfactory bulb when MPTP was intraperitoneally administered at a concentration of 20 mg/kg for four injections [58], suggesting that, in general, data regarding nonmotor symptoms in the MPTP mouse model are in contrast. In MPTP-treated mice, nonmotor symptoms, as well as motor symptoms, are temporary and strictly dependent on the extent of the dopaminergic neuronal loss; this aspect, together with a high variability among different mouse strains, might limit the use of this model for behavioral studies. On the other hand, this model represents a useful tool to investigate molecular mechanisms involved in neuronal cell death and neuro-inflammation, allowing to test the efficacy of neuroprotective interventions [47].

4   Rotenone Rodent Model of PD

Rotenone is another toxin able to induce DA nigrostriatal degeneration. This herbicide and insecticide, extracted from *Leguminosae* plants, is a lipophilic molecule that crosses the brain-blood barrier and has a half-life of 3–5 days. Exposure to rotenone has been linked to a higher risk of PD; in fact, similarly to what occurs with MPTP administration, rotenone enters the DA neurons, it accumulates in the mitochondria, where it blocks the complex I, leading to oxidative stress in several rat brain areas, such as midbrain, olfactory bulb, striatum, and cortex [59] (Fig. 1). In addition to this, rotenone inhibits mitosis and cell proliferation [34]. Of note, rotenone induces not only damage to the CNS, but also at a systemic level, reason why it has a high mortality rate. Interestingly, rats might present intrinsic resistance to the toxin [60].

Among the rodents, rotenone is preferentially used in rats, although some studies have evidenced degeneration in the DA nigrostriatal pathway as well as motor dysfunction in mice that received rotenone orally [61, 62]. Depending on the route of administration, rotenone can induce a different extent of neurotoxicity. Particularly, it has been shown that oral administration is little toxic [62], while chronic administration, via osmotic minipumps, causes α-synuclein aggregation and Lewy body–like formations, as well as degeneration of DA neurons, although with a high extent of variability [59, 60]. Moreover, it has been demonstrated that, when administered intraperitoneally, rotenone induces neurochemical changes and behavioral impairments that are relieved by L-DOPA treatment, although this route of administration has been associated with a high mortality rate [63]. It is important to note that rotenone administration influences not only the dopaminergic system, but also other pathways, including the serotonergic, noradrenergic and cholinergic ones, indicating that this toxin causes extensive heterogeneous lesions.

The use of rotenone to model PD in animals is still debated as it is believed not to have many advantages with respect to the 6-OHDA and MPTP models. However, unlike these two toxins, rotenone is able to induce α-synuclein aggregation and Lewy-body formation [59, 64], offering the opportunity to study the mechanisms of protein aggregation.

4.1 Nonmotor Symptoms in the Rotenone Model

In addition to motor symptoms, it has been demonstrated that rotenone also reproduces the typical nonmotor symptoms observed in PD. In particular, in a first study it was reported that bilateral intranigral infusion of 12 mg of rotenone significantly reduced sense of smell in rats [65]. This model also seems to present some features of gastrointestinal alteration, although results are still debated. Particularly, it was found that intraperitoneal injection of rotenone, at a dose not sufficient to reproduce behavioral impairments (2 mg/kg for 6 weeks), reduced gastrointestinal transit, with concomitant loss of enteric neurons, in rats [66]. Furthermore, administration via the osmotic pump of 3 mg/kg/day delayed gastric empting in rats [67]. Gastrointestinal dysfunctions are also evident in mice, after oral administration of 30 mg/kg of rotenone, where it induces a decrease in fecal pellet, but no differences in gastric empting [68]. Rats treated with rotenone display increased/reduction of slow-wave sleep, during the active and the rest phase of the animals, respectively, reproducing sleep disturbance observed in humans [69]. Moreover, cognitive function was found to be altered in rotenone models, as shown in the novel object task [70]. However, the mechanisms controlling these alterations are not clear and results are mixed; for this reason, further studies would be required. Furthermore, treatment with rotenone was shown to induce both anxiety and depressive-like behavior, in both mice and rats, as shown in elevated plus maze, forced swimming, and sucrose preference tests [71–73].

5   Paraquat Rodent Model of PD

N,N′-dimethyl-4,4′-bipyridinium, known as paraquat, is an herbicide structurally similar to MPP+, that induces DA degeneration in the SNc, through oxidative stress [34] (Fig.1). In particular, paraquat enters the cell causing redox cycling and inhibition of the complex I, which in turn leads to ROS generation [74]. These events cause DA cell death in a Bak-dependent manner, mediated by JNK/c-Jun pathway and caspase-3 [75, 76]. Paraquat is a hydrophilic molecule that crosses the brain blood barrier via the neutral amino acid transporters. The common regimen of intoxication, able to cause a significant reduction in the striatal DA levels, consists of one intraperitoneal injection of paraquat per week, for 3 consecutive weeks, in mice [77]. By contrast, it has been reported that one single injection of paraquat is not sufficient to induce DA cell damage, but it is able to cause microglia activation, which may play a role in triggering the dopaminergic degeneration [78]. The paraquat rodent model of PD has now been widely validated as an experimental model of the disease, although not without debate. In fact, some studies reported high variability in the extent of DA neuronal cell death among animals, or no cell death at all, and also absence of behavioral abnormalities, raising concerns about the reliability of this model to mimic PD [42, 79]. This is possibly due to the fact that the sensitivity to the compound increases over ageing and that different strains of mice respond differently to the toxin [80]. In order to potentiate the toxic effect on DA neurons, paraquat is often used in combination with a fungicide called manganese ethylenebisdithiocarbamate (maneb), which preferentially inhibits complex III in the mitochondrial respiratory chain [79, 81]. It has been suggested that the concomitant exposure to both paraquat and maneb might represent an important environmental risk factor for PD, although further studies would be required [82, 83]. In experimental studies, combined administration of paraquat (10 mg/kg, i.p.) and maneb (30 mg/kg, i.p.) twice weekly for up to 6 weeks, induced a 20–35% of nigrostriatal degeneration in both mice and rats, accompanied by motor deficits [79, 84]. It has also been shown that some animals might experience body weight loss as well as respiration problems, when exposed to paraquat treatment [85]. Unlike 6-OHDA and MPTP models of PD, paraquat intoxication has been associated to an increase in α-synuclein as well as Lewy body formation [86]. For this reason, this toxin is considered a useful tool to study this aspect of the PD pathology [87, 88].

5.1 Nonmotor Symptoms in the Paraquat Model

Paraquat is not often used to model the nonmotor symptoms observed in PD. However, few studies reported that, rats receiving paraquat once a week, for 3 consecutive weeks, display impairment in the nigrovagal pathway that controls gastric functions, with subsequent alteration in the gastric tone and motility [89]. Furthermore, cotreatment with paraquat and maneb cause an impairment in the elevated plus maze test and in the forced swimming test, indicating that the toxin might induce anxiety and depressive-like behavior, respectively [90].

6 Conclusions

Animal models of PD represent valuable tools for the understanding of the pathogenesis of PD, although none of them entirely reproduce the human condition. As reported in this chapter, the 6-OHDA- and MPTP-treated animals remain the most used models to study PD features, thanks to their ability to produce a consistent and reproducible DAergic neurodegeneration, accompanied by motor and nonmotor symptoms, making them useful tools to screen drugs for symptomatic treatment of the disease. However, these models resemble the later stages of PD, thus they are not suitable for neuroprotective investigations. The rotenone and paraquat models, on the opposite, are associated with development of α-synuclein aggregates and Lewy body–like formations, that allow to mimic early stages of PD and study the neurodegenerative process; however, the extent of DA damage produced by these toxins is highly variable, and often not accompanied by motor or nonmotor disturbances.

Thus, although with some limitations, each toxin-based model shows specific peculiarity and represents a valuable tool for preclinical investigation of the mechanisms involved in PD and for studying possible symptomatic or disease modifying strategies.

References

1. Chung KKK, Zhang Y, Lim KL, Tanaka Y, Huang H, Gao J, Ross CA, Dawson VL, Dawson TM (2001) Parkin ubiquitinates the α-synuclein–interacting protein, synphilin-1: implications for Lewy-body formation in Parkinson disease. Nat Med 7(10):1144–1150. https://doi.org/10.1038/nm1001-1144

2. Brichta L, Greengard P, Flajolet M (2013) Advances in the pharmacological treatment of Parkinson's disease: targeting neurotransmitter systems. Trends Neurosci 36(9):543–554. https://doi.org/10.1016/j.tins.2013.06.003

3. Asakawa T, Fang H, Sugiyama K, Nozaki T, Kobayashi S, Hong Z, Suzuki K, Mori N, Yang Y, Hua F, Ding G, Wen G, Namba H, Xia Y (2016) Human behavioral assessments in current research of Parkinson's disease. Neurosci Biobehav Rev 68:741–772. https://doi.org/10.1016/j.neubiorev.2016.06.036

4. Takakusaki K, Saitoh K, Harada H, Kashiwayanagi M (2004) Role of basal ganglia–brainstem pathways in the control of motor behaviors. Neurosci Res 50(2):137–151. https://doi.org/10.1016/j.neures.2004.06.015

5. Takakusaki K (2008) Forebrain control of locomotor behaviors. Brain Res Rev 57(1):192–198. https://doi.org/10.1016/j.brainresrev.2007.06.024

6. McDowell K, Chesselet M-F (2012) Animal models of the non-motor features of Parkinson's disease. Neurobiol Dis 46(3):597–606. https://doi.org/10.1016/j.nbd.2011.12.040

7. Chaudhuri KR, Schapira AHV (2009) Non-motor symptoms of Parkinson's disease: dopaminergic pathophysiology and treatment. Lancet Neurol 8(5):464–474. https://doi.org/10.1016/s1474-4422(09)70068-7

8. Tolosa E, Compta Y, Gaig C (2007) The premotor phase of Parkinson's disease. Parkinsonism Relat Disord 13:S2–S7. https://doi.org/10.1016/j.parkreldis.2007.06.007

9. Schrag A (2004) Psychiatric aspects of Parkinson's disease. J Neurol 251(7). https://doi.org/10.1007/s00415-004-0483-3

10. Tadaiesky MT, Dombrowski PA, Figueiredo CP, Cargnin-Ferreira E, Da Cunha C, Takahashi RN (2008) Emotional, cognitive and neurochemical alterations in a premotor stage model of Parkinson's disease. Neuroscience 156(4):830–840. https://doi.org/10.1016/j.neuroscience.2008.08.035

11. Hawkes CH, Shephard BC, Daniel SE (1997) Olfactory dysfunction in Parkinson's disease. J Neurol Neurosurg Psychiatry 62(5):436–446. https://doi.org/10.1136/jnnp.62.5.436

12. Braak H, Tredici KD, Rüb U, de Vos RAI, Jansen Steur ENH, Braak E (2003) Staging of brain pathology related to sporadic Parkinson's disease. Neurobiol Aging 24(2):197–211. https://doi.org/10.1016/s0197-4580(02)00065-9

13. Bezard E, Fernagut P-O (2014) Premotor parkinsonism models. Parkinsonism Relat Disord 20:S17–S19. https://doi.org/10.1016/s1353-8020(13)70007-5

14. Hauser RA, Rascol O, Korczyn AD, Jon Stoessl A, Watts RL, Poewe W, De Deyn PP, Lang AE (2007) Ten-year follow-up of Parkinson's disease patients randomized to initial therapy with ropinirole or levodopa. Mov Disord 22(16):2409–2417. https://doi.org/10.1002/mds.21743

15. Carlsson A, Lindqvist M, Magnusson TOR (1957) 3,4-dihydroxyphenylalanine and 5-hydroxytryptophan as reserpine antagonists. Nature 180(4596):1200–1200. https://doi.org/10.1038/1801200a0

16. Blesa J, Phani S, Jackson-Lewis V, Przedborski S (2012) Classic and new animal models of Parkinson's disease. J Biomed Biotechnol 2012:1–10. https://doi.org/10.1155/2012/845618

17. Senoh S, Witkop B (1959) Non-enzymatic conversions of dopamine to norepinephrine and trihydroxyphenethylamines. J Am Chem Soc 81(23):6222–6231. https://doi.org/10.1021/ja01532a028

18. Ungerstedt U (1968) 6-hydroxy-dopamine induced degeneration of central monoamine neurons. Eur J Pharmacol 5(1):107–110. https://doi.org/10.1016/0014-2999(68)90164-7

19. da Conceição FSL, Ngo-Abdalla S, Houzel J-C, Rehen SK (2010) Murine model for Parkinson's disease: from 6-OH dopamine lesion to behavioral test. J Vis Exp (35). https://doi.org/10.3791/1376

20. Thiele SL, Warre R, Nash JE (2012) Development of a unilaterally-lesioned 6-OHDA mouse model of Parkinson's disease. J Vis Exp (60). https://doi.org/10.3791/3234

21. Blesa J, Przedborski S (2014) Parkinson's disease: animal models and dopaminergic cell vulnerability. Front Neuroanat 8. https://doi.org/10.3389/fnana.2014.00155

22. Schober A (2004) Classic toxin-induced animal models of Parkinson's disease: 6-OHDA and MPTP. Cell Tissue Res 318(1):215–224. https://doi.org/10.1007/s00441-004-0938-y

23. Kirik D, Rosenblad C, Björklund A (1998) Characterization of behavioral and neurodegenerative changes following partial lesions of the nigrostriatal dopamine system induced by intrastriatal 6-hydroxydopamine in the rat. Exp Neurol 152(2):259–277. https://doi.org/10.1006/exnr.1998.6848

24. Carta M, Carlsson T, Kirik D, Bjorklund A (2007) Dopamine released from 5-HT terminals is the cause of L-DOPA-induced dyskinesia in parkinsonian rats. Brain 130(7):1819–1833. https://doi.org/10.1093/brain/awm082

25. Francardo V, Recchia A, Popovic N, Andersson D, Nissbrandt H, Cenci MA (2011) Impact of the lesion procedure on the profiles of motor impairment and molecular responsiveness to L-DOPA in the 6-hydroxydopamine mouse model of Parkinson's disease. Neurobiol Dis 42(3):327–340. https://doi.org/10.1016/j.nbd.2011.01.024

26. Lundblad M, Andersson M, Winkler C, Kirik D, Wierup N, Cenci MA (2002) Pharmacological validation of behavioural measures of akinesia

and dyskinesia in a rat model of Parkinson's disease. Eur J Neurosci 15(1):120–132. https://doi.org/10.1046/j.0953-816x.2001.01843.x

27. Przedbroski S, Leviver M, Jiang H, Ferreira M, Jackson-Lewis V, Donaldson D, Togasaki DM (1995) Dose-dependent lesions of the dopaminergic nigrostriatal pathway induced by instrastriatal injection of 6-hydroxydopamine. Neuroscience 67(3):631–647. https://doi.org/10.1016/0306-4522(95)00066-r

28. Aarsland D, Påhlhagen S, Ballard CG, Ehrt U, Svenningsson P (2011) Depression in Parkinson disease—epidemiology, mechanisms and management. Nat Rev Neurol 8(1):35–47. https://doi.org/10.1038/nrneurol.2011.189

29. Bonito-Oliva A, Pignatelli M, Spigolon G, Yoshitake T, Seiler S, Longo F, Piccinin S, Kehr J, Mercuri NB, Nisticò R, Fisone G (2014) Cognitive impairment and dentate gyrus synaptic dysfunction in experimental Parkinsonism. Biol Psychiatry 75(9):701–710. https://doi.org/10.1016/j.biopsych.2013.02.015

30. Carvalho MM, Campos FL, Coimbra B, Pêgo JM, Rodrigues C, Lima R, Rodrigues AJ, Sousa N, Salgado AJ (2013) Behavioral characterization of the 6-hydroxidopamine model of Parkinson's disease and pharmacological rescuing of non-motor deficits. Mol Neurodegener 8(1):14. https://doi.org/10.1186/1750-1326-8-14

31. Zhu HC, Zhao J, Luo CY, Li QQ (2011) Gastrointestinal dysfunction in a Parkinson's disease rat model and the changes of dopaminergic, nitric oxidergic, and cholinergic neurotransmitters in myenteric plexus. J Mol Neurosci 47(1):15–25. https://doi.org/10.1007/s12031-011-9560-0

32. Vo Q, Gilmour TP, Venkiteswaran K, Fang J, Subramanian T (2014) Polysomnographic features of sleep disturbances and REM sleep behavior disorder in the unilateral 6-OHDA lesioned hemiparkinsonian rat. Parkinson's Dis 2014:1–8. https://doi.org/10.1155/2014/852965

33. Fahn S (1996) The case of the frozen addicts: how the solution of an extraordinary medical mystery spawned a revolution in the understanding and treatment of Parkinson's disease By J. William Langston and Jon Palfreman. 309 pp. New York, Pantheon, 1996. $25. 0-679-42465-2. N Engl J Med 335(26):2002–2003. https://doi.org/10.1056/nejm199612263352618

34. Jagmag SA, Tripathi N, Shukla SD, Maiti S, Khurana S (2016) Evaluation of models of Parkinson's disease. Front Neurosci 9. https://doi.org/10.3389/fnins.2015.00503

35. Bové J, Perier C (2012) Neurotoxin-based models of Parkinson's disease. Neuroscience 211:51–76. https://doi.org/10.1016/j.neuroscience.2011.10.057

36. Chan P, DeLanney LE, Irwin I, Langston JW, Monte D (1991) Rapid ATP loss caused by 1-methyl-4-phenyl-1,2,3,6-tetrahydropyridine in mouse brain. J Neurochem 57(1):348–351. https://doi.org/10.1111/j.1471-4159.1991.tb02134.x

37. Bezard E, Przedborski S (2011) A tale on animal models of Parkinson's disease. Mov Disord 26(6):993–1002. https://doi.org/10.1002/mds.23696

38. Crossman AR, Mitchell IJ, Sambrook MA (1985) Regional brain uptake of 2-deoxyglucose in N-methyl-4-phenyl-1,2,3,6-tetrahydropyridine (MPTP)—induced parkinsonism in the macaque monkey. Neuropharmacology 24(6):587–591. https://doi.org/10.1016/0028-3908(85)90070-x

39. Bezard E, Imbert C, Deloire X, Bioulac B, Gross CE (1997) A chronic MPTP model reproducing the slow evolution of Parkinson's disease: evolution of motor symptoms in the monkey. Brain Res 766(1–2):107–112. https://doi.org/10.1016/s0006-8993(97)00531-3

40. Fox SH, Brotchie JM (2010) The MPTP-lesioned non-human primate models of Parkinson's disease. Past, present, and future. Prog Brain Res 184:133–157. https://doi.org/10.1016/s0079-6123(10)84007-5

41. Smeyne RJ, Jackson-Lewis V (2005) The MPTP model of Parkinson's disease. Mol Brain Res 134(1):57–66. https://doi.org/10.1016/j.molbrainres.2004.09.017

42. Duty S, Jenner P (2011) Animal models of Parkinson's disease: a source of novel treatments and clues to the cause of the disease. Br J Pharmacol 164(4):1357–1391. https://doi.org/10.1111/j.1476-5381.2011.01426.x

43. Jackson-Lewis V, Jakowec M, Burke RE, Przedborski S (1995) Time course and morphology of dopaminergic neuronal death caused by the neurotoxin 1-methyl-4-phenyl-1,2,3,6-tetrahydropyridine. Neurodegeneration 4(3):257–269. https://doi.org/10.1016/1055-8330(95)90015-2

44. Perier C, Bove J, Wu DC, Dehay B, Choi DK, Jackson-Lewis V, Rathke-Hartlieb S, Bouillet P, Strasser A, Schulz JB, Przedborski S, Vila M (2007) Two molecular pathways initiate mitochondria-dependent dopaminergic neurodegeneration in experimental Parkinson's disease. Proc Natl Acad Sci 104(19):8161–8166. https://doi.org/10.1073/pnas.0609874104

45. Fornai F, Schluter OM, Lenzi P, Gesi M, Ruffoli R, Ferrucci M, Lazzeri G, Busceti CL,

34 Francesca Rossi et al.

Pontarelli F, Battaglia G, Pellegrini A, Nicoletti F, Ruggieri S, Paparelli A, Sudhof TC (2005) Parkinson-like syndrome induced by continuous MPTP infusion: convergent roles of the ubiquitin-proteasome system and -synuclein. Proc Natl Acad Sci 102(9):3413–3418. https://doi.org/10.1073/pnas.0409713102

46. Sonsalla PK, Zeevalk GD, German DC (2008) Chronic intraventricular administration of 1-methyl-4-phenylpyridinium as a progressive model of Parkinson's disease. Parkinsonism Relat Disord 14:S116–S118. https://doi.org/10.1016/j.parkreldis.2008.04.008

47. Schintu N, Frau L, Ibba M, Garau A, Carboni E, Carta AR (2009) Progressive dopaminergic degeneration in the chronic MPTPp mouse model of Parkinson's disease. Neurotox Res 16(2):127–139. https://doi.org/10.1007/s12640-009-9061-x

48. Bezard E, Jaber M, Gonon F, Boireau A, Bloch B, Gross CE (2000) Adaptive changes in the nigrostriatal pathway in response to increased 1-methyl-4-phenyl-1,2,3,6-tetrahydropyridine-induced neurodegeneration in the mouse. Eur J Neurosci 12(8):2892–2900. https://doi.org/10.1046/j.1460-9568.2000.00180.x

49. Vila M, Jackson-Lewis V, Vukosavic S, Djaldetti R, Liberatore G, Offen D, Korsmeyer SJ, Przedborski S (2001) Bax ablation prevents dopaminergic neurodegeneration in the 1-methyl- 4-phenyl-1,2,3,6-tetrahydropyridine mouse model of Parkinson's disease. Proc Natl Acad Sci 98(5):2837–2842. https://doi.org/10.1073/pnas.051633998

50. Dauer W, Kholodilov N, Vila M, Trillat AC, Goodchild R, Larsen KE, Staal R, Tieu K, Schmitz Y, Yuan CA, Rocha M, Jackson-Lewis V, Hersch S, Sulzer D, Przedborski S, Burke R, Hen R (2002) Resistance of -synuclein null mice to the parkinsonian neurotoxin MPTP. Proc Natl Acad Sci 99(22):14524–14529. https://doi.org/10.1073/pnas.172514599

51. Jackson-Lewis V, Przedborski S (2007) Protocol for the MPTP mouse model of Parkinson's disease. Nat Protoc 2(1):141–151. https://doi.org/10.1038/nprot.2006.342

52. Hébert G, Arsaut J, Dantzer R, Demotes-Mainard J (2003) Time-course of the expression of inflammatory cytokines and matrix metalloproteinases in the striatum and mesencephalon of mice injected with 1-methyl-4-phenyl-1,2,3,6-tetrahydropyridine, a dopaminergic neurotoxin. Neurosci Lett 349(3):191–195. https://doi.org/10.1016/s0304-3940(03)00832-2

53. Vučković MG, Wood RI, Holschneider DP, Abernathy A, Togasaki DM, Smith A, Petzinger GM, Jakowec MW (2008) Memory, mood, dopamine, and serotonin in the 1-methyl-4-phenyl-1,2,3,6-tetrahydropyridine-lesioned mouse model of basal ganglia injury. Neurobiol Dis 32(2):319–327. https://doi.org/10.1016/j.nbd.2008.07.015

54. Anderson G, Noorian A, Taylor G, Anitha M, Bernhard D, Srinivasan S, Greene J (2007) Loss of enteric dopaminergic neurons and associated changes in colon motility in an MPTP mouse model of Parkinson's disease. Exp Neurol 207(1):4–12. https://doi.org/10.1016/j.expneurol.2007.05.010

55. Prediger R, Batista L, Medeiros R, Pandolfo P, Florio J, Takahashi R (2006) The risk is in the air: intranasal administration of MPTP to rats reproducing clinical features of Parkinson's disease. Exp Neurol 202(2):391–403. https://doi.org/10.1016/j.expneurol.2006.07.001

56. Prediger RDS, Aguiar AS, Rojas-Mayorquin AE, Figueiredo CP, Matheus FC, Ginestet L, Chevarin C, Bel ED, Mongeau R, Hamon M, Lanfumey L, Raisman-Vozari R (2009) Single intranasal administration of 1-methyl-4-phenyl-1,2,3,6-tetrahydropyridine in C57BL/6 mice models early preclinical phase of Parkinson's disease. Neurotox Res 17(2):114–129. https://doi.org/10.1007/s12640-009-9087-0

57. Prediger RDS, Aguiar AS, Matheus FC, Walz R, Antoury L, Raisman-Vozari R, Doty RL (2011) Intranasal administration of neurotoxicants in animals: support for the olfactory vector hypothesis of Parkinson's disease. Neurotox Res 21(1):90–116. https://doi.org/10.1007/s12640-011-9281-8

58. Dluzen DE (1992) 1-Methyl-4-phenyl-1,2,3,6-tetrahydropyridine (MPTP) reduces norepinephrine concentrations in the olfactory bulbs of male mice. Brain Res 586(1):144–147. https://doi.org/10.1016/0006-8993(92)91385-r

59. Sherer TB, Kim J-H, Betarbet R, Greenamyre JT (2003) Subcutaneous rotenone exposure causes highly selective dopaminergic degeneration and α-synuclein aggregation. Exp Neurol 179(1):9–16. https://doi.org/10.1006/exnr.2002.8072

60. Betarbet R, Sherer TB, MacKenzie G, Garcia-Osuna M, Panov AV, Greenamyre JT (2000) Chronic systemic pesticide exposure reproduces features of Parkinson's disease. Nat Neurosci 3(12):1301–1306. https://doi.org/10.1038/81834

61. Pan-Montojo F, Anichtchik O, Dening Y, Knels L, Pursche S, Jung R, Jackson S, Gille G, Spillantini MG, Reichmann H, Funk RHW (2010) Progression of Parkinson's disease

pathology is reproduced by intragastric administration of rotenone in mice. PLoS One 5(1):e8762. https://doi.org/10.1371/journal.pone.0008762

62. Inden M, Kitamura Y, Abe M, Tamaki A, Takata K, Taniguchi T (2011) Parkinsonian rotenone mouse model: reevaluation of long-term administration of rotenone in C57BL/6 mice. Biol Pharm Bull 34(1):92–96. https://doi.org/10.1248/bpb.34.92

63. Alam M, Schmidt WJ (2004) l-DOPA reverses the hypokinetic behaviour and rigidity in rotenone-treated rats. Behav Brain Res 153(2):439–446. https://doi.org/10.1016/j.bbr.2003.12.021

64. Höglinger GU, Féger J, Prigent A, Michel PP, Parain K, Champy P, Ruberg M, Oertel WH, Hirsch EC (2003) Chronic systemic complex I inhibition induces a hypokinetic multisystem degeneration in rats. J Neurochem 84(3):491–502. https://doi.org/10.1046/j.1471-4159.2003.01533.x

65. Rodrigues LS, Targa ADS, Noseda ACD, Aurich MF, Da Cunha C, Lima MMS (2014) Olfactory impairment in the rotenone model of Parkinson's disease is associated with bulbar dopaminergic D2 activity after REM sleep deprivation. Front Cell Neurosci 8. https://doi.org/10.3389/fncel.2014.00383

66. Drolet RE, Cannon JR, Montero L, Greenamyre JT (2009) Chronic rotenone exposure reproduces Parkinson's disease gastrointestinal neuropathology. Neurobiol Dis 36(1):96–102. https://doi.org/10.1016/j.nbd.2009.06.017

67. Greene JG, Noorian AR, Srinivasan S (2009) Delayed gastric emptying and enteric nervous system dysfunction in the rotenone model of Parkinson's disease. Exp Neurol 218(1):154–161. https://doi.org/10.1016/j.expneurol.2009.04.023

68. Tasselli M, Chaumette T, Paillusson S, Monnet Y, Lafoux A, Huchet-Cadiou C, Aubert P, Hunot S, Derkinderen P, Neunlist M (2013) Effects of oral administration of rotenone on gastrointestinal functions in mice. Neurogastroenterol Motility 25(3):e183–e193. https://doi.org/10.1111/nmo.12070

69. Yi P-L, Tsai C-H, Lu M-K, Liu H-J, Chen Y-C, Chang F-C (2007) Interleukin-1β mediates sleep alteration in rats with rotenone-induced parkinsonism. Sleep 30(4):413–425. https://doi.org/10.1093/sleep/30.4.413

70. Dos Santos ACD, Castro MAV, Jose EAK, Delattre AM, Dombrowski PA, Da Cunha C, Ferraz AC, Lima MMS (2013) REM sleep deprivation generates cognitive and neurochemical disruptions in the intranigral rote-

none model of Parkinson's disease. J Neurosci Res 91(11):1508–1516. https://doi.org/10.1002/jnr.23258

71. Santiago RM, Barbieiro J, Lima MMS, Dombrowski PA, Andreatini R, Vital MABF (2010) Depressive-like behaviors alterations induced by intranigral MPTP, 6-OHDA, LPS and rotenone models of Parkinson's disease are predominantly associated with serotonin and dopamine. Prog Neuro-Psychopharmacol Biol Psychiatry 34(6):1104–1114. https://doi.org/10.1016/j.pnpbp.2010.06.004

72. Bassani TB, Gradowski RW, Zaminelli T, Barbiero JK, Santiago RM, Boschen SL, da Cunha C, Lima MMS, Andreatini R, Vital MABF (2014) Neuroprotective and antidepressant-like effects of melatonin in a rotenone-induced Parkinson's disease model in rats. Brain Res 1593:95–105. https://doi.org/10.1016/j.brainres.2014.09.068

73. Gokul K, Muralidhara (2014) Oral supplements of aqueous extract of tomato seeds alleviate motor abnormality, oxidative impairments and neurotoxicity induced by rotenone in mice: relevance to Parkinson's disease. Neurochem Res 39 (7):1382–1394. https://doi.org/10.1007/s11064-014-1323-1

74. Miller GW (2007) Paraquat: the red herring of Parkinson's disease research. Toxicol Sci 100(1):1–2. https://doi.org/10.1093/toxsci/kfm223

75. Fei Q, McCormack AL, Di Monte DA, Ethell DW (2007) Paraquat neurotoxicity is mediated by a Bak-dependent mechanism. J Biol Chem 283(6):3357–3364. https://doi.org/10.1074/jbc.m708451200

76. Peng J, Mao XO, Stevenson FF, Hsu M, Andersen JK (2004) The herbicide paraquat induces dopaminergic nigral apoptosis through sustained activation of the JNK pathway. J Biol Chem 279(31):32626–32632. https://doi.org/10.1074/jbc.m404596200

77. McCormack AL, Atienza JG, Johnston LC, Andersen JK, Vu S, Di Monte DA (2005) Role of oxidative stress in paraquat-induced dopaminergic cell degeneration. J Neurochem 93(4):1030–1037. https://doi.org/10.1111/j.1471-4159.2005.03088.x

78. Peng J, Stevenson FF, Oo ML, Andersen JK (2009) Iron-enhanced paraquat-mediated dopaminergic cell death due to increased oxidative stress as a consequence of microglial activation. Free Radic Biol Med 46(2):312–320. https://doi.org/10.1016/j.freeradbiomed.2008.10.045

79. Thiruchelvam M, Brockel BJ, Richfield EK, Baggs RB, Cory-Slechta DA (2000) Potentiated and preferential effects of combined paraquat

and maneb on nigrostriatal dopamine systems: environmental risk factors for Parkinson's disease? Brain Res 873(2):225–234. https://doi.org/10.1016/s0006-8993(00)02496-3

80. McCormack AL, Thiruchelvam M, Manning-Bog AB, Thiffault C, Langston JW, Cory-Slechta DA, Di Monte DA (2002) Environmental risk factors and Parkinson's disease: selective degeneration of nigral dopaminergic neurons caused by the herbicide paraquat. Neurobiol Dis 10(2):119–127. https://doi.org/10.1006/nbdi.2002.0507

81. Krackow S, Vannoni E, Codita A, Mohammed AH, Cirulli F, Branchi I, Alleva E, Reichelt A, Willuweit A, Voikar V, Colacicco G, Wolfer DP, Buschmann JUF, Safi K, Lipp HP (2010) Consistent behavioral phenotype differences between inbred mouse strains in the IntelliCage. Genes Brain Behav 9(7):722–731. https://doi.org/10.1111/j.1601-183x.2010.00606.x

82. Tang CC, Poston KL, Dhawan V, Eidelberg D (2010) Abnormalities in metabolic network activity precede the onset of motor symptoms in Parkinson's disease. J Neurosci 30(3):1049–1056. https://doi.org/10.1523/jneurosci.4188-09.2010

83. Tanner CM, Kamel F, Ross GW, Hoppin JA, Goldman SM, Korell M, Marras C, Bhudhikanok GS, Kasten M, Chade AR, Comyns K, Richards MB, Meng C, Priestley B, Fernandez HH, Cambi F, Umbach DM, Blair A, Sandler DP, Langston JW (2011) Rotenone, paraquat, and Parkinson's disease. Environ Health Perspect 119(6):866–872. https://doi.org/10.1289/ehp.1002839

84. Cicchetti F, Lapointe N, Roberge-Tremblay A, Saint-Pierre M, Jimenez L, Ficke BW, Gross RE (2005) Systemic exposure to paraquat and maneb models early Parkinson's disease in young adult rats. Neurobiol Dis 20(2):360–371. https://doi.org/10.1016/j.nbd.2005.03.018

85. Saint-Pierre M, Tremblay M-E, Sik A, Gross RE, Cicchetti F (2006) Temporal effects of paraquat/maneb on microglial activation and dopamine neuronal loss in older rats. J Neurochem 98(3):760–772. https://doi.org/10.1111/j.1471-4159.2006.03923.x

86. Manning-Bog AB, McCormack AL, Li J, Uversky VN, Fink AL, Di Monte DA (2001) The herbicide paraquat causes up-regulation and aggregation of α-synuclein in mice. J Biol Chem 277(3):1641–1644. https://doi.org/10.1074/jbc.c100560200

87. Muthukumaran K, Leahy S, Harrison K, Sikorska M, Sandhu JK, Cohen J, Keshan C, Lopatin D, Miller H, Borowy-Borowski H, Lanthier P, Weinstock S, Pandey S (2014) Orally delivered water soluble Coenzyme Q10 (Ubisol-Q10) blocks on-going neurodegeneration in rats exposed to paraquat: potential for therapeutic application in Parkinson's disease. BMC Neurosci 15(1):21. https://doi.org/10.1186/1471-2202-15-21

88. Singhal NK, Srivastava G, Patel DK, Jain SK, Singh MP (2010) Melatonin or silymarin reduces maneb- and paraquat-induced Parkinson's disease phenotype in the mouse. J Pineal Res 50(2):97–109. https://doi.org/10.1111/j.1600-079x.2010.00819.x

89. Anselmi L, Toti L, Bove C, Hampton J, Travagli RA (2017) A Nigro–Vagal pathway controls gastric motility and is affected in a rat model of parkinsonism. Gastroenterology 153(6):1581–1593. https://doi.org/10.1053/j.gastro.2017.08.069

90. Tinakoua A, Bouabid S, Faggiani E, De Deurwaerdère P, Lakhdar-Ghazal N, Benazzouz A (2015) The impact of combined administration of paraquat and maneb on motor and non-motor functions in the rat. Neuroscience 311:118–129. https://doi.org/10.1016/j.neuroscience.2015.10.021

Chapter 3

Genetic Models of Parkinson's Disease

Philippe Kachidian and Paolo Gubellini

Abstract

Parkinson's disease (PD) is a chronic and evolving neurodegenerative disorder primarily due to the progressive loss of *substantia nigra pars compacta* neurons releasing dopamine. The etiology of idiopathic PD, which represents most of the cases, is still unclear but seems to be multifactorial, associating environmental and/or genetic factors. The major limitation of "classical" toxin-based animal models of PD is that they do not replicate some characterizing features of the pathology. Animal models based on gene mutations connected with PD may overcome, at least partially, this limitation. In this chapter, animal models targeting orthologs of genes linked with PD in humans will be reviewed. These models can represent excellent tools to investigate the specific role of the targeted gene—and its protein product—in PD pathophysiology.

Key words Parkinson's disease, Animal models, Preclinical, Gene mutations, Rodent, Fish

1 Parkinson's Disease

Parkinson's disease (PD) is a chronic and evolving neurodegenerative disorder primarily due to the progressive loss of *substantia nigra pars compacta* (SNc) neurons releasing dopamine (DA). PD is characterized by motor symptoms including bradykinesia, akinesia, rigidity, postural instability, and resting tremor [1–4]. SNc DAergic neurons project to the basal ganglia (BG), a highly organized network of brain nuclei implicated in movement control and motor learning, as well as limbic and cognitive functions, which receives massive projections from the cortex and thalamus. The BG nuclei include the caudate-putamen (or striatum in non primates), which is the main input station and is composed of more than 95% of spiny projection neurons (SPNs) using γ-aminobutyric acid (GABA) as neurotransmitter, and by cholinergic and GABAergic interneurons; the external *globus pallidus* (GPe or GP in nonprimates); the so-called output structures, that is, the *substantia nigra pars reticulata* (SNr) and the internal *globus pallidus* (GPi or entopeduncular nucleus, EP, in nonprimates) that project to the

Santiago Perez-Lloret (ed.), *Clinical Trials In Parkinson's Disease*, Neuromethods, vol. 160,
https://doi.org/10.1007/978-1-0716-0912-5_3, © Springer Science+Business Media, LLC, part of Springer Nature 2021

thalamus and brainstem; and the subthalamic nucleus (STN), which is the second input station and the only intrinsic glutamatergic structure of the BG, all the others being mainly GABAergic. Two distinct efferent pathways (although such segregation is not complete [5]) originate from the striatum: the "direct" pathway, consisting in striatal SPNs expressing mainly D1-like DA receptors and projecting directly to the BG output structures (SNr/GPi), and the "indirect" pathway, composed of SPNs expressing D2-like receptors and projecting to the SNr/GPi indirectly via two intermediate stations: the GPe and the STN [6, 7]. In PD, the loss of SNc DAergic neurons leads to two opposite effects in the BG: the hypoactivity of direct pathway SPNs and, conversely, the hyperactivity of the indirect pathway SPNs resulting in the increased inhibition of the GPe which, in turn, disinhibits the STN [8]. The overall effect is a decreased GABAergic and an increased glutamatergic tone at the level of the SNr/GPi, which become hyperactive leading to excessive inhibition of the thalamus and brainstem, and ultimately to the hypokinetic PD motor symptoms [2, 7, 9]. Moreover, an increased corticostriatal glutamatergic input and an increased striatal cholinergic tone is thought to also contribute to such imbalance in the cortico-BG circuit [10–14]. PD is also associated with several nonmotor neurological symptoms such as autonomic dysfunction, pain, sensory disturbances, mood disorders, and sleep impairment, and it may also lead to psychiatric and cognitive complications such as depression, hallucinations, memory troubles, and, especially (but not only) in advanced cases, dementia [15]. Indeed, the neurodegenerative process of PD affects not only the BG DAergic system but also involves glutamatergic, serotonergic, noradrenergic, GABAergic, and cholinergic systems, as well as other DAergic structures besides the SNc, such as the ventral tegmental area [16]. Another hallmark of PD—with or without dementia—is the presence of intracellular inclusions known as Lewy bodies, which contain aggregates of α-synuclein, ubiquitin and neurofilament proteins [17]. Thus, given the multiple symptoms, the diverse etiologic mechanisms as well as the complex pathophysiological pattern, PD is far more than a motor trouble due to striatal DA depletion. However, despite such complexity, it can be argued that the main hallmarks defining PD are three: (1) the loss of DAergic neurons in the SNc; (2) the presence of Lewy bodies; and (3) hypokinetic motor impairments.

The etiology of idiopathic PD, which represents most of the cases, is still unclear but seems to be multifactorial, associating environmental and genetic factors. Some risk factors have been reported, the major being advanced age, family history, declining estrogen levels, exposure to certain toxic agents, genetic factors, low levels of B9 vitamin (folate), and head trauma [4]. Mitochondrial dysfunctions at the level of complex I of the respiratory chain (NADH dehydrogenase or NADH-coenzyme Q oxidoreductase) have also been described in PD [18, 19]. Finally, genetic forms of PD, also called familial or Mendelian forms, presently explain about 10–20% of all cases at the general population level, and are due to mutations of genes, most of them grouped in the so-called PARK family (*see* refs. 20–22 and https://www.omim.org). Orthologs of most of these genes exist in organisms used as experimental models, in particular the mouse (*Mus musculus*, *see* http://www.informatics.jax.org), but also the rat (*Rattus norvegicus*), the fruit fly (*Drosophila*), the zebrafish (*Danio rerio*), the nematode *Caenorhabditis elegans* (*C. elegans*) and others. More recently, other genes not included in the PARK family have been identified as possible risk factors for PD (*see* following sections).

By cross-checking online databases (mainly https://www.omim.org; https://www.ncbi.nlm.nih.gov/gene; https://www.ncbi.nlm.nih.gov/pubmed, and also others) we found 34 identified genes or loci directly involved in familial PD, or identified as risk factors for developing this pathology, among which 22 (considering PARK1 and PARK4 as the same gene/locus) included in the PARK gene family [23–25]. Table 1 resumes all these genes/loci, their protein product, their mutations identified as PD risk factors, their inheritance mechanism, and whether or not genetic animal models exist. In Table 2, we provide a list of the available animal models targeting orthologs of these genes that could be found in the scientific literature (*see* the abovementioned databases). For each model, we have ranked the ability (or not) to produce the three pathological phenotypes that mostly characterize PD, i.e., the loss of DAergic neurons, the presence of Lewy body–like inclusions and the development of motor troubles, and, notably, that can be detected inter-species. The complete and most relevant bibliography concerning these animal models is also provided in Table 2. Finally, Table 3 resumes other animal models of PD targeting genes that have not been directly involved in PD (or whose linkage needs confirmation) but code for proteins whose mutation can produce a pathological state mimicking one (or more) of those characterizing PD (mitochondrial dysfunction, DAergic neuron loss, etc.).

2   Animal PD Models Targeting PARK Genes

The first mutations responsible for hereditary PD were identified in 1996 in the *SNCA* gene coding for α-synuclein [26, 27]. In the following years, other new heterogeneous chromosomal regions whose mutation leads to familial PD were reported. In the current PD genetics nomenclature, 22 specific chromosomal regions (also called chromosomal loci) are termed "PARK" to denote their confirmed (or putative) link to PD, and numbered in chronological order of their identification. However, this list does not include all

Table 1

Genes whose mutation is involved in familial/inheritable forms of PD, or constitutes a risk factor for developing the pathology. The rightmost column specifies whether animal models targeting ortholog genes exist; details on these models are provided in Table 2

Human condition

Locus	Position (NCBI ID)	Gene	Protein	Mutation type	PD onset	Inh.	References	Genetic model
PARK1	4q21-q22.1 (6622)	SNCA	α-synuclein	Missense; regulatory; gene duplication or triplication	Early (rarely late)	AD	[24–27, 143, 145, 146, 237]	Y
PARK2	6q25.2-q27 (50873)	PRKN	Parkin (or ubiquitin E3 ligase)	Missense or nonsense; regulatory; splicing; small indels; deletions; insertions	Early, sporadic	AR	[24, 42, 43, 238]	Y
PARK3	2p13 (5072)	SPR	Sepiapterin reductase (or 7,8-dihydrobiopterin: NADP+ oxidoreductase)	Single nucleotide polymorphisms	Variable, susceptibility	AD	[25, 50, 51, 239]	Y
PARK4 (see PARK1)	4q21-q22.1 (6622)	SNCA	α-synuclein	Missense; regulatory; gene duplication or triplication	Early (rarely late)	AD	[240, 241] See also PARK1	Y
PARK5	4p13 (7345)	UCH-L1	Ubiquitin C-terminal hydrolase L1 (or PGP9.5)	Missense	Late (controversial)	AD	[25, 56, 57]	Y
PARK6	1p36.12 (65018)	PINK1	PTEN induced putative kinase 1	Missense or nonsense; splicing; small indels; deletions; insertions	Early	AR	[24, 62–64, 238]	Y

Human condition

Locus	Position (NCBI ID)	Gene	Protein	Mutation type	PD onset	Inh.	References	Genetic model
PARK7	1p36.23 (11315)	*DJ-1* (or *DJ1*)	DJ1 (or parkinsonism associated deglycase)	Missense; regulatory; splicing; small indels; deletions; insertions	Early	AR	[24, 72, 73, 238]	Y
PARK8	12q12 (120892)	*LRRK2*	Leucine rich repeat kinase 2 (or dardarin)	Missense; splicing; small deletions	Late, sporadic	AD	[24, 80–83, 143, 145, 238, 242]	Y
PARK9	1p36.13 (23400)	*ATP13A2*	P-type cation transporting ATPase 13A2	Missense; splicing; small indels; deletions; insertions	Early (Kufor Rakeb syndrome, NCL, HSP)	AR	[24, 94–96]	Y
PARK10	1p32 (170534)	*RNF11* *UQCRH* *HIVEP3* *EIF2B3* *USP24* *ELAVL4* *PLPP3* (?)	Several	Single nucleotide polymorphisms	Controversial	?	[25, 102, 103, 243–245]	Y
PARK11	2q35-q37 (26058)	*GIGYF2*	GRB10 interacting GYF protein 2	Missense; small indels	Late (controversial)	AD	[25, 105–107]	Y
PARK12	Xq21-q25 (677662)	*PGK-1* *RAB39B* *ATP1B4* *IL13Rα1*	Phosphoglycerate kinase 1; RAS-associated protein RAB39B; P-type ATPase βm; interlukin 13 receptor α1 subunit	Missense (*PGK-1*); duplication (*RAB39*); single nucleotide variant (*ATP1B4*)	Susceptibility	? AR ? ?	[23, 25, 111–113]	Y

(continued)

Table 1
(continued)

Human condition

Locus	Position (NCBI ID)	Gene	Protein	Mutation type	PD onset	Inh.	References	Genetic model
PARK13	2p13.1 (27429)	*HTRA2*	HtrA serine peptidase 2	Missense; splicing	Late (controversial)	AD	[25, 115, 116]	Y
PARK14	22q13.1 (8398)	*PLA2G6*	Phospholipase A2, group 6	Missense; splicing; deletions; insertions	Early (involved in infantile neuroaxonal dystrophy)	AR	[25, 125, 126]	Y
PARK15	22q12.3 (25793)	*FBXO7* (or *FBX7*)	F-box protein O-type 7	Missense; splicing	Early	AR	[24, 136, 137]	Y
PARK16	1q32 (100359403)	*RAB7L1, SLC41A1,ADORA1*	Several	Single nucleotide polymorphisms	Controversial (seems to lower risk)	AD	[23, 143–146]	N
PARK17	16q11.2 (55737)	*VPS35 GAK*	Vacuolar protein sorting 35	Missense; splicing	Late (controversial)	AD	[24, 146–150]	Y
PARK18	3q27.1 (1981)	*EIF4G1 HLA-DRA*	Eukaryotic translation initiation factor 4 gamma, 1	Missense; deletions; insertions	Late (controversial)	AD	[24, 146, 150, 161]	N
PARK19	1p31.3 (9829)	*DNAJC6*	DnaJ heat shock protein family (Hsp40) member C6 (auxilin)	Missense or nonsense; splicing	Early	AR	[23, 162–165]	Y
PARK20	21q22.11 (8867)	*SYNJ1*	Synaptojanin-1	Missense	Early	AR	[23, 170, 171]	Y*

Human condition

Locus	Position (NCBI ID)	Gene	Protein	Mutation type	PD onset	Inh.	References	Genetic model
PARK21	3q22.1 (23317)	DNAJC13	DnaJ heat shock protein family (Hsp40) member C13	Missense	Early	AD	[23, 182, 183]	Y
PARK22	7p11.2 (51142)	CHCHD2	Coiled-coil helix-coiled-Coil helix domain 2	Missense	Late (controversial)	AD	[23, 187, 188]	Y
PARK23	15q22 (54832)	VPS13C	Vacuolar protein sorting 13C	Missense; small deletion	Early	AR	[23, 192]	Y
–	1q21–22 (2629)	GBA	β-glucocerebrosidase	Missense; regulatory; splicing; small indels; deletions; insertions	Late (involved in Gaucher disease)	AD	[25, 194–196]	Y*
–	12q24.1-q24.12 (6311)	ATXN2 (or SCA2)	Ataxin 2	Expanded CAG repeats	Late, susceptibility (involved in spinocerebellar ataxia type 2)	AD	[25, 197–199]	Y*
–	4q23 (126)	ADHIC (or ADH3)	Alcohol dehydrogenase IC (class I), gamma polypeptide	Single nucleotide polymorphisms; nonsense mutation	Susceptibility	?	[201, 202]	Y
–	6q27 (6908)	TBP	TATA-box binding protein	Expanded CAG/CAA repeats	Susceptibility (involved in spinocerebellar ataxia type 17)	?	[205, 206]	Y*

(continued)

Table 1
(continued)

Human condition

Locus	Position (NCBI ID)	Gene	Protein	Mutation type	PD onset	Inh.	References	Genetic model
–	17q21.31 (4137)	MAPT (or MTBT1)	Microtubule associated protein tau	Single nucleotide polymorphisms	Late, susceptibility (involved in tauopathies)	?	[145, 146, 207]	Y*
–	Xq24 (2747)	GLUD2	Glutamate dehydrogenase-2 (hGDH2)	Single nucleotide polymorphisms	Susceptibility	GR	[208]	Y*
–	20p13-p12.3 (29058)	TMEM230	Transmembrane protein 230	Single nucleotide polymorphisms	Unconfirmed	AD	[23, 209]	N
–	11p15.4 (79608)	RIC3	RIC3 acetylcholine receptor chaperone	Single nucleotide polymorphisms	Unconfirmed	AD	[23, 210]	N
–	2p23.3 (391356)	PTRHD1	Peptidyl-tRNA hydrolase domain containing 1	Single nucleotide polymorphisms	Unconfirmed	AR	[23, 211]	N
–	22q11.2 (undefined)	SMARCB1?	SWI/SNF related, matrix associated, actin dependent regulator of chromatin, subfamily b, member 1?	Deletions	Susceptibility	AD	[23, 212]	N
–	7q32.3 (5420)	PODXL	Podocalyxin like	Insertion	Early	AR	[23, 213]	N

AD autosomal dominant, *AR* autosomal recessive, *GR* gonosomic recessive, *HSP* hereditary spastic paraplegia, *Inh.* inheritance, *N* no genetic animal models reported to date, *NCL* neuronal ceroid lipofuscinosis, *Sp* sporadic, *Y* at least one genetic animal model exists, *Y** models targeting this gene exist, but they have not been studied/characterized as PD models

Table 2
Animal models of PD targeting orthologs of genes implicated in the human pathology (*see* Table 1)

Locus (gene)	Species	Genetic model	DL	LB	MS	References
PARK1, 4 (*SNCA* or *α-synuclein*)	*M. mulatta*	Overexpression of human mutated	+	+	NA	[34]
	Marmoset	Overexpression of human mutated	+	+	+	[35, 36]
	Rat	Overexpression of human WT or mutated, KD	+/−	+/−	+/−	[47, 246–250]
	Mouse	Overexpression of WT (mouse or human), mutated or truncated, KO, overexpression + *Pink1* KO	+/−	+	+/−	[66, 89, 250–259]
	Zebrafish	KO	+	−	+	[37]
	Drosophila	Overexpression of human WT or mutated	+	+	+	[38, 39, 260]
	C. elegans	Overexpression of human WT or mutated	+	+	+	[40, 261]
PARK2 (*PRKN* or *parkin*)	Pig	KO + *Pink1* KO KI *Prkn* + *Pink1* + *DJ-1*	NA	NA	NA	[48, 49]
	Rat	KO, deletions, KI	−	−	−	[47]
	Mouse	Exon deletions, mutations, *Prkn* KO + α-synuclein overexpression, triple KO *Prkn* + *Pink1* + *DJ-1*	+/−	+/−	+/−	[18, 44–46, 89]
	Zebrafish	KD, overexpression	−	NA	NA	[262]
	O. latipes	KO, KO + *Pink1* KO	+	NA	+	[263, 264]
	Drosophila	KO	+	−	+	[38, 39]
	C. elegans	KO, KD	+	+	+	[40, 261, 265, 266]
PARK3 (*SPR*)	Mouse	KO	+	NA	+	[52–54]
	Silkworm	KO	NA	NA	NA	[55]

(continued)

Table 2
(continued)

Locus (gene)	Species	Genetic model	DL	LB	MS	References
PARK5 (*UCH-L1*)	Mouse	*Uch-l1* KO or mutation + α-synuclein expression, overexpression	+	−	+	[58, 59]
	Drosophila	KO	+	NA	+	[60]
PARK6 (*PINK1*)	Pig	KO + KO *Prkn* KI *Prkn* + *Pink1* + *DJ-1*	NA	NA	NA	[48, 49]
	Rat	KO	+/−	NA	+	[47, 67, 68, 267]
	Mouse	KO, KO + α-synuclein overexpression triple KO *Prkn* + *Pink1* + *DJ-1*	+/−	−	+	[18, 45, 65, 66, 268, 269]
	Zebrafish	KO, KD	+/−	NA	NA	[69, 70, 262, 270–272]
	O. latipes	KO, KO + *Prkn* KO	+	NA	+	[263, 264, 273, 274]
	Drosophila	KO, mutations, overexpression	+	−	+	[38, 39, 269, 275–280]
	C. elegans	KO	−	NA	+	[40, 265, 281]
PARK7 (*DJ-1*)	Pig	KO, KI *Prkn* + *Pink1* + *DJ-1*	NA	NA	NA	[49, 78]
	Rat	KO	−	−	NA	[47, 282]
	Mouse	KO, triple KO *Prkn* + *Pink1* + *DJ-1*	+/−	−	+/−	[45, 74–77, 282–284]
	Zebrafish	KD	−	NA	NA	[262]
	O. latipes	KO	NA	NA	NA	[285]
	Drosophila	KO, loss of function, overexpression	+/−	NA	+/−	[38, 39, 122]
	C. elegans	KO, KD	NA	NA	NA	[40, 261, 265]
PARK8 (*LRKK2*)	Rat	KO, point mutations, expression of human (WT or mutated)	+/−	NA	+/−	[47, 286–289]
	Mouse	KO, KI, expression of human (mutated), *Lrrk2* overexpression + mutated *SNCA*	+/−	NA	−	[87–93, 282, 290, 291]
	Zebrafish	KD, mutation	+/−	NA	+/−	[262, 292–294]
	Drosophila	KO, overexpression of human (WT or mutated)	+	−	+	[18, 38, 295–298]
	C. elegans	KO, expression human (WT or mutated)	+	NA	+	[40, 281]

Table 2
(continued)

Locus (gene)	Species	Genetic model	DL	LB	MS	References
PARK9 (*ATP13A2*)	Dog	Spontaneous mutation	–	NA	+	[97, 98]
	Rat	Overexpression of mutated	NA	NA	NA	[99, 282]
	Mouse	KO, heterozygous	–	+	+	[100, 101, 282, 299]
	Zebrafish	KO	Lethal			[300]
	O. latipes	Mutation	+	NA	+	[264, 301]
	C. elegans	Mutation	+	NA	+	[40]
PARK10 (*see* Table 1)	Mouse	KO	–	NA	+	[104]
PARK11 (*GIGYF2*)	Mouse	KO (lethal) and heterozygous	–	+	+	[108]
	Zebrafish	KD	–	NA	NA	[110]
	Drosophila	KD, KO	+	+	+	[109]
PARK12 (*see* Table 1)	Mouse	KO	–	NA	NA	[114]
PARK13 (*HTRA2*)	Mouse	KO, overexpression, KI, mutation	+/–	NA	+	[117–121]
	Drosophila	KD, KO	NA	NA	+	[122–124]
PARK14 (*PLA2G6*)	Dog	Spontaneous mutation	NA	NA	NA	[135]
	Mouse	KO, KI	+	+	+	[127–131]
	Drosophila	KO, KD	NA	NA	+	[133, 134]
	Zebrafish	KO	+	+	+	[132]
PARK15 (*FBXO7*)	Mouse	KO, reduced expression	–	–	+	[138, 139]
	Zebrafish	KD	+	NA	+	[142]
	Drosophila	Overexpression ± *Prkn* KO	+/–	NA	NA	[140, 141]
PARK17 (*VPS35*)	Rat	KI	+	NA	–	[157]
	Mouse	KI, KO, heterozygous	+/–	+	NA	[151–156]
	Drosophila	KI, KD, expression of human (mutated)	+	+	+	[158–160]
PARK19 (*DNAJC6* or auxilin)	Mouse	KO	–	NA	–	[166]
	Zebrafish	KD	NA	NA	NA	[167]
	Drosophila	KD	+	NA	+	[168, 169]

(continued)

Table 2
(continued)

Locus (gene)	Species	Genetic model	DL	LB	MS	References
PARK20 (*SYNJ1*)	Mouse	KO, KD	NA	NA	+	[172–174, 181]
	Zebrafish	Mutation	NA	NA	NA	[175, 176]
	Drosophila	Mutation	NA	NA	NA	[177, 178]
	C. elegans	Mutation	NA	NA	NA	[179, 180]
PARK21 (*DNAJC13*)	*Drosophila*	KI + human *SNCA* overexpression	+	+	+	[184]
	C. elegans	KO	NA	NA	NA	[185, 186]
PARK22 (*CHCHD2*)	Mouse	KO	−	−	−	[189]
	Drosophila	KO, KI	+	NA	+	[190, 191]
PARK23 (*VPS13C*)	*Drosophila*	KD + human *SNCA* overexpression	NA	NA	NA	[193]
(*GBA*)	Rat	KO	NA	+	NA	[302]
	Mouse	KI, KO and heterozygous	NA	+	NA	[303–310]
	Zebrafish	KO, heterozygous	−	+	+	[311]
	O. latipes	KO	+	+	+	[312]
	Drosophila	KI, KD	+/−	NA	+	[38]
(*ATX2* or *ATXN2* or *SCA2*)	Mouse	Transgenic, KI, KO, expression of human (mutated)	NA	NA	+	[313–317]
(*ADH1C* or *ADH3*)	Mouse	KO	NA	NA	NA	[203]
(*TBP*)	Rat	Transgenic	NA	NA	NA	[318, 319]
	Mouse	Transgenic, KI	NA	NA	+/−	[319–322]
	X. laevis	KD	NA	NA	NA	[323]
(*MAPT* or *MTBT1*)	Mouse	Transgenic, KO	NA	NA	NA	[324–327]
	Zebrafish	Expression of human	NA	NA	NA	[328]
	C. elegans	Transgenic	NA	NA	+	[329, 330]
	Drosophila	Misexpression	+	NA	NA	[331, 332]
(*GLUD2*)	Mouse	Transgenic, KO	NA	NA	NA	[333–338]

DL DAergic neuron loss, *KD* knockdown, *KI* knockin (mostly of a human mutated form of the gene), *KO* knockout, *LB* Lewy body–like inclusions (or α-synuclein-positive aggregates), *MS* motor signs, *WT* wild-type
Scale: (−) absent; (+/−) debated or depending on the genetic approach; (+) typically present; NA, not assessed, not applies or not tested in the context of PD

Table 3
Animal PD models targeting other genes

Targeted gene	Organism	Genetic approach	DL	LB	MS	References
Sonic hedgehog (*Shh*)	Mouse	KO in DAergic neurons	+	−	+/−	[214]
Nuclear receptor related transcription factor 1 (*Nurr1*)	Mouse	KO or heterozygous	+	−	+/−	[215, 219, 222, 339]
Paired-like homeodomain transcription factor 3 (*Pitx3*)	"Aphakia" mouse	KO	+	−	+/−	[216, 225–227, 340–343]
	Zebrafish	KD	NA	NA	NA	[228]
Engrailed-1 (*En1*)	Mouse	Heterozygous (KO lethal)	+	−	+/−	[217, 344, 345]
Mitochondrial transcription factor A (*Tfam*)	"MitoPark" mouse	KO in DA neurons	+	+/−	+	[218, 232, 346]

DL DAergic neuron loss, *KD* knockdown, *KO* knockout, *LB* Lewy body–like inclusions (or α-synuclein-positive aggregates), *MS* motor signs
Scale: (−) absent; (+/−) debated or depending on the genetic approach; (+) typically present; NA, not assessed, not applies or not tested in the context of PD

genes that have been linked to PD and, unfortunately, has a number of inconsistencies: it comprises confirmed loci as well as those for which the linkage or association with PD could not be replicated or is controversial (PARK3, 5, 10, 11, 16–18), and some loci include multiple genes among which the PD-causative one is not yet identified (PARK10, 12). Moreover, variations in some of these genes are considered as risk factors for PD, rather than being a sufficient cause. Finally, one locus, PARK4, was wrongly designated as a novel chromosomal region associated with PD, but later was found corresponding to PARK1 [22]. In this section we will focus on these 22 genes/loci of the PARK gene family and the corresponding animal models.

2.1 SNCA or α-Synuclein (PARK1/ PARK4)

PARK1/PARK4 (or *SNCA*) codes for α-synuclein (Table 1), a protein whose exact functions are still largely unknown. It has been reported that α-synuclein interacts with tubulin [28] and may participate in synaptic vesicle trafficking by interacting with the SNARE complex [29, 30]. PD linked to autosomal dominant *SNCA* point mutations was first characterized in Italian and Greek families [26]. These mutations, most commonly A30P or A53T, seem to accelerate α-synuclein aggregation and thus the formation of Lewy bodies, resulting in early-onset PD [31], while *SNCA* duplication or triplication is supposed to increase the risk of sporadic PD [32]. In addition, phosphorylation of α-synuclein at the ser129 site seems to play a role in its toxicity and ability to form pathogenic aggregates [33].

SNCA is obviously one of the most targeted genes for generating animal PD models (Table 2), and the only one to date that has been utilized in non-human primates, that is, *Macaca mulatta* (rhesus macaque) [34] and *Callithrix jacchus* (marmoset) [35, 36]. By a viral vector-mediated approach, the animals expressed mutant (A53T) α-synuclein in the brain and showed significant DAergic neuron loss accompanied by Lewy body–like inclusions and, in the marmoset, progressive motor impairment was described. In order to study the pathophysiological role of α-synuclein, common genetic approaches consist in overexpressing wild-type or mutated forms of this protein, but some works (e.g., in mouse and zebrafish) also report the use of knockout (KO) or knockdown (KD) models. In rats and mice models, which are abundant (*see* Table 2 for references), the most reproducible feature seems the presence Lewy body–like inclusions, while DAergic neuron loss and motor troubles are absent or controversial. Conversely, zebrafish [37], *Drosophila* [38, 39] and *C. elegans* [40] show systematically DAergic neuron loss and motor impairments, the two latter also producing Lewy body–like inclusions. For these reasons, and considering the progressive propagation of α-synuclein in these models, rodents seem more suited to study α-synucleopathies rather than strictly PD pathophysiology, while primate models seem much better for modeling PD. Strikingly, zebrafish and invertebrates also seem to be powerful models due to the rich pathological phenotype.

2.2 PRKN or Parkin (PARK2)

Parkin (Table 1) is an ubiquitin E3 ligase that participates in clearing damaged mitochondria [41]. Autosomal recessive, juvenile forms of PD linked to the *PRKN* gene were first identified in Japanese families [42, 43]. Mutations generally consist of deletions (in exons 2, 3, and 8), duplications (in exons 2–4 and 9), or point mutations (P437L).

There are several mouse models typically produced by fully or partially (exon 2, 3 and 7) deleting the *Prkn* gene, and one paper reports combined *Prkn* deletion coupled to the expression of mutated a-synuclein [44], while another consist in a triple KO of *Prkn*, *Pink1*, and *DJ-1* [45]. None of these models shows DAergic cell loss, Lewy body–like inclusions and motor deficits, but some exhibit changes in striatal glutamatergic activity and mitochondrial impairment in the brain (Table 2). Another mouse model consists in the expression of a C-terminal truncated human mutant parkin (Parkin-Q311X) and is the only one presenting DAergic neuron loss, α-synuclein inclusions in the SNc, and late-onset and progressive motor impairments [46]. Rat *Prkn* models using KO, knockin (KI) and partial deletion approaches also do not show PD-like features [47]. Conversely, *Drosophila* and *C. elegans* KO or KD models targeting *PRKN* homolog genes show DAergic cell loss associated to motor impairments, accompanied by Lewy body–like inclusions in the latter organism (Table 2). Recently, using the CRISPR/Cas9 technique, pigs (*Sus scrofa domesticus*) with double deletion of *PRKN* and *PINK1* [48] or triple deletion of *PRKN*, *PINK1*, and *DJ-1*, [49] have been generated, but were not characterized for PD signs. Invertebrate *PRKN* models seem thus the more effective in mimicking PD pathophysiology.

2.3   SPR (PARK3)

SPR or PARK3 codes for sepiapterin reductase (7,8-dihydrobiopterin: NADP+ oxidoreductase), an enzyme that catalyzes the NADPH-dependent reduction of various carbonyl substances (Table 1). Single nucleotide polymorphisms in the PARK3 locus were identified as an autosomal dominant risk factor for both sporadic and familial PD [50, 51]. However, the molecular and cellular mechanisms by which these *SPR* mutations lead to PD remain to be elucidated.

Mice with *Spr* deletion show significant reduction of DA and other monoamines in the brain, impaired motor activity, as well as other dysfunctions of biochemical processes depending on tetrahydrobiopterin (BH4), which requires sepiapterin reductase for its synthesis [52–54]. An interesting insight concerning the possible role of SPR and BH4 in brain pathology could arise from a silkworm (*Bombyx mori*) model carrying mutations in the *SPR* gene homolog (*BmSpr*), which result in a lethal phenotype that can be rescued by oral administration of DA and BH4 [55]. Overall, these *SPR* models (Table 2) could thus be useful for investigating the possible role of sepiapterin reductase-dependent biochemical processes in PD as well as in physiological condition.

2.4   UCH-L1 or PGP9.5 (PARK5)

A heterozygous dominant missense mutation (I93M) of *UCH-L1* or PARK5 (Table 1), coding for ubiquitin C-terminal hydrolase L1 (also known as PGP9.5), was found in members of a German family with hereditary PD, and therefore considered as a possible risk factor for late-onset PD [56]. However, these findings were never replicated, and a study on larger cohorts of patients failed to associate *UCH-L1* polymorphism to PD, casting doubts on the role of this gene as a risk factor for this pathology [57].

Intriguingly, *UCH-L1* KO or KI (I93M) mice show central and peripheral neurodegeneration including SNc DAergic cell loss, that are paralleled by motor impairments. Double mutants that also overexpress α-synuclein show earlier-onset motor deficits, a shorter lifespan and forebrain astrogliosis, but no pathological α-synuclein and ubiquitin deposits [58, 59]. It is worth mentioning that KD of the *UCH-L1* homolog (*dUCH*) in *Drosophila* also results in DAergic neuron loss and impaired movement [60]. Thus, despite controversial data in humans, these animal models (Table 2) seem somehow leading to a PD-relevant pathological state, and can be useful for investigating DAergic neuron degeneration and the subsequent motor impairments.

2.5 PINK1 (PARK6)

PINK1 or PARK6 (Table 1) codes for the PTEN-induced putative kinase 1 (PINK1), a protein that interacts with parkin and participates in mitophagy, a process required for eliminating damaged and nonfunctional mitochondria, which is crucial for cell survival [61]. Homozygous *PINK1* mutations were first identified in a large Italian family with several cases of early-onset PD [62], then other autosomal recessive mutations (missense, nonsense) of this gene were found by separate studies in other Italian and Asian patients [63, 64].

According to the role of PINK1, *Pink1* KO mice show mitochondrial deficits, increased vulnerability to oxidative stress and mild motor impairment, but they present neither DAergic neuron loss nor Lewy body–like inclusions [65]. One study using *Pink1* KO mice also overexpressing α-synuclein showed, however, DAergic neuron loss [66], although the already mentioned triple KO (*Prkn*, *Pink1*, and *DJ-1*) did not [45]. Similarly, *Pink1* KO rats show several abnormalities of energy and oxygen metabolism, but no other signs indicating a PD-relevant pathological process, while some locomotor impairments have been reported [47, 67, 68]. Zebrafish KO and KD models also exist where DAergic neuron degeneration is sometimes observed, but they express neither morphologic nor motor changes [69, 70]. We mentioned already pigs carrying either double *PINK1* and *PRKN* deletion [48] or triple *PINK1*, *PRKN* and *DJ-1* deletion [49], which however were not characterized for PD signs. Finally, *Drosophila* [38, 39] and *C. elegans* models [40] have been generated (mostly KO), the former showing DAergic neuron loss and both resulting in motor troubles (Table 2). Overall, these animal models with *PINK1* deletion do not replicate the drastic early-onset PD observed in patients, but can be useful for studying pathological processes due to mitochondrial dysfunction, related or not to PD.

2.6 DJ-1 (PARK7)

The protein coded by the *parkinsonism associated deglycase* gene or *DJ-1* (PARK7) has various functions, including transcription regulation, oxidative stress sensor, protease and mitochondrial regulation [71]. Initially found in Dutch and Italian families, missense *DJ-1* mutations (Table 1) are linked to a form of autosomal recessive, juvenile PD [72, 73].

DJ-1 KO mice show no loss of SNc DAergic neurons, but they have reduced striatal DA release and responsiveness to D2 autoreceptor stimulation, as well as decreased locomotor activity and altered corticostriatal synaptic plasticity [74]. These mice were also more vulnerable to MPTP treatment [75]. In *DJ-1* "gene trapped" mice (basically, KI/KO), other authors reported upregulation of mitochondrial respiratory activity, decreased DAergic neurons in the ventral tegmental area but no decreased DAergic terminals in the striatum, paralleled by nonmotor deficits, suggesting the implication of *DJ-1* in nonmotor symptoms of early-phase PD [76]. Surprisingly, one paper reports that a subset of *DJ-1* KO mice (backcrossed with C57BL/6) showed early-onset and unilateral nigrostriatal degeneration, which progressed and became bilateral with age, leading to mild motor deficits [77]: this model could provide a possible tool to study the progression of PD from early presymptomatic to symptomatic stages, and from unilateral to bilateral. *DJ-1* KO rats were recently added to the available models and seem a promising tool for studying PD pathophysiology and possible neuroprotective strategies, since they show significant loss of nigral DAergic neurons accompanied by locomotor impairments [47]. In the pig, a biallelic *DJ-1* KO and site-specific KI mediated by TALENs (transcription activator–like effector nucleases) has also been generated [78], and the abovementioned triple KO (*DJ-1, PRKN,* and *PINK1*) also exist [49], but none of them was characterized as a PD model. Finally, other species have been utilized to produce *DJ-1* mutants (mostly KO), namely, the zebrafish, the medaka fish (*O. latipes*), the *Drosophila* and the flatworm *C. elegans*, but they were not studied as PD models (Table 2).

2.7 LRRK2 (PARK8)

Leucine-rich repeat kinase 2, also known as dardarin, is a large protein coded by the *LRKK2* gene (PARK8) that has several domains including a kinase, a RAS and a GTPase domain, and which, among several functions, interacts with parkin [79] (Table 1). The implication of *LRRK2* gene in familial PD was first established in a Japanese family [80], and lately confirmed as autosomal dominant missense and point mutations in this gene for late-onset and autosomal dominant cases of PD [81–83], which are significantly enriched (~20% and ~ 40%, respectively) in PD patients among Ashkenazi Jews [84] and with Berber Arab ancestors [85]. Mutations of *LRRK2* gene mainly involve the kinase (G2019S) [85] or the GTPase (R1441C/G) [86] domain.

Lrrk2 mutations have been produced in mouse models (Table 2), which may show mild DAergic degeneration, as well as alterations in DA release and uptake [87–89]. Interestingly, *Lrrk2* KO mice do not show DAergic deficits or other neuropathological features [90], but they present α-synuclein inclusions in the kidneys [91]. BAC transgenic mice expressing mutated *Lrrk2* show age-dependent and progressive motor impairment, and slight reduction of DA release in the striatum, but no nigrostriatal degeneration [92, 93]. Overall, this mild pathological phenotype in *Lrrk2* mouse models makes them unlikely to be really valuable for studying PD pathophysiology or testing therapeutic agents. It is worth noting that *Lrrk2* KO mice are less affected than those carrying a mutated form of *Lrrk2*. In consideration of the multiple functions of LRKK2 protein, it could be possible that losing only part of the protein's functions could be more harmful that suppressing the whole of them, possibly due to an unbalance in one or more LRRK2-dependent processes of cell metabolism. On the other hand, rat and zebrafish *Lrrk2* models seem slightly more promising because they often (but not always) show progressive degeneration of DAergic neurons and motor symptoms (*see* Table 2 also for references). Finally, although phylogenetically distant from humans, *Drosophila* and *C. elegans* models expressing human *LRRK2* (normal or mutated) or are *lrrk2* KO, seem those which produce more reliably DAergic neuron loss and motor impairments (Table 2). This further underlies the complex and multiple function of this protein, and thus the difficulty to interpret these phenotypes and utilize these animal models for exploring PD pathological mechanisms.

2.8 ATP13A2 (PARK9)

ATP13A2 (locus 1p36.13) codes for lysosomal type P cation transporting ATPase 13A2. Loss-of-function mutations of this gene lead to a very rare, autosomal recessive, and early-onset form of PD called Kufor Rakeb syndrome (PARK9), with additional symptoms including supranuclear gaze palsy, spasticity, and dementia [94–96]. These mutations are also associated to a lysosomal storage disorder named neuronal ceroid lipofuscinosis (NCL), and to a form of hereditary spastic paraplegia (HSP) (Table 1). Moreover, heterozygous mutations of *ATP13A2* may constitute a risk factor for PD development of classical PD.

Interestingly, homozygous truncating mutation (1623delC) in *ATP13A2* gene was found in dogs (Tibetan terriers) with NCL [97, 98], but these animals can hardly be defined as PD models due to the specificity of NCL symptoms and pathology. Overexpression of mutated *ATP13A2* alone in rats was not tested for PD features, but it was reported that it could not reduce the formation of protein aggregates after α-synuclein injection [99]. Conversely, in mice, *ATP13A2* KO produced motor troubles and α-synuclein positive inclusions [100, 101]. However, in none of these rodent models, DAergic neuron loss was observed. Finally, *ATP13A2* KO resulted lethal in the zebrafish, while mutations in *O. latipes* and *C. elegans* resulted in DAergic cell loss and motor deficits, but the presence of Lewy body–like inclusions was not assessed (Table 2). Due to the different and complex pathological phenotypes resulting from *ATP13A2* mutation (Kufor Rakeb syndrome, NCL, and HSP), the abovementioned models can hardly be considered as "strictly" PD models.

2.9 PARK10 (Several Genes)

The PARK10 locus (1p32) contains several genes (*see* Table 1), and polymorphisms in this region have been associated to familial PD since 2002 [102]. However, more recent studies suggest that common variations in the PARK10 region are not associated with PD risk [25, 103].

However, one mouse model exists, targeting the *PLPP3* gene (formerly *PPAP2B* or *LPP3*) that maps within the PARK10 locus (Table 2). *Plpp3* KO mice show motor impairments possibly due to decreased DA release in the striatum, but no DAergic neuronal loss [104]. However, although this gene is actually located in the PARK10 locus (more exactly, at 1p32.2), it does not appear in the scientific literature as being associated with PD.

2.10 GIGYF2 (PARK11)

Autosomal dominant mutation in the *GRB10 interacting GYF protein 2* gene (*GIGYF2*), coding for a transcription regulator, result in a higher risk of late-onset PD [25, 105, 106]. However, at least one study casted doubts about the actual link between PD and PARK11, at least in European families [107] (Table 1).

Heterozygous *Gigyf2$^{+/-}$* mice show neuronal loss in the cerebellum and midbrain as well as in the spinal cord, but no evident loss of tyrosine hydroxylase staining in the SN. KO mice present ~95% lethality at 3 weeks after birth, making impossible to define their growth and pathological characteristics [108]. *Drosophila* models (KD or KO) exhibit the three main PD signs [109], while *gigyf2* deficiency in zebrafish [110] did not produce any of them (Table 2). Given the debated involvement of PARK11 in familial PD and the contrasting results of these models, *GIGYF2* does not seem the best gene to target for producing a PD model.

2.11 PARK12 (Locus Xq21-q25)

Linkage to the X chromosome for familial PD was evidenced in the early 2000 [111, 112], but given the large size of the locus the exact gene(s) involved are multiple and they need to be confirmed (Table 1). To date, the only PARK12 gene associated to familial PD is *RAB39B* (RAS-associated protein 39B), which is responsible of the Waisman syndrome, an X-linked neurologic disorder characterized by delayed psychomotor development, intellectual disability, and early-onset PD [113].

To our knowledge, only one mouse model targets PARK12, namely, bearing a deletion of the *Il-13Ralpha1* mouse gene, whose ortholog in human actually maps in the Xq21-q25 region. However, as mentioned above, there is no evidence to date linking this gene to familial PD (Table 2), and these KO mice (*Il13ra$^{-/Y}$*) do not show DAergic neuron loss and were not further tested as a PD model [114].

2.12 HTRA2 (PARK13)

The development of an autosomal dominant form of late-onset PD has been associated to heterozygous mutation in the *HtrA serine peptidase 2* gene (*HTRA2*). Several cases have been reported in Germany [115], but the linkage between PARK13 and familial PD was not confirmed by a successive study that failed to find an association between *HTRA2* polymorphism (G399S or A141S) and PD [116] (Table 1).

Several mouse [117–121] and *Drosophila* [122–124] models targeting *HTRA2* gene orthologs exist, utilizing heterogeneous genetic approaches (KO, KI, KD, overexpression, mutation): they exhibit some motor signs, but no clear cellular alterations were evidenced or assessed (Table 2).

2.13 PLA2G6 (PARK14)

The *PLA2G6* gene codes for the phospholipase A2 group 6, which catalyzes the S_N2 hydrolysis of acyl-ester bonds in phospholipids, leading to the release of arachidonic acid and other fatty acids. *PLA2G6* mutations are associated with two pathologies, namely, infantile neuroaxonal dystrophy and neurodegeneration with brain iron accumulation [125]. Autosomal recessive mutation in the *PLA2G6* gene have also been evidenced in a form of adult-onset dystonia-parkinsonism (Table 1) without the other symptoms of the two pathologies mentioned above [126].

Mouse [127–131] and zebrafish [132] with deletion or suppression of the *PLA2G6* ortholog gene show all the three mains PD signs, namely, DAergic neuron degeneration, presence of Lewy body–like inclusions rich in α-synuclein (with or without neuron-axonal dystrophy) and motor deficits. Conversely, *Drosophila* models [133, 134] seem less promising in producing PD signs since they present only motor impairment (Table 2). Finally, neuronaxonal dystrophy was found in "papillon dogs" with *PLA2G6* missense mutations, but no information is provided about PD-relevant pathology [135].

2.14 FBXO7 (PARK15)

Autosomal recessive mutation in the *FBXO7* gene, coding for F-box protein O-type 7 (a component of E3 ubiquitin protein ligases), constitutes a risk factor of a rare, early-onset parkinsonian-pyramidal syndrome (Table 1), also known as pallidopyramidal syndrome [24, 136, 137].

The existing mouse models (KO or reduced *Fbxo7* expression) only exhibit motor deficits but no other PD-relevant signs [138, 139]. *Drosophila* models show only little (if any) DAergic neuron loss [140, 141], while a zebrafish model seems more promising since it produces both cell loss and DA-dependent motor defects [142] (Table 2).

2.15 RAB7L1 (PARK16)

PARK16 corresponds to locus 1q32 and contains multiple genes including *RAB7L1*, *SLC41A1* and *ADORA1* [23, 143, 144]. Mutations in this locus seem to constitute a risk factor for developing PD [145], but data are controversial and it is reported that they even may lower the risk of developing the pathology [146] (Table 1). To date, no genetic animal PD models are available.

2.16 VPS35 (PARK17)

VPS35 gene codes for the vacuolar protein sorting 35, which is involved in endosome trans-Golgi trafficking and recycling of membrane-associated proteins. Autosomal dominant mutation in the *VPS35* gene was associated to a higher risk of developing PD in several studies [147–150], but this finding was unconfirmed by another report that found no association between PARK16–18 and PD [146] (Table 1).

Several mouse models are available (KO, KI, heterozygous), showing Lewy body–like inclusions but no motor sign; DAergic neuron loss is sometimes reported [151–156] (Table 2). A rat model has been generated [157], and exhibits only DAergic neuron loss. Strikingly, three different *Drosophila* genetic models (KI, KD and expressing the human mutated protein) produce the three PD-relevant phenotypes that we consider in this review [158–160].

2.17   EIF4G1 (PARK18)

EIF4G1 codes for the eukaryotic translation initiation factor 4 gamma 1, and autosomal dominant mutations are supposed to increase the risk for PD [150, 161], but this finding was unconfirmed by another report that found no association between PARK16–18 and PD [146] (Table 1). No genetic animal models were found in the scientific literature.

2.18   DNAJC6 (PARK19)

Autosomal recessive mutation in the *DNAJC6* gene [*DnaJ heat shock protein family (Hsp40) member C6*] coding for auxilin (Table 1), has been associated to two forms of PD, namely, PARK19A, due to loss-of-function mutations, leading to a very early-onset (10–20 years) atypical form of PD with pyramidal signs [162–164], and PARK19B, due to point mutations, resulting in an early-onset (30–50 years) form of "classical" PD [165].

Three genetic animal models targeting auxilin were produced (Table 2). Auxilin KO mice, showing a high rate of early postnatal mortality, with surviving animals generally having a normal life span but lower body weight than wild type; however, they show neither DAergic neuron loss nor motor impairments [166]. Auxilin KD Zebrafish were not tested for PD signs, but they show an elevated level of apoptosis in neural tissues, resulting in severe degeneration of neural structures [167]. In *Drosophila*, auxilin KD results in progressive locomotor deficits and loss of DAergic neurons [168, 169], confirming that this model organism is often more affected than vertebrates by genetic manipulation of PARK genes.

2.19   SYNJ1 (PARK20)

PARK20 is caused by autosomal recessive mutation in the *SYNJ1* gene coding for synaptojanin-1 (a polyphosphoinositide phosphatase involved in synaptic vesicle dynamics), an is characterized by a higher risk of juvenile onset, atypical form of PD with other symptoms including seizures, cognitive decline, abnormal eye movements, and dystonia [170, 171] (Table 1).

Several genetic animal models targeting *SYNJ1* ortholog genes exist (KD, KO, mutations), namely, in mouse [172–174], zebrafish [175, 176], *Drosophila* [177, 178], and *C. elegans* [179, 180], but none of them has been characterized for PD-relevant pathology (Table 2). However, one study reported that mice carrying sinaptojanin-1 R258Q mutation (the same as in humans) show ~40% mortality, and the surviving animals presenting seizures, motor deficits, and dystrophic DAergic axons in the striatum [181].

2.20 DNAJC13 (PARK21)

DNAJC13 is a protein of the endosomal system involved in vesicle formation and trafficking. Missense, autosomal dominant mutations in the *DNAJC13* gene (Table 1) have been linked to late-onset PD [182, 183].

Strikingly, a *Drosophila* model expressing human mutated *DNAJC13* (N855S) and overexpressing human *SNCA* presents the three main PD features, that is, DAergic neuron loss, α-synuclein inclusions and motor deficits [184]. Some *C. elegans* KO models also exist [185, 186] but they were not characterized as PD models (Table 2).

2.21 CHCHD2 (PARK22)

Autosomal dominant mutation in the *CHCHD2* gene (Table 1), coding for the coiled-coil helix–coiled-coil helix domain–containing protein 2 (a transcription factor), have been linked to a higher risk of classical PD [187], but this finding was questioned [188].

The only available mouse model to date (KO) shows no PD signs [189]. Conversely, *Drosophila* models (KO and KI) exhibit both DAergic neuron loss and motor deficits [190, 191], highlighting again this invertebrate model as the most efficient in producing a PD-relevant pathological state (Table 2).

2.22 VPS13C (PARK23)

Homozygous or compound heterozygous mutation in the *VPS13C* gene, coding for vacuolar protein sorting 13C (Table 1), results in young-adult PD onset, associated with a progressive cognitive impairment leading to dementia and dysautonomia. Some individuals also present additional motor abnormalities [192].

To date, only one genetic animal model (Table 2) is available in *Drosophila* (KD of the *VPS13C* ortholog gene and overexpressing human *SNCA*), but it does not show any PD-related signs [193].

3   Animal PD Models Targeting "non-PARK" Genes Involved in Familial PD

Several other genes have been identified as potentially involved in inheritable PD, but they are not included in the PARK family because their role in this pathology needs further confirmation, and/or because they are primarily involved in other diseases. Here we provide a list of the 11 most studied and characterized, so far (Table 1). Not all of these genes have been targeted to generate animal models, and for those which such models exist, they often address other pathologies rather than PD (*see* below). However, we felt to include this group of genes and their corresponding animal models because the literature refers to them as "PD susceptibility genes" or as risk factors (*see* for example https://www.omim.org/entry/168600). Thus, these models could possibly be of some relevance and use in the future.

Mutation in the glucocerebrosidase (*GBA*) gene leads to type 1 Gaucher disease, and patients with this mutation who also show PD symptoms have been identified [194–196]. Several genetic animal models have been generated. In rodent and *Drosophila* models only α-synuclein inclusions or motor symptoms, respectively, were identified. The zebrafish model exhibits both these signs and, intriguingly, the best *GBA* model seems to be *O. latipes*, in which the three pathological hallmarks are present in *GBA*$^{-/-}$ individuals compared to wild-type and *GBA*$^{+/-}$ (*see* Table 2 for references).

CAG expansion in the *SCA* gene coding for ataxin 2 (also known as *ATX2* or *ATXN2*) have been described in spinocerebellar ataxia type 2. However, such mutation (or CAA interruption of such expansion) have been evidenced in some patients with autosomal dominant parkinsonism [197–199]. Thus, *SCA* mutations likely constitutes a risk factor for developing PD, and the configuration of repeat expansions seems to play an important role in phenotypic variability. Although several mouse models targeting *Atxn2* were generated, they have been characterized as spinocerebellar ataxia models: they actually exhibit motor deficits, but PD-relevant cellular alterations were not assessed (*see* Table 2 for references).

The *ADH1C* gene codes for the gamma subunit of class I alcohol dehydrogenase, and polymorphisms in this gene have been associated to type I and II alcoholism [200]. A linkage between a truncating polymorphism (G78Stop) in this gene and the development of PD has been evidenced [201, 202]. Only one mouse model targeting *Adh1* and *4* (*Adh1*$^{-/-}$) is available so far (Table 2) but, besides some changes in DA-related behavior, no PD-relevant signs were reported, and in was suggested that these mice might be possible models to study presymptomatic PD [203].

Mutations (expansion of CAG repeats) in the *TBP* gene coding for the TATA-box binding protein have been found in patients with spinocerebellar ataxia type 17 or SCA17 [204]. Abnormal trinucleotide repeat expansion (46 repeats) of the same gene have also been identified in a PD patient [205, 206]. Several rodent transgenic models exist as well as a KD model in the frog (*Xenopus laevis*), but they have not been studied for PD-relevant pathophysiology. Only mice show some motor impairments (*see* Table 2 for references).

The *MAPT* gene codes for the microtubule-associated protein tau, a protein that is involved in several neurological disorders generally termed tauopathies that include dementia, Pick disease, and supranuclear palsy. Autosomal dominant frontotemporal dementia with parkinsonism linked to chromosome 17 has been associated with missense mutations in the *MAPT* gene [145, 146, 207]. Mouse and zebrafish models are available, but they have not been characterized for PD-relevant parameters. Transgenic *C. elegans* expressing mutated tau protein show motor deficits, and loss of DAergic neurons has been reported in *Drosophila* models (*see* Table 2 for references).

The *GLUD2* gene codes for glutamate dehydrogenase 2, whose polymorphism (S445A) was associated with earlier age of onset in 2 cohorts of PD patients. This substitution was demonstrated to confer a gain of function to the enzyme, which may confer it the capability of increasing glutamate oxidation and the production of reactive oxygen species in the brain, lowering the age of PD onset [208]. Transgenic and KO mice models were produced, but they were not investigated for PD-relevant features (*see* Table 2 for references).

Finally, mutations in other genes, including *TMEM230* [209], *RIC3* [210], *PTRHD1* [211], *SMARCB1* [212], and *PODXL* [213] have been suspected as potential risk factors for developing PD, but none of them has been confirmed (Table 1). Animal models targeting orthologs of these genes do not exist to our knowledge.

4   Other Genetic Models of PD

The PD models listed in this section are based on genes whose mutation is not directly linked to genetic/familial PD, or for which the linkage is suspected but needs to be confirmed (*see* below). The peculiarity of these models, all engineered in mouse, is that their rationale is to disrupt genes coding for proteins involved in metabolic pathways that are impaired in PD, or that are required for the survival of DAergic neurons but are not (necessarily) mutated in PD. These target genes include, for example, transcription factors such as sonic hedgehog (*Shh*) [214], the nuclear receptor related protein-1 (*Nurr1*) [215], the pituitary homeobox 3 (*Pitx3*) [216] and engrailed-1 (*En1*) [217]. in general, all these models show some DAergic neuron loss but no Lewy body–like inclusions, and the presence of motor trouble is sometimes reported (Table 3). Another original model is based on the selective inactivation of a mitochondrial transcription factor in DAergic neurons [218].

The conditional ablation of *Shh* in DAergic neurons has been reported to result in progressive, adult-onset degeneration of DAergic, cholinergic, and fast-spiking GABAergic neurons of the mesostriatal circuit. Imbalance of cholinergic and DAergic neurotransmission, and mild motor deficits were also reported [214].

Ablation of *Nurr1* in midbrain DAergic neurons resulted in rapid loss of striatal DA and DAergic neuron markers, and striato-nigral degeneration, while a more slowly progressing process was observed when conditional ablation was performed in adults [219]. Interestingly, increased *NURR1* (or *NR4A2*) expression has also been reported in the midbrain of PD patients (particularly, in neurons containing Lewy bodies), and *NURR1* polymorphism has been reported in association with familial or sporadic PD [220–222], but this linkage is still debated [223, 224]. Overall, these data confirm that this transcription factor is essential for the development and survival of mesencephalic DAergic neurons.

The *Pitx3* KO mouse, also called "aphakia mouse," shows a complete loss of nigrostriatal DAergic neurons at early postnatal age and motor deficits corrected by L-DOPA [225–227]. In addition, a zebrafish *pitx3* KD model exists, but it was not investigated for PD signs [228].

En1[+/-] mice have been shown to develop a progressive retrograde degeneration of nigrostriatal DAergic neurons [229]. Interestingly, *PITX3* and *EN1* polymorphism has been associated with an increased risk for PD, but these findings need to be confirmed [230].

Finally, the "MitoPark" mouse model targets the mitochondrial transcription factor A (*Tfam*), which is required for the transcription of the mitochondrial DNA coding for essential subunits of the mitochondrial respiratory chain [231]. The deletion of this gene in DAergic neurons results in adult-onset and slowly progressive motor impairment, degeneration of nigrostriatal DAergic neurons, and intraneuronal inclusions not containing α-synuclein [218, 232]. Among these "atypical" PD models, the MitoPark is probably the most promising due to its strong phenotype, pathology and progressivity, and to the involvement of mitochondrial deficits.

5   General Considerations on Genetic PD Models

When modeling a gene-linked pathology such as familial PD in animals, several questions arise. First of all, what is needed from a PD model in general, and a genetic one in particular? The relevance to PD of the model's pathological outcome is of primary importance: an "ideal" model should be capable of producing (or, rather, mimicking) the whole PD pathology, in order to be as close as possible to the human condition. In particular, it should produce a progressive, unilateral to bilateral SNc DAergic neuron loss, paralleled by the accumulation of Lewy bodies in neurons, leading to progressive motor impairments [18, 233, 234]. Ideally, these motor symptoms should be preceded by nonmotor ones, which could be difficult (or impossible) to assess depending on the model organism. This raises the crucial question about the choice of the animal model, which, for translational purposes, should be as close as possible to human. From this point of view, vertebrate models are the most relevant despite the possible ethical considerations. However, the outcome of rodent genetic PD models is generally poor (*see* Tables 2 and 3), in contrast to the two primate α-synuclein models that provided the best results in terms of mimicking PD pathophysiology (but for whom the ethical issues are even more relevant). Strikingly, despite their phylogenetic distance, invertebrate models seem the most powerful in terms of pathological phenotype, and could be of particular interest for investigating specific molecular/cellular dysfunctions that characterize PD pathophysiology due to a given mutation, and/or the general pathogenicity of such mutation. However, on the one hand, the phylogenic distance from human can cast doubts on the pertinence of these models that could be considered as barely relevant to PD. On the other hand, it could be argued that at molecular/cellular level, the pathological processes triggered by manipulating PD-related genes are actually comparable between phylogenetically distant organisms. Thus, while vertebrate could be considered as "as close as possible" PD models, invertebrate can be seen as a tool for investigating more specific and fundamental pathological mechanisms.

Another issue that should be considered at the same level of importance as the points above, is the reproducibility of the model, which should be capable of producing a reliable phenotype and data that can be replicated from one laboratory to another.

Finally, concerning PD, it must be reminded that new therapeutical strategies are currently required and multiple lines of investigation are ongoing in this field, in order to ameliorate patient management and, possibly, counteract the progression of the disease. To date, dopamine replacement treatments, which are now more than 60 years old [235], or deep brain stimulation [236], are the only available alternatives, with their well-known limitations in terms of side effects and, for the latter, eligibility. More importantly, they are both symptomatic treatments, thus the main challenge for PD still remains a disease-modifying, curative therapy. In this context, genetic animal models could provide the mechanistic, molecular, and cellular bases for such new treatments for both familial as well as sporadic PD.

As a general conclusion, it seems that the "ideal" animal model of PD (genetic or other) does not exist, with the possible exception of the primate one that matches with all the expectations but has been largely underexploited. Thus, maybe each of these genetic models should be taken into consideration not "strictly" as a PD model aimed a mimicking as much as possible the pathology, but as a tool to investigate the specific role of the targeted gene—and its protein product—in PD pathophysiology.

References

1. Olanow CW (2007) The pathogenesis of cell death in Parkinson's disease. Mov Disord 22 Suppl 17:S335–S342
2. Obeso JA, Rodriguez-Oroz MC, Rodriguez M, Lanciego JL, Artieda J, Gonzalo N, Olanow CW (2000) Pathophysiology of the basal ganglia in Parkinson's disease. Trends Neurosci 23(10 Suppl):S8–S19
3. Jankovic J (2008) Parkinson's disease: clinical features and diagnosis. J Neurol Neurosurg Psychiatry 79(4):368–376
4. Schapira AH, Jenner P (2011) Etiology and pathogenesis of Parkinson's disease. Mov Disord 26(6):1049–1055
5. Calabresi P, Picconi B, Tozzi A, Ghiglieri V, Di FM (2014) Direct and indirect pathways of basal ganglia: a critical reappraisal. Nat Neurosci 17(8):1022–1030
6. Levy R, Hazrati LN, Herrero MT, Vila M, Hassani OK, Mouroux M, Ruberg M, Asensi H, Agid Y, Feger J, Obeso JA, Parent A, Hirsch EC (1997) Re-evaluation of the functional anatomy of the basal ganglia in normal and Parkinsonian states. Neuroscience 76(2):335–343
7. DeLong MR, Wichmann T (2007) Circuits and circuit disorders of the basal ganglia. Arch Neurol 64(1):20–24
8. Blandini F, Nappi G, Tassorelli C, Martignoni E (2000) Functional changes of the basal ganglia circuitry in Parkinson's disease. Prog Neurobiol 62(1):63–88
9. Wichmann T, Dostrovsky JO (2011) Pathological basal ganglia activity in movement disorders. Neuroscience 198:232–244
10. Chase TN, Oh JD (2000) Striatal dopamine- and glutamate-mediated dysregulation in experimental parkinsonism. Trends Neurosci 23(10 Suppl):S86–S91
11. Chase TN, Bibbiani F, Oh JD (2003) Striatal glutamatergic mechanisms and extrapyramidal movement disorders. Neurotox Res 5(1–2):139–146
12. Bamford NS, Robinson S, Palmiter RD, Joyce JA, Moore C, Meshul CK (2004) Dopamine modulates release from corticostriatal terminals. J Neurosci 24(43):9541–9552
13. Pisani A, Bernardi G, Ding J, Surmeier DJ (2007) Re-emergence of striatal cholinergic interneurons in movement disorders. Trends Neurosci 30(10):545–553
14. Obeso JA, Rodriguez-Oroz MC, Benitez-Temino B, Blesa FJ, Guridi J, Marin C, Rodriguez M (2008) Functional organization of the basal ganglia: therapeutic implications for Parkinson's disease. Mov Disord 23(Suppl 3):S548–S559
15. Chaudhuri KR, Odin P, Antonini A, Martinez-Martin P (2011) Parkinson's disease: the non-motor issues. Parkinsonism Relat Disord 17(10):717–723
16. Brichta L, Greengard P, Flajolet M (2013) Advances in the pharmacological treatment of Parkinson's disease: targeting neurotransmitter systems. Trends Neurosci 36(9):543–554
17. Spillantini MG, Schmidt ML, Lee VM, Trojanowski JQ, Jakes R, Goedert M (1997) Alpha-synuclein in Lewy bodies. Nature 388(6645):839–840. https://doi.org/10.1038/42166
18. Dawson TM, Ko HS, Dawson VL (2010) Genetic animal models of Parkinson's disease. Neuron 66(5):646–661. https://doi.org/10.1016/j.neuron.2010.04.034
19. Gubellini P, Picconi B, Di FM, Calabresi P (2010) Downstream mechanisms triggered by mitochondrial dysfunction in the basal ganglia: from experimental models to neurodegenerative diseases. Biochim Biophys Acta 1802(1):151–161
20. Dauer W, Przedborski S (2003) Parkinson's disease: mechanisms and models. Neuron 39(6):889–909
21. Gasser T, Hardy J, Mizuno Y (2011) Milestones in PD genetics. Mov Disord 26(6):1042–1048
22. Klein C, Westenberger A (2012) Genetics of Parkinson's disease. Cold Spring Harb Perspect Med 2(1):a008888. https://doi.org/10.1101/cshperspect.a008888
23. Puschmann A (2017) New genes causing hereditary Parkinson's disease or parkinsonism. Curr Neurol Neurosci Rep 17(9):66. https://doi.org/10.1007/s11910-017-0780-8
24. Puschmann A (2013) Monogenic Parkinson's disease and parkinsonism: clinical phenotypes and frequencies of known mutations. Parkinsonism Relat Disord 19(4):407–415. https://doi.org/10.1016/j.parkreldis.2013.01.020
25. Karimi-Moghadam A, Charsouei S, Bell B, Jabalameli MR (2018) Parkinson disease from Mendelian forms to genetic susceptibility: new molecular insights into the neurodegeneration process. Cell Mol Neurobiol 38(6):1153–1178. https://doi.org/10.1007/s10571-018-0587-4
26. Polymeropoulos MH, Lavedan C, Leroy E, Ide SE, Dehejia A, Dutra A, Pike B, Root H, Rubenstein J, Boyer R, Stenroos ES, Chandrasekharappa S, Athanassiadou A, Papapetropoulos T, Johnson WG, Lazzarini AM, Duvoisin RC, Di Iorio G,

Golbe LI, Nussbaum RL (1997) Mutation in the alpha-synuclein gene identified in families with Parkinson's disease. Science 276(5321):2045–2047

27. Polymeropoulos MH, Higgins JJ, Golbe LI, Johnson WG, Ide SE, Di Iorio G, Sanges G, Stenroos ES, Pho LT, Schaffer AA, Lazzarini AM, Nussbaum RL, Duvoisin RC (1996) Mapping of a gene for Parkinson's disease to chromosome 4q21-q23. Science 274(5290):1197–1199

28. Alim MA, Hossain MS, Arima K, Takeda K, Izumiyama Y, Nakamura M, Kaji H, Shinoda T, Hisanaga S, Ueda K (2002) Tubulin seeds alpha-synuclein fibril formation. J Biol Chem 277(3):2112–2117. https://doi.org/10.1074/jbc.M102981200

29. Bonini NM, Giasson BI (2005) Snaring the function of alpha-synuclein. Cell 123(3):359–361. https://doi.org/10.1016/j.cell.2005.10.017

30. Burre J, Sharma M, Tsetsenis T, Buchman V, Etherton MR, Sudhof TC (2010) Alpha-synuclein promotes SNARE-complex assembly in vivo and in vitro. Science 329(5999):1663–1667. https://doi.org/10.1126/science.1195227

31. Narhi L, Wood SJ, Steavenson S, Jiang Y, Wu GM, Anafi D, Kaufman SA, Martin F, Sitney K, Denis P, Louis JC, Wypych J, Biere AL, Citron M (1999) Both familial Parkinson's disease mutations accelerate alpha-synuclein aggregation. J Biol Chem 274(14):9843–9846

32. Ross OA, Braithwaite AT, Skipper LM, Kachergus J, Hulihan MM, Middleton FA, Nishioka K, Fuchs J, Gasser T, Maraganore DM, Adler CH, Larvor L, Chartier-Harlin MC, Nilsson C, Langston JW, Gwinn K, Hattori N, Farrer MJ (2008) Genomic investigation of alpha-synuclein multiplication and parkinsonism. Ann Neurol 63(6):743–750. https://doi.org/10.1002/ana.21380

33. Fujiwara H, Hasegawa M, Dohmae N, Kawashima A, Masliah E, Goldberg MS, Shen J, Takio K, Iwatsubo T (2002) alpha-Synuclein is phosphorylated in synucleinopathy lesions. Nat Cell Biol 4(2):160–164. https://doi.org/10.1038/ncb748

34. Yang W, Wang G, Wang CE, Guo X, Yin P, Gao J, Tu Z, Wang Z, Wu J, Hu X, Li S, Li XJ (2015) Mutant alpha-synuclein causes age-dependent neuropathology in monkey brain. J Neurosci 35(21):8345–8358. https://doi.org/10.1523/JNEUROSCI.0772-15.2015

35. Eslamboli A, Romero-Ramos M, Burger C, Bjorklund T, Muzyczka N, Mandel RJ, Baker H, Ridley RM, Kirik D (2007) Long-term consequences of human alpha-synuclein overexpression in the primate ventral mid-brain. Brain 130(Pt3):799–815. https://doi.org/10.1093/brain/awl382

36. Kirik D, Annett LE, Burger C, Muzyczka N, Mandel RJ, Bjorklund A (2003) Nigrostriatal alpha-synucleinopathy induced by viral vector-mediated overexpression of human alpha-synuclein: a new primate model of Parkinson's disease. Proc Natl Acad Sci U S A 100(5):2884–2889. https://doi.org/10.1073/pnas.0536383100

37. Milanese C, Sager JJ, Bai Q, Farrell TC, Cannon JR, Greenamyre JT, Burton EA (2012) Hypokinesia and reduced dopamine levels in zebrafish lacking beta- and gamma1-synucleins. J Biol Chem 287(5):2971–2983. https://doi.org/10.1074/jbc.M111.308312

38. Hewitt VL, Whitworth AJ (2017) Mechanisms of Parkinson's disease: lessons from Drosophila. Curr Top Dev Biol 121:173–200. https://doi.org/10.1016/bs.ctdb.2016.07.005

39. Vanhauwaert R, Verstreken P (2015) Flies with Parkinson's disease. Exp Neurol 274(Pt A):42–51. https://doi.org/10.1016/j.expneurol.2015.02.020

40. Cooper JF, Van Raamsdonk JM (2018) Modeling Parkinson's disease in C. elegans. J Parkinsons Dis 8(1):17–32. https://doi.org/10.3233/JPD-171258

41. Yoshii SR, Kishi C, Ishihara N, Mizushima N (2011) Parkin mediates proteasome-dependent protein degradation and rupture of the outer mitochondrial membrane. J Biol Chem 286(22):19630–19640. https://doi.org/10.1074/jbc.M110.209338

42. Kitada T, Asakawa S, Hattori N, Matsumine H, Yamamura Y, Minoshima S, Yokochi M, Mizuno Y, Shimizu N (1998) Mutations in the parkin gene cause autosomal recessive juvenile parkinsonism. Nature 392(6676):605–608

43. Matsumine H, Saito M, Shimoda-Matsubayashi S, Tanaka H, Ishikawa A, Nakagawa-Hattori Y, Yokochi M, Kobayashi T, Igarashi S, Takano H, Sanpei K, Koike R, Mori H, Kondo T, Mizutani Y, Schaffer AA, Yamamura Y, Nakamura S, Kuzuhara S, Tsuji S, Mizuno Y (1997) Localization of a gene for an autosomal recessive form of juvenile Parkinsonism to chromosome 6q25.2-27. Am J Hum Genet 60(3):588–596

44. Stichel CC, Zhu XR, Bader V, Linnartz B, Schmidt S, Lubbert H (2007) Mono- and double-mutant mouse models of Parkinson's disease display severe mitochondrial damage. Hum Mol Genet 16(20):2377–2393. https://doi.org/10.1093/hmg/ddm083

45. Kitada T, Tong Y, Gautier CA, Shen J (2009) Absence of nigral degeneration in aged parkin/DJ-1/PINK1 triple knockout mice. J

Neurochem 111(3):696–702. https://doi.org/10.1111/j.1471-4159.2009.06350.x

46. Lu XH, Fleming SM, Meurers B, Ackerson LC, Mortazavi F, Lo V, Hernandez D, Sulzer D, Jackson GR, Maidment NT, Chesselet MF, Yang XW (2009) Bacterial artificial chromosome transgenic mice expressing a truncated mutant parkin exhibit age-dependent hypokinetic motor deficits, dopaminergic neuron degeneration, and accumulation of proteinase K-resistant alpha-synuclein. J Neurosci 29(7):1962–1976

47. Creed RB, Goldberg MS (2018) New developments in genetic rat models of Parkinson's disease. Mov Disord 33(5):717–729. https://doi.org/10.1002/mds.27296

48. Zhou X, Xin J, Fan N, Zou Q, Huang J, Ouyang Z, Zhao Y, Zhao B, Liu Z, Lai S, Yi X, Guo L, Esteban MA, Zeng Y, Yang H, Lai L (2015) Generation of CRISPR/Cas9-mediated gene-targeted pigs via somatic cell nuclear transfer. Cell Mol Life Sci 72(6):1175–1184. https://doi.org/10.1007/s00018-014-1744-7

49. Wang X, Cao C, Huang J, Yao J, Hai T, Zheng Q, Wang X, Zhang H, Qin G, Cheng J, Wang Y, Yuan Z, Zhou Q, Wang H, Zhao J (2016) One-step generation of triple gene-targeted pigs using CRISPR/Cas9 system. Sci Rep 6:20620. https://doi.org/10.1038/srep20620

50. Karamohamed S, DeStefano AL, Wilk JB, Shoemaker CM, Golbe LI, Mark MH, Lazzarini AM, Suchowersky O, Labelle N, Guttman M, Currie LJ, Wooten GF, Stacy M, Saint-Hilaire M, Feldman RG, Sullivan KM, Xu G, Watts R, Growdon J, Lew M, Waters C, Vieregge P, Pramstaller PP, Klein C, Racette BA, Perlmutter JS, Parsian A, Singer C, Montgomery E, Baker K, Gusella JF, Fink SJ, Myers RH, Herbert A, Gene PD (2003) A haplotype at the PARK3 locus influences onset age for Parkinson's disease: the GenePD study. Neurology 61(11):1557–1561

51. Klein C, Vieregge P, Hagenah J, Sieberer M, Doyle E, Jacobs H, Gasser T, Breakefield XO, Risch NJ, Ozelius LJ (1999) Search for the PARK3 founder haplotype in a large cohort of patients with Parkinson's disease from northern Germany. Ann Hum Genet 63(Pt4):285–291

52. Yang S, Lee YJ, Kim JM, Park S, Peris J, Laipis P, Park YS, Chung JH, Oh SP (2006) A murine model for human sepiapterin-reductase deficiency. Am J Hum Genet 78(4):575–587. https://doi.org/10.1086/501372

53. Takazawa C, Fujimoto K, Homma D, Sumi-Ichinose C, Nomura T, Ichinose H, Katoh S (2008) A brain-specific decrease of the tyrosine hydroxylase protein in sepiapterin reductase-null mice--as a mouse model for Parkinson's disease. Biochem Biophys Res Commun 367(4):787–792. https://doi.org/10.1016/j.bbrc.2008.01.028

54. Sumi-Ichinose C, Suganuma Y, Kano T, Ihira N, Nomura H, Ikemoto K, Hata T, Katoh S, Ichinose H, Kondo K (2017) Sepiapterin reductase gene-disrupted mice suffer from hypertension with fluctuation and bradycardia. Physiol Rep 5(6). https://doi.org/10.14814/phy2.13196

55. Meng Y, Katsuma S, Daimon T, Banno Y, Uchino K, Sezutsu H, Tamura T, Mita K, Shimada T (2009) The silkworm mutant lemon (lemon lethal) is a potential insect model for human sepiapterin reductase deficiency. J Biol Chem 284(17):11698–11705. https://doi.org/10.1074/jbc.M900485200

56. Leroy E, Boyer R, Auburger G, Leube B, Ulm G, Mezey E, Harta G, Brownstein MJ, Jonnalagada S, Chernova T, Dehejia A, Lavedan C, Gasser T, Steinbach PJ, Wilkinson KD, Polymeropoulos MH (1998) The ubiquitin pathway in Parkinson's disease. Nature 395(6701):451–452. https://doi.org/10.1038/26652

57. Healy DG, Abou-Sleiman PM, Casas JP, Ahmadi KR, Lynch T, Gandhi S, Muqit MM, Foltynie T, Barker R, Bhatia KP, Quinn NP, Lees AJ, Gibson JM, Holton JL, Revesz T, Goldstein DB, Wood NW (2006) UCHL-1 is not a Parkinson's disease susceptibility gene. Ann Neurol 59(4):627–633. https://doi.org/10.1002/ana.20757

58. Yasuda T, Nihira T, Ren YR, Cao XQ, Wada K, Setsuie R, Kabuta T, Wada K, Hattori N, Mizuno Y, Mochizuki H (2009) Effects of UCH-L1 on alpha-synuclein over-expression mouse model of Parkinson's disease. J Neurochem 108(4):932–944. https://doi.org/10.1111/j.1471-4159.2008.05827.x

59. Shimshek DR, Schweizer T, Schmid P, van der Putten PH (2012) Excess alpha-synuclein worsens disease in mice lacking ubiquitin carboxy-terminal hydrolase L1. Sci Rep 2:262. https://doi.org/10.1038/srep00262

60. Nguyen TT, Vuu MD, Huynh MA, Yamaguchi M, Tran LT, Dang TPT (2018) Curcumin effectively rescued Parkinson's disease-like phenotypes in a novel Drosophila melanogaster model with dUCH knockdown. Oxidative Med Cell Longev 2018:2038267. https://doi.org/10.1155/2018/2038267

61. Chu CT (2018) Mechanisms of selective autophagy and mitophagy: implications for neurodegenerative diseases. Neurobiol Dis. https://doi.org/10.1016/j.nbd.2018.07.015

62. Valente EM, Abou-Sleiman PM, Caputo V, Muqit MM, Harvey K, Gispert S, Ali Z, Del Turco D, Bentivoglio AR, Healy

DG, Albanese A, Nussbaum R, Gonzalez-Maldonado R, Deller T, Salvi S, Cortelli P, Gilks WP, Latchman DS, Harvey RJ, Dallapiccola B, Auburger G, Wood NW (2004) Hereditary early-onset Parkinson's disease caused by mutations in PINK1. Science 304(5674):1158–1160. https://doi.org/10.1126/science.1096284

63. Bonifati V, Rohe CF, Breedveld GJ, Fabrizio E, De Mari M, Tassorelli C, Tavella A, Marconi R, Nicholl DJ, Chien HF, Fincati E, Abbruzzese G, Marini P, De Gaetano A, Horstink MW, Maat-Kievit JA, Sampaio C, Antonini A, Stocchi F, Montagna P, Toni V, Guidi M, Dalla Libera A, Tinazzi M, De Pandis F, Fabbrini G, Goldwurm S, de Klein A, Barbosa E, Lopiano L, Martignoni E, Lamberti P, Vanacore N, Meco G, Oostra BA, Italian Parkinson Genetics N (2005) Early-onset parkinsonism associated with PINK1 mutations: frequency, genotypes, and phenotypes. Neurology 65(1):87–95. https://doi.org/10.1212/01.wnl.0000167546.39375.82

64. Hatano Y, Sato K, Elibol B, Yoshino H, Yamamura Y, Bonifati V, Shinotoh H, Asahina M, Kobayashi S, Ng AR, Rosales RL, Hassin-Baer S, Shinar Y, Lu CS, Chang HC, Wu-Chou YH, Atac FB, Kobayashi T, Toda T, Mizuno Y, Hattori N (2004) PARK6-linked autosomal recessive early-onset parkinsonism in Asian populations. Neurology 63(8):1482–1485

65. Gispert S, Ricciardi F, Kurz A, Azizov M, Hoepken HH, Becker D, Voos W, Leuner K, Muller WE, Kudin AP, Kunz WS, Zimmermann A, Roeper J, Wenzel D, Jendrach M, Garcia-Arencibia M, Fernandez-Ruiz J, Huber L, Rohrer H, Barrera M, Reichert AS, Rub U, Chen A, Nussbaum RL, Auburger G (2009) Parkinson phenotype in aged PINK1-deficient mice is accompanied by progressive mitochondrial dysfunction in absence of neurodegeneration. PLoS One 4(6):e5777. https://doi.org/10.1371/journal.pone.0005777

66. Oliveras-Salva M, Macchi F, Coessens V, Deleersnijder A, Gerard M, Van der Perren A, Van den Haute C, Baekelandt V (2014) Alpha-synuclein-induced neurodegeneration is exacerbated in PINK1 knockout mice. Neurobiol Aging 35(11):2625–2636. https://doi.org/10.1016/j.neurobiolaging.2014.04.032

67. Villeneuve LM, Purnell PR, Boska MD, Fox HS (2016) Early expression of Parkinson's disease-related mitochondrial abnormalities in PINK1 knockout rats. Mol Neurobiol 53(1):171–186. https://doi.org/10.1007/s12035-014-8927-y

68. Ferris CF, Morrison TR, Iriah S, Malmberg S, Kulkarni P, Hartner JC, Trivedi M (2018) Evidence of neurobiological changes in the presymptomatic PINK1 knockout rat. J Parkinsons Dis 8(2):281–301. https://doi.org/10.3233/JPD-171273

69. Zhang Y, Nguyen DT, Olzomer EM, Poon GP, Cole NJ, Puvanendran A, Phillips BR, Hesselson D (2017) Rescue of Pink1 deficiency by stress-dependent activation of autophagy. Cell Chem Biol 24(4):471–480 e474. https://doi.org/10.1016/j.chembiol.2017.03.005

70. Priyadarshini M, Tuimala J, Chen YC, Panula P (2013) A zebrafish model of PINK1 deficiency reveals key pathway dysfunction including HIF signaling. Neurobiol Dis 54:127–138. https://doi.org/10.1016/j.nbd.2013.02.002

71. Ariga H, Takahashi-Niki K, Kato I, Maita H, Niki T, Iguchi-Ariga SM (2013) Neuroprotective function of DJ-1 in Parkinson's disease. Oxidative Med Cell Longev 2013:683920

72. Hedrich K, Djarmati A, Schafer N, Hering R, Wellenbrock C, Weiss PH, Hilker R, Vieregge P, Ozelius LJ, Heutink P, Bonifati V, Schwinger E, Lang AE, Noth J, Bressman SB, Pramstaller PP, Riess O, Klein C (2004) DJ-1 (PARK7) mutations are less frequent than Parkin (PARK2) mutations in early-onset Parkinson disease. Neurology 62(3):389–394

73. Bonifati V, Rizzu P, van Baren MJ, Schaap O, Breedveld GJ, Krieger E, Dekker MC, Squitieri F, Ibanez P, Joosse M, van Dongen JW, Vanacore N, van Swieten JC, Brice A, Meco G, van Duijn CM, Oostra BA, Heutink P (2003) Mutations in the DJ-1 gene associated with autosomal recessive early-onset parkinsonism. Science 299(5604):256–259. https://doi.org/10.1126/science.1077209

74. Goldberg MS, Pisani A, Haburcak M, Vortherms TA, Kitada T, Costa C, Tong Y, Martella G, Tscherter A, Martins A, Bernardi G, Roth BL, Pothos EN, Calabresi P, Shen J (2005) Nigrostriatal dopaminergic deficits and hypokinesia caused by inactivation of the familial Parkinsonism-linked gene DJ-1. Neuron 45(4):489–496

75. Kim RH, Smith PD, Aleyasin H, Hayley S, Mount MP, Pownall S, Wakeham A, You-Ten AJ, Kalia SK, Horne P, Westaway D, Lozano AM, Anisman H, Park DS, Mak TW (2005) Hypersensitivity of DJ-1-deficient mice to 1-methyl-4-phenyl-1,2,3,6-tetrahydropyrindine (MPTP) and oxidative stress. Proc Natl Acad Sci U S A 102(14):5215–5220. https://doi.org/10.1073/pnas.0501282102

76. Pham TT, Giesert F, Rothig A, Floss T, Kallnik M, Weindl K, Holter SM, Ahting U, Prokisch H, Becker L, Klopstock T, Hrabe de AM, Beyer K, Gorner K, Kahle PJ, Vogt Weisenhorn DM, Wurst W (2010) DJ-1-deficient mice show less TH-positive neurons in the ventral tegmental area and exhibit non-motoric behavioural impairments. Genes Brain Behav 9(3):305–317

77. Rousseaux MW, Marcogliese PC, Qu D, Hewitt SJ, Seang S, Kim RH, Slack RS, Schlossmacher MG, Lagace DC, Mak TW, Park DS (2012) Progressive dopaminergic cell loss with unilateral-to-bilateral progression in a genetic model of Parkinson disease. Proc Natl Acad Sci U S A 109(39):15918–15923. https://doi.org/10.1073/pnas.1205102109

78. Yao J, Huang J, Hai T, Wang X, Qin G, Zhang H, Wu R, Cao C, Xi JJ, Yuan Z, Zhao J (2014) Efficient bi-allelic gene knockout and site-specific knock-in mediated by TALENs in pigs. Sci Rep 4:6926. https://doi.org/10.1038/srep06926

79. Smith WW, Pei Z, Jiang H, Moore DJ, Liang Y, West AB, Dawson VL, Dawson TM, Ross CA (2005) Leucine-rich repeat kinase 2 (LRRK2) interacts with parkin, and mutant LRRK2 induces neuronal degeneration. Proc Natl Acad Sci U S A 102(51):18676–18681

80. Funayama M, Hasegawa K, Kowa H, Saito M, Tsuji S, Obata F (2002) A new locus for Parkinson's disease (PARK8) maps to chromosome 12p11.2-q13.1. Ann Neurol 51(3):296–301

81. Zimprich A, Biskup S, Leitner P, Lichtner P, Farrer M, Lincoln S, Kachergus J, Hulihan M, Uitti RJ, Calne DB, Stoessl AJ, Pfeiffer RF, Patenge N, Carbajal IC, Vieregge P, Asmus F, Muller-Myhsok B, Dickson DW, Meitinger T, Strom TM, Wszolek ZK, Gasser T (2004) Mutations in LRRK2 cause autosomal-dominant parkinsonism with pleomorphic pathology. Neuron 44(4):601–607. https://doi.org/10.1016/j.neuron.2004.11.005

82. Healy DG, Falchi M, O'Sullivan SS, Bonifati V, Durr A, Bressman S, Brice A, Aasly J, Zabetian CP, Goldwurm S, Ferreira JJ, Tolosa E, Kay DM, Klein C, Williams DR, Marras C, Lang AE, Wszolek ZK, Berciano J, Schapira AH, Lynch T, Bhatia KP, Gasser T, Lees AJ, Wood NW (2008) Phenotype, genotype, and worldwide genetic penetrance of LRRK2-associated Parkinson's disease: a case-control study. Lancet Neurol 7(7):583–590

83. Gilks WP, bou-Sleiman PM, Gandhi S, Jain S, Singleton A, Lees AJ, Shaw K, Bhatia KP, Bonifati V, Quinn NP, Lynch J, Healy DG, Holton JL, Revesz T, Wood NW (2005) A common LRRK2 mutation in idiopathic Parkinson's disease. Lancet 365(9457):415–416

84. Ozelius LJ, Senthil G, Saunders-Pullman R, Ohmann E, Deligtisch A, Tagliati M, Hunt AL, Klein C, Henick B, Hailpern SM, Lipton RB, Soto-Valencia J, Risch N, Bressman SB (2006) LRRK2 G2019S as a cause of Parkinson's disease in Ashkenazi Jews. New Engl J Med 354(4):424–425

85. Hulihan MM, Ishihara-Paul L, Kachergus J, Warren L, Amouri R, Elango R, Prinjha RK, Upmanyu R, Kefi M, Zouari M, Sassi SB, Yahmed SB, El Euch-Fayeche G, Matthews PM, Middleton LT, Gibson RA, Hentati F, Farrer MJ (2008) LRRK2 Gly2019Ser penetrance in Arab-Berber patients from Tunisia: a case-control genetic study. Lancet Neurol 7(7):591–594

86. Rudenko IN, Cookson MR (2014) Heterogeneity of leucine-rich repeat kinase 2 mutations: genetics, mechanisms and therapeutic implications. Neurotherapeutics 11(4):738–750

87. Ramonet D, Daher JP, Lin BM, Stafa K, Kim J, Banerjee R, Westerlund M, Pletnikova O, Glauser L, Yang L, Liu Y, Swing DA, Beal MF, Troncoso JC, McCaffery JM, Jenkins NA, Copeland NG, Galter D, Thomas B, Lee MK, Dawson TM, Dawson VL, Moore DJ (2011) Dopaminergic neuronal loss, reduced neurite complexity and autophagic abnormalities in transgenic mice expressing G2019S mutant LRRK2. PLoS One 6(4):e18568

88. Tong Y, Pisani A, Martella G, Karouani M, Yamaguchi H, Pothos EN, Shen J (2009) R1441C mutation in LRRK2 impairs dopaminergic neurotransmission in mice. Proc Natl Acad Sci U S A 106(34):14622–14627

89. Antony PM, Diederich NJ, Balling R (2011) Parkinson's disease mouse models in translational research. Mamm Genome 22(7–8):401–419. https://doi.org/10.1007/s00335-011-9330-x

90. Hinkle KM, Yue M, Behrouz B, Dachsel JC, Lincoln SJ, Bowles EE, Beevers JE, Dugger B, Winner B, Prots I, Kent CB, Nishioka K, Lin WL, Dickson DW, Janus CJ, Farrer MJ, Melrose HL (2012) LRRK2 knockout mice have an intact dopaminergic system but display alterations in exploratory and motor co-ordination behaviors. Mol Neurodegener 7:25

91. Tong Y, Yamaguchi H, Giaime E, Boyle S, Kopan R, Kelleher RJ III, Shen J (2010) Loss of leucine-rich repeat kinase 2 causes impairment of protein degradation pathways, accumulation of alpha-synuclein, and apoptotic cell death in aged mice. Proc Natl Acad Sci U S A 107(21):9879–9884

92. Li Y, Liu W, Oo TF, Wang L, Tang Y, Jackson-Lewis V, Zhou C, Geghman K, Bogdanov M, Przedborski S, Beal MF, Burke RE, Li C (2009) Mutant LRRK2(R1441G) BAC transgenic mice recapitulate cardinal features of Parkinson's disease. Nat Neurosci 12(7):826–828

93. Melrose HL, Dachsel JC, Behrouz B, Lincoln SJ, Yue M, Hinkle KM, Kent CB, Korvatska E, Taylor JP, Witten L, Liang YQ, Beevers JE, Boules M, Dugger BN, Serna VA, Gaukhman A, Yu X, Castanedes-Casey M, Braithwaite AT, Ogholikhan S, Yu N, Bass D, Tyndall G, Schellenberg GD, Dickson DW, Janus C, Farrer MJ (2010) Impaired dopaminergic neurotransmission and microtubule-associated protein tau alterations in human LRRK2 transgenic mice. Neurobiol Dis 40(3):503–517

94. Ramirez A, Heimbach A, Grundemann J, Stiller B, Hampshire D, Cid LP, Goebel I, Mubaidin AF, Wriekat AL, Roeper J, Al-Din A, Hillmer AM, Karsak M, Liss B, Woods CG, Behrens MI, Kubisch C (2006) Hereditary parkinsonism with dementia is caused by mutations in ATP13A2, encoding a lysosomal type 5 P-type ATPase. Nat Genet 38(10):1184–1191. https://doi.org/10.1038/ng1884

95. Williams DR, Hadeed A, al-Din AS, Wreikat AL, Lees AJ (2005) Kufor Rakeb disease: autosomal recessive, levodopa-responsive parkinsonism with pyramidal degeneration, supranuclear gaze palsy, and dementia. Mov Disord 20(10):1264–1271. https://doi.org/10.1002/mds.20511

96. Najim al-Din AS, Wriekat A, Mubaidin A, Dasouki M, Hiari M (1994) Pallido-pyramidal degeneration, supranuclear upgaze paresis and dementia: Kufor-Rakeb syndrome. Acta Neurol Scand 89(5):347–352

97. Farias FH, Zeng R, Johnson GS, Wininger FA, Taylor JF, Schnabel RD, McKay SD, Sanders DN, Lohi H, Seppala EH, Wade CM, Lindblad-Toh K, O'Brien DP, Katz ML (2011) A truncating mutation in ATP13A2 is responsible for adult-onset neuronal ceroid lipofuscinosis in Tibetan terriers. Neurobiol Dis 42(3):468–474. https://doi.org/10.1016/j.nbd.2011.02.009

98. Wohlke A, Philipp U, Bock P, Beineke A, Lichtner P, Meitinger T, Distl O (2011) A one base pair deletion in the canine ATP13A2 gene causes exon skipping and late-onset neuronal ceroid lipofuscinosis in the Tibetan terrier. PLoS Genet 7(10):e1002304. https://doi.org/10.1371/journal.pgen.1002304

99. Daniel G, Musso A, Tsika E, Fiser A, Glauser L, Pletnikova O, Schneider BL, Moore DJ (2015) alpha-Synuclein-induced dopaminergic neurodegeneration in a rat model of Parkinson's disease occurs independent of ATP13A2 (PARK9). Neurobiol Dis 73:229–243. https://doi.org/10.1016/j.nbd.2014.10.007

100. Kett LR, Stiller B, Bernath MM, Tasset I, Blesa J, Jackson-Lewis V, Chan RB, Zhou B, Di Paolo G, Przedborski S, Cuervo AM, Dauer WT (2015) alpha-Synuclein-independent histopathological and motor deficits in mice lacking the endolysosomal Parkinsonism protein Atp13a2. J Neurosci 35(14):5724–5742. https://doi.org/10.1523/JNEUROSCI.0632-14.2015

101. Schultheis PJ, Fleming SM, Clippinger AK, Lewis J, Tsunemi T, Giasson B, Dickson DW, Mazzulli JR, Bardgett ME, Haik KL, Ekhator O, Chava AK, Howard J, Gannon M, Hoffman E, Chen Y, Prasad V, Linn SC, Tamargo RJ, Westbroek W, Sidransky E, Krainc D, Shull GE (2013) Atp13a2-deficient mice exhibit neuronal ceroid lipofuscinosis, limited alpha-synuclein accumulation and age-dependent sensorimotor deficits. Hum Mol Genet 22(10):2067–2082. https://doi.org/10.1093/hmg/ddt057

102. Hicks AA, Petursson H, Jonsson T, Stefansson H, Johannsdottir HS, Sainz J, Frigge ML, Kong A, Gulcher JR, Stefansson K, Sveinbjornsdottir S (2002) A susceptibility gene for late-onset idiopathic Parkinson's disease. Ann Neurol 52(5):549–555. https://doi.org/10.1002/ana.10324

103. Wan JY, Edwards KL, Hutter CM, Mata IF, Samii A, Roberts JW, Agarwal P, Checkoway H, Farin FM, Yearout D, Zabetian CP (2014) Association mapping of the PARK10 region for Parkinson's disease susceptibility genes. Parkinsonism Relat Disord 20(1):93–98. https://doi.org/10.1016/j.parkreldis.2013.10.001

104. Gomez-Lopez S, Martinez-Silva AV, Montiel T, Osorio-Gomez D, Bermudez-Rattoni F, Massieu L, Escalante-Alcalde D (2016) Neural ablation of the PARK10 candidate Plpp3 leads to dopaminergic transmission deficits without neurodegeneration. Sci Rep 6:24028. https://doi.org/10.1038/srep24028

105. Lautier C, Goldwurm S, Durr A, Giovannone B, Tsiaras WG, Pezzoli G, Brice A, Smith RJ (2008) Mutations in the GIGYF2 (TNRC15) gene at the PARK11 locus in familial Parkinson disease. Am J Hum Genet 82(4):822–833. https://doi.org/10.1016/j.ajhg.2008.01.015

106. Pankratz N, Nichols WC, Uniacke SK, Halter C, Rudolph A, Shults C, Conneally PM, Foroud T, Parkinson Study G (2003) Significant linkage of Parkinson disease

to chromosome 2q36-37. Am J Hum Genet 72(4):1053–1057. https://doi.org/10.1086/374383

107. Prestel J, Sharma M, Leitner P, Zimprich A, Vaughan JR, Durr A, Bonifati V, De Michele G, Hanagasi HA, Farrer M, Hofer A, Asmus F, Volpe G, Meco G, Brice A, Wood NW, Muller-Myhsok B, Gasser T, European Consortium on Genetic Susceptibility in Parkinson's D (2005) PARK11 is not linked with Parkinson's disease in European families. Eur J Hum Genet 13(2):193–197. https://doi.org/10.1038/sj.ejhg.5201317

108. Giovannone B, Tsiaras WG, de la Monte S, Klysik J, Lautier C, Karashchuk G, Goldwurm S, Smith RJ (2009) GIGYF2 gene disruption in mice results in neurodegeneration and altered insulin-like growth factor signaling. Hum Mol Genet 18(23):4629–4639. https://doi.org/10.1093/hmg/ddp430

109. Kim M, Semple I, Kim B, Kiers A, Nam S, Park HW, Park H, Ro SH, Kim JS, Juhasz G, Lee JH (2015) Drosophila Gyf/GRB10 interacting GYF protein is an autophagy regulator that controls neuron and muscle homeostasis. Autophagy 11(8):1358–1372. https://doi.org/10.1080/15548627.2015.1063766

110. Guella I, Pistocchi A, Asselta R, Rimoldi V, Ghilardi A, Sironi F, Trotta L, Primignani P, Zini M, Zecchinelli A, Coviello D, Pezzoli G, Del Giacco L, Duga S, Goldwurm S (2011) Mutational screening and zebrafish functional analysis of GIGYF2 as a Parkinson-disease gene. Neurobiol Aging 32(11):1994–2005. https://doi.org/10.1016/j.neurobiolaging.2009.12.016

111. Pankratz N, Nichols WC, Uniacke SK, Halter C, Murrell J, Rudolph A, Shults CW, Conneally PM, Foroud T, Parkinson Study G (2003) Genome-wide linkage analysis and evidence of gene-by-gene interactions in a sample of 362 multiplex Parkinson disease families. Hum Mol Genet 12(20):2599–2608. https://doi.org/10.1093/hmg/ddg270

112. Pankratz N, Nichols WC, Uniacke SK, Halter C, Rudolph A, Shults C, Conneally PM, Foroud T, Parkinson Study G (2002) Genome screen to identify susceptibility genes for Parkinson disease in a sample without parkin mutations. Am J Hum Genet 71(1):124–135. https://doi.org/10.1086/341282

113. Wilson GR, Sim JC, McLean C, Giannandrea M, Galea CA, Riseley JR, Stephenson SE, Fitzpatrick E, Haas SA, Pope K, Hogan KJ, Gregg RG, Bromhead CJ, Wargowski DS, Lawrence CH, James PA, Churchyard A, Gao Y, Phelan DG, Gillies G, Salce N, Stanford L, Marsh AP, Mignogna ML, Hayflick SJ,

Leventer RJ, Delatycki MB, Mellick GD, Kalscheuer VM, D'Adamo P, Bahlo M, Amor DJ, Lockhart PJ (2014) Mutations in RAB39B cause X-linked intellectual disability and early-onset Parkinson disease with alpha-synuclein pathology. Am J Hum Genet 95(6):729–735. https://doi.org/10.1016/j.ajhg.2014.10.015

114. Morrison BE, Marcondes MC, Nomura DK, Sanchez-Alavez M, Sanchez-Gonzalez A, Saar I, Kim KS, Bartfai T, Maher P, Sugama S, Conti B (2012) Cutting edge: IL-13Ralpha1 expression in dopaminergic neurons contributes to their oxidative stress-mediated loss following chronic peripheral treatment with lipopolysaccharide. J Immunol 189(12):5498–5502. https://doi.org/10.4049/jimmunol.1102150

115. Strauss KM, Martins LM, Plun-Favreau H, Marx FP, Kautzmann S, Berg D, Gasser T, Wszolek Z, Muller T, Bornemann A, Wolburg H, Downward J, Riess O, Schulz JB, Kruger R (2005) Loss of function mutations in the gene encoding Omi/HtrA2 in Parkinson's disease. Hum Mol Genet 14(15):2099–2111. https://doi.org/10.1093/hmg/ddi215

116. Simon-Sanchez J, Singleton AB (2008) Sequencing analysis of OMI/HTRA2 shows previously reported pathogenic mutations in neurologically normal controls. Hum Mol Genet 17(13):1988–1993. https://doi.org/10.1093/hmg/ddn096

117. Jones JM, Datta P, Srinivasula SM, Ji W, Gupta S, Zhang Z, Davies E, Hajnoczky G, Saunders TL, Van Keuren ML, Fernandes-Alnemri T, Meisler MH, Alnemri ES (2003) Loss of Omi mitochondrial protease activity causes the neuromuscular disorder of mnd2 mutant mice. Nature 425(6959):721–727. https://doi.org/10.1038/nature02052

118. Martins LM, Morrison A, Klupsch K, Fedele V, Moisoi N, Teismann P, Abuin A, Grau E, Geppert M, Livi GP, Creasy CL, Martin A, Hargreaves I, Heales SJ, Okada H, Brandner S, Schulz JB, Mak T, Downward J (2004) Neuroprotective role of the Reaper-related serine protease HtrA2/Omi revealed by targeted deletion in mice. Mol Cell Biol 24(22):9848–9862. https://doi.org/10.1128/MCB.24.22.9848-9862.2004

119. Moisoi N, Klupsch K, Fedele V, East P, Sharma S, Renton A, Plun-Favreau H, Edwards RE, Teismann P, Esposti MD, Morrison AD, Wood NW, Downward J, Martins LM (2009) Mitochondrial dysfunction triggered by loss of HtrA2 results in the activation of a brain-specific transcriptional stress response. Cell Death Differ 16(3):449–464. https://doi.org/10.1038/cdd.2008.166

120. Casadei N, Sood P, Ulrich T, Fallier-Becker P, Kieper N, Helling S, May C, Glaab E, Chen J, Nuber S, Wolburg H, Marcus K, Rapaport D, Ott T, Riess O, Kruger R, Fitzgerald JC (2016) Mitochondrial defects and neurodegeneration in mice overexpressing wild-type or G399S mutant HtrA2. Hum Mol Genet 25(3):459–471. https://doi.org/10.1093/hmg/ddv485

121. Patterson VL, Zullo AJ, Koenig C, Stoessel S, Jo H, Liu X, Han J, Choi M, DeWan AT, Thomas JL, Kuan CY, Hoh J (2014) Neural-specific deletion of Htra2 causes cerebellar neurodegeneration and defective processing of mitochondrial OPA1. PLoS One 9(12):e115789. https://doi.org/10.1371/journal.pone.0115789

122. Dwivedi V, Lakhotia SC (2016) Ayurvedic Amalaki Rasayana promotes improved stress tolerance and thus has anti-aging effects in Drosophila melanogaster. J Biosci 41(4):697–711

123. Yun J, Cao JH, Dodson MW, Clark IE, Kapahi P, Chowdhury RB, Guo M (2008) Loss-of-function analysis suggests that Omi/HtrA2 is not an essential component of the PINK1/PARKIN pathway in vivo. J Neurosci 28(53):14500–14510. https://doi.org/10.1523/JNEUROSCI.5141-08.2008

124. M'Angale PG, Staveley BE (2017) The HtrA2 Drosophila model of Parkinson's disease is suppressed by the pro-survival Bcl-2 Buffy. Genome 60(1):1–7. https://doi.org/10.1139/gen-2016-0069

125. Morgan NV, Westaway SK, Morton JE, Gregory A, Gissen P, Sonek S, Cangul H, Coryell J, Canham N, Nardocci N, Zorzi G, Pasha S, Rodriguez D, Desguerre I, Mubaidin A, Bertini E, Trembath RC, Simonati A, Schanen C, Johnson CA, Levinson B, Woods CG, Wilmot B, Kramer P, Gitschier J, Maher ER, Hayflick SJ (2006) PLA2G6, encoding a phospholipase A2, is mutated in neurodegenerative disorders with high brain iron. Nat Genet 38(7):752–754. https://doi.org/10.1038/ng1826

126. Paisan-Ruiz C, Bhatia KP, Li A, Hernandez D, Davis M, Wood NW, Hardy J, Houlden H, Singleton A, Schneider SA (2009) Characterization of PLA2G6 as a locus for dystonia-parkinsonism. Ann Neurol 65(1):19–23. https://doi.org/10.1002/ana.21415

127. Blanchard H, Taha AY, Cheon Y, Kim HW, Turk J, Rapoport SI (2014) iPLA2beta knockout mouse, a genetic model for progressive human motor disorders, develops age-related neuropathology. Neurochem Res 39(8):1522–1532. https://doi.org/10.1007/s11064-014-1342-y

128. Wada H, Kojo S, Seino K (2013) Mouse models of human INAD by Pla2g6 deficiency. Histol Histopathol 28(8):965–969. https://doi.org/10.14670/HH-28.965

129. Chiu CC, Lu CS, Weng YH, Chen YL, Huang YZ, Chen RS, Cheng YC, Huang YC, Liu YC, Lai SC, Lin KJ, Lin YW, Chen YJ, Chen CL, Yeh TH, Wang HL (2018) PARK14 (D331Y) PLA2G6 causes early-onset degeneration of substantia nigra dopaminergic neurons by inducing mitochondrial dysfunction, ER stress, mitophagy impairment and transcriptional dysregulation in a knockin mouse model. Mol Neurobiol. https://doi.org/10.1007/s12035-018-1118-5

130. Zhou Q, Yen A, Rymarczyk G, Asai H, Trengrove C, Aziz N, Kirber MT, Mostoslavsky G, Ikezu T, Wolozin B, Bolotina VM (2016) Impairment of PARK14-dependent Ca(2+) signalling is a novel determinant of Parkinson's disease. Nat Commun 7:10332 https://doi.org/10.1038/ncomms10332

131. Sumi-Akamaru H, Beck G, Kato S, Mochizuki H (2015) Neuroaxonal dystrophy in PLA2G6 knockout mice. Neuropathology 35(3):289–302. https://doi.org/10.1111/neup.12202

132. Sanchez E, Azcona LJ, Paisan-Ruiz C (2018) Pla2g6 deficiency in zebrafish leads to dopaminergic cell death, axonal degeneration, increased beta-synuclein expression, and defects in brain functions and pathways. Mol Neurobiol 55(8):6734–6754. https://doi.org/10.1007/s12035-017-0846-2

133. Iliadi KG, Gluscencova OB, Iliadi N, Boulianne GL (2018) Mutations in the Drosophila homolog of human PLA2G6 give rise to age-dependent loss of psychomotor activity and neurodegeneration. Sci Rep 8(1):2939. https://doi.org/10.1038/s41598-018-21343-8

134. Kinghorn KJ, Castillo-Quan JI, Bartolome F, Angelova PR, Li L, Pope S, Cocheme HM, Khan S, Asghari S, Bhatia KP, Hardy J, Abramov AY, Partridge L (2015) Loss of PLA2G6 leads to elevated mitochondrial lipid peroxidation and mitochondrial dysfunction. Brain 138(Pt7):1801–1816. https://doi.org/10.1093/brain/awv132

135. Tsuboi M, Watanabe M, Nibe K, Yoshimi N, Kato A, Sakaguchi M, Yamato O, Tanaka M, Kuwamura M, Kushida K, Ishikura T, Harada T, Chambers JK, Sugano S, Uchida K, Nakayama H (2017) Identification of the PLA2G6 c.1579G>A missense mutation in papillon dog neuroaxonal dystrophy using whole exome sequencing analysis. PLoS One 12(1):e0169002. https://doi.org/10.1371/journal.pone.0169002

136. Di Fonzo A, Dekker MC, Montagna P, Baruzzi A, Yonova EH, Correia Guedes L,

Szczerbinska A, Zhao T, Dubbel-Hulsman LO, Wouters CH, de Graaff E, Oyen WJ, Simons EJ, Breedveld GJ, Oostra BA, Horstink MW, Bonifati V (2009) FBXO7 mutations cause autosomal recessive, early-onset parkinsonian-pyramidal syndrome. Neurology 72(3):240–245. https://doi.org/10.1212/01.wnl.0000338144.10967.2b

137. Shojaee S, Sina F, Banihosseini SS, Kazemi MH, Kalhor R, Shahidi GA, Fakhrai-Rad H, Ronaghi M, Elahi E (2008) Genome-wide linkage analysis of a Parkinsonian-pyramidal syndrome pedigree by 500 K SNP arrays. Am J Hum Genet 82(6):1375–1384. https://doi.org/10.1016/j.ajhg.2008.05.005

138. Vingill S, Brockelt D, Lancelin C, Tatenhorst L, Dontcheva G, Preisinger C, Schwedhelm-Domeyer N, Joseph S, Mitkovski M, Goebbels S, Nave KA, Schulz JB, Marquardt T, Lingor P, Stegmuller J (2016) Loss of FBXO7 (PARK15) results in reduced proteasome activity and models a parkinsonism-like phenotype in mice. EMBO J 35(18):2008–2025. https://doi.org/10.15252/embj.201593585

139. Randle SJ, Nelson DE, Patel SP, Laman H (2015) Defective erythropoiesis in a mouse model of reduced Fbxo7 expression due to decreased p27 expression. J Pathol 237(2):263–272. https://doi.org/10.1002/path.4571

140. Merzetti EM, Staveley BE (2016) Altered expression of CG5961, a putative Drosophila melanogaster homologue of FBXO9, provides a new model of Parkinson disease. Genet Mol Res 15(2). https://doi.org/10.4238/gmr.15028579

141. Burchell VS, Nelson DE, Sanchez-Martinez A, Delgado-Camprubi M, Ivatt RM, Pogson JH, Randle SJ, Wray S, Lewis PA, Houlden H, Abramov AY, Hardy J, Wood NW, Whitworth AJ, Laman H, Plun-Favreau H (2013) The Parkinson's disease-linked proteins Fbxo7 and Parkin interact to mediate mitophagy. Nat Neurosci 16(9):1257–1265. https://doi.org/10.1038/nn.3489

142. Zhao T, Zondervan-van der Linde H, Severijnen LA, Oostra BA, Willemsen R, Bonifati V (2012) Dopaminergic neuronal loss and dopamine-dependent locomotor defects in Fbxo7-deficient zebrafish. PLoS One 7(11):e48911. https://doi.org/10.1371/journal.pone.0048911

143. Satake W, Nakabayashi Y, Mizuta I, Hirota Y, Ito C, Kubo M, Kawaguchi T, Tsunoda T, Watanabe M, Takeda A, Tomiyama H, Nakashima K, Hasegawa K, Obata F, Yoshikawa T, Kawakami H, Sakoda S, Yamamoto M, Hattori N, Murata M, Nakamura Y, Toda T (2009) Genome-wide association study identifies common variants at four loci as genetic risk factors for Parkinson's disease. Nat Genet 41(12):1303–1307. https://doi.org/10.1038/ng.485

144. Jaberi E, Rohani M, Shahidi GA, Nafissi S, Arefian E, Soleimani M, Moghadam A, Arzenani MK, Keramatian F, Klotzle B, Fan JB, Turk C, Steemers F, Elahi E (2016) Mutation in ADORA1 identified as likely cause of early-onset parkinsonism and cognitive dysfunction. Mov Disord 31(7):1004–1011. https://doi.org/10.1002/mds.26627

145. Simon-Sanchez J, Schulte C, Bras JM, Sharma M, Gibbs JR, Berg D, Paisan-Ruiz C, Lichtner P, Scholz SW, Hernandez DG, Kruger R, Federoff M, Klein C, Goate A, Perlmutter J, Bonin M, Nalls MA, Illig T, Gieger C, Houlden H, Steffens M, Okun MS, Racette BA, Cookson MR, Foote KD, Fernandez HH, Traynor BJ, Schreiber S, Arepalli S, Zonozi R, Gwinn K, van der Brug M, Lopez G, Chanock SJ, Schatzkin A, Park Y, Hollenbeck A, Gao J, Huang X, Wood NW, Lorenz D, Deuschl G, Chen H, Riess O, Hardy JA, Singleton AB, Gasser T (2009) Genome-wide association study reveals genetic risk underlying Parkinson's disease. Nat Genet 41(12):1308–1312. https://doi.org/10.1038/ng.487

146. Mata IF, Yearout D, Alvarez V, Coto E, de Mena L, Ribacoba R, Lorenzo-Betancor O, Samaranch L, Pastor P, Cervantes S, Infante J, Garcia-Gorostiaga I, Sierra M, Combarros O, Snapinn KW, Edwards KL, Zabetian CP (2011) Replication of MAPT and SNCA, but not PARK16-18, as susceptibility genes for Parkinson's disease. Mov Disord 26(5):819–823. https://doi.org/10.1002/mds.23642

147. Zimprich A, Benet-Pages A, Struhal W, Graf E, Eck SH, Offman MN, Haubenberger D, Spielberger S, Schulte EC, Lichtner P, Rossle SC, Klopp N, Wolf E, Seppi K, Pirker W, Presslauer S, Mollenhauer B, Katzenschlager R, Foki T, Hotzy C, Reinthaler E, Harutyunyan A, Kralovics R, Peters A, Zimprich F, Brucke T, Poewe W, Auff E, Trenkwalder C, Rost B, Ransmayr G, Winkelmann J, Meitinger T, Strom TM (2011) A mutation in VPS35, encoding a subunit of the retromer complex, causes late-onset Parkinson disease. Am J Hum Genet 89(1):168–175. https://doi.org/10.1016/j.ajhg.2011.06.008

148. Vilarino-Guell C, Wider C, Ross OA, Dachsel JC, Kachergus JM, Lincoln SJ, Soto-Ortolaza AI, Cobb SA, Wilhoite GJ, Bacon JA, Behrouz B, Melrose HL, Hentati E, Puschmann A, Evans DM, Conibear E, Wasserman WW, Aasly JO, Burkhard PR, Djaldetti R, Ghika J, Hentati F, Krygowska-Wajs A, Lynch T,

Melamed E, Rajput A, Rajput AH, Solida A, Wu RM, Uitti RJ, Wszolek ZK, Vingerhoets F, Farrer MJ (2011) VPS35 mutations in Parkinson disease. Am J Hum Genet 89(1):162–167. https://doi.org/10.1016/j.ajhg.2011.06.001

149. Pankratz N, Wilk JB, Latourelle JC, DeStefano AL, Halter C, Pugh EW, Doheny KF, Gusella JF, Nichols WC, Foroud T, Myers RH, Psg P, GenePd Investigators C, Molecular Genetic L (2009) Genomewide association study for susceptibility genes contributing to familial Parkinson disease. Hum Genet 124(6):593–605. https://doi.org/10.1007/s00439-008-0582-9

150. Hamza TH, Zabetian CP, Tenesa A, Laederach A, Montimurro J, Yearout D, Kay DM, Doheny KF, Paschall J, Pugh E, Kusel VI, Collura R, Roberts J, Griffith A, Samii A, Scott WK, Nutt J, Factor SA, Payami H (2010) Common genetic variation in the HLA region is associated with late-onset sporadic Parkinson's disease. Nat Genet 42(9):781–785. https://doi.org/10.1038/ng.642

151. Ishizu N, Yui D, Hebisawa A, Aizawa H, Cui W, Fujita Y, Hashimoto K, Ajioka I, Mizusawa H, Yokota T, Watase K (2016) Impaired striatal dopamine release in homozygous Vps35 D620N knock-in mice. Hum Mol Genet 25(20):4507–4517. https://doi.org/10.1093/hmg/ddw279

152. Cataldi S, Follett J, Fox JD, Tatarnikov I, Kadgien C, Gustavsson EK, Khinda J, Milnerwood AJ, Farrer MJ (2018) Altered dopamine release and monoamine transporters in Vps35 p.D620N knock-in mice. NPJ Parkinsons Dis 4:27. https://doi.org/10.1038/s41531-018-0063-3

153. Tang FL, Liu W, Hu JX, Erion JR, Ye J, Mei L, Xiong WC (2015) VPS35 deficiency or mutation causes dopaminergic neuronal loss by impairing mitochondrial fusion and function. Cell Rep 12(10):1631–1643. https://doi.org/10.1016/j.celrep.2015.08.001

154. Tang FL, Erion JR, Tian Y, Liu W, Yin DM, Ye J, Tang B, Mei L, Xiong WC (2015) VPS35 in dopamine neurons is required for endosome-to-golgi retrieval of Lamp2a, a receptor of chaperone-mediated autophagy that is critical for alpha-synuclein degradation and prevention of pathogenesis of Parkinson's disease. J Neurosci 35(29):10613–10628. https://doi.org/10.1523/JNEUROSCI.0042-15.2015

155. Wang W, Wang X, Fujioka H, Hoppel C, Whone AL, Caldwell MA, Cullen PJ, Liu J, Zhu X (2016) Parkinson's disease-associated mutant VPS35 causes mitochondrial dysfunction by recycling DLP1 complexes. Nat Med 22(1):54–63. https://doi.org/10.1038/nm.3983

156. Tian Y, Tang FL, Sun X, Wen L, Mei L, Tang BS, Xiong WC (2015) VPS35-deficiency results in an impaired AMPA receptor trafficking and decreased dendritic spine maturation. Mol Brain 8(1):70. https://doi.org/10.1186/s13041-015-0156-4

157. Tsika E, Glauser L, Moser R, Fiser A, Daniel G, Sheerin UM, Lees A, Troncoso JC, Lewis PA, Bandopadhyay R, Schneider BL, Moore DJ (2014) Parkinson's disease-linked mutations in VPS35 induce dopaminergic neurodegeneration. Hum Mol Genet 23(17):4621–4638. https://doi.org/10.1093/hmg/ddu178

158. Wang HS, Toh J, Ho P, Tio M, Zhao Y, Tan EK (2014) In vivo evidence of pathogenicity of VPS35 mutations in the Drosophila. Mol Brain 7:73. https://doi.org/10.1186/s13041-014-0073-y

159. Malik BR, Godena VK, Whitworth AJ (2015) VPS35 pathogenic mutations confer no dominant toxicity but partial loss of function in Drosophila and genetically interact with parkin. Hum Mol Genet 24(21):6106–6117. https://doi.org/10.1093/hmg/ddv322

160. Miura E, Hasegawa T, Konno M, Suzuki M, Sugeno N, Fujikake N, Geisler S, Tabuchi M, Oshima R, Kikuchi A, Baba T, Wada K, Nagai Y, Takeda A, Aoki M (2014) VPS35 dysfunction impairs lysosomal degradation of alpha-synuclein and exacerbates neurotoxicity in a Drosophila model of Parkinson's disease. Neurobiol Dis 71:1–13. https://doi.org/10.1016/j.nbd.2014.07.014

161. Chartier-Harlin MC, Dachsel JC, Vilarino-Guell C, Lincoln SJ, Lepretre F, Hulihan MM, Kachergus J, Milnerwood AJ, Tapia L, Song MS, Le Rhun E, Mutez E, Larvor L, Duflot A, Vanbesien-Mailliot C, Kreisler A, Ross OA, Nishioka K, Soto-Ortolaza AI, Cobb SA, Melrose HL, Behrouz B, Keeling BH, Bacon JA, Hentati E, Williams L, Yanagiya A, Sonenberg N, Lockhart PJ, Zubair AC, Uitti RJ, Aasly JO, Krygowska-Wajs A, Opala G, Wszolek ZK, Frigerio R, Maraganore DM, Gosal D, Lynch T, Hutchinson M, Bentivoglio AR, Valente EM, Nichols WC, Pankratz N, Foroud T, Gibson RA, Hentati F, Dickson DW, Destee A, Farrer MJ (2011) Translation initiator EIF4G1 mutations in familial Parkinson disease. Am J Hum Genet 89(3):398–406. https://doi.org/10.1016/j.ajhg.2011.08.009

162. Edvardson S, Cinnamon Y, Ta-Shma A, Shaag A, Yim YI, Zenvirt S, Jalas C, Lesage S, Brice A, Taraboulos A, Kaestner KH, Greene LE, Elpeleg O (2012) A deleterious mutation in DNAJC6 encoding the neuronal-specific clathrin-uncoating co-chaperone

auxilin, is associated with juvenile parkinsonism. PLoS One 7(5):e36458. https://doi.org/10.1371/journal.pone.0036458

163. Koroglu C, Baysal L, Cetinkaya M, Karasoy H, Tolun A (2013) DNAJC6 is responsible for juvenile parkinsonism with phenotypic variability. Parkinsonism Relat Disord 19(3):320–324. https://doi.org/10.1016/j.parkreldis.2012.11.006

164. Elsayed LE, Drouet V, Usenko T, Mohammed IN, Hamed AA, Elseed MA, Salih MA, Koko ME, Mohamed AY, Siddig RA, Elbashir MI, Ibrahim ME, Durr A, Stevanin G, Lesage S, Ahmed AE, Brice A (2016) A novel nonsense mutation in DNAJC6 expands the phenotype of autosomal-recessive juvenile-onset Parkinson's disease. Ann Neurol 79(2):335–337. https://doi.org/10.1002/ana.24591

165. Olgiati S, Quadri M, Fang M, Rood JP, Saute JA, Chien HF, Bouwkamp CG, Graafland J, Minneboo M, Breedveld GJ, Zhang J, International Parkinsonism Genetics N, Verheijen FW, Boon AJ, Kievit AJ, Jardim LB, Mandemakers W, Barbosa ER, Rieder CR, Leenders KL, Wang J, Bonifati V (2016) DNAJC6 mutations associated with early-onset Parkinson's disease. Ann Neurol 79(2):244–256. https://doi.org/10.1002/ana.24553

166. Yim YI, Sun T, Wu LG, Raimondi A, De Camilli P, Eisenberg E, Greene LE (2010) Endocytosis and clathrin-uncoating defects at synapses of auxilin knockout mice. Proc Natl Acad Sci U S A 107(9):4412–4417. https://doi.org/10.1073/pnas.1000738107

167. Bai T, Seebald JL, Kim KE, Ding HM, Szeto DP, Chang HC (2010) Disruption of zebrafish cyclin G-associated kinase (GAK) function impairs the expression of Notch-dependent genes during neurogenesis and causes defects in neuronal development. BMC Dev Biol 10:7. https://doi.org/10.1186/1471-213X-10-7

168. Song L, He Y, Ou J, Zhao Y, Li R, Cheng J, Lin CH, Ho MS (2017) Auxilin underlies progressive locomotor deficits and dopaminergic neuron loss in a Drosophila model of Parkinson's disease. Cell Rep 18(5):1132–1143. https://doi.org/10.1016/j.celrep.2017.01.005

169. Banks SM, Cho B, Eun SH, Lee JH, Windler SL, Xie X, Bilder D, Fischer JA (2011) The functions of auxilin and Rab11 in Drosophila suggest that the fundamental role of ligand endocytosis in notch signaling cells is not recycling. PLoS One 6(3):e18259. https://doi.org/10.1371/journal.pone.0018259

170. Quadri M, Fang M, Picillo M, Olgiati S, Breedveld GJ, Graafland J, Wu B, Xu F, Erro R, Amboni M, Pappata S, Quarantelli M, Annesi G, Quattrone A, Chien HF,

Barbosa ER, International Parkinsonism Genetics N, Oostra BA, Barone P, Wang J, Bonifati V (2013) Mutation in the SYNJ1 gene associated with autosomal recessive, early-onset Parkinsonism. Hum Mutat 34 (9):1208–1215. https://doi.org/10.1002/humu.22373

171. Krebs CE, Karkheiran S, Powell JC, Cao M, Makarov V, Darvish H, Di Paolo G, Walker RH, Shahidi GA, Buxbaum JD, De Camilli P, Yue Z, Paisan-Ruiz C (2013) The Sac1 domain of SYNJ1 identified mutated in a family with early-onset progressive Parkinsonism with generalized seizures. Hum Mutat 34(9):1200–1207. https://doi.org/10.1002/humu.22372

172. Zhu L, Zhong M, Zhao J, Rhee H, Caesar I, Knight EM, Volpicelli-Daley L, Bustos V, Netzer W, Liu L, Lucast L, Ehrlich ME, Robakis NK, Gandy SE, Cai D (2013) Reduction of synaptojanin 1 accelerates Abeta clearance and attenuates cognitive deterioration in an Alzheimer mouse model. J Biol Chem 288(44):32050–32063. https://doi.org/10.1074/jbc.M113.504365

173. Cremona O, Di Paolo G, Wenk MR, Luthi A, Kim WT, Takei K, Daniell L, Nemoto Y, Shears SB, Flavell RA, McCormick DA, De Camilli P (1999) Essential role of phosphoinositide metabolism in synaptic vesicle recycling. Cell 99(2):179–188

174. Kim WT, Chang S, Daniell L, Cremona O, Di Paolo G, De Camilli P (2002) Delayed reentry of recycling vesicles into the fusion-competent synaptic vesicle pool in synaptojanin 1 knockout mice. Proc Natl Acad Sci U S A 99(26):17143–17148. https://doi.org/10.1073/pnas.222657399

175. Van Epps HA, Hayashi M, Lucast L, Stearns GW, Hurley JB, De Camilli P, Brockerhoff SE (2004) The zebrafish nrc mutant reveals a role for the polyphosphoinositide phosphatase synaptojanin 1 in cone photoreceptor ribbon anchoring. J Neurosci 24(40):8641–8650. https://doi.org/10.1523/JNEUROSCI.2892-04.2004

176. George AA, Hayden S, Stanton GR, Brockerhoff SE (2016) Arf6 and the 5'phosphatase of synaptojanin 1 regulate autophagy in cone photoreceptors. BioEssays 38(Suppl 1):S119–S135. https://doi.org/10.1002/bies.201670913

177. Dickman DK, Horne JA, Meinertzhagen IA, Schwarz TL (2005) A slowed classical pathway rather than kiss-and-run mediates endocytosis at synapses lacking synaptojanin and endophilin. Cell 123(3):521–533. https://doi.org/10.1016/j.cell.2005.09.026

178. Verstreken P, Koh TW, Schulze KL, Zhai RG, Hiesinger PR, Zhou Y, Mehta SQ, Cao

Y, Roos J, Bellen HJ (2003) Synaptojanin is recruited by endophilin to promote synaptic vesicle uncoating. Neuron 40(4):733–748

179. Harris TW, Hartwieg E, Horvitz HR, Jorgensen EM (2000) Mutations in synaptojanin disrupt synaptic vesicle recycling. J Cell Biol 150(3):589–600

180. Schuske KR, Richmond JE, Matthies DS, Davis WS, Runz S, Rube DA, van der Bliek AM, Jorgensen EM (2003) Endophilin is required for synaptic vesicle endocytosis by localizing synaptojanin. Neuron 40(4):749–762

181. Cao M, Wu Y, Ashrafi G, McCartney AJ, Wheeler H, Bushong EA, Boassa D, Ellisman MH, Ryan TA, De Camilli P (2017) Parkinson sac domain mutation in synaptojanin 1 impairs clathrin uncoating at synapses and triggers dystrophic changes in dopaminergic axons. Neuron 93(4):882–896.e885. https://doi.org/10.1016/j.neuron.2017.01.019

182. Vilarino-Guell C, Rajput A, Milnerwood AJ, Shah B, Szu-Tu C, Trinh J, Yu I, Encarnacion M, Munsie LN, Tapia L, Gustavsson EK, Chou P, Tatarnikov I, Evans DM, Pishotta FT, Volta M, Beccano-Kelly D, Thompson C, Lin MK, Sherman HE, Han HJ, Guenther BL, Wasserman WW, Bernard V, Ross CJ, Appel-Cresswell S, Stoessl AJ, Robinson CA, Dickson DW, Ross OA, Wszolek ZK, Aasly JO, Wu RM, Hentati F, Gibson RA, McPherson PS, Girard M, Rajput M, Rajput AH, Farrer MJ (2014) DNAJC13 mutations in Parkinson disease. Hum Mol Genet 23(7):1794–1801. https://doi.org/10.1093/hmg/ddt570

183. Gustavsson EK, Trinh J, Guella I, Vilarino-Guell C, Appel-Cresswell S, Stoessl AJ, Tsui JK, McKeown M, Rajput A, Rajput AH, Aasly JO, Farrer MJ (2015) DNAJC13 genetic variants in parkinsonism. Mov Disord 30(2):273–278. https://doi.org/10.1002/mds.26064

184. Yoshida S, Hasegawa T, Suzuki M, Sugeno N, Kobayashi J, Ueyama M, Fukuda M, Ido-Fujibayashi A, Sekiguchi K, Ezura M, Kikuchi A, Baba T, Takeda A, Mochizuki H, Nagai Y, Aoki M (2018) Parkinson's disease-linked DNAJC13 mutation aggravates alpha-synuclein-induced neurotoxicity through perturbation of endosomal trafficking. Hum Mol Genet 27(5):823–836. https://doi.org/10.1093/hmg/ddy003

185. Norris A, Tammineni P, Wang S, Gerdes J, Murr A, Kwan KY, Cai Q, Grant BD (2017) SNX-1 and RME-8 oppose the assembly of HGRS-1/ESCRT-0 degradative microdomains on endosomes. Proc Natl Acad Sci U S A 114(3):E307–E316. https://doi.org/10.1073/pnas.1612730114

186. Shi A, Sun L, Banerjee R, Tobin M, Zhang Y, Grant BD (2009) Regulation of endosomal clathrin and retromer-mediated endosome to Golgi retrograde transport by the J-domain protein RME-8. EMBO J 28(21):3290–3302. https://doi.org/10.1038/emboj.2009.272

187. Funayama M, Ohe K, Amo T, Furuya N, Yamaguchi J, Saiki S, Li Y, Ogaki K, Ando M, Yoshino H, Tomiyama H, Nishioka K, Hasegawa K, Saiki H, Satake W, Mogushi K, Sasaki R, Kokubo Y, Kuzuhara S, Toda T, Mizuno Y, Uchiyama Y, Ohno K, Hattori N (2015) CHCHD2 mutations in autosomal dominant late-onset Parkinson's disease: a genome-wide linkage and sequencing study. Lancet Neurol 14(3):274–282. https://doi.org/10.1016/S1474-4422(14)70266-2

188. Jansen IE, Bras JM, Lesage S, Schulte C, Gibbs JR, Nalls MA, Brice A, Wood NW, Morris H, Hardy JA, Singleton AB, Gasser T, Heutink P, Sharma M, IPDGC (2015) CHCHD2 and Parkinson's disease. Lancet Neurol 14(7):678–679. https://doi.org/10.1016/S1474-4422(15)00094-0

189. Burstein SR, Valsecchi F, Kawamata H, Bourens M, Zeng R, Zuberi A, Milner TA, Cloonan SM, Lutz C, Barrientos A, Manfredi G (2018) In vitro and in vivo studies of the ALS-FTLD protein CHCHD10 reveal novel mitochondrial topology and protein interactions. Hum Mol Genet 27(1):160–177. https://doi.org/10.1093/hmg/ddx397

190. Tio M, Wen R, Lim YL, Zukifli ZHB, Xie S, Ho P, Zhou Z, Koh TW, Zhao Y, Tan EK (2017) Varied pathological and therapeutic response effects associated with CHCHD2 mutant and risk variants. Hum Mutat 38(8):978–987. https://doi.org/10.1002/humu.23234

191. Meng H, Yamashita C, Shiba-Fukushima K, Inoshita T, Funayama M, Sato S, Hatta T, Natsume T, Umitsu M, Takagi J, Imai Y, Hattori N (2017) Loss of Parkinson's disease-associated protein CHCHD2 affects mitochondrial crista structure and destabilizes cytochrome c. Nat Commun 8:15500. https://doi.org/10.1038/ncomms15500

192. Lesage S, Drouet V, Majounie E, Deramecourt V, Jacoupy M, Nicolas A, Cormier-Dequaire F, Hassoun SM, Pujol C, Ciura S, Erpapazoglou Z, Usenko T, Maurage CA, Sahbatou M, Liebau S, Ding J, Bilgic B, Emre M, Erginel-Unaltuna N, Guven G, Tison F, Tranchant C, Vidailhet M, Corvol JC, Krack P, Leutenegger AL, Nalls MA, Hernandez DG, Heutink P, Gibbs JR, Hardy J, Wood NW, Gasser T, Durr A, Deleuze JF, Tazir M, Destee A, Lohmann E, Kabashi E, Singleton A, Corti O, Brice A, French Parkinson's Disease Genetics S, International Parkinson's Disease Genomics C (2016) Loss of VPS13C function in autosomal-recessive parkinsonism causes

mitochondrial dysfunction and increases PINK1/Parkin-dependent mitophagy. Am J Hum Genet 98(3):500–513. https://doi.org/10.1016/j.ajhg.2016.01.014

193. Jansen IE, Ye H, Heetveld S, Lechler MC, Michels H, Seinstra RI, Lubbe SJ, Drouet V, Lesage S, Majounie E, Gibbs JR, Nalls MA, Ryten M, Botia JA, Vandrovcova J, Simon-Sanchez J, Castillo-Lizardo M, Rizzu P, Blauwendraat C, Chouhan AK, Li Y, Yogi P, Amin N, van Duijn CM, International Parkinson's Disease Genetics C, Morris HR, Brice A, Singleton AB, David DC, Nollen EA, Jain S, Shulman JM, Heutink P (2017) Discovery and functional prioritization of Parkinson's disease candidate genes from large-scale whole exome sequencing. Genome Biol 18 (1):22. https://doi.org/10.1186/s13059-017-1147-9

194. Aharon-Peretz J, Rosenbaum H, Gershoni-Baruch R (2004) Mutations in the glucocerebrosidase gene and Parkinson's disease in Ashkenazi Jews. N Engl J Med 351(19):1972–1977. https://doi.org/10.1056/NEJMoa033277

195. Lesage S, Brice A (2009) Parkinson's disease: from monogenic forms to genetic susceptibility factors. Hum Mol Genet 18(R1):R48–R59

196. Ron I, Rapaport D, Horowitz M (2010) Interaction between parkin and mutant glucocerebrosidase variants: a possible link between Parkinson disease and Gaucher disease. Hum Mol Genet 19(19):3771–3781. https://doi.org/10.1093/hmg/ddq292

197. Gwinn-Hardy K, Chen JY, Liu HC, Liu TY, Boss M, Seltzer W, Adam A, Singleton A, Koroshetz W, Waters C, Hardy J, Farrer M (2000) Spinocerebellar ataxia type 2 with parkinsonism in ethnic Chinese. Neurology 55(6):800–805

198. Shan DE, Soong BW, Sun CM, Lee SJ, Liao KK, Liu RS (2001) Spinocerebellar ataxia type 2 presenting as familial levodopa-responsive parkinsonism. Ann Neurol 50(6):812–815

199. Charles P, Camuzat A, Benammar N, Sellal F, Destee A, Bonnet AM, Lesage S, Le Ber I, Stevanin G, Durr A, Brice A (2007) Are interrupted SCA2 CAG repeat expansions responsible for parkinsonism? Neurology 69(21):1970–1975. https://doi.org/10.1212/01.wnl.0000269323.21969.db

200. Chai YG, Oh DY, Chung EK, Kim GS, Kim L, Lee YS, Choi IG (2005) Alcohol and aldehyde dehydrogenase polymorphisms in men with type I and Type II alcoholism. Am J Psychiatry 162(5):1003–1005. https://doi.org/10.1176/appi.ajp.162.5.1003

201. Buervenich S, Carmine A, Galter D, Shahabi HN, Johnels B, Holmberg B, Ahlberg J, Nissbrandt H, Eerola J, Hellstrom O, Tienari PJ, Matsuura T, Ashizawa T, Wullner U, Klockgether T, Zimprich A, Gasser T, Hanson M, Waseem S, Singleton A, McMahon FJ, Anvret M, Sydow O, Olson L (2005) A rare truncating mutation in ADH1C (G78Stop) shows significant association with Parkinson disease in a large international sample. Arch Neurol 62(1):74–78. https://doi.org/10.1001/archneur.62.1.74

202. Westerlund M, Belin AC, Felder MR, Olson L, Galter D (2007) High and complementary expression patterns of alcohol and aldehyde dehydrogenases in the gastrointestinal tract: implications for Parkinson's disease. FEBS J 274(5):1212–1223. https://doi.org/10.1111/j.1742-4658.2007.05665.x

203. Anvret A, Ran C, Westerlund M, Gellhaar S, Lindqvist E, Pernold K, Lundstromer K, Duester G, Felder MR, Galter D, Belin AC (2012) Adh1 and Adh1/4 knockout mice as possible rodent models for presymptomatic Parkinson's disease. Behav Brain Res 227(1):252–257. https://doi.org/10.1016/j.bbr.2011.10.040

204. Koide R, Kobayashi S, Shimohata T, Ikeuchi T, Maruyama M, Saito M, Yamada M, Takahashi H, Tsuji S (1999) A neurological disease caused by an expanded CAG trinucleotide repeat in the TATA-binding protein gene: a new polyglutamine disease? Hum Mol Genet 8(11):2047–2053

205. Wu YR, Fung HC, Lee-Chen GJ, Gwinn-Hardy K, Ro LS, Chen ST, Hsieh-Li HM, Lin HY, Lin CY, Li SN, Chen CM (2005) Analysis of polyglutamine-coding repeats in the TATA-binding protein in different neurodegenerative diseases. J Neural Transm (Vienna) 112(4):539–546. https://doi.org/10.1007/s00702-004-0197-9

206. Wu YR, Lin HY, Chen CM, Gwinn-Hardy K, Ro LS, Wang YC, Li SH, Hwang JC, Fang K, Hsieh-Li HM, Li ML, Tung LC, Su MT, Lu KT, Lee-Chen GJ (2004) Genetic testing in spinocerebellar ataxia in Taiwan: expansions of trinucleotide repeats in SCA8 and SCA17 are associated with typical Parkinson's disease. Clin Genet 65(3):209–214

207. Hutton M, Lendon CL, Rizzu P, Baker M, Froelich S, Houlden H, Pickering-Brown S, Chakraverty S, Isaacs A, Grover A, Hackett J, Adamson J, Lincoln S, Dickson D, Davies P, Petersen RC, Stevens M, de Graaff E, Wauters E, van Baren J, Hillebrand M, Joosse M, Kwon JM, Nowotny P, Che LK, Norton J, Morris JC, Reed LA, Trojanowski J, Basun H, Lannfelt L, Neystat M, Fahn S, Dark F, Tannenberg T, Dodd PR, Hayward N, Kwok JB, Schofield PR, Andreadis A, Snowden J, Craufurd D, Neary D, Owen F, Oostra BA,

Hardy J, Goate A, van Swieten J, Mann D, Lynch T, Heutink P (1998) Association of missense and 5'-splice-site mutations in tau with the inherited dementia FTDP-17. Nature 393(6686):702–705. https://doi.org/10.1038/31508

208. Plaitakis A, Latsoudis H, Kanavouras K, Ritz B, Bronstein JM, Skoula I, Mastorodemos V, Papapetropoulos S, Borompokas N, Zaganas I, Xiromerisiou G, Hadjigeorgiou GM, Spanaki C (2010) Gain-of-function variant in GLUD2 glutamate dehydrogenase modifies Parkinson's disease onset. Eur J Hum Genet 18(3):336–341. https://doi.org/10.1038/ejhg.2009.179

209. Deng HX, Shi Y, Yang Y, Ahmeti KB, Miller N, Huang C, Cheng L, Zhai H, Deng S, Nuytemans K, Corbett NJ, Kim MJ, Deng H, Tang B, Yang Z, Xu Y, Chan P, Huang B, Gao XP, Song Z, Liu Z, Fecto F, Siddique N, Foroud T, Jankovic J, Ghetti B, Nicholson DA, Krainc D, Melen O, Vance JM, Pericak-Vance MA, Ma YC, Rajput AH, Siddique T (2016) Identification of TMEM230 mutations in familial Parkinson's disease. Nat Genet 48(7):733–739. https://doi.org/10.1038/ng.3589

210. Sudhaman S, Muthane UB, Behari M, Govindappa ST, Juyal RC, Thelma BK (2016) Evidence of mutations in RIC3 acetylcholine receptor chaperone as a novel cause of autosomal-dominant Parkinson's disease with non-motor phenotypes. J Med Genet 53(8):559–566. https://doi.org/10.1136/jmedgenet-2015-103616

211. Khodadadi H, Azcona LJ, Aghamollaii V, Omrani MD, Garshasbi M, Taghavi S, Tafakhori A, Shahidi GA, Jamshidi J, Darvish H, Paisan-Ruiz C (2017) PTRHD1 (C2orf79) mutations lead to autosomal-recessive intellectual disability and parkinsonism. Mov Disord 32(2):287–291. https://doi.org/10.1002/mds.26824

212. Krahn LE, Maraganore DM, Michels VV (1998) Childhood-onset schizophrenia associated with parkinsonism in a patient with a microdeletion of chromosome 22. Mayo Clin Proc 73(10):956–959. https://doi.org/10.4065/73.10.956

213. Sudhaman S, Prasad K, Behari M, Muthane UB, Juyal RC, Thelma BK (2016) Discovery of a frameshift mutation in podocalyxin-like (PODXL) gene, coding for a neural adhesion molecule, as causal for autosomal-recessive juvenile Parkinsonism. J Med Genet 53(7):450–456. https://doi.org/10.1136/jmedgenet-2015-103459

214. Gonzalez-Reyes LE, Verbitsky M, Blesa J, Jackson-Lewis V, Paredes D, Tillack K, Phani S, Kramer ER, Przedborski S, Kottmann AH (2012) Sonic hedgehog maintains cellular and neurochemical homeostasis in the adult nigrostriatal circuit. Neuron 75(2):306–319. https://doi.org/10.1016/j.neuron.2012.05.018

215. Zetterstrom RH, Solomin L, Jansson L, Hoffer BJ, Olson L, Perlmann T (1997) Dopamine neuron agenesis in Nurr1-deficient mice. Science 276(5310):248–250

216. Semina EV, Murray JC, Reiter R, Hrstka RF, Graw J (2000) Deletion in the promoter region and altered expression of Pitx3 homeobox gene in aphakia mice. Hum Mol Genet 9(11):1575–1585

217. Sgado P, Alberi L, Gherbassi D, Galasso SL, Ramakers GM, Alavian KN, Smidt MP, Dyck RH, Simon HH (2006) Slow progressive degeneration of nigral dopaminergic neurons in postnatal Engrailed mutant mice. Proc Natl Acad Sci U S A 103(41):15242–15247. https://doi.org/10.1073/pnas.0602116103

218. Ekstrand MI, Terzioglu M, Galter D, Zhu S, Hofstetter C, Lindqvist E, Thams S, Bergstrand A, Hansson FS, Trifunovic A, Hoffer B, Cullheim S, Mohammed AH, Olson L, Larsson NG (2007) Progressive parkinsonism in mice with respiratory-chain-deficient dopamine neurons. Proc Natl Acad Sci U S A 104(4):1325–1330. https://doi.org/10.1073/pnas.0605208103

219. Kadkhodaei B, Ito T, Joodmardi E, Mattsson B, Rouillard C, Carta M, Muramatsu S, Sumi-Ichinose C, Nomura T, Metzger D, Chambon P, Lindqvist E, Larsson NG, Olson L, Bjorklund A, Ichinose H, Perlmann T (2009) Nurr1 is required for maintenance of maturing and adult midbrain dopamine neurons. J Neurosci 29(50):15923–15932. https://doi.org/10.1523/JNEUROSCI.3910-09.2009

220. Le WD, Xu P, Jankovic J, Jiang H, Appel SH, Smith RG, Vassilatis DK (2003) Mutations in NR4A2 associated with familial Parkinson disease. Nat Genet 33(1):85–89. https://doi.org/10.1038/ng1066

221. Xu PY, Liang R, Jankovic J, Hunter C, Zeng YX, Ashizawa T, Lai D, Le WD (2002) Association of homozygous 7048G7049 variant in the intron six of Nurr1 gene with Parkinson's disease. Neurology 58(6):881–884

222. Jankovic J, Chen S, Le WD (2005) The role of Nurr1 in the development of dopaminergic neurons and Parkinson's disease. Prog Neurobiol 77(1–2):128–138. https://doi.org/10.1016/j.pneurobio.2005.09.001

223. Zimprich A, Asmus F, Leitner P, Castro M, Bereznai B, Homann N, Ott E, Rutgers AW, Wieditz G, Trenkwalder C, Gasser T (2003) Point mutations in exon 1 of

the NR4A2 gene are not a major cause of familial Parkinson's disease. Neurogenetics 4(4):219–220. https://doi.org/10.1007/s10048-003-0156-x

224. Wellenbrock C, Hedrich K, Schafer N, Kasten M, Jacobs H, Schwinger E, Hagenah J, Pramstaller PP, Vieregge P, Klein C (2003) NR4A2 mutations are rare among European patients with familial Parkinson's disease. Ann Neurol 54(3):415. https://doi.org/10.1002/ana.10736

225. Hwang DY, Ardayfio P, Kang UJ, Semina EV, Kim KS (2003) Selective loss of dopaminergic neurons in the substantia nigra of Pitx3-deficient aphakia mice. Brain Res Mol Brain Res 114(2):123–131

226. Nunes I, Tovmasian LT, Silva RM, Burke RE, Goff SP (2003) Pitx3 is required for development of substantia nigra dopaminergic neurons. Proc Natl Acad Sci U S A 100(7):4245–4250. https://doi.org/10.1073/pnas.0230529100

227. van den Munckhof P, Luk KC, Ste-Marie L, Montgomery J, Blanchet PJ, Sadikot AF, Drouin J (2003) Pitx3 is required for motor activity and for survival of a subset of midbrain dopaminergic neurons. Development 130(11):2535–2542

228. Shi X, Bosenko DV, Zinkevich NS, Foley S, Hyde DR, Semina EV, Vihtelic TS (2005) Zebrafish pitx3 is necessary for normal lens and retinal development. Mech Dev 122(4):513–527. https://doi.org/10.1016/j.mod.2004.11.012

229. Le Pen G, Sonnier L, Hartmann A, Bizot JC, Trovero F, Krebs MO, Prochiantz A (2008) Progressive loss of dopaminergic neurons in the ventral midbrain of adult mice heterozygote for Engrailed1: a new genetic model for Parkinson's disease? Parkinsonism Relat Disord 14(Suppl 2):S107–S111. https://doi.org/10.1016/j.parkreldis.2008.04.007

230. Haubenberger D, Reinthaler E, Mueller JC, Pirker W, Katzenschlager R, Froehlich R, Bruecke T, Daniel G, Auff E, Zimprich A (2011) Association of transcription factor polymorphisms PITX3 and EN1 with Parkinson's disease. Neurobiol Aging 32(2):302–307

231. Falkenberg M, Larsson NG, Gustafsson CM (2007) DNA replication and transcription in mammalian mitochondria. Annu Rev Biochem 76:679–699

232. Ekstrand MI, Galter D (2009) The MitoPark Mouse - an animal model of Parkinson's disease with impaired respiratory chain function in dopamine neurons. Parkinsonism Relat Disord 15(Suppl 3):S185–S188. https://doi.org/10.1016/S1353-8020(09)70811-9

233. Gubellini P, Kachidian P (2015) Animal models of Parkinson's disease: an updated overview. Rev Neurol (Paris) 171(11):750–761

234. Jiang P, Dickson DW (2018) Parkinson's disease: experimental models and reality. Acta Neuropathol 135(1):13–32. https://doi.org/10.1007/s00401-017-1788-5

235. LeWitt PA, Fahn S (2016) Levodopa therapy for Parkinson disease: a look backward and forward. Neurology 86(14 Suppl 1):S3–S12. https://doi.org/10.1212/WNL.0000000000002509

236. Wichmann T, DeLong MR (2016) Deep brain stimulation for movement disorders of basal ganglia origin: restoring function or functionality? Neurotherapeutics 13(2):264–283

237. Deng H, Yuan L (2014) Genetic variants and animal models in SNCA and Parkinson disease. Ageing Res Rev 15:161–176. https://doi.org/10.1016/j.arr.2014.04.002

238. Kilarski LL, Pearson JP, Newsway V, Majounie E, Knipe MD, Misbahuddin A, Chinnery PF, Burn DJ, Clarke CE, Marion MH, Lewthwaite AJ, Nicholl DJ, Wood NW, Morrison KE, Williams-Gray CH, Evans JR, Sawcer SJ, Barker RA, Wickremaratchi MM, Ben-Shlomo Y, Williams NM, Morris HR (2012) Systematic review and UK-based study of PARK2 (parkin), PINK1, PARK7 (DJ-1) and LRRK2 in early-onset Parkinson's disease. Mov Disord 27(12):1522–1529. https://doi.org/10.1002/mds.25132

239. Gasser T, Muller-Myhsok B, Wszolek ZK, Oehlmann R, Calne DB, Bonifati V, Bereznai B, Fabrizio E, Vieregge P, Horstmann RD (1998) A susceptibility locus for Parkinson's disease maps to chromosome 2p13. Nat Genet 18(3):262–265. https://doi.org/10.1038/ng0398-262

240. Farrer M, Gwinn-Hardy K, Muenter M, DeVrieze FW, Crook R, Perez-Tur J, Lincoln S, Maraganore D, Adler C, Newman S, MacElwee K, McCarthy P, Miller C, Waters C, Hardy J (1999) A chromosome 4p haplotype segregating with Parkinson's disease and postural tremor. Hum Mol Genet 8(1):81–85

241. Singleton AB, Farrer M, Johnson J, Singleton A, Hague S, Kachergus J, Hulihan M, Peuralinna T, Dutra A, Nussbaum R, Lincoln S, Crawley A, Hanson M, Maraganore D, Adler C, Cookson MR, Muenter M, Baptista M, Miller D, Blancato J, Hardy J, Gwinn-Hardy K (2003) alpha-Synuclein locus triplication causes Parkinson's disease. Science 302(5646):841. https://doi.org/10.1126/science.1090278

242. Paisan-Ruiz C, Jain S, Evans EW, Gilks WP, Simon J, van der BM, Lopez de MA, Aparicio S, Gil AM, Khan N, Johnson J, Martinez JR, Nicholl D, Carrera IM, Pena AS, de

SR, Lees A, Marti-Masso JF, Perez-Tur J, Wood NW, Singleton AB (2004) Cloning of the gene containing mutations that cause PARK8-linked Parkinson's disease. Neuron 44(4):595–600

243. Simon-Sanchez J, Heutink P, Gasser T, International Parkinson's Disease Genomics C (2015) Variation in PARK10 is not associated with risk and age at onset of Parkinson's disease in large clinical cohorts. Neurobiol Aging 36(10):2907.e2913. https://doi.org/10.1016/j.neurobiolaging.2015.07.008

244. Beecham GW, Dickson DW, Scott WK, Martin ER, Schellenberg G, Nuytemans K, Larson EB, Buxbaum JD, Trojanowski JQ, Van Deerlin VM, Hurtig HI, Mash DC, Beach TG, Troncoso JC, Pletnikova O, Frosch MP, Ghetti B, Foroud TM, Honig LS, Marder K, Vonsattel JP, Goldman SM, Vinters HV, Ross OA, Wszolek ZK, Wang L, Dykxhoorn DM, Pericak-Vance MA, Montine TJ, Leverenz JB, Dawson TM, Vance JM (2015) PARK10 is a major locus for sporadic neuropathologically confirmed Parkinson disease. Neurology 84(10):972–980. https://doi.org/10.1212/WNL.0000000000001332

245. Oliveira SA, Li YJ, Noureddine MA, Zuchner S, Qin X, Pericak-Vance MA, Vance JM (2005) Identification of risk and age-at-onset genes on chromosome 1p in Parkinson disease. Am J Hum Genet 77(2):252–264. https://doi.org/10.1086/432588

246. Khodr CE, Sapru MK, Pedapati J, Han Y, West NC, Kells AP, Bankiewicz KS, Bohn MC (2011) An alpha-synuclein AAV gene silencing vector ameliorates a behavioral deficit in a rat model of Parkinson's disease, but displays toxicity in dopamine neurons. Brain Res 1395:94–107. https://doi.org/10.1016/j.brainres.2011.04.036

247. Cannon JR, Geghman KD, Tapias V, Sew T, Dail MK, Li C, Greenamyre JT (2013) Expression of human E46K-mutated alpha-synuclein in BAC-transgenic rats replicates early-stage Parkinson's disease features and enhances vulnerability to mitochondrial impairment. Exp Neurol 240:44–56. https://doi.org/10.1016/j.expneurol.2012.11.007

248. Bido S, Soria FN, Fan RZ, Bezard E, Tieu K (2017) Mitochondrial division inhibitor-1 is neuroprotective in the A53T-alpha-synuclein rat model of Parkinson's disease. Sci Rep 7(1):7495. https://doi.org/10.1038/s41598-017-07181-0

249. Musacchio T, Rebenstorff M, Fluri F, Brotchie JM, Volkmann J, Koprich JB, Ip CW (2017) Subthalamic nucleus deep brain stimulation is neuroprotective in the A53T alpha-synuclein Parkinson's disease rat model. Ann Neurol 81(6):825–836. https://doi.org/10.1002/ana.24947

250. Dehay B, Fernagut PO (2016) alpha-Synuclein-based models of Parkinson's disease. Rev Neurol (Paris) 172(6–7):371–378. https://doi.org/10.1016/j.neurol.2016.04.003

251. Kurz A, Double KL, Lastres-Becker I, Tozzi A, Tantucci M, Bockhart V, Bonin M, Garcia-Arencibia M, Nuber S, Schlaudraff F, Liss B, Fernandez-Ruiz J, Gerlach M, Wullner U, Luddens H, Calabresi P, Auburger G, Gispert S (2010) A53T-alpha-synuclein overexpression impairs dopamine signaling and striatal synaptic plasticity in old mice. PLoS One 5(7):e11464. https://doi.org/10.1371/journal.pone.0011464

252. Abeliovich A, Schmitz Y, Farinas I, Choi-Lundberg D, Ho WH, Castillo PE, Shinsky N, Verdugo JM, Armanini M, Ryan A, Hynes M, Phillips H, Sulzer D, Rosenthal A (2000) Mice lacking alpha-synuclein display functional deficits in the nigrostriatal dopamine system. Neuron 25(1):239–252

253. Sharon R, Bar-Joseph I, Mirick GE, Serhan CN, Selkoe DJ (2003) Altered fatty acid composition of dopaminergic neurons expressing alpha-synuclein and human brains with alpha-synucleinopathies. J Biol Chem 278(50):49874–49881. https://doi.org/10.1074/jbc.M309127200

254. Masliah E, Rockenstein E, Veinbergs I, Mallory M, Hashimoto M, Takeda A, Sagara Y, Sisk A, Mucke L (2000) Dopaminergic loss and inclusion body formation in alpha-synuclein mice: implications for neurodegenerative disorders. Science 287(5456):1265–1269

255. Emmer KL, Waxman EA, Covy JP, Giasson BI (2011) E46K human alpha-synuclein transgenic mice develop Lewy-like and tau pathology associated with age-dependent, detrimental motor impairment. J Biol Chem 286(40):35104–35118. https://doi.org/10.1074/jbc.M111.247965

256. Gomez-Isla T, Irizarry MC, Mariash A, Cheung B, Soto O, Schrump S, Sondel J, Kotilinek L, Day J, Schwarzschild MA, Cha JH, Newell K, Miller DW, Ueda K, Young AB, Hyman BT, Ashe KH (2003) Motor dysfunction and gliosis with preserved dopaminergic markers in human alpha-synuclein A30P transgenic mice. Neurobiol Aging 24(2):245–258

257. Kahle PJ, Neumann M, Ozmen L, Muller V, Jacobsen H, Schindzielorz A, Okochi M, Leimer U, van Der Putten H, Probst A, Kremmer E, Kretzschmar HA, Haass C (2000) Subcellular localization of wild-type and Parkinson's disease-associated mutant

alpha -synuclein in human and transgenic mouse brain. J Neurosci 20(17):6365–6373

258. Mbefo MK, Fares MB, Paleologou K, Oueslati A, Yin G, Tenreiro S, Pinto M, Outeiro T, Zweckstetter M, Masliah E, Lashuel HA (2015) Parkinson disease mutant E46K enhances alpha-synuclein phosphorylation in mammalian cell lines, in yeast, and in vivo. J Biol Chem 290(15):9412–9427. https://doi.org/10.1074/jbc.M114.610774

259. Zhou W, Milder JB, Freed CR (2008) Transgenic mice overexpressing tyrosine-to-cysteine mutant human alpha-synuclein: a progressive neurodegenerative model of diffuse Lewy body disease. J Biol Chem 283(15):9863–9870. https://doi.org/10.1074/jbc.M710232200

260. Feany MB, Bender WW (2000) A Drosophila model of Parkinson's disease. Nature 404(6776):394–398. https://doi.org/10.1038/35006074

261. Ved R, Saha S, Westlund B, Perier C, Burnam L, Sluder A, Hoener M, Rodrigues CM, Alfonso A, Steer C, Liu L, Przedborski S, Wolozin B (2005) Similar patterns of mitochondrial vulnerability and rescue induced by genetic modification of alpha-synuclein, parkin, and DJ-1 in Caenorhabditis elegans. J Biol Chem 280(52):42655–42668. https://doi.org/10.1074/jbc.M505910200

262. Xi Y, Noble S, Ekker M (2011) Modeling neurodegeneration in zebrafish. Curr Neurol Neurosci Rep 11(3):274–282. https://doi.org/10.1007/s11910-011-0182-2

263. Matsui H, Gavinio R, Asano T, Uemura N, Ito H, Taniguchi Y, Kobayashi Y, Maki T, Shen J, Takeda S, Uemura K, Yamakado H, Takahashi R (2013) PINK1 and Parkin complementarily protect dopaminergic neurons in vertebrates. Hum Mol Genet 22(12):2423–2434. https://doi.org/10.1093/hmg/ddt095

264. Matsui H, Uemura N, Yamakado H, Takeda S, Takahashi R (2014) Exploring the pathogenetic mechanisms underlying Parkinson's disease in medaka fish. J Parkinsons Dis 4(2):301–310. https://doi.org/10.3233/JPD-130289

265. Bornhorst J, Chakraborty S, Meyer S, Lohren H, Brinkhaus SG, Knight AL, Caldwell KA, Caldwell GA, Karst U, Schwerdtle T, Bowman A, Aschner M (2014) The effects of pdr1, djr1.1 and pink1 loss in manganese-induced toxicity and the role of alpha-synuclein in C. elegans. Metallomics 6(3):476–490. https://doi.org/10.1039/c3mt00325f

266. Chege PM, McColl G (2014) Caenorhabditis elegans: a model to investigate oxidative stress and metal dyshomeostasis in Parkinson's disease. Front Aging Neurosci 6:89. https://doi.org/10.3389/fnagi.2014.00089

267. Orr AL, Rutaganira FU, de Roulet D, Huang EJ, Hertz NT, Shokat KM, Nakamura K (2017) Long-term oral kinetin does not protect against alpha-synuclein-induced neurodegeneration in rodent models of Parkinson's disease. Neurochem Int 109:106–116. https://doi.org/10.1016/j.neuint.2017.04.006

268. Gandhi S, Vaarmann A, Yao Z, Duchen MR, Wood NW, Abramov AY (2012) Dopamine induced neurodegeneration in a PINK1 model of Parkinson's disease. PLoS One 7(5):e37564. https://doi.org/10.1371/journal.pone.0037564

269. Yang Y, Gehrke S, Imai Y, Huang Z, Ouyang Y, Wang JW, Yang L, Beal MF, Vogel H, Lu B (2006) Mitochondrial pathology and muscle and dopaminergic neuron degeneration caused by inactivation of Drosophila Pink1 is rescued by Parkin. Proc Natl Acad Sci U S A 103(28):10793–10798. https://doi.org/10.1073/pnas.0602493103

270. Soman S, Keatinge M, Moein M, Da Costa M, Mortiboys H, Skupin A, Sugunan S, Bazala M, Kuznicki J, Bandmann O (2017) Inhibition of the mitochondrial calcium uniporter rescues dopaminergic neurons in pink1(−/−) zebrafish. Eur J Neurosci 45(4):528–535. https://doi.org/10.1111/ejn.13473

271. Anichtchik O, Diekmann H, Fleming A, Roach A, Goldsmith P, Rubinsztein DC (2008) Loss of PINK1 function affects development and results in neurodegeneration in zebrafish. J Neurosci 28(33):8199–8207. https://doi.org/10.1523/JNEUROSCI.0979-08.2008

272. Flinn LJ, Keatinge M, Bretaud S, Mortiboys H, Matsui H, De Felice E, Woodroof HI, Brown L, McTighe A, Soellner R, Allen CE, Heath PR, Milo M, Muqit MM, Reichert AS, Koster RW, Ingham PW, Bandmann O (2013) TigarB causes mitochondrial dysfunction and neuronal loss in PINK1 deficiency. Ann Neurol 74(6):837–847. https://doi.org/10.1002/ana.23999

273. Matsui H, Gavinio R, Takahashi R (2012) Medaka fish Parkinson's disease model. Exp Neurobiol 21(3):94–100. https://doi.org/10.5607/en.2012.21.3.94

274. Matsui H, Taniguchi Y, Inoue H, Kobayashi Y, Sakaki Y, Toyoda A, Uemura K, Kobayashi D, Takeda S, Takahashi R (2010) Loss of PINK1 in medaka fish (Oryzias latipes) causes late-onset decrease in spontaneous movement. Neurosci Res 66(2):151–161. https://doi.org/10.1016/j.neures.2009.10.010

275. Clark IE, Dodson MW, Jiang C, Cao JH, Huh JR, Seol JH, Yoo SJ, Hay BA, Guo M

(2006) Drosophila pink1 is required for mito-chondrial function and interacts genetically with parkin. Nature 441(7097):1162–1166. https://doi.org/10.1038/nature04779

276. Todd AM, Staveley BE (2008) Pink1 sup-presses alpha-synuclein-induced phenotypes in a Drosophila model of Parkinson's disease. Genome 51(12):1040–1046. https://doi.org/10.1139/G08-085

277. Park J, Lee SB, Lee S, Kim Y, Song S, Kim S, Bae E, Kim J, Shong M, Kim JM, Chung J (2006) Mitochondrial dysfunction in Drosophila PINK1 mutants is complemented by parkin. Nature 441(7097):1157–1161. https://doi.org/10.1038/nature04788

278. Cornelissen T, Vilain S, Vints K, Gounko N, Verstreken P, Vandenberghe W (2018) Deficiency of parkin and PINK1 impairs age-dependent mitophagy in Drosophila. elife 7. https://doi.org/10.7554/eLife.35878

279. Song S, Jang S, Park J, Bang S, Choi S, Kwon KY, Zhuang X, Kim E, Chung J (2013) Characterization of PINK1 (PTEN-induced putative kinase 1) mutations associated with Parkinson disease in mammalian cells and Drosophila. J Biol Chem 288(8):5660–5672. https://doi.org/10.1074/jbc.M112.430801

280. Koh H, Kim H, Kim MJ, Park J, Lee HJ, Chung J (2012) Silent information regula-tor 2 (Sir2) and Forkhead box O (FOXO) complement mitochondrial dysfunction and dopaminergic neuron loss in Drosophila PTEN-induced kinase 1 (PINK1) null mutant. J Biol Chem 287(16):12750–12758. https://doi.org/10.1074/jbc.M111.337907

281. Samann J, Hegermann J, von Gromoff E, Eimer S, Baumeister R, Schmidt E (2009) Caenorhabditits elegans LRK-1 and PINK-1 act antagonistically in stress response and neu-rite outgrowth. J Biol Chem 284(24):16482–16491. https://doi.org/10.1074/jbc.M808255200

282. Vingill S, Connor-Robson N, Wade-Martins R (2018) Are rodent models of Parkinson's disease behaving as they should? Behav Brain Res 352:133–141. https://doi.org/10.1016/j.bbr.2017.10.021

283. Lev N, Barhum Y, Ben-Zur T, Melamed E, Steiner I, Offen D (2013) Knocking out DJ-1 attenuates astrocytes neuroprotection against 6-hydroxydopamine toxicity. J Mol Neurosci 50(3):542–550. https://doi.org/10.1007/s12031-013-9984-9

284. Chandran JS, Lin X, Zapata A, Hoke A, Shimoji M, Moore SO, Galloway MP, Laird FM, Wong PC, Price DL, Bailey KR, Crawley JN, Shippenberg T, Cai H (2008) Progressive behavioral deficits in DJ-1-deficient mice are associated with normal nigrostriatal function. Neurobiol Dis 29(3):505–514. https://doi.org/10.1016/j.nbd.2007.11.011

285. Ansai S, Sakuma T, Yamamoto T, Ariga H, Uemura N, Takahashi R, Kinoshita M (2013) Efficient targeted mutagenesis in medaka using custom-designed transcription activator-like effector nucleases. Genetics 193(3):739–749. https://doi.org/10.1534/genetics.112.147645

286. Lee JW, Tapias V, Di Maio R, Greenamyre JT, Cannon JR (2015) Behavioral, neuro-chemical, and pathologic alterations in bacte-rial artificial chromosome transgenic G2019S leucine-rich repeated kinase 2 rats. Neurobiol Aging 36(1):505–518. https://doi.org/10.1016/j.neurobiolaging.2014.07.011

287. Zhou H, Huang C, Tong J, Hong WC, Liu YJ, Xia XG (2011) Temporal expression of mutant LRRK2 in adult rats impairs dopa-mine reuptake. Int J Biol Sci 7(6):753–761

288. Tsika E, Nguyen AP, Dusonchet J, Colin P, Schneider BL, Moore DJ (2015) Adenoviral-mediated expression of G2019S LRRK2 induces striatal pathology in a kinase-dependent manner in a rat model of Parkinson's disease. Neurobiol Dis 77:49–61

289. Dusonchet J, Kochubey O, Stafa K, Young SM Jr, Zufferey R, Moore DJ, Schneider BL, Aebischer P (2011) A rat model of progres-sive nigral neurodegeneration induced by the Parkinson's disease-associated G2019S muta-tion in LRRK2. J Neurosci 31(3):907–912

290. Tagliaferro P, Kareva T, Oo TF, Yarygina O, Kholodilov N, Burke RE (2015) An early axo-nopathy in a hLRRK2(R1441G) transgenic model of Parkinson disease. Neurobiol Dis 82:359–371. https://doi.org/10.1016/j.nbd.2015.07.009

291. Lin X, Parisiadou L, Gu XL, Wang L, Shim H, Sun L, Xie C, Long CX, Yang WJ, Ding J, Chen ZZ, Gallant PE, Tao-Cheng JH, Rudow G, Troncoso JC, Liu Z, Li Z, Cai H (2009) Leucine-rich repeat kinase 2 regulates the progression of neuropathology induced by Parkinson's-disease-related mutant alpha-synuclein. Neuron 64(6):807–827. https://doi.org/10.1016/j.neuron.2009.11.006

292. Sheng D, Qu D, Kwok KH, Ng SS, Lim AY, Aw SS, Lee CW, Sung WK, Tan EK, Lufkin T, Jesuthasan S, Sinnakaruppan M, Liu J (2010) Deletion of the WD40 domain of LRRK2 in Zebrafish causes Parkinsonism-like loss of neurons and locomotive defect. PLoS Genet 6(4):e1000914. https://doi.org/10.1371/journal.pgen.1000914

293. Prabhudesai S, Bensabeur FZ, Abdullah R, Basak I, Baez S, Alves G, Holtzman NG, Larsen JP, Moller SG (2016) LRRK2 knock-down in zebrafish causes developmental

defects, neuronal loss, and synuclein aggregation. J Neurosci Res 94(8):717–735. https://doi.org/10.1002/jnr.23754

294. Ren G, Xin S, Li S, Zhong H, Lin S (2011) Disruption of LRRK2 does not cause specific loss of dopaminergic neurons in zebrafish. PLoS One 6(6):e20630. https://doi.org/10.1371/journal.pone.0020630

295. Afsari F, Christensen KV, Smith GP, Hentzer M, Nippe OM, Elliott CJ, Wade AR (2014) Abnormal visual gain control in a Parkinson's disease model. Hum Mol Genet 23(17):4465–4478. https://doi.org/10.1093/hmg/ddu159

296. Hindle S, Afsari F, Stark M, Middleton CA, Evans GJ, Sweeney ST, Elliott CJ (2013) Dopaminergic expression of the Parkinsonian gene LRRK2-G2019S leads to non-autonomous visual neurodegeneration, accelerated by increased neural demands for energy. Hum Mol Genet 22(11):2129–2140. https://doi.org/10.1093/hmg/ddt061

297. Hindle SJ, Elliott CJ (2013) Spread of neuronal degeneration in a dopaminergic, Lrrk-G2019S model of Parkinson disease. Autophagy 9(6):936–938. https://doi.org/10.4161/auto.24397

298. Venderova K, Kabbach G, Abdel-Messih E, Zhang Y, Parks RJ, Imai Y, Gehrke S, Ngsee J, Lavoie MJ, Slack RS, Rao Y, Zhang Z, Lu B, Haque ME, Park DS (2009) Leucine-Rich Repeat Kinase 2 interacts with Parkin, DJ-1 and PINK-1 in a Drosophila melanogaster model of Parkinson's disease. Hum Mol Genet 18(22):4390–4404. https://doi.org/10.1093/hmg/ddp394

299. Rayaprolu S, Seven YB, Howard J, Duffy C, Altshuler M, Moloney C, Giasson BI, Lewis J (2018) Partial loss of ATP13A2 causes selective gliosis independent of robust lipofuscinosis. Mol Cell Neurosci 92:17–26. https://doi.org/10.1016/j.mcn.2018.05.009

300. Lopes da Fonseca T, Correia A, Hasselaar W, van der Linde HC, Willemsen R, Outeiro TF (2013) The zebrafish homologue of Parkinson's disease ATP13A2 is essential for embryonic survival. Brain Res Bull 90:118–126. https://doi.org/10.1016/j.brainresbull.2012.09.017

301. Matsui H, Sato F, Sato S, Koike M, Taruno Y, Saiki S, Funayama M, Ito H, Taniguchi Y, Uemura N, Toyoda A, Sakaki Y, Takeda S, Uchiyama Y, Hattori N, Takahashi R (2013) ATP13A2 deficiency induces a decrease in cathepsin D activity, fingerprint-like inclusion body formation, and selective degeneration of dopaminergic neurons. FEBS Lett 587(9):1316–1325. https://doi.org/10.1016/j.febslet.2013.02.046

302. Du TT, Wang L, Duan CL, Lu LL, Zhang JL, Gao G, Qiu XB, Wang XM, Yang H (2015) GBA deficiency promotes SNCA/alpha-synuclein accumulation through autophagic inhibition by inactivated PPP2A. Autophagy 11(10):1803–1820. https://doi.org/10.1080/15548627.2015.1086055

303. Sinclair GB, Jevon G, Colobong KE, Randall DR, Choy FY, Clarke LA (2007) Generation of a conditional knockout of murine glucocerebrosidase: utility for the study of Gaucher disease. Mol Genet Metab 90(2):148–156. https://doi.org/10.1016/j.ymgme.2006.09.008

304. Kim D, Hwang H, Choi S, Kwon SH, Lee S, Park JH, Kim S, Ko HS (2018) D409H GBA1 mutation accelerates the progression of pathology in A53T alpha-synuclein transgenic mouse model. Acta Neuropathol Commun 6(1):32. https://doi.org/10.1186/s40478-018-0538-9

305. Enquist IB, Nilsson E, Ooka A, Mansson JE, Olsson K, Ehinger M, Brady RO, Richter J, Karlsson S (2006) Effective cell and gene therapy in a murine model of Gaucher disease. Proc Natl Acad Sci U S A 103(37):13819–13824. https://doi.org/10.1073/pnas.0606016103

306. Taguchi YV, Liu J, Ruan J, Pacheco J, Zhang X, Abbasi J, Keutzer J, Mistry PK, Chandra SS (2017) Glucosylsphingosine promotes alpha-synuclein pathology in mutant GBA-associated Parkinson's disease. J Neurosci 37(40):9617–9631. https://doi.org/10.1523/JNEUROSCI.1525-17.2017

307. Sardi SP, Clarke J, Kinnecom C, Tamsett TJ, Li L, Stanek LM, Passini MA, Grabowski GA, Schlossmacher MG, Sidman RL, Cheng SH, Shihabuddin LS (2011) CNS expression of glucocerebrosidase corrects alpha-synuclein pathology and memory in a mouse model of Gaucher-related synucleinopathy. Proc Natl Acad Sci U S A 108(29):12101–12106. https://doi.org/10.1073/pnas.1108197108

308. Yun SP, Kim D, Kim S, Kim S, Karuppagounder SS, Kwon SH, Lee S, Kam TI, Lee S, Ham S, Park JH, Dawson VL, Dawson TM, Lee Y, Ko HS (2018) alpha-Synuclein accumulation and GBA deficiency due to L444P GBA mutation contributes to MPTP-induced parkinsonism. Mol Neurodegener 13(1):1. https://doi.org/10.1186/s13024-017-0233-5

309. Papadopoulos VE, Nikolopoulou G, Antoniadou I, Karachaliou A, Arianoglou G, Emmanouilidou E, Sardi SP, Stefanis L, Vekrellis K (2018) Modulation of beta-glucocerebrosidase increases alpha-synuclein secretion and exosome release in mouse models of Parkinson's disease. Hum Mol

Genet 27(10):1696–1710. https://doi.org/10.1093/hmg/ddy075

310. Ginns EI, Mak SK, Ko N, Karlgren J, Akbarian S, Chou VP, Guo Y, Lim A, Samuelsson S, LaMarca ML, Vazquez-DeRose J, Manning-Bog AB (2014) Neuroinflammation and alpha-synuclein accumulation in response to glucocerebrosidase deficiency are accompanied by synaptic dysfunction. Mol Genet Metab 111(2):152–162. https://doi.org/10.1016/j.ymgme.2013.12.003

311. Keatinge M, Bui H, Menke A, Chen YC, Sokol AM, Bai Q, Ellett F, Da Costa M, Burke D, Gegg M, Trollope L, Payne T, McTighe A, Mortiboys H, de Jager S, Nuthall H, Kuo MS, Fleming A, Schapira AH, Renshaw SA, Highley JR, Chacinska A, Panula P, Burton EA, O'Neill MJ, Bandmann O (2015) Glucocerebrosidase 1 deficient Danio rerio mirror key pathological aspects of human Gaucher disease and provide evidence of early microglial activation preceding alpha-synuclein-independent neuronal cell death. Hum Mol Genet 24(23):6640–6652. https://doi.org/10.1093/hmg/ddv369

312. Uemura N, Koike M, Ansai S, Kinoshita M, Ishikawa-Fujiwara T, Matsui H, Naruse K, Sakamoto N, Uchiyama Y, Todo T, Takeda S, Yamakado H, Takahashi R (2015) Viable neuronopathic Gaucher disease model in Medaka (Oryzias latipes) displays axonal accumulation of alpha-synuclein. PLoS Genet 11(4):e1005065. https://doi.org/10.1371/journal.pgen.1005065

313. Liu J, Tang TS, Tu H, Nelson O, Herndon E, Huynh DP, Pulst SM, Bezprozvanny I (2009) Deranged calcium signaling and neurodegeneration in spinocerebellar ataxia type 2. J Neurosci 29(29):9148–9162. https://doi.org/10.1523/JNEUROSCI.0660-09.2009

314. Dansithong W, Paul S, Figueroa KP, Rinehart MD, Wiest S, Pflieger LT, Scoles DR, Pulst SM (2015) Ataxin-2 regulates RGS8 translation in a new BAC-SCA2 transgenic mouse model. PLoS Genet 11(4):e1005182. https://doi.org/10.1371/journal.pgen.1005182

315. Meierhofer D, Halbach M, Sen NE, Gispert S, Auburger G (2016) Ataxin-2 (Atxn2)-knock-out mice show branched chain amino acids and fatty acids pathway alterations. Mol Cell Proteomics 15(5):1728–1739. https://doi.org/10.1074/mcp.M115.056770

316. Pfeffer M, Gispert S, Auburger G, Wicht H, Korf HW (2017) Impact of Ataxin-2 knock out on circadian locomotor behavior and PER immunoreaction in the SCN of mice. Chronobiol Int 34(1):129–137. https://doi.org/10.1080/07420528.2016.1245666

317. Alves-Cruzeiro JM, Mendonca L, Pereira de Almeida L, Nobrega C (2016) Motor dysfunc-

tions and neuropathology in mouse models of spinocerebellar ataxia type 2: a comprehensive review. Front Neurosci 10:572. https://doi.org/10.3389/fnins.2016.00572

318. Kelp A, Koeppen AH, Petrasch-Parwez E, Calaminus C, Bauer C, Portal E, Yu-Taeger L, Pichler B, Bauer P, Riess O, Nguyen HP (2013) A novel transgenic rat model for spinocerebellar ataxia type 17 recapitulates neuropathological changes and supplies in vivo imaging biomarkers. J Neurosci 33(21):9068–9081. https://doi.org/10.1523/JNEUROSCI.5622-12.2013

319. Cui Y, Yang S, Li XJ, Li S (2017) Genetically modified rodent models of SCA17. J Neurosci Res 95(8):1540–1547. https://doi.org/10.1002/jnr.23984

320. Chang YC, Lin CY, Hsu CM, Lin HC, Chen YH, Lee-Chen GJ, Su MT, Ro LS, Chen CM, Hsieh-Li HM (2011) Neuroprotective effects of granulocyte-colony stimulating factor in a novel transgenic mouse model of SCA17. J Neurochem 118(2):288–303. https://doi.org/10.1111/j.1471-4159.2011.07304.x

321. Chen ZZ, Wang CM, Lee GC, Hsu HC, Wu TL, Lin CW, Ma CK, Lee-Chen GJ, Huang HJ, Hsieh-Li HM (2015) Trehalose attenuates the gait ataxia and gliosis of spinocerebellar ataxia type 17 mice. Neurochem Res 40(4):800–810. https://doi.org/10.1007/s11064-015-1530-4

322. Huang DS, Lin HY, Lee-Chen GJ, Hsieh-Li HM, Wu CH, Lin JY (2016) Treatment with a Ginkgo biloba extract, EGb 761, inhibits excitotoxicity in an animal model of spinocerebellar ataxia type 17. Drug Des Devel Ther 10:723–731. https://doi.org/10.2147/DDDT.S98156

323. Myslinski E, Schuster C, Huet J, Sentenac A, Krol A, Carbon P (1993) Point mutations 5′ to the tRNA selenocysteine TATA box alter RNA polymerase III transcription by affecting the binding of TBP. Nucleic Acids Res 21(25):5852–5858

324. Przybyla M, Stevens CH, van der Hoven J, Harasta A, Bi M, Ittner A, van Hummel A, Hodges JR, Piguet O, Karl T, Kassiou M, Housley GD, Ke YD, Ittner LM, Eersel J (2016) Disinhibition-like behavior in a P301S mutant tau transgenic mouse model of frontotemporal dementia. Neurosci Lett 631:24–29. https://doi.org/10.1016/j.neulet.2016.08.007

325. Tan DCS, Yao S, Ittner A, Bertz J, Ke YD, Ittner LM, Delerue F (2018) Generation of a new tau knockout (tauDeltaex1) line using CRISPR/Cas9 genome editing in mice. J Alzheimers Dis 62(2):571–578. https://doi.org/10.3233/JAD-171058

326. Wobst HJ, Denk F, Oliver PL, Livieratos A, Taylor TN, Knudsen MH, Bengoa-Vergniory N, Bannerman D, Wade-Martins R (2017) Increased 4R tau expression and behavioural changes in a novel MAPT-N296H genomic mouse model of tauopathy. Sci Rep 7:43198. https://doi.org/10.1038/srep43198

327. Vargas-Caballero M, Denk F, Wobst HJ, Arch E, Pegasiou CM, Oliver PL, Shipton OA, Paulsen O, Wade-Martins R (2017) Wild-type, but not mutant N296H, human tau restores abeta-mediated inhibition of LTP in Tau(−/−) mice. Front Neurosci 11:201. https://doi.org/10.3389/fnins.2017.00201

328. Newman M, Ebrahimie E, Lardelli M (2014) Using the zebrafish model for Alzheimer's disease research. Front Genet 5:189. https://doi.org/10.3389/fgene.2014.00189

329. Choudhary B, Mandelkow E, Mandelkow EM, Pir GJ (2018) Glutamatergic nervous system degeneration in a C. elegans Tau(A152T) tauopathy model involves pathways of excitotoxicity and Ca(2+) dysregulation. Neurobiol Dis 117:189–202. https://doi.org/10.1016/j.nbd.2018.06.005

330. Morelli F, Romeo M, Barzago MM, Bolis M, Mattioni D, Rossi G, Tagliavini F, Bastone A, Salmona M, Diomede L (2018) V363I and V363A mutated tau affect aggregation and neuronal dysfunction differently in C. elegans. Neurobiol Dis 117:226–234. https://doi.org/10.1016/j.nbd.2018.06.018

331. Roy B, Jackson GR (2014) Interactions between Tau and alpha-synuclein augment neurotoxicity in a Drosophila model of Parkinson's disease. Hum Mol Genet 23(11):3008–3023. https://doi.org/10.1093/hmg/ddu011

332. Chouhan AK, Guo C, Hsieh YC, Ye H, Senturk M, Zuo Z, Li Y, Chatterjee S, Botas J, Jackson GR, Bellen HJ, Shulman JM (2016) Uncoupling neuronal death and dysfunction in Drosophila models of neurodegenerative disease. Acta Neuropathol Commun 4(1):62. https://doi.org/10.1186/s40478-016-0333-4

333. Nishiyama J, Matsuda K, Kakegawa W, Yamada N, Motohashi J, Mizushima N, Yuzaki M (2010) Reevaluation of neurodegeneration in lurcher mice: constitutive ion fluxes cause cell death with, not by, autophagy. J Neurosci 30(6):2177–2187. https://doi.org/10.1523/JNEUROSCI.6030-09.2010

334. Li Q, Guo S, Jiang X, Bryk J, Naumann R, Enard W, Tomita M, Sugimoto M, Khaitovich P, Paabo S (2016) Mice carrying a human GLUD2 gene recapitulate aspects of human transcriptome and metabolome development. Proc Natl Acad Sci U S A 113(19):5358–5363. https://doi.org/10.1073/pnas.1519261113

335. Kotajima-Murakami H, Narumi S, Yuzaki M, Yanagihara D (2016) Involvement of GluD2 in fear-conditioned bradycardia in mice. PLoS One 11(11):e0166144. https://doi.org/10.1371/journal.pone.0166144

336. Plaitakis A, Kotzamani D, Petraki Z, Delidaki M, Rinotas V, Zaganas I, Douni E, Sidiropoulou K, Spanaki C (2018) Transgenic mice carrying GLUD2 as a tool for studying the expressional and the functional adaptation of this positive selected gene in human brain evolution. Neurochem Res. https://doi.org/10.1007/s11064-018-2546-3

337. Hashizume M, Miyazaki T, Sakimura K, Watanabe M, Kitamura K, Kano M (2013) Disruption of cerebellar microzonal organization in GluD2 (GluRdelta2) knockout mouse. Front Neural Circuit 7:130. https://doi.org/10.3389/fncir.2013.00130

338. Yuzaki M (2013) Cerebellar LTD vs. motor learning-lessons learned from studying GluD2. Neural Netw 47:36–41. https://doi.org/10.1016/j.neunet.2012.07.001

339. Zhang L, Le W, Xie W, Dani JA (2012) Age-related changes in dopamine signaling in Nurr1 deficient mice as a model of Parkinson's disease. Neurobiol Aging 33(5):1001 e1007–1001 e1016. https://doi.org/10.1016/j.neurobiolaging.2011.03.022

340. Le W, Zhang L, Xie W, Li S, Dani JA (2015) Pitx3 deficiency produces decreased dopamine signaling and induces motor deficits in Pitx3(−/−) mice. Neurobiol Aging 36(12):3314–3320. https://doi.org/10.1016/j.neurobiolaging.2015.08.012

341. Hwang DY, Fleming SM, Ardayfio P, Moran-Gates T, Kim H, Tarazi FI, Chesselet MF, Kim KS (2005) 3,4-dihydroxyphenylalanine reverses the motor deficits in Pitx3-deficient aphakia mice: behavioral characterization of a novel genetic model of Parkinson's disease. J Neurosci 25(8):2132–2137

342. Suarez LM, Alberquilla S, Garcia-Montes JR, Moratalla R (2018) Differential synaptic remodeling by dopamine in direct and indirect striatal projection neurons in Pitx3(−/−) mice, a genetic model of Parkinson's disease. J Neurosci 38(15):3619–3630. https://doi.org/10.1523/JNEUROSCI.3184-17.2018

343. Graw J (2017) From eyeless to neurological diseases. Exp Eye Res 156:5–9. https://doi.org/10.1016/j.exer.2015.11.006

344. Nordstroma U, Beauvais G, Ghosh A, Pulikkaparambil Sasidharan BC, Lundblad M, Fuchs J, Joshi RL, Lipton JW, Roholt A, Medicetty S, Feinstein TN, Steiner JA, Escobar Galvis ML, Prochiantz A, Brundin P (2015) Progressive nigrostriatal terminal dys-

function and degeneration in the engrailed1 heterozygous mouse model of Parkinson's disease. Neurobiol Dis 73:70–82. https://doi.org/10.1016/j.nbd.2014.09.012

345. Hanks MC, Loomis CA, Harris E, Tong CX, Anson-Cartwright L, Auerbach A, Joyner A (1998) Drosophila engrailed can substitute for mouse Engrailed1 function in mid-hindbrain, but not limb development. Development 125(22):4521–4530

346. Galter D, Pernold K, Yoshitake T, Lindqvist E, Hoffer B, Kehr J, Larsson NG, Olson L (2010) MitoPark mice mirror the slow progression of key symptoms and L-DOPA response in Parkinson's disease. Genes Brain Behav 9(2):173–181. https://doi.org/10.1111/j.1601-183X.2009.00542.x

Part II

Clinical Trials for Motor Symptoms in Parkinson's Disease

Chapter 4

A Review of Randomized Phase III Pharmacological Clinical Trials for Motor Symptoms in Parkinson's Disease Patients and Quality of Evidence Recommendations

Mónica M. Kurtis, Carmen Rodriguez-Blazquez, and Isabel Pareés

Abstract

The phase III randomized clinical trials designed to improve movement symptomatology in initial and advanced stages of Parkinson disease are critically reviewed. This summary provides the reader with an overall clinical picture of the oral therapies that have been tested to treat motor symptoms in PD in the past decade. The responsiveness and interpretability of reported results are considered and a quality of evidence rating scale is proposed to guide the critical review of randomized clinical trials. These recommendations, along with consideration of the large placebo effect in Parkinson disease patients, will hopefully aid in the design of future clinical trials.

Key words Randomized clinical trials, Phase III, Parkinson disease, Motor symptoms, Minimally important clinical change, Quality of evidence, Placebo

1 Introduction

Motor parkinsonism, based on the detection of bradykinesia with rest tremor or rigidity, is currently the core feature defining Parkinson's disease [1]. Motor symptoms can greatly impact patient quality of life and lead to disability thus treatment efforts are continuously expanding. A number of interventions have proven to be efficacious and a recent review by the Movement Disorder Society Evidence-Based Medicine Committee establishes recommendations for the clinician [2]. Ongoing basic research continues to explore new molecules that will go from bench to bedside only after following a series of sequential trials that prove safety and efficacy.

In the initial phase O, trials are designed to gather preliminary data on the pharmacodynamics and pharmacokinetics of the tested drug and are not always performed in humans. Phase I trials are the first to be performed in human subjects, generally small in number to screen for drug safety. Phase II trials are performed with

Santiago Perez-Lloret (ed.), *Clinical Trials In Parkinson's Disease*, Neuromethods, vol. 160,
https://doi.org/10.1007/978-1-0716-0912-5_4, © Springer Science+Business Media, LLC, part of Springer Nature 2021

patients and powered to demonstrate efficacy of the treatment being investigated against placebo and to assess side effect profile. Phase III clinical trials have the objective of confirming drug safety and efficacy. This type of randomized clinical trials (RCTs) imply testing large populations in order to compare the potential treatment's effect with placebo and possibly other established treatments, to monitor side effects and to collect information that will aid in the future use of the drug. Phase IV trials are postcommercialization trials aimed to delineate treatment benefits, risks, and optimal use in routine clinical practice.

This review focuses on the phase III RCTs published in the past decade investigating therapies for the treatment of motor symptoms in PD as these trials are pivotal for determining drug safety, efficacy and gaining approval by medical agencies. Thus, a critical review of RCTs' patient selection, measurement and statistical methods, and reporting of results is crucial in determining the quality of evidence presented. Guidelines to assess (and design) quality of RCTs are provided through a modified version of a previously published scale [3].

The purpose of randomization, masking, and placebo control in RCTs is to control for the placebo effect. However, placebo intake triggers a physiological response directed by the activation of dopaminergic circuits involved reward and learning mechanisms. Thus, the great importance of advancing in this area of research in PD as it will also have implications in the design of future clinical trials.

2   Methodology

The authors searched bibliographic databases, including MEDLINE, PUBMED and ClinicalTrials.gov for phase III randomized clinical trials regarding conventional treatments of motor symptoms in Parkinson disease (PD) for all clinical stages (both early and levodopa treated patients) published in the past decade (2007–2017). The following search terms were used: PD AND randomized clinical trial AND pharmacotherapy (limits: past 10 years). Papers were screened for relevance and references checked for further trials. Only randomized, double blinded, controlled, phase III trials with oral pharmacological therapies for PD patients with motor improvement as the primary endpoint were analyzed. Trials included patients in both early (Hoehn and Yahr stages 1–2) and later stages of disease (Hoehn and Yahr stages 3–4) and patients with motor complications such as wearing-off or/and dyskinesias.

Data extraction was performed by two of the authors (MK and IP), noting important variables, similarly to a recent review of clinical trials in Multiple System Atrophy promoted by the Movement

Disorders Society MSA Study Group [4]. The following data was recorded: demographic and disease characteristics of participants (age, gender, disease stage), diagnostic criteria used (Brain Bank, MDS or other), therapeutic intervention, primary outcomes, study design characteristics (sample size, participating centers, parallel or crossover groups, cointervention, blinding, allocation and randomization, trial design, duration of treatment and follow-up), assessments used (scales, diaries), statistical analysis (clearly stated, software used, sample size calculation, intention to treat analysis), reported results, and limitations specified by the investigators. The results are summarized in Table 1 and Subheading 3. Results of the analysis of the clinical trials included are reported as mean (standard deviation) and as percentages. Trials were classified as positive if the tested drug showed superior motor effects when compared to placebo or noninferiority versus other standard treatment, while negative trials were those in which no significant differences were found between the trial drug and placebo.

The second part of the chapter's objective is to provide the reader with the necessary background to critically assess the results of a clinical trial. A modified version of a previously published rating scale for evaluating clinical trial quality [3] is proposed. A practical example of how to apply the scale is provided (Table 2). Two raters (CB and MK) performed the assessment independently and discussed differing scores to reach consensus. Some recommended statistical methods such as the responsiveness of clinical scales and their interpretation are discussed and concepts of effect size, relative change, and minimally important clinical change are defined. Furthermore, publication bias in PD and the large placebo response in PD are discussed.

3 Results

3.1 Publications

A total of 30 publications published between 2007 and 2017 and describing the results of phase III RCTs for treatment of motor symptoms in PD were included (Table 1). Twenty-nine (96.7%) studies were multicenter. The number of centers involved in one clinical trial was not specified [5]. The studies were most often published in neurology journals (n = 16, 53.3%), specific movement disorders journals (n = 11, 36.7%), pharmacology journals (n = 2, 6.7%), and a general medicine journal (n = 1, 3.3%). They were all written in English and published as original research articles. Mean number of publications per year was 2.72 (±1.9), being 2012 and 2017 the years with the highest number of published studies in the field (5 and 7 publications per year respectively).

3.2 Study Participants

A total of 13,542 participants with PD were randomized in the 30 clinical trials included. Mean age of participants ranged between 56.5 and 68.8 years according to the demographic data provided

Table 1
Phase III randomized double-blind controlled studies with motor improvement as primary outcome in PD

Study	Drug	Control	Randomized patients (N)	Male %	Age (mean)*	H&Y inclusion	Design	Intervention duration	Primary outcome	Results
Watts 2007 [17]	Rotigotine	P	T = 181 P = 96	T = 68 P = 60	T = 62 P = 64.5	<3	Parallel group	28w	Change in UPDRS (III)	+
Pahwa 2007 [7]	Ropinirole PR	P	T = 202 P = 191	T = 58 P = 68	T = 66.3 P = 66	2–4	Parallel group	24w	Change in OFF time in Diary	+
Zhang 2013 [11]	Ropinirole PR	P	T = 175 P = 170	T = 63.3 P = 61.8	T = 64.1 P = 63.6	2–4	Parallel group	24w	Change in OFF time in Diary	+
Hauser 2008 [24]	Istradefylline	P	T = 116 P = 115	T = 66.1 P = 67	T = 63 P = 64	2–4	Parallel group	12w	Change in OFF time in Diary	+
Mizuno 2010 [25]	Istradefylline	P	T20 = 119 T40 = 125 P = 119	T20 = 43 T40 = 44 P = 38	T20 = 65 T40 = 63 P = 65	2–4	Parallel group	12w	Change in OFF time in Diary	+
Pourcher 2012 [26]	Istradefylline	P	T10 = 155 T20 = 149 T40 = 152 P = 154	T10 = 67.3 T20 = 69.1 T40 = 65.8 P = 64.2	T10 = 63 T20 = 64 T40 = 63 P = 63	2–4	Parallel group	12w	Change in OFF time in Diary	–
Olanow 2009 [19]	Rasagiline	P	ET1 = 288 ET2 = 293 DT1 = 300 DT2 = 295	ET1 = 60.8 ET2 = 59.7 DT1 = 62.0 DT2 = 61.7	ET1 = 62.4 ET2 = 62.3 DT1 = 61.9 DT2 = 62.4	<3	Delayed-start	18m	Change in total UPDRS	+ For ET1
Hauser 2010 [8]	Pramipexole PR	P and IR pramipexole	T = 106 P = 50 IR = 103	T = 58.5 P = 46 IR = 57.3	T = 61.6 P = 63.2 IR = 62.6	1–3	Parallel group	18w	Change in UPDRS (II + III)	+
Poewe 2011 [9]	Pramipexole PR	P and IR pramipexole	T = 223 P = 213 IR = 103	T = 57 P = 56.8 IR = 49.5	T = 61.3 P = 61.7 IR = 62	1–3	Parallel group	33w	Change in UPDRS (II + III)	+

Schapira 2011 [10]	Pramipexole PR	P and IR pramipexole	T = 165 P = 178 IR = 175	T = 56.1 P = 52.8 IR = 56	T = 61.6 P = 60.9 IR = 62	2–4	Parallel group	33w	Change in UPDRS (II + III)	+
Mizuno 2012 [11]	Pramipexol PR	IR pramipexole	T = 56 IR = 56	T = 37.5 IR = 37.5	T = 68.8 IR = 66.1	2–3	Parallel group	12w	No	+
Shapira 2013 [18]	Pramipexole	P	ET = 261 DT = 274	ET = 68 DT = 61	ET = 62.1 DT = 62.9	1–2	Delayed-start	15m	Change in total UPDRS	-
Lees 2012 [31]	Perampanel	P	T2 = 258 T4 = 251 P = 254	T2 = 60 T4 = 61 P = 60	T2 = 63.8 T4 = 64.3 P = 64.1	Not specified	Parallel group	30w	Change in OFF time in Diary	−
Lees 2012 [31]	Perampanel	P	T2 = 251 T4 = 250 P = 250	T2 65 T4 66 P = 66	T2 = 62.8 T4 = 62.9 P = 62.2	Not specified	Parallel group	20w	Change in OFF time in Diary	-
Stocchi 2012 [20]	Safinamide	P	T100 = 90 T200 = 90 P = 90	T100 = 66 T200 = 61 P = 62	T100 = 56.5 T200 = 58.5 P = 57.3	1–3	Parallel group	24w	Change in UPDRS (III)	−
Shapira 2013 [21]	Safinamide	P	T100 = 80 T200 = 69 P = 78	T100 = 67.5 T200 = 62.3 P = 60.3	T100 = 56.6 T200 = 59.8 P = 57	1–3	Parallel group	12m	Time from baseline to need for increase in dopamine agonist dose	−
Borgohain 2014 [22]	Safinamide	P	T50 = 223 T100 = 224 P = 222	T50 = 70.4 T100 = 72.8 P = 72.1	T50 = 60.1 T100 = 60.1 P = 59.4	1–4	Parallel group	24w	Change in ON time with no troublesome dyskinesia in DRS	+
Schapira 2017 [23]	Safinamide	P	T = 274 P = 275	T = 62.4 P = 59.3	T = 61.7 P = 62.1	1–4	Parallel group	24w	Change in ON time with no troublesome dyskinesia in Diary	+

(continued)

Table 1
(continued)

Study	Drug	Control	Randomized patients (N)	Male %	Age (mean)*	H&Y inclusion	Design	Intervention duration	Primary outcome	Results
Hauser 2013 [13]	Carbidopa-levodopa (IPX066) PR	IR carbidopa levodopa	T = 201 IR = 192	T = 64 IR = 65	T = 63.1 IR = 63.4	1–4	Parallel group	13w	Change in OFF time in Diary	+
Stocchi 2014 [14]	carbidopa-levodopa (IPX066) PR	P and Carbidopa-levodopa-entacapone	91	74.7	64.1	1–4	Crossover	11w	Change in OFF time in Diary	+
PSG QE3 Investigators 2014 [32]	Coenzyme Q10	P	T1200 = 201 T2400 = 196 P = 203	T1200 = 69.2 T2400 = 65.3 P = 64	T1200 = 63.3 T2400 = 62.8 P = 61.3	≤2.5	Parallel group	16m	Change in total UPDRS	–
Hauser 2015 [27]	Preladenant	P and Rasagiline	T2 = 156 T5 = 155 T10 = 156 P = 155 R = 156	T2 = 63.0 T5 = 51.0 T10 = 60.1 P = 50.3 R = 61.7	T2 = 61.6 T5 = 62.6 T10 = 63.5 P = 63.0 R = 63.3	2.5–4	Parallel group	12w	Change in OFF time in Diary	–
Hauser 2015 [27]	Preladenant	P	T2 = 158 T5 = 159 P = 159	T2 = 68.8 T5 = 54.8 P = 59.7	T2 = 62.9 T5 = 64.2 P = 64.2	2.5–4	Parallel group	12w	Change in OFF time in Diary	–
Stocchi 2017 [28]	Preladenant	P and rasagiline	T2 = 204 T5 = 204 T10 = 206 P = 204 R = 204	T2 = 62 T5 = 56 T10 = 56 P = 60 R = 58	T2 = 63.0 T5 = 62.3 T10 = 63.8 P = 63.3 R = 62.9	≤ 3	Parallel group	26w	Change in UPDRS (II + III)	–

Study	Drug	Comparator	n			Dose	Design	Duration	Outcome	Result
Ferreira 2016 [29]	Opicapone	P and entacapone	T5 = 122, T25 = 119, T50 = 116, P = 121, E = 122	T5 = 58, T25 = 56, T50 = 60, P = 59, E = 62	T5 = 63.6, T25 = 64.4, T50 = 63.5, P = 64.3, E = 63.7	1–3	Parallel group	14–15w	Change in OFF time in Diary	+ for T50
Lees 2017 [30]	Opicapone	P	T25 = 129, T50=154, P = 144	T25 = 65.6, T50 = 60.5	T25 = 62.5, T50 = 65.5	1–3	Parallel group	14–15w	Change in OFF time in Diary	+
Pawha 2017 [15]	ADS-5102 Amantadine PR	P	T = 60, P = 61	T = 55.6, P = 60.3	63.9, 65.5	NS	Parallel group	25w	Change in total UDysRS	+
Oertel 2017 [16]	ADS-5102 Amantadine PR	P	T = 38, P = 39	T = 51.4, P = 52.6	T = 64.7, P = 64.9	NS	Parallel group	13w	Change in total UDysRS	+
Mizuno 2017 [5]	Selegiline	P	T = 146, P = 146	T = 42, P = 47	T = 64, P = 64	1–3	Parallel group	12w	Change in UPDRS (I + II + III)	+
Postuma 2017 [33]	Caffeine	P	T = 60, P = 61	T = 68, P = 60	T = 62, P = 62	1–3	Parallel group	6–12m	Change in MDS-UPDRS (III)	–

DRS Dystonia Rating Scale, DT delayed treatment, ET early treatment, IR immediate-release, P placebo, PR prolonged release, R rasagiline, T treatment group, E entacapone, numbers following T, R, E refers to dose in mg in each group; w weeks, NS not specified, UDysRS United Dyskinesia Rating Scale, w weeks, m months, + positive trial, – negative trial

* years old

Table 2
Clinical trial quality of evidence rating scale: a practical example

Quality of evidence rating scale	Score	Preladenant trials [27]
Validity: selection		
1. Was diagnosis well defined?	2	Brain Bank criteria
2. Spectrum of patients well defined?	2	>2 h off time, of H & Y: 2.5–4, >5 year-disease
3. If explanatory, were eligibility criteria suitably narrow?	2	Yes. *See* item 2
4. If pragmatic, was suitability broad eligible criteria used?	NA	Applies to phase IV trials
Measurement		
5. Was assignment to treatment random?	2	Specified in overview section
6. If yes, was the method explained	2	Specified in overview section
7. Were all patients accounted for after randomization	2	*See* flow chart (Fig. 1)
8. Were losses to follow up low (<10%)?	0	Trial 1 11% lost to follow up Trial 2 15% lost to follow up
9. Were treatment groups similar?	2	"Treatment groups were similar"
10. Were all patients otherwise treated alike?	2	By protocol (in supplements)
11. Were all involved blind to treatment	2	Clearly specified in overview section
12. Was assessment of outcome blind?	1	Unclear
13. Was the occurrence of side effects looked for?	2	Specified in safety assessments
Statistical analysis		
14. Was main analysis on ITT?	2	Define efficacy population: randomized and at least 1 dose of study medication
15. If no, was sensitivity analysis performed?	NA	
16. Were appropriate and sufficient methods used?	1	Constrained longitudinal data analysis (missing responsiveness statistics)
17. Were additional clinically relevant factors analyzed?	2	UPDRS results in supplementary material
18. If subgroup analysis were done, were they explicitly presented as such?	NA	

(continued)

Table 2
(continued)

Quality of evidence rating scale	Score	Preladenant trials [27]
Results		
19. Is estimate of treatment effect given?	2	Differences in off time vs. placebo were not significant for dose 2 mg, 5 mg 10 mg or rasagiline 1 mg
20. Is it of clinical importance?	1	MIC/MID is not provided. Clinical relevance is unclear
21. Is estimate of treatment effect sufficiently precise?	0	ES and RC cannot be calculated SD at baseline are not provided
22. If side effects noted, were frequency/ severity given?	2	Provided in Table III of results
Utility		
23. Do results help me chose treatment?	0	Negative results for rasagiline are unclear. High placebo response
Score		
Total (add scores) (*A*)	31	
No. of questions which apply (max = 23) (*B*)	20	
Max possible score (2 × *B*) (*C*)	40	
Overall rating (A/C)	*77.5%*	

SD standard deviation, ES effect size, MIC minimally important change, max maximum, RC relative change

by each article. There was a predominance of males in 27 (90%) of the 30 studies. Most clinical trials (*n* = 26, 86.7%) used the Hoehn and Yahr scale to specify motor status as part of the inclusion criteria for participants. In 14 studies Hoehn and Yahr stage for participants was of three or less. Only 11 (36.7%) of the clinical trials specified whether the Hoehn and Yahr stage was determined during an ON or OFF period. Participants' UPDRS score at baseline was reported in 27 (90%) studies (one used the MDS-UPDRS score), whereas this scale score was not specified in 3 (10%). Of those studies that reported participants' baseline UPDRS score, seven only reported part III subscale score, four reported part II and III subscale scores, and nine studies reported total UPDRS score (only three of these specified the score on each UPDRS subscale). Only 12 (40%) clinical trials specified whether UPDRS score was calculated in the ON or OFF medication state.

3.3 Diagnostic Criteria

Twelve (40%) of the 30 studies did not report the specific criteria used to make a diagnosis of PD. Of the 18 (60%) studies that reported diagnostic inclusion criteria, 15 of them used the Brain Bank criteria for PD [6] and 3 studies reported that diagnosis of PD was based on the presence of at least two of the three cardinal signs of PD.

3.4 Therapeutic Interventions

A total of 15 different pharmacological interventions were studied in the 30 phase III clinical trials included in this chapter (*see* Table 1). Ten of them assessed the efficacy and safety of the prolonged release version of well-established and commercially available medications [7–16]; eight RCT studied intervention with dopamine agonists [7–12, 17, 18]; six trials tested monoamine oxidase B (MAO) inhibitors [5, 8–12]; five studied adenosine A2a receptor antagonists [13–17]; two tested catechol-o-methyltransferase (COMT) inhibitors [18, 19], and three studied other molecules such as a 5-AMPA antagonist [20], Coenz Q10 [21], and caffeine [22].

3.5 Study Design and Comparators

A total of 27 clinical trials had a parallel group design, 2 studies used a delayed start design [8, 23] and 1 of the studies had a crossover design [24]. Parallel groups involve two or more arms comparing different treatment groups from baseline until the end of the trial. In crossover studies, all participants receive all treatments in consecutive time periods and outcomes are measured at the end of every period. In delayed start trials, the delayed-start group will follow the same treatment protocol as the early start group after an established time period.

Of the 27 parallel group-designed clinical trials, 19 of them were only placebo controlled, 6 studies used both placebo and other control interventions [16–18, 25–27], and 2 clinical trials assessed the efficacy and safety of an extended release version of a drug by comparing it with the immediate release version [28, 29]. Fourteen studies compared different dosages of the same drug for the management of motor aspects of PD as can be seen in Table 2.

3.6 Duration of the Pharmacological Intervention

The duration of the pharmacological intervention for each study ranged between 11 weeks and 18 months. Most commonly, the pharmacological intervention lasted 12 weeks, as seen in 7 (23.3%) studies [5, 13–16, 28];followed by 5 (16.7%) studies with a duration of 24 weeks [7, 9, 11, 12, 30]. Only 4 (13.3%) of the 30 studies had a duration of intervention of 1 year or longer [8, 10, 21, 23]. Follow-up after stopping the tested drug was not performed or was not clearly specified in the methods section of 16 (53.9%) of the RCTs. When a follow up or wash out period was reported, duration ranged from 4 days to 4 weeks.

3.7 Outcomes and Assessment Tools

A clear definition of the primary outcomes of the study was reported in 29 (96.7%) clinical trials. Only one study assessing the efficacy and safety of extended vs immediate release pramipexol solely defined secondary outcomes [28]. Of the 29 clinical trials with well-defined motor primary outcomes, the UPDRS was used as the measuring tool for evaluating motor change in 11 of them. Of these, 3 reported change in total UPDRS scores as the primary outcome, 3 studies reported change only in the part III subscore, 4 studies reported change in part II + III subscores and 1 study used change in parts I through III subscores. Only one study used change in part III subscoring of the MDS-UPDRS as the primary outcome. Thirteen studies involving drugs designed to improve motor fluctuations reported changes in daily OFF periods assessed by patients' daily diaries as the primary outcome. Two studies used the duration of daily ON state and time with no or nontroublesome dyskinesia reported in diaries or Dyskinesia Rating Scale [11, 12] as a primary outcome. Finally, two studies used change in total Unified Dyskinesia Rating Scale [31, 32] as primary outcomes. Ten trials were considered negative as the intervention did not prove superiority to placebo.

3.8 Adverse Events

The extent of adverse events reporting in the reviewed RCTs was not consistent across studies. Most trials reported total adverse effects and those that lead to discontinuation of the pharmacological intervention. However, the adverse events leading to study drug discontinuation in four trials were not clearly specified [5, 14, 20]. A total of 43 deaths were reported in the articles included in this chapter. One of them occurred 8 days after study completion [21] and five in the open-label phase of the study [19]. The other 37 occurred in the double blinded phase of the clinical trials and only 1 was considered to be possibly related to the pharmacological intervention by the investigator.

3.9 Statistical Analysis

A description of the statistical methods used to assess results was available in all clinical trials. Analysis of covariance (ANCOVA) was the most prevalent method used ($n = 21$, 70%), followed by a constrained longitudinal data analysis approach ($n = 3$, 10%). Only 4 (13.3%) studies specified the software used to perform statistical analysis. Analysis in the full analysis set was reported in 10 (33.3%) studies, whereas intention-to-treat analysis was reported in 13 (43.3%). A per-protocol analysis was also specified in 5 (16.7%) studies. Finally, safety population analysis was mentioned in 17 (56.7%) of the reviewed studies. Post hoc analysis were performed and clearly explained in 8 (26.7%) studies.

3.10 Limitations

Description of the methodological details varied across the clinical trials and a number of quality indicators were not completely reflected in many publications. Sample size calculation process was described in 24 (80%) studies. All studies were reported to be randomized and participants were blinded to research intervention. However, investigators clearly described the process of random sequence generation in only 18 (60%) studies, most of them based on computer generation sequencing, thus minimizing the risk of selection bias. Drop-outs and reason for missing data were reported in most cases. Thirteen (43.32%) clinical trials specified the way they dealt with missing data to perform the analysis (11 of them used a last observation carried forward method). However, a section clearly acknowledging the limitations of each clinical trial could only be found in 13 (43.3%) of the studies. Moreover, the effect size (ES) and relative change (RC) of the resulting benefit and whether this change was clinically meaningful and relevant to the patient was not specifically reported in any of the clinical trials. Fifteen (30%) studies noted the Global Impression Scale score but did not calculate the minimally important change (MIC).

4　Interpreting Results of Clinical Trials

Assessment of RCT quality is essential for the clinician to interpret reported results and plays a crucial role for the design of future trials and for drug approval. For this purpose, we propose a modified version of a previously published Rating Scale for Quality of Evidence designed by the Movement Disorder Society (MDS) Task Force on Evidence-Based Medicine [3]. This scale provides a checklist when evaluating the quality of a trial with regards to five important domains: patient selection, appropriate measurement, statistical analysis, reporting of results and utility. This quality of evidence scale has 23 items, each rated on a Likert scale from 0 to 2 (0 = no, 1 = unclear and 2 = yes). Some items may not apply to the study under evaluation and should be answered "not applicable." The overall quality score is expressed as a percentage (total score/maximum possible score), thus taking into account items that may not apply.

The first domain of the Rating Scale for Quality of Evidence evaluates appropriateness of diagnostic criteria and definition of patient population included in the trial. For explanatory (phase II and III trials), the "ideal" patient population should have sufficiently narrow inclusion criteria to test for drug efficacy [33]. For pragmatic trials (phase IV postmarketing trials) the eligibility criteria should be sufficiently broad to reflect "real world" clinical practice patients. In this review, most trials adequately explained their inclusion criteria, frequently using Hoehn and Yahr staging or a certain amount of off time during the day based on patient diaries. However, although a seemingly straightforward standard, up to 40% of trials did not explain how PD diagnosis was made. Those that did clarify this point, mostly used the traditional Brain Bank criteria [6].

The measurement domain includes nine items on patient randomization, appropriate blinding, baseline patient characteristics, clarity of treatment protocol and side effect reporting. As an example of a common error, we noted that the randomization process was not described in 40% of reviewed trials. Also in this section, in the reported protocol, most trials did not clearly explain the follow-up period after drug suspension, washout periods or whether continued compassionate use was allowed.

The statistical analysis domain of the scale includes five items that score whether an intention to treat or some kind of sensitivity analysis was performed, whether appropriate and sufficient statistical methods were used to analyze the endpoint data and other clinically relevant factors and if ad hoc analysis were performed and clearly explained. ANCOVA, an adequate test of statistical significance was the most widely used, however, none of the trials used sufficient statistical methods as ES and RC, and MIC was not calculated. These statistical techniques are crucial in reporting high quality evidence and are further discussed below. Moreover, less than half the RCTs used intention to treat analysis.

The results domain is made up of four items considering the effect of treatment, how precisely it is reported (again, are effect size or other responsiveness statistics provided?) and whether it is clinically relevant (is MIC or MID provided?). Negative results should not be penalized in this section as it is the quality of the evidence that is being evaluated and not the direction of results. Side effect reporting is also evaluated in this section. This is a basic point for any potential treatment and most but not all reviewed trials reported adverse events adequately as reviewed above. Finally, a utility item evaluates whether the trial provides the clinician with enough information to make a treatment choice.

5   Responsiveness and Interpretability of Rating Scales

To interpret the estimated benefit of tested drugs in RCTs, methods to measure change, differentiate real change from measurement error and identify what amount of change is clinically relevant for the patient have been developed [36, 37]. To measure change, tests of statistical significance such as ANOVA or Student's t are traditionally used as exemplified in the reviewed studies. However, these tests do not inform about the direction or magnitude of change and they are related to sample size and variance: a small and nonrelevant change can be significant in studies with large samples. To overcome these limitations, different responsiveness statistics have been developed in recent years. Responsiveness, or sensitivity to change, is the ability of an instrument to detect change [34], and is related to the reliability and the precision of the instrument. Precision, defined as the ability of the instrument to detect small differences, is assessed with the standard error of measurement (SEM), calculated with the formula SEM = SD $\times \sqrt{(1 - r_{xx})}$, where SD is the standard deviation and r_{xx} is the reliability coefficient of the rating scale [34].

5.1 Effect Size and Relative Change

Based on the SEM, several methods to detect a real change (a change higher than the measurement error) [35] are available, although they provide nonequivalent results [36]. The SEM value, the upper limit of the 95% interval of confidence (1.96 SEM), the smallest real difference (1.96 $\times \sqrt{[2 \times \text{SEM}]}$) and the reliable change index (1.96 $\times \sqrt{[\text{SEM}_1^2 + \text{SEM}_2^2]}$) are some of these methods.

Responsiveness statistics can also be used for calculating the magnitude of change: the relative change (RC), the effect size (ES), and the standardized response mean (SRM) are some of the most used indices for this purpose [37]. The RC is the average change divided by the mean at baseline and is expressed as a percentage. The ES is the average change divided by the standard deviation at baseline, whereas the SRM is the average change divided by the standard deviation of change. Although the last two formulas produce different results, the proposed benchmark values for the interpretation of these indices are the same (<0.20, negligible change; 0.20–0.49, small; 0.50–0.79, moderate; and ≥0.80, large) [38].

The UPDRS and the MDS-UPDRS have been widely studied and have shown to be responsive to changes over time and in response to treatment, with moderate and large ES in some studies. For example, in their observational study comparing subcutaneous apomorphine and intrajejunal levodopa infusions, Martinez-Martin et al. found large RC and ES for both these interventions for the UPDRS. For levodopa and apomorphine infusions respectively: the RC of part III was 44.79% and 43.26%, of part IV it was 55.06% and 40.84%, while the ES was 1.00 and 1.28 for part III and 1.69 and 0.87 for part IV [39]. With these calculations, the reader can conclude that both advanced treatments greatly improve motor signs, while levodopa infusions may be more beneficial for motor fluctuations. Similarly, there are other examples of reported RC and ES on the MDS-UPDRS after subthalamic nucleus deep brain stimulation [40, 41]. As described above, these statistical methods were lacking in the reviewed studies, although some reported enough information (mean scores and SD at baseline and follow-up) for their calculation.

5.2 Minimally Important Change or Difference

Clinical trials should also determine if the estimated change with interventions is clinically relevant and meaningful for the patient. In this sense, the extent to which a qualitative meaning can be assigned to quantitative scores of change is referred to as interpretability [42]. The calculation of minimally important change (MIC) or difference (MID), defined as the smallest difference in score in the outcome of interest that patients or proxies perceive as important, either beneficial or harmful, and that would lead the patient or clinician to consider a change in therapeutic management, are the most common tools of interpreting the change [43]. Two main approaches can be used to calculate the MIC or MID: anchor-based and distribution-based methods [44]. Anchor-based methods use an external criterion or anchor (a clinician or patient-based measure) to operationalize the change. Distribution-based approaches estimate the MIC/MID by means of the distribution of observed scores. Some authors propose the use of precision (SEM, ½ SD) and responsiveness (ES, SRM) statistics to determine the MIC or MID, based on the idea that a large change may be an important one [44]. Determining MIC or MID with anchor or distribution-based methods leads to different results, therefore some authors have proposed calculating a range of values (triangulation) that presumably contains the real MIC value. Moreover, the MIC/MID values also depend on the selected anchor, as reflected in the study by Hauser et al. [45], that determined this data based on two pivotal RCTs of ET and IR pramipexole reviewed above [25, 26]. This group compared the MIC obtained with the Clinician-Rated Global Impression of Improvement (CGI-I) scale and the Patient-Rated Global Impression of Improvement (PGI-I) [45]. Patient or clinician CGI-I scale allows for MIC calculations based on anchor methods. Table 3 displays several examples of studies on the MIC/MID of the most common motor rating scales used for PD and may be used as a reference to interpret the clinical meaningfulness of the results reported in the reviewed RCTs.

6   Publication Bias

Trial discontinuation and trial publication bias is also of crucial importance when reviewing clinical trials. Similarly to other diseases, in PD too, about 40% of trials do not reach publication [46]. Research on predictors of trial discontinuation have reported an association with the number of patients (more likely in trials with <100 patients) and trial phase (more likely in phase IV than phase III, and phase II more than phase IV). Trial nonpublication is also associated with number of participating centers (more likely in single center trials), blinding status (single blinded more than double blinded), and sponsorship (industry-sponsored trials versus university-sponsored). [46].

Publication bias can also determine that negative trials not reach publication [47]. One third of the reviewed studies showed negative results [9, 10, 15–17, 20–23, 28].

Table 3
Responsiveness and interpretability of several motor rating scales for Parkinson's disease

Rating scale	References	MIC/MID	Method
UPDRS	[65]	UPDRS total: 4.3 UPDRS III: 2.5	Anchor-and-distribution-based
	[66]	UPDRS total: 3.0 for improvement; 0.5 for worsening. UPDRS II: 0.5 for improvement; 1 for worsening. UPDRS III: 2 for improvement; 0.5 for worsening.	Anchor-based
	[45]	UPDRS II: 1.8 to 2.3 UPDRS III: 5.8 to 6.5 UPDRS II + III: 7.1 to 8.8 (early vs advanced PD and depending on the anchor)	Anchor-based
MDS-UPDRS	[67]	Part 3: −3.25 for improvement; 4.63 for worsening	Anchor-and-distribution-based
	[68]	Part 1: 2.64 for improvement; 2.45 for worsening. Part 2: 3.05 for improvement; 2.51 for worsening. Parts 1 + 2: 5.73 for improvement; 4.70 for worsening.	Anchor-based
UDysRS	[69]	Part 3: 2.32 for improvement.	Anchor-based
Hauser diary	[60]	21.5% (day 1) to 8.8% (day 6)	Distribution-based

Figures are expressed in absolute values for comparative purposes
MIC/MID minimally important change or difference, *UPDRS* Unified Parkinson's Disease Rating Scale, *MDS* Movement Disorders Society, *UDysRS* Unified Dyskinesia Rating Scale

As the authors of the Café-PD trial explain, it is important that negative results be disseminated and shared with the scientific community [22]. This trial, for example, provides class I evidence that caffeine does not significantly improve PD motor symptoms. Similarly, the high-dose CoQ10 study published by the Parkinson Study Group [21] puts an undisputed close to the controversy of whether this antioxidant is neuroprotective for PD. Disappointing results must be published so that research can focus on more promising efforts.

7   Placebo Effect and PD Trials

The placebo effect is particularly important in disorders affecting dopaminergic circuits. RCTs are planned to control for this effect and most of the reviewed trials chose a parallel group design, but crossover or delayed start designs were also seen. There are advantages and disadvantages for each design. Crossover trials assume stability of disease and have period and carry over effects that require randomization and sufficient washout periods. Parallel group designs require more patients and are more costly but offer the versatility that phase III trials generally require [48]. Delayed start designs are used to demonstrate a treatment's modifying effect disease progression. Some investigators suggest adding an "untreated group" to RCTs [49] that study PD since the placebo response in clinical trials in PD is well established and reported to be between 16 and 21% [50–52]. Almost two decades ago, de la Fuente-Fernandez et al. investigated how placebo and the expectation of clinical benefit triggers the activation of reward mechanisms [49]. Through positron emission tomography (PET) neuroimaging with raclopride (binds to D2 receptors) this group found that after placebo intake, dopamine is released in the ventral and dorsal striatum. In the ventral caudate and pallidum, dopamine activation is associated with reward expectancy; while in the dorsal striatum, dopamine is intrinsically related to motor circuitry. This research group proposed an expectation model of placebo effect based on probability of reward. Activated dopamine neurons can display two types of response: phasic activation, which occurs after the reward and depends on how surprising the reward is; and tonic activation, which precedes the reward and depends on reward uncertainty.

Reported clinical improvements seen with placebo are not methodological errors but a direct consequence of dopaminergic activation in the neural circuitry involved in reward mechanisms. Ongoing research is furthering our understanding of the predictors of placebo response and will have profound implications in clinical trial design [53]. Basic research suggests that low expectation of clinical benefit may underestimate the placebo effect, while the contrary will be seen with high expectations [54]. A recent negative phase IIb trial with inhaled glutathione may be an example of how high expectations in the participating community may lead to large placebo effects [55]. Thus, the management of expectations is important for clinical trial results.

Other groups have dwelled on the key role of learning (conditioning) in the placebo effect. Frisaldi et al. studied the effects of apomorphine versus placebo for bradykinesia in a small sample of patients and found that placebo only improved bradykinesia after previous exposure to the active drug [56]. With this study and a previous one on rigidity [57] they suggest that pharmacological conditioning modulates the placebo effect and suggest that unblinded exposure to active agents should thus be avoided before randomization in clinical trials.

Recently, a meta-analysis of 48 RCTs with improvement in motor function or motor fluctuations as a primary endpoint investigated the placebo response according to PD population. Chae

Won Shin et al. concluded that baseline higher UPDRS III scores, lower levodopa daily dose and presence of motor fluctuations may show higher placebo effects [58].

Furthermore, geographical region may have an effect on the placebo effect as the investigators of the negative phase III preladenant trial discuss in their publication [16]. In future trials, they suggest treatment stratification to have balanced assignment of arms by region, as well as specialized training to mitigate the placebo effect. This trial also found differences in the placebo effect according to time period of enrollment, as the mean placebo response markedly increased in the second half of the enrollment period. They suggest this may be a result of pressuring sites to enroll more patients and at a quicker pace to reach sponsor deadlines and thus site may loosen their standards, choosing less than ideal patients that may exhibit a larger placebo effect.

8   Motor PD Treatment Implications Based on RCTs in the Past Decade

The 30 reviewed phase III RCTs testing drugs to improve motor symptoms in PD have established high quality evidence for new treatments that allow better control of the initial and later stages of disease. These trials have established the superiority or noninferiority of extended release forms of well-established dopamine agonists such as ropinirole and pramipexole. New extended release formulations of levodopa are the focus of ongoing clinical trials and the phase III trials with IPX066 have allowed drug approval and commercialization in some countries [24, 29]. For patients with motor fluctuations, two new drugs have demonstrated to reduce off time considerably: opicapone by mean 2.0 h [19] and safinamide by mean 1.4 h [12] which amounts to 3.5 h more on time for these patients, well above the MIC established for the Hauser diary [59]. There is currently no approved drug for the treatment of dyskinesias in PD and the recent trials with Amantadine ER [31, 32] showing a decrease in the total UDysRS of 20.7 and 15.9 points respectively, should facilitate steps in that direction.

On the other hand, the past decade has seen a number of negative trials published, overcoming the possible publication bias. After positive Phase II RCTs [60, 61], the phase III trials for two adenosine A2a receptor antagonists: istradefylline and preladenant, showed disappointing results. The phase II RCT investigating the effect of perampanel (a 5-AMPA receptor antagonist) did not show encouraging results [62] and neither did the phase III trial [20]. Two large RCTs with CoEnzQ10 [21] and caffeine [22] respectively did not show motor benefit of either intervention in PD. To date, no therapies have been tested at the phase III level to treat axial symptoms such as balance and gait impairment which remain unresolved problems and a challenge for future investigation.

9 Conclusions and Recommendations for Future Clinical Trials

When designing or reading a RCT study the clinician should peruse every step of the methodology, bearing in mind the following recommendations: (1) Is the selection of patients adequate for the drug that is being tested? We encourage future trials to take advantage of the updated MDS diagnostic criteria for PD to reduce error in patient enrollment as much as possible [1]. (2) Are appropriate methods used to reduce measurement bias, with adequate randomization and masking of investigators, patients as well as outcome analyzers. (3) Does the trial use the most adequate design (parallel group, delayed start, crossover) and is the time of intervention and washout/follow up period sufficient and adequately explained. (4) Is the treatment protocol available and does it establish equal treatment for all patients? (5) Are the appropriate scales used to measure the primary outcome? In order to determine clinical relevance, we suggest to use a clinician-based scale such as the MDS UPDRS and patient reported outcomes such as quality of life scales. Future RCTs may also benefit from the use of electronic diaries and wearable devices made up of sensors/accelerometers to provide objective data on patient movement that will certainly aid precise measurement of motor symptomatology [64]. (6) Is the placebo effect considered with appropriate randomization, masking, management of expectations, and conditioning as well as balanced patient and center selection to minimize this physiological response? Further research is needed in this area to improve clinical trial designs and control for this powerful effect in PD. Furthermore, a critical appraisal of statistical methods and reporting of results is also recommendable and the reader can use the checklist provided in the Quality of Evidence Scale provided in Table 2. (7) Are appropriate statistical methods used to avoid selection biases with intention to treat or some kind of sensitivity analysis performed? (8) Are results properly reported, including the treatment effect size, relative change, and MIC that will determine whether the benefit is clinically relevant? And finally, after careful assessment of all the above items, (9) does the RCT help to make a clinical treatment decision? The direction of outcomes should not penalize the quality of evidence as communication of negative results is also important for the advancement of pharmacological therapies in PD.

References

1. Postuma RB, Berg D, Stern M et al (2015) MDS clinical diagnostic criteria for Parkinson's disease. Mov Disord 30:1591–1601
2. Fox SH, Katzenschlager R, Lim S-Y et al (2018) International Parkinson and movement disorder society evidence-based medicine review: update on treatments for the motor symptoms of Parkinson's disease. Mov Disord 33(8):1248–1266
3. Task Force of the Movement Disorders Society, Lang AE, Lees A (2002) Management of

Parkinson's disease: an evidence-based review. Mov Disord 17(Suppl 4):S1–S6

4. Castro Caldas A, Levin J, Djaldetti R et al (2017) Critical appraisal of clinical trials in multiple system atrophy: toward better quality. Mov Disord 32:1–9

5. Mizuno Y, Hattori N, Kondo T et al (2017) A randomized double-blind placebo-controlled phase III trial of selegiline monotherapy for early Parkinson disease. Clin Neuropharmacol 40:201–207

6. Hughes AJ, Daniel SE, Kilford L, Lees AJ (1992) Accuracy of clinical diagnosis of idiopathic Parkinson's disease: a clinico-pathological study of 100 cases. J Neurol Neurosurg Psychiatry 55:181–184

7. Pahwa R, Stacy MA, Factor SA et al (2007) Ropinirole 24-hour prolonged release: randomized, controlled study in advanced Parkinson disease. Neurology 68:1108–1115

8. Hauser RA, Schapira AHV, Rascol O et al (2010) Randomized, double-blind, multicenter evaluation of pramipexole extended release once daily in early Parkinson's disease. Mov Disord 25:2542–2549

9. Poewe W, Rascol O, Barone P et al (2011) Extended-release pramipexole in early Parkinson disease: a 33-week randomized controlled trial. Neurology 77:759–766

10. Schapira AHV, Barone P, Hauser RA et al (2011) Extended-release pramipexole in advanced Parkinson disease: a randomized controlled trial. Neurology 77:759–766

11. Mizuno Y, Yamamoto M, Kuno S et al (2012) Efficacy and safety of extended-versus immediate-release pramipexole in Japanese patients with advanced and L-dopa-undertreated Parkinson disease: a double-blind, randomized trial. Clin Neuropharmacol 35:174–181

12. Zhang Z, Wang J, Zhang X et al (2013) The efficacy and safety of ropinirole prolonged release tablets as adjunctive therapy in Chinese subjects with advanced Parkinson's disease: a multicenter, double-blind, randomized, placebo-controlled study. Park Relat Disord 19:1022–1026

13. Hauser RA, Hsu A, Kell S et al (2013) Extended-release carbidopa-levodopa (IPX066) compared with immediate-release carbidopa-levodopa in patients with Parkinson's disease and motor fluctuations: a phase 3 randomised, double-blind trial. Lancet Neurol 12:346–356

14. Stocchi F, Hsu A, Khanna S et al (2014) Comparison of IPX066 with carbidopa-levodopa plus entacapone in advanced PD patients. Park Relat Disord 20:1335–1340

15. Pahwa R, Tanner CM, Hauser RA et al (2017) ADS-5102 (Amantadine) extended-release capsules for levodopa-induced dyskinesia in Parkinson disease (EASE LID study). JAMA Neurol 74:941–949

16. Oertel W, Eggert K, Pahwa R et al (2017) Randomized, placebo-controlled trial of ADS-5102 (amantadine) extended-release capsules for levodopa-induced dyskinesia in Parkinson's disease (EASE LID 3). Mov Disord 32:1701–1709

17. Watts RL, Jankovic J, Waters C et al (2007) Randomized, blind, controlled trial of transdermal rotigotine in early Parkinson disease. Neurology 68:272–276

18. Schapira AHV, McDermott MP, Barone P et al (2013) Pramipexole in patients with early Parkinson's disease (PROUD): a randomised delayed-start trial. Lancet Neurol 12:747–755

19. Olanow CW, Rascol O, Hauser R et al (2009) A double-blind, delayed-start trial of rasagiline in Parkinson's disease. N Engl J Med 361:1268–1278

20. Stocchi F, Borgohain R, Onofrj M et al (2012) A randomized, double-blind, placebo-controlled trial of safinamide as add-on therapy in early Parkinson's disease patients. Mov Disord 27:106–112

21. Schapira AHV, Stocchi F, Borgohain R et al (2013) Long-term efficacy and safety of safinamide as add-on therapy in early Parkinson's disease. Eur J Neurol 20:271–280

22. Borgohain R, Szasz J, Stanzione P et al (2014) Two-year, randomized, controlled study of safinamide as add-on to levodopa in mid to late Parkinson's disease. Mov Disord 29:1273–1280

23. Schapira AHV, Fox SH, Hauser RA et al (2017) Assessment of safety and efficacy of safinamide as a levodopa adjunct in patients with Parkinson disease and motor fluctuations. JAMA Neurol 74:216–224

24. Hauser RA, Shulman LM, Trugman JM et al (2008) Study of istradefylline in patients with Parkinson's disease on levodopa with motor fluctuations. Mov Disord 23:2177–2185

25. Mizuno Y, Hasegawa K, Kondo T et al (2010) Clinical efficacy of istradefylline (KW-6002) in Parkinson's disease: a randomized, controlled study. Mov Disord 25:1437–1443

26. Pourcher E, Fernandez HH, Stacy M et al (2012) Istradefylline for Parkinson's disease patients experiencing motor fluctuations: results of the KW-6002-US-018 study. Park Relat Disord 18:178–184

27. Hauser RA, Stocchi F, Rascol O et al (2015) Preladenant as an adjunctive therapy with

levodopa in Parkinson disease: two randomized clinical trials and lessons learned. JAMA Neurol 72:1491–1500

28. Stocchi F, Rascol O, Hauser RA et al (2017) Randomized trial of preladenant, given as monotherapy, in patients with early Parkinson disease. Neurology 88:2198–2206

29. Ferreira JJ, Lees A, Rocha JF et al (2016) Opicapone as an adjunct to levodopa in patients with Parkinson's disease and end-of-dose motor fluctuations: a randomised, double-blind, controlled trial. Lancet Neurol 15:154–165

30. Lees AJ, Ferreira J, Rascol O et al (2017) Opicapone as adjunct to levodopa therapy in patients with Parkinson disease and motor fluctuations a randomized clinical trial. JAMA Neurol 74:197–206

31. Lees A, Fahn S, Eggert KM et al (2012) Perampanel, an AMPA antagonist, found to have no benefit in reducing "off" time in Parkinson's disease. Mov Disord 27:284–288

32. Investigators PSGQ, Beal MF, Oakes D et al (2014) A randomized clinical trial of high-dosage coenzyme Q10 in early Parkinson disease: no evidence of benefit. JAMA Neurol 71:543–552

33. Postuma RB, Anang J, Pelletier A et al (2017) Caffeine as symptomatic treatment for Parkinson disease (Café-PD): a randomized trial. Neurology 89:1795–1803. https://doi.org/10.1212/WNL.0000000000004568

34. Sedgwick P (2014) Explanatory trials versus pragmatic trials. BMJ 349:g6694

35. Scientific Advisory Committee of the Medical Outcomes Trust (2002) Assessing health status and quality of life instruments: attributes and review criteria. Qual Life Res 11:193–205

36. Wyrwich KW, Wolinsky FD (2000) Identifying meaningful intra-individual change standards for health- related quality of life measures. J Eval Clin Pract 6:39–49

37. Revicki D, Hays RD, Cella D, Sloan J (2008) Recommended methods for determining responsiveness and minimally important differences for patient-reported outcomes. J Clin Epidemiol 61:102–109

38. Beaton DE, Bombardier C, Katz JN, Wright JG (2001) A taxonomy for responsiveness. J Clin Epidemiol 54:1204–1217

39. Cohen J (1988) Statistical power analysis for the behavioral sciences, 2nd edn. Erlbaum, Hillsdale, NJ

40. Martinez-Martin P, Reddy P, Katzenschlager R et al (2015) EuroInf: a multicenter comparative observational study of apomorphine and levodopa infusion in Parkinson's disease. Mov Disord 30:510–516

41. Dafsari HS, Reker P, Stalinski L et al (2018) Quality of life outcome after subthalamic stimulation in Parkinson's disease depends on age. Mov Disord 33:99–107

42. Juhász A, Deli G, Aschermann Z et al (2017) How efficient is subthalamic deep brain stimulation in reducing dyskinesia in Parkinson's disease? Eur Neurol 77:281–287

43. Mokkink LB, Terwee CB, Patrick DL et al (2010) The COSMIN study reached international consensus on taxonomy, terminology, and definitions of measurement properties for health-related patient-reported outcomes. J Clin Epidemiol 63:737–745

44. Jaeschke R, Singer J, Guyatt G (1989) Measurement of health status. Ascertaining the minimal clinically important difference. Control Clin Trials 10:407–415

45. Wyrwich KW, Norquist JM, Lenderking WR, Acaster S (2013) Methods for interpreting change over time in patient-reported outcome measures. Qual Life Res 22:475–483

46. Hauser RA, Gordon MF, Mizuno Y et al (2014) Minimal clinically important difference in Parkinson's disease as assessed in pivotal trials of pramipexole extended release. Park Dis 2014:467131

47. Stefaniak JD, Lam TCH, Sim NE et al (2017) Discontinuation and non-publication of neurodegenerative disease trials: a cross-sectional analysis. Eur J Neurol 24:1071–1076

48. Hopewell S, Loudon K, Clarke MJ et al (2009) Publication bias in clinical trials due to statistical significance or direction of trial results. Cochrane Database Syst Rev: MR000006

49. Richens A (2001) Proof of efficacy trials: cross-over versus parallel-group. Epilepsy Res 45:43–47

50. De La Fuente-Fernández R, Schulzer M, Stoessl AJ (2004) Placebo mechanisms and reward circuitry: clues from Parkinson's disease. Biol Psychiatry 56:67–71

51. Shetty N, Friedman JH, Kieburtz K et al (1999) The placebo response in Parkinson's disease. Parkinson Study Group. Clin Neuropharmacol 22:207–212

52. Goetz CG, Leurgans S, Raman R, Stebbins GT (2000) Objective changes in motor function during placebo treatment in PD. Neurology 54:710–714

53. Goetz CG, Wuu J, McDermott MP et al (2008) Placebo response in Parkinson's disease: comparisons among 11 trials covering medical and surgical interventions. Mov Disord 23:690–699

54. Goetz CG (2005) Movement disorders: understanding clinical trials. Lancet Neurol 4:5–6

55. de la Fuente-Fernández R, Schulzer M, Stoessl AJ (2002) The placebo effect in neurological disorders. Lancet Neurol 1:85–91

56. Mischley LK, Lau RC, Shankland EG et al (2017) Phase IIb study of intranasal glutathione in Parkinson's disease. J Parkinsons Dis 7:289–299

57. Frisaldi E, Carlino E, Zibetti M et al (2017) The placebo effect on bradykinesia in Parkinson's disease with and without prior drug conditioning. Mov Disord 32:1474–1478

58. Benedetti F, Frisaldi E, Carlino E et al (2016) Teaching neurons to respond to placebos. J Physiol 594:5647–5660

59. Shin CW, Hahn S, Park BJ et al (2016) Predictors of the placebo response in clinical trials on Parkinson's disease: a meta-analysis. Park Relat Disord 29:83–89

60. Hauser RA, Deckers F, Lehert P (2004) Parkinson's disease home diary: further validation and implications for clinical trials. Mov Disord 19:1409–1413

61. Hauser RA, Cantillon M, Pourcher E et al (2011) Preladenant in patients with Parkinson's disease and motor fluctuations: a phase 2, double-blind, randomised trial. Lancet Neurol 10:221–229

62. LeWitt PA, Guttman M, Tetrud JW et al (2008) Adenosine A $_{2A}$ receptor antagonist is tradefylline (KW-6002) reduces "off" time in Parkinson's disease: a double-blind,

randomized, multicenter clinical trial (6002-US-005). Ann Neurol 63:295–302

63. Eggert K, Squillacote D, Barone P et al (2010) Safety and efficacy of perampanel in advanced Parkinson's disease: a randomized, placebo-controlled study. Mov Disord 25:896–905

64. Perry B, Herrington W, Goldsack JC et al (2018) Use of mobile devices to measure outcomes in clinical research, 2010–2016: a systematic literature review. Digit Biomarkers 2:11–30

65. Shulman LM, Gruber-Baldini AL, Anderson KE et al (2010) The clinically important difference on the unified Parkinson's disease rating scale. Arch Neurol 67:64–70

66. Hauser RA, Auinger P, Parkinson Study Group (2011) Determination of minimal clinically important change in early and advanced Parkinson's disease. Mov Disord 26:813–818

67. Horváth K, Aschermann Z, Ács P et al (2015) Minimal clinically important difference on the motor examination part of MDS-UPDRS. Parkinsonism Relat Disord 21:1421–1426

68. Horváth K, Aschermann Z, Kovács M et al (2017) Minimal clinically important differences for the experiences of daily living parts of movement disorder society-sponsored unified Parkinson's disease rating scale. Mov Disord 32:789–793

69. Mestre TA, Beaulieu-Boire I, Aquino CC et al (2015) What is a clinically important change in the Unified Dyskinesia Rating Scale in Parkinson's disease? Parkinsonism Relat Disord 21:1349–1354

Chapter 5

Clinical Trials for Motor Complications in Parkinson's Disease

Tiago A. Mestre, Joaquim J. Ferreira, and Olivier Rascol

Abstract

The development of motor fluctuations with the dopaminergic treatment response together with treatment-associated dyskinesia is inevitable in Parkinson's disease (PD) and determine disability and reduction in quality of life. Currently, there are many options for the management of motor complications in PD namely oral pharmacological interventions for both motor fluctuations (various formulation of levodopa/carbidopa, catechol-O-methyl transferase inhibitors, use of agents with a longer half-life such as dopamine agonists or the inhibitors of the monoaminoxidase type B) and levodopa-associated dyskinesia (amantadine, clozapine). Device-aided therapies (deep brain stimulation, apomorphine subcutaneous pump, and levodopa/carbidopa intestinal gel infusion) are options generally used for more severe motor complications refractory to oral pharmacological options. Therapeutic development in PD for the treatment of motor complications has focused mostly in the symptomatic treatment of established motor complications, namely, for the reduction of daily off-time duration, peak-dose dyskinesia, followed by the acute rescue of an off-time. A significant number of clinical trials has been conducted to inform the evidence-based management of motor complications in PD. We describe the pivotal clinical trials that led to the current evidence-based recommendations for the treatment of motor complications and present the clinical trials conducted to explore the therapeutic benefit of other neurotransmitter systems for the treatment of motor fluctuations such as adenosine 2A antagonist, AMPA glutamate receptor antagonists, compounded dopamine, noradrenergic and serotonin 5-HT1A agonism, and for levodopa-associated dyskinesia, metabotropic glutamate receptor antagonists, among others. An important aspect of the management of motor complications is the ability to delay motor fluctuations or levodopa-related dyskinesia in PD, which has been evaluated with different outcomes using dopamine agonists and different formulations of levodopa/carbidopa with or without entacapone. The development of disease-modifying therapies or non-dopaminergic symptomatic treatment may change the nature, history, and management of motor complications in PD.

Key words Parkinson's disease, Motor fluctuations, Dyskinesia, Motor complications, Treatment

1 Motor Complications in Parkinson's Disease

Parkinson's disease (PD) is a neurodegenerative disorder diagnosed clinically based on the mandatory presence of bradykinesia together with resting tremor or rigidity complemented by support-

Santiago Perez-Lloret (ed.), *Clinical Trials In Parkinson's Disease*, Neuromethods, vol. 160,
https://doi.org/10.1007/978-1-0716-0912-5_5, © Springer Science+Business Media, LLC, part of Springer Nature 2021

ive and exclusionary features [1]. Symptomatic pharmacological treatments target for the most part the hypodopaminergic state associated with the presence of parkinsonism, although nonmotor symptoms have gained more interest and relevance in the care of patients with PD [2]. Classically, dopaminergic treatments are associated with an initial dramatic and improvement of the various features of parkinsonism, namely, for levodopa. The hallmark of this initial honeymoon period is a stable and long-lasting response to levodopa, the long-duration response (LDR), despite the half-life of levodopa being 60–90 min [3]. With the progressive dopaminergic neuronal loss, there is a reduced buffering capacity between doses of levodopa and greater dependency on the timing of dopaminergic medications. These changes together with a chronic pulsatile dopaminergic stimulation are the main factors for patients with PD starting to experience changes in the antiparkinsonian response to individual doses of levodopa generating a fluctuation of the therapeutic response throughout the day (motor fluctuations) [4]. Also, patients start to experience involuntary movements (mostly in the form of chorea and/or dystonia) most commonly time-locked with the occurrence of an antiparkinsonian response, the levodopa-related dyskinesia (LRD). The term "motor complications" is used to include both motor fluctuations and LRD.

1.1 Epidemiology

Virtually, all PD patients experience motor complications at some point of their clinical disease [5]. In a meta-analysis, the cumulative incidence of motor fluctuations and LRD was 40.8% and 36.2%, respectively, at 4–6 years of clinical disease [6]. The duration of motor fluctuations and severity of LRD correlates with health-related quality of life [7]. The occurrence of either motor fluctuations or LRD determines the earlier occurrence of the other [8]. Clinical trials have provided insight into the development of motor complications. The first trial that tested formally levodopa in a placebo-controlled study design for PD patients at an early stage without need for symptomatic treatment showed that motor complications can develop as early as 5–6 months after starting medication, and a higher dose (e.g., 600 mg daily *vs.* 300 mg/d) was associated with a higher incidence of wearing-off (29.7% vs. 18.2%) and LRD (16.5% vs. 2.3%) after only 40 weeks of treatment [9]. More recently, a placebo-controlled study assessed a potential disease-modifying effect of levodopa/carbidopa with a delayed start design with negative efficacy outcome measures, namely, change in UPDRS score from baseline to end of treatment (primary) and rate of progression using the UPDRS [10]. In a clinical trial conducted in early PD and comparing levodopa/carbidopa with levodopa/carbidopa/entacapone [11] as initial monotherapy, young age at onset, higher levodopa daily dose, low body weight, female gender, and more severe baseline parkinsonian disability were predictive factors of LRD incidence [12]. More recently, a case–control study of levodopa initiation in PD patients showed that both a higher daily dose and longer disease duration was associated with a higher incidence of motor complications, but not the duration of treatment with levodopa [13].

1.2 Phenomenology

Motor complications have a rich and complex clinical presentation. The most common form of motor fluctuations is the predictable wearing-off occurring at the end of a levodopa dose. A variation of a predictable wearing-off is the reemergence of parkinsonian symptoms during the night (nocturnal akinesia) or in the morning after waking from sleep (morning akinesia). Other forms include the unpredictable wearing-off characterized by the emergence of parkinsonian symptoms unrelated with the timing of levodopa intake, extremely sudden and severe "super-offs," a delayed-on response, or dose failure when the intake of levodopa is not associated with a clinical benefit. More rarely, patients can repeatedly switch between an on- and off-state, the "yo-yoing" phenomenon (for review, *see* ref. 14). Dyskinesia are classified according to its occurrence in relation with the dosing of levodopa and its phenomenology. The most common are peak-dose dyskinesia occurring at a period of maximum antiparkinsonian effect of levodopa. Other types of dyskinesia include diphasic dyskinesia occurring closer to the timing of levodopa dosing (either before or soon after) and off-time dystonia occurring concomitantly with the presence of parkinsonism [14], and frequently involving the feet.

1.3 Current Treatment Principles for Motor Complications in Parkinson's Disease

The current pharmacological approach to motor fluctuations is based on the principle of "continuous" dopamine stimulation attempting to compensate for a shortening LDR. This therapeutic principle is implemented in clinical practice in multiple ways: (1) increase in frequency of dosing of levodopa; (2) improving absorption of levodopa by taking it on an empty stomach, using carbonated water; (3) lengthening the duration of an individual response of levodopa using a catechol-O-methyl transferase inhibitor (COMT-i), (4) use of novel formulation of levodopa that allows for a slower release of levodopa and a longer half-life such as the extended-release levodopa/carbidopa capsule IPX066. Other options are (1) the use of less powerful dopaminergic agents with a longer half-life and distinct side effect profiles such as dopamine agonists (DAs) or the inhibitors of the monoaminoxidase type B (MAOB-i). It is important to recognize that the addition of a dopaminergic drug carries the risk of emergence or worsening of peak-dose dyskinesia. The pharmacological treatment of peak-dose dyskinesia is limited almost to the use of amantadine. Other strategies can be considered, such as the reduction of levodopa together with an increase of a dopamine agonist and, less frequently, the use of clozapine due to the required regular hematological workup for the risk of neutropenia. When pharmacological options are insufficient to manage motor complications, device-aided therapies are considered and include deep brain stimulation (DBS), apomorphine subcutaneous pump [15] and levodopa/carbidopa intestinal gel infusion [16]. A recent evidence-based medicine (EBM) review of treatment options available for Parkinson's disease has been released by the International Parkinson's disease and Movement Disorders Society (MDS) [17]. The EBM-MDS review appraised the data from randomized controlled trials (RCTs) in which the primary endpoint measured motor symptoms and used an established rating scale or well-described outcome. The following pharmacological interventions were considered "*efficacious*" for motor fluctuations: the nonergot DA pramipexole, ropinirole, rotigotine, apomorphine, the ergot DA pergolide, the standard and extended release formulation of levodopa/carbidopa, the levodopa/carbidopa intestinal gel infusion, the COMT-is entacapone, tolcapone and opicapone, the MAOB-i rasagiline, and the MAOB-i *plus* zonisamide and safinamide (For an overview of surgical interventions, *see* Subheading 5). In the same review, the following pharmacological interventions were considered "*efficacious*" for LRD: amantadine and clozapine. Also, pramipexole and cabergoline are "*efficacious*" in delaying motor fluctuations compared to levodopa, while pramipexole, ropinirole, and cabergoline have also been considered "*efficacious*" for delaying LRD. These conclusions reflect the results of about 120 higher-quality RCTs since last century up to 2017, demonstrating the vitality of therapeutic development in PD.

The optimization of quality of life ultimately guides symptomatic management in early PD. When treating the individual patient, the potential delaying effect of some DAs for motor complications needs to be balanced with the fact that levodopa is better at improving parkinsonian disability and quality of life, and is better tolerated. Currently, a combined approach of levodopa and DA avoiding the use of high doses of either drug is preferred. In a pragmatic, open-label RCT 1620 patients newly diagnosed PD were randomly assigned to a levodopa-sparing therapy (DAs or MAOB-i) with the option of levodopa supplementation, or levodopa alone. At a 3-year median follow-up, only minimal benefits (lasting up to 7 years) were observed for mobility scores in a patient-reported health-related quality of life scale developed for PD (PDQ39) in the group of patients initially assigned to levodopa monotherapy. This difference was below the minimal clinically difference initially defined in the study.

2 Overview of Trial Design for Motor Complications in Parkinson's Disease

Drug development in PD for the treatment of motor complications has focused mostly in the symptomatic treatment of established motor complications and most frequently target the reduction of daily off-time duration, followed by the acute rescue of an off-time. The latter can include morning akinesia or isolated disabling wearing-off periods. Regarding LRD, the few clinical trials directly assessing this indication have targeted mostly peak-dose dyskinesia.

The clinical trials for motor fluctuations are more commonly parallel in design and controlled with placebo. The most commonly used efficacy outcome measure is the reduction of daily off-time, for the most part, measured using patient-report either through a PD diary or the sect. IV of the Unified Parkinson Disease Rating Scale (UPDRS) [18] or its newer version the Movement Disorder Society—UPDRS [19]. PD diaries typically require for the patient to rate its functional state every 30 min regarding awake vs. sleep, and if awake whether in an off-time, on-time, and on-time with LRD. It is also required to record the timing of the PD medications intake. PD diaries are demanding for patients, and its validity as a measurement tool to quantify motor fluctuations has been questioned [20, 21]. The minimal clinically important change for off-time reduction has been calculated to be of 1 hour using PD diaries and an anchor-based approach [22]. In the MDS-UPDRS, both predictable and unpredictable wearing-off is quantified as a percentage of the waking time, and its functional impact is rated.

The trials conducted to assess the efficacy of an experimental intervention for the acute rescue of off-time are usually placebo-controlled, and there is no active comparison with levodopa, for example. Both parallel and crossover study design have been adopted. The most commonly used efficacy outcomes are the temporal change in parkinsonian disability after the administration of the study drug, time from drug administration to on-time, percentage of rescued off-time episodes, duration and severity of off-time. Patients can be tested in a controlled environment where an acute challenge of the intervention is done in a patient in a pragmatically defined off-time or naturally occurring off-time in the patient's routine daily life and assessed using on-off diaries. Repeat dosing is allowed.

In clinical trials conducted for LRD, a few rating scales have been used as outcomes. These include "older" scales such as the Abnormal Involuntary Movement Scale (AIMS), the Rush Dyskinesia Rating Scale (RDRS), the Lang–Fahn Activities of Daily Living Dyskinesia Rating Scale (LF), the 26-Item Parkinson's Disease Dyskinesia scale (PDD-26), the part IV of the UPDRS and the more recent Unified Dyskinesia Rating Scale (UDysRS) [23]. The UDysRS, the LF, and the PDD-26 have shown to be responsive to treatment [24]. The UDysRS had the highest effect size [24]. These clinical rating scales have been comprehensively assessed by the MDS [23].

The placebo effect is recognizably strong and prevalent in PD [25]. Even in PD patients not requiring symptomatic treatment, a placebo group reported LRD and wearing-off after 40 weeks, with an incidence similar to 300 mg/day of levodopa [9]. Older age, a lower baseline parkinsonism score, and a lower total daily levodopa dose were associated with placebo-associated improvement for LRD [26]. The assessment of the efficacy and safety of device-aided therapies is more challenging to control for the placebo effect, due to ethical concerns for a sham intervention and high risk of unblinding. For example, in DBS an ideal placebo would be sham stimulation but the easy "unblinding" related with acute side effects of neurostimulation and the complex management of patients with PD postoperatively with concomitant modifications of neurostimulation and dopaminergic medication are of difficult (if not impossible) implementation in a clinical trial [27]. Consequently, clinical trials in DBS have used as comparator a best medical treatment group and the studies have been open label. Another example of complexity in the administration of the intervention is the evaluation of levodopa/carbidopa intestinal gel infusion [16]. To reduce the risk of unblinding, the authors adopted a double-dummy double-titration design, in which patients were randomly allocated to levodopa/carbidopa intestinal gel infusion/ oral placebo or placebo infusion/oral levodopa/carbidopa, with subsequent titration of each active drug/placebo interventions. In transplantation surgical trials, a "sham" intervention has been used with more successful masking of study participants patients and raters, but still was associated with a significant placebo effect [28].

Another set of trials assesses the delaying effect of an intervention for motor complications. These trials are conducted in early PD patients starting symptomatic treatment for parkinsonian disability and time to onset of motor complications is the commonly adopted efficacy outcome measure. The rationale for these trials is that the nonphysiological pulsatile dopaminergic stimulation with levodopa leads to the development of motor complications via postsynaptic dopamine receptor sensitization [29]. Consequently, the use of dopaminergic drugs with a longer half-life than levodopa will provide a more continuous dopaminergic stimulation [4] and delay the onset of motor complications. Typically, these trials have compared DAs with levodopa and have a long duration of follow-up, usually 3–5 years. This hypothesis also led to other continuous drug delivery methods including patches, subcutaneous/jejunal pumps and novel formulations of dopamine agonists with prolonged release.

3   Overview of Pharmacological Clinical Trials for Motor Fluctuations in Parkinson's Disease

3.1   Levodopa

Although changes in the regimen of regular levodopa/carbidopa or levodopa/benserazide are used in clinical practice, there are no clinical trials conducted to specifically assess the efficacy and tolerability of levodopa in the context of motor fluctuations. Available data were collected in clinical trials controlled with levodopa. The development of novel oral formulations of levodopa is an expanding field in the pharmacological management of PD. Examples of alternative formulation of levodopa approved for clinical use are the controlled-release levodopa/carbidopa [30], the modified-release levodopa/benserazide [31], and the extended-release levodopa/carbidopa capsule IPX066 [32].

Examples of other alternative formulations evaluated in the past or under development include (1) melevodopa, a rapid-acting methyl-ester of levodopa [33, 34], (2) a solution formulation of L-dopa, levodopa ethyl ester [35], the gastro-retentive formulation DM-1992 [36] with bilayer of immediate-release and extended-release form of levodopa/carbidopa, (3) the dispersible formulation of levodopa/benserazide [37], (4) the Accordion Pill with gastroretentive features [38], (5) XP21279, a prodrug of levodopa that is absorbed from the gut by high-capacity nutrient transporters available along the gut including the colon [39, 40], (6) ODM-101 is a carbidopa-enriched formulation of levodopa/carbidopa/entacapone that demonstrated an improved pharmacokinetic profile of levodopa compared with Stalevo, (7) ONO-2160 is a recently developed levodopa pro-drug designed to reduce the fluctuations of levodopa plasma levels [41].

3.2   Dopamine Agonists

DAs were extensively studied for the treatment of motor fluctuations and a delaying effect for the onset of motor complications when used preferentially to levodopa in earlier stages of the clinical disease (see below). DAs are divided in ergot and nonergot compounds. The ergot compounds have fallen out of clinical use due to safety concerns of retroperitoneal and cardiac valve fibrosis identified in the postmarketing phase [42]. For this reason, in this chapter we will focus on the demonstrated the efficacy and safety for the most commonly used DAs: pramipexole, ropinirole and rotigotine. A Cochrane review on pramipexole for motor fluctuations in PD [43] reported a significant reduction in off-time with a weighted mean difference (WMD) of 1.8 h; 95% CI = 1.2, 2.3 between pramipexole and placebo, after evaluating the data of four double-blind RCT trials with 669 patients with PD and motor complications. Although no significant changes were found in a dyskinesia rating scale, LRD were reported more often with pramipexole. A levodopa dose reduction was also documented with pramipexole (WMD: 115 mg; 87, 143 95% CI) [43]. In the case of ropinirole, there was a concern for a lack of consistent benefit in reducing off-time and an increased risk of LRD in an early 2001 Cochrane review [44]. Since then, additional trial results demonstrated that ropinirole has a role in the reduction of off-time [17]. More recently, extended-release formulations of pramipexole [45, 46] and ropinirole [47, 48] became available that allow for a single daily administration with an overall control of motor fluctuations and tolerability similar to the corresponding immediate-release formulations. Higher dose of prolonged release ropinirole may allow for the maximum tolerated doses to be higher compared with immediate release formulation (18.6 ± 6.5 mg/day vs. 10.4 ± 6.4 mg/day), which in turn was associated with a favorable odds-ratio (OR) for ≥20% reduction in off-time (adjusted OR: 1.82; 95% CI: 1.16, 2.86; p = 0.009) [49]. Rotigotine is the most recent DAs approved for the treatment of motor fluctuations. It has the advantage of a single daily transdermal administration. In a placebo-controlled RCT, rotigotine was associated with a reduction in off-time of 1.8 h/day (8 mg/day) and 1.2 h/day (12 mg/day). Application site reactions are specific to rotigotine and include erythema and pruritus [50]. The extended-release formulation of ropinirole and the rotigotine 24 h-patch has been associated with better sleep quality [51, 52] and early morning motor function before any PD medication dosing, in placebo-controlled trials. Overnight switch between DAs is possible, and equivalence ratios have been developed to aid in clinical practice [53–55].

3.3 MAO Inhibitors   Selegiline and rasagiline are the two MAOB-is assessed for the treatment of motor fluctuations, thought there is insufficient evidence in the case of selegiline [17]. Rasagiline has the most substantial evidence with the two pivotal LARGO [56] and PRESTO [57] trials. Both trials were placebo-controlled double-blind RCTs assessing the efficacy of rasagiline in PD patients with motor fluctuations not optimally treated with levodopa, with or without a DA or COMT-i for a period of 18 [56] and 26 [57] weeks (total number of participants = 1159). The reduction in mean daily off-time was of 0.78–0.94 h with 1 mg/day of rasagiline, with a corresponding increase in on-time without troublesome LRD. LRD were significantly increased with rasagiline 1 mg/day, but not with rasagiline 0.5 mg/day [57]. Common adverse events were weight loss, vomiting, and anorexia in the rasagiline 1 mg/day group [57]. In the past, other MAOB-is have been evaluated such as lazabemide but yielded negative efficacy results [58].

More recently, zonisamide and safinamide have been assessed for motor fluctuations in PD. These compounds are considered MAOB-i *plus* compounds due to ancillary actions in other neurotransmitter systems including glutamate release inhibition, antagonism to calcium and sodium channels, although the clinical relevance of the latter two putative mechanisms of safin-amide in PD are less certain. Zonisamide has been approved in Japan for the treatment of motor complications [59] after a phase III multicenter placebo-controlled RCT involving 422 patients and testing zonisamide 25 and 50 mg/day. Only zonisamide 50 mg reduced the daily off-time by about 0.7 h as per patient diary. Somnolence was more frequent with zonisamide (zonisamide 25 mg/day: 3.1%; zonisamide 50 mg/day: 6.3%) than in the pla-cebo group. EMA and FDA approved safinamide for the adjunc-tive treatment to levodopa/carbidopa for off-time following the results of two RCTs [60, 61]. One large multicenter, multinational trial recruited 549 patients with PD and a daily off-time > 1.5 h (excluding morning akinesia) with a target safinamide dose of 100 mg/day. The mean change from baseline to week 24 in daily on-time without troublesome LRD was 0.96 h better than com-pared with placebo [60]. The most frequently reported adverse event was LRD: 14.6% (safinamide group) vs. 5.5% (placebo group) [60]. Tesofensine, a triple MAO inhibitor, was tested in a phase 2 placebo-controlled RCT with modest improvements in off-time (reduction in 68 min daily) at a dose of 0.25 mg/day only without a dose–response relationship. Adverse drug reactions tended to be more frequent at higher dosages [62].

3.4 COMT-Inhibitors

Tolcapone, entacapone and opicapone are the three COMT-is that have been approved for the management of motor fluctuations in PD. Tolcapone was the first COMT-i available. In a Cochrane review, [63] a total of 6 placebo-controlled RCTs (1006 patients) were included assessing the safety and efficacy of tolcapone 50–400 mg in a three-daily dose regimen. Efficacy results were in general consistent across doses. At the most commonly used dose of 200 mg/day three times daily, there was a reduction of mean daily off-time of approximately 1.5 h, and a reduction of the levodopa dose by approximately 150 mg/day. The most common side effects were LRD, nausea, vomiting, and diarrhea. Although in these trials only a few participants had an elevation of liver enzyme levels, the report of potentially fatal, acute fulminant liver failure in the postmarketing phase, led to tolcapone being discon-tinued. The very few patients still taking this medication require ongoing monitoring of liver function. Entacapone is the most widely used COMT-i. In the same Cochrane review [63], a total of 8 placebo-controlled RCTs (1560 patients) were included assess-ing the safety and efficacy of entacapone 200 mg with each dose of levodopa/carbidopa. Available efficacy results show there was a reduction of mean daily off-time of approximately 41 min and an on-time increase of approximately 1 h. The side effect profile in clinical trials was similar to tolcapone, without the risk of a serious hepatic adverse event. More recently, a new COMT-i, opicapone, has been approved for the treatment of motor fluctuations in PD.

Opicapone is the only COMT-i available with single daily dosing. The confirmatory study was a randomized, double-blind, both placebo- and active-controlled trial of opicapone as an adjunct to levodopa/carbidopa in patients with PD with end-of-dose motor fluctuations, comparing oral treatment with opicapone (5 mg, 25 mg, or 50 mg/day), entacapone (200 mg with every levodopa intake) and placebo for 14–15 weeks [64]. Treatment with opicapone 50 mg was the only dose that was both superior to placebo (mean difference in change from baseline: −60.8 min, 95% CI: −97.2; −24.4; p = 0.0015) and noninferior to entacapone (−26.2 min, 95% CI: −63.8; 11.4; p = 0.0051) [64]. The most common adverse events were LRD (opicapone 50 mg: 15.6%) [64].

3.5 Exploring the Therapeutic Benefit of Other Neurotransmitter Systems for the Treatment of Motor Fluctuations

Nondopaminergic neurotransmitter systems have been explored for its therapeutic potential in the treatment of motor fluctuations and include drugs that act on adenosine, glutamate, and serotonin systems.

3.5.1 Adenosine 2A Antagonist

Three A_{2A} receptor antagonist have been developed for PD: istradefylline, preladenant, and tozadenant. Istradefylline has been studied more extensively [65] and is the only compound in this drug class approved for treatment of PD patients experiencing wearing-off phenomena although only in Japan. One of the studies that supported regulatory approval was a double-blind, placebo-controlled RCT conducted in 373 patients with motor complications that showed a significant reduction in off-time of 0.99 h (p < 0.003, Istradefylline 20 mg/day) and 0.96 h (p < 0.003, Istradefylline 40 mg/day) compared with placebo (−0.23 h). The most common adverse event was LRD (placebo: 4.0%; istradefylline 20 mg/day: 13.0%; istradefylline 40 mg/day: 12.1%) [66]. Studies performed outside Japan have reported mixed efficacy results [67, 68]. The FDA issued a letter of nonapproved in the past. A new application is expected with new results of a recently completed study (NCT01968031). Preladenant is the second A_{2A} receptor antagonist developed for PD. The initial suggestion of efficacy for the reduction of off-time in a placebo-controlled phase IIb trial in PD patients [69] was not confirmed in two 12-week, phase III, randomized, placebo-controlled, double-blind trials assessing preladenant in multiple doses from 2 mg to 10 mg twice daily in 778 PD patients [70]. Interestingly, in one of these trials, rasagiline 1 mg/day was not superior to placebo for the reduction of off-time. In virtue of the known efficacy of rasagiline for the treatment of motor fluctuations, the investigators suggested that a high magnitude placebo-associated improvement, recruitment of less ideal patients in competitive recruitment, and the less than optimal reliability of completion of PD diaries could have contributed for these findings [70]. There are no ongoing or registered trials with Preladenant. Tozadenant is the third A_{2A} receptor antagonist developed for PD and a single phase 2b, double-blind, placebo-controlled RCT has been published with a reduction of approximately 1 h in mean daily off-time compared with placebo for tozadenant 240–360 mg/day [71]. The most common side effect was LRD. A phase III study has been completed but results have not been published (NCT02453386). Agranulocytosis resulting in death led to the suspension of drug development.

3.5.2 Other Compounds

Pardoprunox, a mixed effect compound with partial dopamine agonism, and noradrenergic and serotonin 5-HT_{1A} agonism, was recently evaluated for the treatment of motor fluctuations. Pardoprunox was studied in two placebo-controlled RCTs [72, 73] in PD patients with motor fluctuations with mixed efficacy results. A high drop-out rate was also observed due to adverse events. Perampanel, an antagonist of the AMPA glutamate receptors, was studied for the reduction of daily off-time in three RCTs [74–76] that documented a lack of a therapeutic benefit for this indication.

Another important indication assessed in the clinical trials for motor complications is the acute rescue of off-time either in the context of a predictable wearing-off or on-off episodes. Interventions assessed in clinical trials include the DAs apomorphine and piribedil [77], levodopa (CVT-301) [78] using routes of administration with faster absorption. Apomorphine has been tested more consistently in subcutaneous, intranasal, sublingual [79, 80], and rectal formulations [81]. The FDA and EMA approved the subcutaneous formulation of apomorphine for off-time rescue. For approval, three studies were considered. Participants had a mean duration of PD of approximately 11 years, and at least 2 h of off-time. All patients took levodopa in a variable combination most often with an oral dopaminergic agonist, and less frequently with a MAOB-i or COMT-i. The initial study was conducted in PD patients naïve to subcutaneous apomorphine and included an inpatient and 1-month outpatient phases. The percentage reduction in the UPDRS motor score was 62% with apomorphine compared with 1% after a placebo injection. Interestingly, the percentage of rescue of off-time episodes in an outpatient setting was 95% for apomorphine and 23% for placebo [82]. In the two additional trials that included patients with at least a 3-month experience of apomorphine before study enrollment, efficacy results were similar in terms of improvement from baseline in the UPDRS motor score. The most common adverse events were LRD, yawning, drowsiness, nausea, and orthostatic hypotension, and injection site reactions. To prevent nausea, patients used trimethobenzamide typically 3 days before starting the subcutaneous apomorphine, and half of them discontinued treatment on average 2 months later [83]. More recently, a sublingual formulation of apomorphine is under development, with results of initial phase II study published [84]. Based on results not yet published of a phase III study (CTH-300, NCT02469090), a new drug application was submitted to the FDA. The development of the inhaled formulation of levodopa (CVT-301) led to the approval by the FDA and it is currently available in the USA. The confirmatory study was a randomized, double-blind, placebo-controlled, phase III study of 351 patients assigned to receive two different doses of CVT-301 (60 mg: $n = 115$; 84 mg: $n = 120$) or placebo ($n = 116$). The mean between-group difference in the UPDRS motor score change from predose to 30 min postdose (primary efficacy outcome measure) was -3.92 (95% CI: -6.84; 1.00; $p = 0.0088$) between the CVT-301 84 mg and placebo groups [85]. There are no active studies of piribedil for this indication.

4   Overview of Pharmacological Clinical Trials in Levodopa-Related Dyskinesia in Parkinson's Disease

4.1 NMDA
Antagonists

Amantadine is a well-known nonselective NMDA antagonist currently recommended for LRD. There have been multiple small clinical trials to assess the efficacy and tolerability of Amantadine for the treatment of LRD [24, 86–92]. A systematic review conducted in 2003 that included most of these studies denoted significant methodological weaknesses including lack of randomization or a comparator group, a small sample size, cross-over design without a washout period [93]. The largest placebo-controlled RCT of amantadine (up to a daily dose of 300 mg) involved 68 participants and demonstrated a therapeutic benefit in various dyskinesia rating scales [24]. More recently, two extended-release formulations of amantadine with single daily dosing were approved by the FDA. ADS-5102 has been assessed for efficacy and tolerability in clinical trials up to 12 weeks [94–96] with open-label extensions with a mean follow-up of about 1 year [97]. This formulation allows for one daily dosing at bedtime with a slow initial rise and delayed time to maximum concentration (12–16 h) of amantadine in plasma [98]. Interestingly, the extended-release formulation of amantadine showed a consistent effect in reducing LRD severity in the Unified Dyskinesia Rating Scale (10.1 points, 95% CI: 13.8; 6.5) and a reduction in off-time with a treatment difference for placebo of 1.00 h daily (95% CI: 1.57, 0.44) [99]. Amantadine HCl extended-release (combined with an immediate-release) with a single daily dose has been evaluated in two Phase III RCTs (NCT02153632, NCT02153645). The results led to FDA approval, but have not yet been published in a peer-review journal. Other NMDA antagonists assessed for the treatment of LRD include Dextromethorphan [100, 101], an NR2B subunit selective MK-0657 [102], memantine [103, 104] with either negative or inconsistent efficacy results.

4.2 Metabotropic Glutamate Receptor Antagonists

Mavoglurant (AFQ056) is a selective metabotropic glutamate receptor 5 (mGLuR5) antagonist assessed in PD patients with LRD. A total of four studies have been conducted with this compound. Despite initial signs of efficacy with a dose–antidyskinetic response relationship and a more robust effect with Mavoglurant 200 mg daily [105], two placebo-controlled double-blind RCTs failed to document an improvement in LRD compared with placebo in the modified AIMS. An immediate-release (100 mg twice daily, Study 1) and modified-release (150–200 mg once daily, Study 2) were evaluated. Dipraglurant (ADX48621) is another mGLu5 antagonist assessed in a phase IIa double-blind, placebo-controlled RCT involving 76 PD patients. Dipraglurant reduced peak dose dyskinesia (modified AIMS) on day 14 (100 mg, 32%; $p = 0\ 0.04$) and across a 3-h postdose period on day 14 ($p = 0.04$), without worsening parkinsonism [106]. There are no ongoing studies with Dipraglurant.

4.3 Dopamine Antagonism

Clozapine is considered the most atypical of neuroleptic drugs, with action on D_1 and D_2 dopamine receptors, but also D_4 dopamine receptors and antagonism to serotonin, muscarinic, adrenergic, and histamine receptor systems. A single double-blind, placebo-controlled RCT confirmed prior observations in open-label studies. In this trial [107], clozapine (mean daily dose: 39.4 mg) was associated with a reduction in the daily duration of LRD of about 2 h compared with placebo, without a significant worsening of parkinsonism. There was no case of agranulocytosis. Three cases of eosinophilia were reported with reversibility after discontinuing clozapine. Quetiapine has been assessed in a placebo-controlled blinded study for LRD [108] with negative results.

4.4 Other Compounds

Many other compounds have been tested for its efficacy in treating LRD targeting different neurotransmitters including 5-HT_{1A} receptors, α2-adrenergic receptor antagonist; cannabinoids receptors, histamine receptors, ionotropic glutamate receptors, GABAergic receptors and opiate receptors, among others (*see* Table 1 for further details).

5 Clinical Trials of Device-Aided Therapies for Motor Complications in Parkinson's Disease

The progression of the disease in PD may lead to disabling motor complications for which available medical oral treatments are insufficient for its management. In the last decades, more effective (but more invasive) treatment modalities were successfully developed and considered an option in clinical practice at tertiary PD centers. These treatment options are grouped as device-aided therapies

Table 1
Compounds investigated for levodopa-related dyskinesia (LRD) in Parkinson's disease currently not approved (only randomized controlled trials with published results are included)

	Drug class	Studies	Efficacy (LRD)	Safety
Sarizotan	*Selective 5-HT$_{1A}$ R agonist, D$_2$ R antagonist*	DP BC [123, 124]	Potencial improvement of LRD [123]	Increase in off-time duration [123]
Fipamezole	*α2-adrenergic R antagonist*	DB PC [125]	Negative	Blood pressure elevation
Idazoxan		DB PC [126], DB PC CO [127]	No effect on LRD [126] Improvement of LRD [127]	Blood pressure elevation
Eltoprazine	*Partial 5-HT$_{1A/B}$ R agonists, nicotinic Ach R agonists*	DB PC [128]	Improvement of LRD	Nausea, dizziness
Buspirone	*5-HT$_{1A}$ R agonist, dopamine mixed agonist/antagonist*	DB PC CO [129]	Improvement of LRD	No concerns
Famotidine	*Histamine R antagonist*	DB PC CO [130]	Negative	No concerns
Cannabis sativa extract (THC/ cannabidiol)	*Cannabinoid R agonist*	DB PC CO [131]	Negative [131]	No concerns
Nabilone	*Cannabinoid R agonist*	DB PC CO [132]	Improvement in LRD [132]	No concerns
Levetiracetam	*Synaptic vesicle glycoprotein modulator*	DB PC CO [133], DB PC [134]	Improvement in LRD [133, 134]	No concerns
Topiramate	*Modulator of AMPA receptors*	DB PC CO [135]	Worsening of LRD	Cognition, dry eyes/mouth
Sodium valproate	*Inhibit GABAergic tone*	DB PC CO [136]	No change in LRD	No concerns
AQW051	*Nicotinic Ach α7 R agonist*	DB PC [137]	No change in LRD	Dyskinesia
NP002	*Nicotinic Ach R agonist*	DB PC	No change in LRD	Nausea
Deanol	*Acetylcholine precursor*	DB PC CO [138]	No change in LRD	No concerns
Naltrexone	*Opiate R antagonist*	DB PC CO [139]	No change in LRD	No concerns

(continued)

Table 1
(continued)

	Drug class	Studies	Efficacy (LRD)	Safety
Naloxone	*Nonselective opioid R antagonist*	DB PC CO [140]	No change in LRD Longer on-time	No concerns
Naftazone	*Glutamate release inhibitor*	Multiple n-of-1 [141] DB PC CO [142]	Reduction in troublesome LRD [141] No change in LRD [142]	No concerns
Remacemide	*NMDA R antagonist*	DB PC [143]	No change in LRD	Dizziness, gastrointestinal discomfort
SR 27897B	*Cholecystokinin-A receptor inhibitor*	DB PC [144]	No change in LRD	No concerns
Perindopril	*ACE inhibitor*	DB PC CO [145]	Improvement in LRD	No concerns
17Beta-estradiol	*Estrogens*	DB PC CO [146]	No change in LRD	No concerns

DB double-blind, *PC* placebo-controlled, *CO* crossover, *CS* cross-sectional, *Ach* acetylcholine, *GABA* γ-aminobutyric acid, *R* receptor, *ACE* angiotensin-converting enzyme

since all of them require the use of a device, either a pump (levodopa/carbidopa continuous intestinal gel infusion, and apomorphine subcutaneous infusion), or an impulse generator coupled to a lead system placed in the brain in the case of DBS. Currently, a subcutaneous infusion of levodopa/carbidopa is being evaluated in clinical trials (NCT04006210, NCT03033498). Other experimental therapies have been assessed such as pallidotomy, cell transplantation treatment with human fetal mesencephalic cells or porcine fetal cells. Bilateral DBS of the subthalamic nucleus (STN), bilateral Globus pallidus *pars interna* (GPi), levodopa/carbidopa continuous intestinal gel infusion have been recommended as *efficacious* and the apomorphine subcutaneous infusion *likely efficacious* [17].

5.1 Deep Brain Stimulation

Multiple trials have been conducted to assess the safety and efficacy of DBS to manage motor complications in PD [17, 109]. Commonly, two targets are considered to treat parkinsonism and/ or LRD: STN and GPi. Another target, the nucleus *ventralis intermedius* of the thalamus (Vim) is used exclusively to treat refractory limb tremor in PD. Among the studies with higher methodological quality is the German Study group trial of bilateral STN-DBS compared with best medical treatment for 6 months [110] in patients under 75 years of age with severe motor complications of PD (mean age, 60 years; mean duration of disease, 13.4 years). The primary endpoints of the trial were the changes from baseline to end of treatment regarding quality of life assessed by the Parkinson's Disease Questionnaire (PDQ-39), and severity of symptoms without medication measured by the UPDRS-III. Bilateral STN-DBS was associated with greater improvements in the PDQ-39 and UPDRS-III with a mean improvement of 9.5 and 19.6 points, respectively [110]. Serious adverse events were more common with neurostimulation arm than with medication alone arm (13% vs. 4%, $p < 0.04$) and included a fatal intracerebral hemorrhage. The overall frequency of adverse events was higher in the medication group (64% vs. 50%, $p = 0.08$). The effect of bilateral STN-DBS has been explored at an earlier stage of PD patients (mean age, 52 years; mean duration of disease, 7.5 years) when patient start to experience impairment in their daily function related with motor complications occurring for less than 3 years [111]. After 2 years, bilateral STN-DBS was superior to best medical treatment concerning quality of life ($p = 0.0002$), activities of daily living ($p < 0.001$), for time with good mobility and no LRD ($p = 0.01$), and levodopa-induced motor complications ($p < 0.001$), among other outcomes. Although, we recognize that these studies are amongst the ones with the better methodologically rigor, they cannot address concerns about the placebo effect and the ancillary effect related with the open-label use of best medical treatment (lessebo-like effect) which have been discussed elsewhere [27, 112, 113].

5.2 Levodopa/Carbidopa Continuous Intestinal Gel Infusion

The development of levodopa/carbidopa continuous intestinal gel infusion generated the adoption of an interesting study design that could control for placebo and nocebo effect related both with the experimental intervention and the reduction/stoppage of oral daytime dopaminergic medication after the start of this device-aided therapy. In the confirmatory north-american trial of levodopa/carbidopa continuous intestinal gel infusion [16], a randomized, double-blind, double-dummy, double-titration design was adopted. In this particular study design, each participant was allocated to the "real" experimental drug, the levodopa/carbidopa intestinal gel plus placebo pills (dummy #1), or to immediate-release oral levodopa/carbidopa and placebo intestinal gel infusion (dummy #2). The adopted primary endpoint was the "classical" change from baseline to end of treatment in off-time, and the pre-specified key secondary outcome was changed in on-time without troublesome LRD. After 12 weeks of treatment, the mean off-time reduction difference was −1.91 h [95% CI: −3.05 to −0.76], $p = 0.0015$) with advantage for the levodopa/carbidopa intestinal gel plus placebo pills study arm, which had converted almost entirely to on-time without troublesome LRD. Safety issues were mainly associated with the percutaneous gastrojejunostomy tube and tended to occur more frequently in the first month. The two most commonly used targets for DBS have been tested in a few trials. The most recent NSTAPS study [114] enrolled 128 patients (GPi DBS, $n = 65$; STN DBS, $n = 63$) After 12 months, there was no statistically significant difference the quality of life, and in the number of patients with cognitive, mood, and behavioral side effects. There was a more favorable outcome for bilateral STN-DBS in the off-medication stimulation-on state and levodopa equivalent dose reduction [114].

5.3   Apomorphine Subcutaneous Continuous Infusion

Apomorphine subcutaneous pump has been available in a few countries for decades, with multiple clinical studies conducted showing an effect in reduction of off time but also LRD [115]. Notwithstanding, the results of the single double-blind, parallel RCT of this treatment was only recently reported [15]. The study involved 107 PD patients with more than 3 years of disease and refractory motor fluctuations [15]. Apomorphine infusion (mean final dose: 4.68 mg/h) significantly reduced mean daily off-time compared with placebo by 1.9 h (95% CI: 3.16 to 0.62; $p = 0.0025$) after 12 weeks with a corresponding increase in on-time without troublesome LRD ($p = 0.00008$) [15]. The report of adverse events corresponded to the known safety profile of apomorphine subcutaneous administration. The most common side effects were skin nodules (44%), followed by nausea and somnolence (both 22%) [15]. In the apomorphine group, treatment-related adverse events were the cause for withdrawal in five out of the six patients that dropped out [15]. The head-to-head comparisons of device-aided therapies are rare. In an open-label, nonrandomized trial, apomorphine subcutaneous continuous infusion and levodopa/carbidopa intestinal gel infusion were associated to similar benefits for motor symptoms, quality of life but the improvement was different in the various nonmotor features [116].

6   Clinical Trials to Delay Motor Fluctuations or Levodopa-Related Dyskinesia in Parkinson's Disease

Clinical trials have been conducted having as outcome the onset of motor fluctuations and/or LRD. In the pivotal trial of pramipexole using as comparator levodopa, initial pramipexole treatment resulted in a significant reduction in the risk of developing LRD (24.5% vs 54%; $p < 0.001$) and wearing-off (47% vs 62.7%; $p = 0.002$) [117]. Carbegoline is the DAs with the longest half-life and requires a single daily administration. In a double-blind levodopa-controlled RCT [118] after 5 years, motor complications were less frequent in cabergoline-treated patients than in levodopa-treated patients (22.3% vs. 33.7%, $p = 0.0175$). The difference between cabergoline-treated and levodopa-treated patients was more striking for the onset of LRD (cabergoline group: 9.5% vs. levodopa group: 21.2%, $p < 0.001$). Levodopa-treated patients had a greater improvement in the UPDRS motor score over time than cabergoline-treated patients. For example, at 5 years the score was 16.3 vs. 19.2, ($p < 0.001$) respectively. The landmark trial that allowed for ropinirole to be recommended for delaying LRD was a 5-year trial double-blind RCT controlled with levodopa monotherapy [119]. This trial demonstrated that ropinirole monotherapy was associated with a lower risk for LRD compared with levodopa (hazard ratio: 6.67; 95% confidence interval [CI], 3.23–14.29; $p < 0.001$). After the addition of supplemental levodopa, the incidence of LRD was the same in ropinirole and levodopa groups (HR, 0.80; 95% CI, 0.48–1.33; $p = 0.39$). At the end of the study, the mean ± SD daily doses were 16.5 ± 6.6 mg of ropinirole (about 330 ± 132 mg of levodopa equivalent daily dose) plus 427 ± 221 mg of levodopa in the ropinirole group, and 753 ± 398 mg of levodopa in the levodopa group. The ultimate advantage to first start with ropinirole monotherapy was a delay of onset of LRD by 3 years compared with levodopa monotherapy. A differential impact on delaying LRD or motor fluctuations was also observed in PD-MED describe earlier in this chapter, with a hazard ratio of 1.52, 95% CI 1.16–2.00, $p = 0.003$) in favor of the levodopa monotherapy group compared with the levodopa sparing group. Other DAs that have been assessed for the same indication with less robust results include bromocriptine, pergolide, piribedil, and lisuride [120].

Different formulations of levodopa/carbidopa with variable pharmacokinetic of levodopa delivery have been tested for the same indication. As described earlier in the chapter, the ELLDOPA study [9] provided clear evidence that the daily dose of levodopa is important to determine the time to development of motor complications. The study STRIDE-PD [11] was a placebo-controlled RCT of entacapone combined with levodopa/carbidopa that assessed the time to onset of LRD over 134 weeks. The group taking entacapone had a shorter time to onset of LRD (HR, 1.29; $p = 0.04$) and increased frequency (42% vs 32%; $p = 0.02$). This group had a greater levodopa daily equivalent dose compared with the levodopa/carbidopa group (396.8 and 306.8 mg, respectively, $p < 0.001$). The modified-release formulation of levodopa/benserazide also failed to show a delaying effect on motor complications [31]. A pilot study of DBS enrolled patients with PD without motor fluctuations or LRD [121]. After 2 years, while the duration of off-time and/or LRD worsened in the best medical treatment group compared with the group undergoing STN DBS, the mean difference was above the reported minimally important difference but lacked statistical significance, most likely due to a high intragroup variability in outcome data [122].

7   Future Views

Motor complications in Parkinson's disease continue to be a landmark in the natural history of the disease that leads frequently to disability and the need for invasive therapies such as DBS or Levodopa/carbidopa continuous intestinal gel infusion. In the future, the development of disease-modifying therapies represents the greatest challenge in PD that can theoretically reduce the incidence of motor complications. Until we reach this new era in PD management, efforts should be made to prevent the development of motor complications, with an increase in translational knowledge of the pathophysiology of motor fluctuations and dyskinesia. The development of methods that allow for a continuous administration of a dopaminergic treatment that are less invasive and less costly than the device-aided therapies currently available may allow for its use earlier in the disease.

References

1. Postuma RB, Berg D, Stern M, Poewe W, Olanow CW, Oertel W, Obeso J, Marek K, Litvan I, Lang AE, Halliday G, Goetz CG, Gasser T, Dubois B, Chan P, Bloem BR, Adler CH, Deuschl G (2015) MDS clinical diagnostic criteria for Parkinson's disease. Mov Disord 30(12):1591–1601. https://doi.org/10.1002/mds.26424

2. Seppi K, Weintraub D, Coelho M, Perez-Lloret S, Fox SH, Katzenschlager R, Hametner EM, Poewe W, Rascol O, Goetz CG, Sampaio C (2011) The movement disorder society evidence-based medicine review update: treatments for the non-motor symptoms of Parkinson's disease. Mov Disord 26 Suppl 3:S42–S80. https://doi.org/10.1002/mds.23884

3. Nutt JG, Holford NH (1996) The response to levodopa in Parkinson's disease: imposing pharmacological law and order. Ann Neurol 39(5):561–573. https://doi.org/10.1002/ana.410390504

4. Nutt JG (2007) Continuous Dopaminergic stimulation: is it the answer to the motor complications of levodopa. Mov Disord 22(1):1–9. https://doi.org/10.1002/mds.21060

5. Hely MA, Morris JGL, Reid WGJ, Trafficante R (2005) Sydney multicenter study of Parkinson's disease: non-L-dopa-responsive problems dominate at 15 years. Mov Disord 20(2):190–199

6. Ahlskog JE, Muenter MD (2001) Frequency of levodopa-related dyskinesias and motor fluctuations as estimated from the cumulative literature. Mov Disord 16(3):448–458

7. Perez-Lloret S, Negre-Pages L, Damier P, Delval A, Derkinderen P, Destee A, Meissner WG, Tison F, Rascol O (2017) L-DOPA-induced dyskinesias, motor fluctuations and health-related quality of life: the COPARK survey. Eur J Neurol 24(12):1532–1538. https://doi.org/10.1111/ene.13466

8. Hauser RA, McDermott MP, Messing S (2006) Factors associated with the development of motor fluctuations and dyskinesias in Parkinson disease. Arch Neurol 63(12):1756–1760. https://doi.org/10.1001/archneur.63.12.1756

9. Fahn S, Oakes D, Shoulson I, Kieburtz K, Rudolph A, Lang A, Olanow CW, Tanner C, Marek K, Group PS (2004) Levodopa and the progression of Parkinson's disease. N Engl J Med 351(24):2498–2508. https://doi.org/10.1056/NEJMoa033447

10. CVM V, Suwijn SR, Boel JA, Post B, Bloem BR, van Hilten JJ, van Laar T, Tissingh G, Munts AG, Deuschl G, Lang AE, MGW D, de Haan RJ, de Bie RMA, Group LS (2019) Randomized delayed-start trial of levodopa in Parkinson's disease. N Engl J Med 380(4):315–324. https://doi.org/10.1056/NEJMoa1809983

11. Stocchi F, Rascol O, Kieburtz K, Poewe W, Jankovic J, Tolosa E, Barone P, Lang AE, Olanow CW (2010) Initiating levodopa/carbidopa therapy with and without entacapone in early Parkinson disease: the STRIDE-PD study. Ann Neurol 68(1):18–27. https://doi.org/10.1002/ana.22060

12. Warren Olanow C, Kieburtz K, Rascol O, Poewe W, Schapira AH, Emre M, Nissinen H, Leinonen M, Stocchi F (2013) Factors predictive of the development of Levodopa-induced

dyskinesia and wearing-off in Parkinson's disease. Mov Disord 28(8):1064–1071. https://doi.org/10.1002/mds.25364

13. Cilia R, Akpalu A, Sarfo FS, Cham M, Amboni M, Cereda E, Fabbri M, Adjei P, Akassi J, Bonetti A, Pezzoli G (2014) The modern pre-levodopa era of Parkinson's disease: insights into motor complications from sub-Saharan Africa. Brain 137(Pt 10):2731–2742. https://doi.org/10.1093/brain/awu195

14. Aquino CC, Fox SH (2015) Clinical spectrum of levodopa-induced complications. Mov Disord 30(1):80–89. https://doi.org/10.1002/mds.26125

15. Katzenschlager R, Poewe W, Rascol O, Trenkwalder C, Deuschl G, Chaudhuri KR, Henriksen T, van Laar T, Spivey K, Vel S, Staines H, Lees A (2018) Apomorphine subcutaneous infusion in patients with Parkinson's disease with persistent motor fluctuations (TOLEDO): a multicentre, double-blind, randomised, placebo-controlled trial. Lancet Neurol 17(9):749–759. https://doi.org/10.1016/s1474-4422(18)30239-4

16. Olanow CW, Kieburtz K, Odin P, Espay AJ, Standaert DG, Fernandez HH, Vanagunas A, Othman AA, Widnell KL, Robieson WZ, Pritchett Y, Chatamra K, Benesh J, Lenz RA, Antonini A, Group LHS (2014) Continuous intrajejunal infusion of levodopa-carbidopa intestinal gel for patients with advanced Parkinson's disease: a randomised, controlled, double-blind, double-dummy study. Lancet Neurol 13(2):141–149. https://doi.org/10.1016/S1474-4422(13)70293-X

17. Fox SH, Katzenschlager R, Lim SY, Barton B, de Bie RMA, Seppi K, Coelho M, Sampaio C (2018) International Parkinson and movement disorder society evidence-based medicine review: update on treatments for the motor symptoms of Parkinson's disease. Mov Disord. https://doi.org/10.1002/mds.27372

18. Fahn S, Elton R, Committee. MotUD (1987) The unified Parkinson's disease rating scale. In: Fahn SMC, Calne DB, Goldstein M (eds) Recent developments in Parkinson's disease, vol 2. Macmillan Health Care Information, Florham Park, NJ, pp 153–163

19. Goetz CG, Tilley BC, Shaftman SR, Stebbins GT, Fahn S, Martinez-Martin P, Poewe W, Sampaio C, Stern MB, Dodel R, Dubois B, Holloway R, Jankovic J, Kulisevsky J, Lang AE, Lees A, Leurgans S, LeWitt PA, Nyenhuis D, Olanow CW, Rascol O, Schrag A, Teresi JA, van Hilten JJ, LaPelle N, Force MDSURT (2008) Movement disorder society-sponsored revision of the unified Parkinson's disease rating scale (MDS-UPDRS): scale presentation and clinimetric testing results. Mov

20. Reimer J, Grabowski M, Lindvall O, Hagell P (2004) Use and interpretation of on/off diaries in Parkinson's disease. J Neurol Neurosurg Psychiatry 75(3):396–400

21. Papapetropoulos S (2012) Patient diaries as a clinical endpoint in Parkinson's disease clinical trials. CNS Neurosci Ther 18(5):380–387. https://doi.org/10.1111/j.1755-5949.2011.00253.x

22. Hauser RA, Auinger P, Group PS (2011) Determination of minimal clinically important change in early and advanced Parkinson's disease. Mov Disord 26(5):813–818. https://doi.org/10.1002/mds.23638

23. Colosimo C, Martínez-Martín P, Fabbrini G, Hauser RA, Merello M, Miyasaki J, Poewe W, Sampaio C, Rascol O, Stebbins GT, Schrag A, Goetz CG (2010) Task force report on scales to assess dyskinesia in Parkinson's disease: critique and recommendations. Mov Disord 25(9):1131–1142. https://doi.org/10.1002/mds.23072

24. Goetz CG, Stebbins GT, Chung KA, Hauser RA, Miyasaki JM, Nicholas AP, Poewe W, Seppi K, Rascol O, Stacy MA, Nutt JG, Tanner CM, Urkowitz A, Jaglin JA, Ge S (2013) Which dyskinesia scale best detects treatment response? Mov Disord 28(3):341–346. https://doi.org/10.1002/mds.25321

25. Goetz CG, Wuu J, McDermott MP, Adler CH, Fahn S, Freed CR, Hauser RA, Olanow WC, Shoulson I, Tandon PK, Leurgans S, Group PS (2008) Placebo response in Parkinson's disease: comparisons among 11 trials covering medical and surgical interventions. Mov Disord 23(5):690–699. https://doi.org/10.1002/mds.21894

26. Goetz CG, Laska E, Hicking C, Damier P, Müller T, Nutt J, Warren Olanow C, Rascol O, Russ H (2008) Placebo influences on dyskinesia in Parkinson's disease. Mov Disord 23(5):700–707. https://doi.org/10.1002/mds.21897

27. Mestre TA, Lang AE, Okun MS (2016) Factors influencing the outcome of deep brain stimulation: placebo, nocebo, lessebo, and lesion effects. Mov Disord 31(3):290–296. https://doi.org/10.1002/mds.26500

28. Gross RE, Watts RL, Hauser RA, Bakay RA, Reichmann H, von Kummer R, Ondo WG, Reissig E, Eisner W, Steiner-Schulze H, Siedentop H, Fichte K, Hong W, Cornfeldt M, Beebe K, Sandbrink R (2011) Intrastriatal transplantation of microcarrier-bound human retinal pigment epithelial cells versus sham surgery in patients with advanced Parkinson's disease: a double-blind, randomised, controlled trial. Lancet Neurol 10(6):509–519. https://doi.org/10.1016/s1474-4422(11)70097-7

29. Olanow W, Obeso JA, Stocchi F (2006) Drug insight: continuous dopaminergic stimulation in the treatment of Parkinson's disease. Nat Clin Pract Neurol 2(7):382–392. https://doi.org/10.1038/ncpneuro0222

30. Juncos JL, Fabbrini G, Mouradian MM, Chase TN (1987) Controlled release levodopa-carbidopa (CR-5) in the management of parkinsonian motor fluctuations. Arch Neurol 44(10):1010–1012. https://doi.org/10.1001/archneur.1987.00520220016008

31. Dupont E, Andersen A, Boas J, Boisen E, Borgmann R, Helgetveit AC, Kjaer MO, Kristensen TN, Mikkelsen B, Pakkenberg H, Presthus J, Stien R, Worm-Petersen J, Buch D (1996) Sustained-release Madopar HBS compared with standard Madopar in the long-term treatment of de novo parkinsonian patients. Acta Neurol Scand 93(1):14–20

32. Hauser RA, Hsu A, Kell S, Espay AJ, Sethi K, Stacy M, Ondo W, O'Connell M, Gupta S, investigators IA-P (2013) Extended-release carbidopa-levodopa (IPX066) compared with immediate-release carbidopa-levodopa in patients with Parkinson's disease and motor fluctuations: a phase 3 randomised, double-blind trial. Lancet Neurol 12(4):346–356. https://doi.org/10.1016/S1474-4422(13)70025-5

33. Stocchi F, Zappia M, Dall'Armi V, Kulisevsky J, Lamberti P, Obeso JA (2010) Melevodopa/carbidopa effervescent formulation in the treatment of motor fluctuations in advanced Parkinson's disease. Mov Disord 25(12):1881–1887. https://doi.org/10.1002/mds.23206

34. Stocchi F, Vacca L, Grassini P, Pawsey S, Whale H, Marconi S, Torti M (2015) L-Dopa pharmacokinetic profile with effervescent melevodopa/carbidopa versus standard-release levodopa/carbidopa tablets in Parkinson's disease: a randomised study. Parkinson's Dis 2015:369465. https://doi.org/10.1155/2015/369465

35. Djaldetti R, Inzelberg R, Giladi N, Korczyn AD, Peretz-Aharon Y, Rabey MJ, Herishano Y, Honigman S, Badarny S, Melamed E (2002) Oral solution of levodopa ethylester for treatment of response fluctuations in patients with advanced Parkinson's disease. Mov Disord 17(2):297–302

36. Verhagen Metman L, Stover N, Chen C, Cowles VE, Sweeney M (2015) Gastroretentive carbidopa/levodopa, DM-1992, for the treatment of advanced Parkinson's disease. Mov Disord 30(9):1222–1228. https://doi.org/10.1002/mds.26219

37. Monge A, Barbato L, Nordera G, Stocchi F (1997) An acute and long-term study with a dispersible formulation of levodopa/benserazide (Madopar(R)) in Parkinson's disease. Eur J Neurol 4(5):485–490. https://doi.org/10.1111/j.1468-1331.1997.tb00388.x

38. LeWitt PA, Giladi N, Navon N (2019) Pharmacokinetics and efficacy of a novel formulation of carbidopa-levodopa (Accordion Pill). Parkinsonism Relat Disord. https://doi.org/10.1016/j.parkreldis.2019.05.032

39. Lewitt PA, Ellenbogen A, Chen D, Lal R, McGuire K, Zomorodi K, Luo W, Huff FJ (2012) Actively transported levodopa prodrug XP21279: a study in patients with Parkinson disease who experience motor fluctuations. Clin Neuropharmacol 35(3):103–110. https://doi.org/10.1097/WNF.0b013e31824e4d7d

40. LeWitt PA, Huff FJ, Hauser RA, Chen D, Lissin D, Zomorodi K, Cundy KC (2014) Double-blind study of the actively transported levodopa prodrug XP21279 in Parkinson's disease. Mov Disord 29(1):75–82. https://doi.org/10.1002/mds.25742

41. Nomoto M, Nagai M, Nishikawa N, Ando R, Kagamiishi Y, Yano K, Saito S, Takeda A (2018) Pharmacokinetics and safety/efficacy of levodopa pro-drug ONO-2160/carbidopa for Parkinson's disease. eNeurologicalSci 13:8–13. https://doi.org/10.1016/j.ensci.2018.09.003

42. Junghanns S, Fuhrmann JT, Simonis G, Oelwein C, Koch R, Strasser RH, Reichmann H, Storch A (2007) Valvular heart disease in Parkinson's disease patients treated with dopamine agonists: a reader-blinded monocenter echocardiography study. Mov Disord 22(2):234–238. https://doi.org/10.1002/mds.21225

43. Clarke CE, Speller JM, Clarke JA (2000) Pramipexole for levodopa-induced complications in Parkinson's disease. Cochrane Database Syst Rev (3):Cd002261. https://doi.org/10.1002/14651858.cd002261

44. Clarke CE, Deane KH (2001) Ropinirole for levodopa-induced complications in Parkinson's disease. Cochrane Database Syst Rev (1):CD001516. https://doi.org/10.1002/14651858.CD001516

45. Schapira AH, Barone P, Hauser RA, Mizuno Y, Rascol O, Busse M, Salin L, Juhel N, Poewe W (2011) Extended-release pramipexole in advanced Parkinson disease: a randomized controlled trial. Neurology 77(8):767–774. https://doi.org/10.1212/WNL.0b013e31822affdb

46. Shen T, Ye R, Zhang B (2017) Efficacy and safety of pramipexole extended-release in Parkinson's disease: a review based on meta-analysis of randomized controlled trials. Eur J Neurol 24(6):835–843. https://doi.org/10.1111/ene.13303

47. Pahwa R, Stacy MA, Factor SA, Lyons KE, Stocchi F, Hersh BP, Elmer LW, Truong DD, Earl NL, Investigators E-PAS (2007) Ropinirole 24-hour prolonged release: randomized, controlled study in advanced Parkinson disease. Neurology 68(14):1108–1115. https://doi.org/10.1212/01.wnl.0000258660.74391.c1

48. Stocchi F, Giorgi L, Hunter B, Schapira AHV (2011) PREPARED: comparison of prolonged and immediate release ropinirole in advanced Parkinson's disease. Mov Disord 26(7):1259–1265. https://doi.org/10.1002/mds.23498

49. Stocchi F, Giorgi L, Hunter B, Schapira AH (2011) PREPARED: comparison of prolonged and immediate release ropinirole in advanced Parkinson's disease. Mov Disord 26(7):1259–1265. https://doi.org/10.1002/mds.23498

50. LeWitt PA, Lyons KE, Pahwa R, Group SS (2007) Advanced Parkinson disease treated with rotigotine transdermal system: PREFER study. Neurology 68(16):1262–1267. https://doi.org/10.1212/01.wnl.0000259516.61938.bb

51. Chaudhuri KR, Martinez-Martin P, Rolfe KA, Cooper J, Rockett CB, Giorgi L, Ondo WG (2012) Improvements in nocturnal symptoms with ropinirole prolonged release in patients with advanced Parkinson's disease. Eur J Neurol 19(1):105–113. https://doi.org/10.1111/j.1468-1331.2011.03442.x

52. Trenkwalder C, Kies B, Rudzinska M, Fine J, Nikl J, Honczarenko K, Dioszeghy P, Hill D, Anderson T, Myllyla V, Kassubek J, Steiger M, Zucconi M, Tolosa E, Poewe W, Surmann E, Whitesides J, Boroojerdi B, Chaudhuri KR (2011) Rotigotine effects on early morning motor function and sleep in Parkinson's disease: a double-blind, randomized, placebo-controlled study (RECOVER). Mov Disord 26(1):90–99. https://doi.org/10.1002/mds.23441

53. Linazasoro G, Group SDAS (2004) Conversion from dopamine agonists to pramipexole. An open-label trial in 227 patients with advanced Parkinson's disease. J Neurol 251(3):335–339. https://doi.org/10.1007/s00415-004-0328-0

54. Tomlinson CL, Stowe R, Patel S, Rick C, Gray R, Clarke CE (2010) Systematic review of levodopa dose equivalency reporting in Parkinson's disease. Mov Disord 25(15):2649–2653. https://doi.org/10.1002/mds.23429

55. Goetz CG, Blasucci L, Stebbins GT (1999) Switching dopamine agonists in advanced Parkinson's disease: is rapid titration preferable to slow? Neurology 52(6):1227–1229

56. Rascol O, Brooks DJ, Melamed E, Oertel W, Poewe W, Stocchi F, Tolosa E, Group LS (2005) Rasagiline as an adjunct to levodopa in patients with Parkinson's disease and motor fluctuations (LARGO, Lasting effect in Adjunct therapy with Rasagiline Given Once daily, study): a randomised, double-blind, parallel-group trial. Lancet 365(9463):947–954. https://doi.org/10.1016/S0140-6736(05)71083-7

57. Parkinson Study Group (2005) A randomized placebo-controlled trial of rasagiline in levodopa-treated patients with Parkinson disease and motor fluctuations: the PRESTO study. Arch Neurol 62(2):241–248. https://doi.org/10.1001/archneur.62.2.241

58. Parkinson Study Group (1994) A controlled trial of lazabemide (Ro 19-6327) in levodopa-treated Parkinson's disease. Parkinson Study Group. Arch Neurol 51(4):342–347

59. Murata M, Hasegawa K, Kanazawa I, Fukasaka J, Kochi K, Shimazu R (2015) Zonisamide improves wearing-off in Parkinson's disease: a randomized, double-blind study. Mov Disord 30(10):1343–1350. https://doi.org/10.1002/mds.26286

60. Schapira AH, Fox SH, Hauser RA, Jankovic J, Jost WH, Kenney C, Kulisevsky J, Pahwa R, Poewe W, Anand R (2017) Assessment of safety and efficacy of safinamide as a levodopa adjunct in patients with parkinson disease and motor fluctuations: a randomized clinical trial. JAMA Neurol 74(2):216–224. https://doi.org/10.1001/jamaneurol.2016.4467

61. Borgohain R, Szasz J, Stanzione P, Meshram C, Bhatt M, Chirilineau D, Stocchi F, Lucini V, Giuliani R, Forrest E, Rice P, Anand R (2014) Randomized trial of safinamide add-on to levodopa in Parkinson's disease with motor fluctuations. Mov Disord 29(2):229–237. https://doi.org/10.1002/mds.25751

62. Rascol O, Poewe W, Lees A, Aristin M, Salin L, Juhel N, Waldhauser L, Schindler T (2008) Tesofensine (NS 2330), a monoamine reuptake inhibitor, in patients with advanced Parkinson disease and motor fluctuations: the ADVANS Study. Arch Neurol 65(5):577–583. https://doi.org/10.1001/archneur.65.5.577

63. Deane KH, Spieker S, Clarke CE (2004) Catechol-O-methyltransferase inhibitors for levodopa-induced complications in Parkinson's disease. Cochrane Database Syst Rev (4):CD004554. https://doi.org/10.1002/14651858.CD004554.pub2

64. Ferreira JJ, Lees A, Rocha JF, Poewe W, Rascol O, Soares-da-Silva P, investigators B-P (2016) Opicapone as an adjunct to levodopa in patients with Parkinson's disease and end-of-dose motor fluctuations: a randomised,

double-blind, controlled trial. Lancet Neurol 15(2):154–165. https://doi.org/10.1016/S1474-4422(15)00336-1

65. Torti M, Vacca L, Stocchi F (2018) Istradefylline for the treatment of Parkinson's disease: is it a promising strategy? Expert Opin Pharmacother:1–8. https://doi.org/10.1080/14656566.2018.1524876

66. Mizuno Y, Kondo T (2013) Adenosine A2A receptor antagonist istradefylline reduces daily OFF time in Parkinson's disease. Mov Disord 28(8):1138–1141. https://doi.org/10.1002/mds.25418

67. Hauser RA, Hubble JP, Truong DD (2003) Randomized trial of the adenosine A(2A) receptor antagonist istradefylline in advanced PD. Neurology 61(3):297–303

68. Pourcher E, Fernandez HH, Stacy M, Mori A, Ballerini R, Chaikin P (2012) Istradefylline for Parkinson's disease patients experiencing motor fluctuations: results of the KW-6002-US-018 study. Parkinsonism Relat Disord 18(2):178–184. https://doi.org/10.1016/j.parkreldis.2011.09.023

69. Hauser RA, Cantillon M, Pourcher E, Micheli F, Mok V, Onofrj M, Huyck S, Wolski K (2011) Preladenant in patients with Parkinson's disease and motor fluctuations: a phase 2, double-blind, randomised trial. Lancet Neurol 10(3):221–229. https://doi.org/10.1016/s1474-4422(11)70012-6

70. Hauser RA, Stocchi F, Rascol O, Huyck SB, Capece R, Ho TW, Sklar P, Lines C, Michelson D, Hewitt D (2015) Preladenant as an adjunctive therapy with levodopa in Parkinson disease: two randomized clinical trials and lessons learned. JAMA Neurol 72(12):1491–1500. https://doi.org/10.1001/jamaneurol.2015.2268

71. Hauser RA, Olanow CW, Kieburtz KD, Pourcher E, Docu-Axelerad A, Lew M, Kozyolkin O, Neale A, Resburg C, Meya U, Kenney C, Bandak S (2014) Tozadenant (SYN115) in patients with Parkinson's disease who have motor fluctuations on levodopa: a phase 2b, double-blind, randomised trial. Lancet Neurol 13(8):767–776. https://doi.org/10.1016/s1474-4422(14)70148-6

72. Hauser RA, Bronzova J, Sampaio C, Lang AE, Rascol O, Theeuwes A, van de Witte SV, Group PS (2009) Safety and tolerability of pardoprunox, a new partial dopamine agonist, in a randomized, controlled study of patients with advanced Parkinson's disease. Eur Neurol 62(1):40–48. https://doi.org/10.1159/000216839

73. Rascol O, Bronzova J, Hauser RA, Lang AE, Sampaio C, Theeuwes A, van de Witte SV (2012) Pardoprunox as adjunct therapy to levodopa in patients with Parkinson's dis-

ease experiencing motor fluctuations: results of a double-blind, randomized, placebo-controlled, trial. Parkinsonism Relat Disord 18(4):370–376. https://doi.org/10.1016/j.parkreldis.2011.12.006

74. Lees A, Fahn S, Eggert KM, Jankovic J, Lang A, Micheli F, Mouradian MM, Oertel WH, Olanow CW, Poewe W, Rascol O, Tolosa E, Squillacote D, Kumar D (2012) Perampanel, an AMPA antagonist, found to have no benefit in reducing "off" time in Parkinson's disease. Mov Disord 27(2):284–288. https://doi.org/10.1002/mds.23983

75. Rascol O, Barone P, Behari M, Emre M, Giladi N, Olanow CW, Ruzicka E, Bibbiani F, Squillacote D, Patten A, Tolosa E (2012) Perampanel in Parkinson disease fluctuations: a double-blind randomized trial with placebo and entacapone. Clin Neuropharmacol 35(1):15–20. https://doi.org/10.1097/WNF.0b013e318241520b

76. Eggert K, Squillacote D, Barone P, Dodel R, Katzenschlager R, Emre M, Lees AJ, Rascol O, Poewe W, Tolosa E, Trenkwalder C, Onofrj M, Stocchi F, Nappi G, Kostic V, Potic J, Ruzicka E, Oertel W (2010) Safety and efficacy of perampanel in advanced Parkinson's disease: a randomized, placebo-controlled study. Mov Disord 25(7):896–905. https://doi.org/10.1002/mds.22974

77. Rascol O, Azulay JP, Blin O, Bonnet AM, Brefel-Courbon C, Cesaro P, Damier P, Debilly B, Durif F, Galitzky M, Grouin JM, Pennaforte S, Villafane G, Yaici S, Agid Y (2010) Orodispersible sublingual piribedil to abort OFF episodes: a single dose placebo-controlled, randomized, double-blind, crossover study. Mov Disord 25(3):368–376. https://doi.org/10.1002/mds.22922

78. LeWitt PA, Hauser RA, Grosset DG, Stocchi F, Saint-Hilaire MH, Ellenbogen A, Leinonen M, Hampson NB, DeFeo-Fraulini T, Freed MI, Kieburtz KD (2016) A randomized trial of inhaled levodopa (CVT-301) for motor fluctuations in Parkinson's disease. Mov Disord 31(9):1356–1365. https://doi.org/10.1002/mds.26611

79. van Laar T, Neef C, Danhof M, Roon KI, Roos RA (1996) A new sublingual formulation of apomorphine in the treatment of patients with Parkinson's disease. Mov Disord 11(6):633–638. https://doi.org/10.1002/mds.870110607

80. Ondo W, Hunter C, Almaguer M, Gancher S, Jankovic J (1999) Efficacy and tolerability of a novel sublingual apomorphine preparation in patients with fluctuating Parkinson's disease. Clin Neuropharmacol 22(1):1–4

81. van Laar T, Jansen EN, Neef C, Danhof M, Roos RA (1995) Pharmacokinetics and

clinical efficacy of rectal apomorphine in patients with Parkinson's disease: a study of five different suppositories. Mov Disord 10(4):433–439. https://doi.org/10.1002/mds.870100405

82. Dewey RB Jr, Hutton JT, LeWitt PA, Factor SA (2001) A randomized, double-blind, placebo-controlled trial of subcutaneously injected apomorphine for parkinsonian off-state events. Arch Neurol 58(9):1385–1392

83. Hauser RA, Isaacson S, Clinch T (2014) Randomized, placebo-controlled trial of trimethobenzamide to control nausea and vomiting during initiation and continued treatment with subcutaneous apomorphine injection. Parkinsonism Relat Disord 20(11):1171–1176. https://doi.org/10.1016/j.parkreldis.2014.08.010

84. Hauser RA, Olanow CW, Dzyngel B, Bilbault T, Shill H, Isaacson S, Dubow J, Agro A (2016) Sublingual apomorphine (APL-130277) for the acute conversion of OFF to ON in Parkinson's disease. Mov Disord 31(9):1366–1372. https://doi.org/10.1002/mds.26697

85. LeWitt PA, Hauser RA, Pahwa R, Isaacson SH, Fernandez HH, Lew M, Saint-Hilaire M, Pourcher E, Lopez-Manzanares L, Waters C, Rudzínska M, Sedkov A, Batycky R, Oh C, Investigators S-PS (2019) Safety and efficacy of CVT-301 (levodopa inhalation powder) on motor function during off periods in patients with Parkinson's disease: a randomised, double-blind, placebo-controlled phase 3 trial. Lancet Neurol 18(2):145–154. https://doi.org/10.1016/S1474-4422(18)30405-8

86. Wolf E, Seppi K, Katzenschlager R, Hochschorner G, Ransmayr G, Schwingenschuh P, Ott E, Kloiber I, Haubenberger D, Auff E, Poewe W (2010) Long-term antidyskinetic efficacy of amantadine in Parkinson's disease. Mov Disord 25(10):1357–1363. https://doi.org/10.1002/mds.23034

87. Zeldowicz LR, Hubermann J (1973) Long-term therapy of Parkinson's disease with amantadine, alone and combined with levodopa. Can Med Assoc J 109(7):588–593

88. Fahn S, Isgreen WP (1975) Long-term evaluation of amantadine and levodopa combination in parkinsonism by double-blind corssover analyses. Neurology 25(8):695–700

89. Luginger E, Wenning GK, Bösch S, Poewe W (2000) Beneficial effects of amantadine on L-dopa-induced dyskinesias in Parkinson's disease. Mov Disord 15(5):873–878

90. Snow BJ, Macdonald L, Mcauley D, Wallis W (2000) The effect of amantadine on levodopa-induced dyskinesias in Parkinson's disease: a double-blind, placebo-controlled study. Clin Neuropharmacol 23(2):82–85

91. Luginger E, Wenning GK, Bosch S, Poewe W (2000) Beneficial effects of amantadine on L-Dopa-induced dyskinesias in Parkinson's disease. Mov Disord 15(5):873–878

92. Verhagen Metman L, Del Dotto P, van den Munckhof P, Fang J, Mouradian MM, Chase TN (1998) Amantadine as treatment for dyskinesias and motor fluctuations in Parkinson's disease. Neurology 50(5):1323–1326

93. Crosby NJ, Deane KH, Clarke CE (2003) Amantadine for dyskinesia in Parkinson's disease. Cochrane Database Syst Rev (2):CD003467. https://doi.org/10.1002/14651858.CD003467

94. Pahwa R, Tanner CM, Hauser RA, Isaacson SH, Nausieda PA, Truong DD, Agarwal P, Hull KL, Lyons KE, Johnson R, Stempien MJ (2017) ADS-5102 (Amantadine) extended-release capsules for levodopa-induced dyskinesia in Parkinson disease (EASE LID study): a randomized clinical trial. JAMA Neurol 74(8):941–949. https://doi.org/10.1001/jamaneurol.2017.0943

95. Pahwa R, Tanner CM, Hauser RA, Sethi K, Isaacson S, Truong D, Struck L, Ruby AE, McClure NL, Went GT, Stempien MJ (2015) Amantadine extended release for levodopa-induced dyskinesia in Parkinson's disease (EASED study). Mov Disord 30(6):788–795. https://doi.org/10.1002/mds.26159

96. Oertel W, Eggert K, Pahwa R, Tanner CM, Hauser RA, Trenkwalder C, Ehret R, Azulay JP, Isaacson S, Felt L, Stempien MJ (2017) Randomized, placebo-controlled trial of ADS-5102 (amantadine) extended-release capsules for levodopa-induced dyskinesia in Parkinson's disease (EASE LID 3). Mov Disord 32(12):1701–1709. https://doi.org/10.1002/mds.27131

97. Hauser RA, Pahwa R, Tanner CM, Oertel W, Isaacson SH, Johnson R, Felt L, Stempien MJ (2017) ADS-5102 (Amantadine) extended-release capsules for levodopa-induced dyskinesia in Parkinson's disease (EASE LID 2 study): interim results of an open-label safety study. J Parkinsons Dis 7(3):511–522. https://doi.org/10.3233/jpd-171134

98. Hauser RA, Pahwa R, Wargin WA, Souza-Prien CJ, McClure N, Johnson R, Nguyen JT, Patni R, Went GT (2018) Pharmacokinetics of ADS-5102 (Amantadine) extended release capsules administered once daily at bedtime for the treatment of dyskinesia. Clin Pharmacokinet. https://doi.org/10.1007/s40262-018-0663-4

99. Elmer LW, Juncos JL, Singer C, Truong DD, Criswell SR, Parashos S, Felt L, Johnson R, Patni R (2018) Pooled analyses of phase III

studies of ADS-5102 (Amantadine) extended-release capsules for dyskinesia in Parkinson's disease. CNS Drugs 32(4):387–398. https://doi.org/10.1007/s40263-018-0498-4

100. Fox SH, Metman LV, Nutt JG, Brodsky M, Factor SA, Lang AE, Pope LE, Knowles N, Siffert J (2017) Trial of dextromethorphan/quinidine to treat levodopa-induced dyskinesia in Parkinson's disease. Mov Disord 32(6):893–903. https://doi.org/10.1002/mds.26976

101. Metman LV, Del Dotto P, Natte R, van den Munckhof P, Chase TN (1998) Dextromethorphan improves levodopa-induced dyskinesias in Parkinson's disease. Neurology 51(1):203–206

102. Herring WJ, Assaid C, Budd K, Vargo R, Mazenko RS, Lines C, Ellenbogen A, Verhagen Metman L (2017) A phase Ib randomized controlled study to evaluate the effectiveness of a single-dose of the NR2B selective N-methyl-D-aspartate antagonist MK-0657 on levodopa-induced dyskinesias and motor symptoms in patients with parkinson disease. Clin Neuropharmacol 40(6):255–260. https://doi.org/10.1097/wnf.0000000000000241

103. Merello M, Nouzeilles MI, Cammarota A, Leiguarda R (1999) Effect of memantine (NMDA antagonist) on Parkinson's disease: a double-blind crossover randomized study. Clin Neuropharmacol 22(5):273–276

104. Wictorin K, Widner H (2016) Memantine and reduced time with dyskinesia in Parkinson's disease. Acta Neurol Scand 133(5):355–360. https://doi.org/10.1111/ane.12468

105. Stocchi F, Rascol O, Destee A, Hattori N, Hauser RA, Lang AE, Poewe W, Stacy M, Tolosa E, Gao H, Nagel J, Merschhemke M, Graf A, Kenney C, Trenkwalder C (2013) AFQ056 in Parkinson patients with levodopa-induced dyskinesia: 13-week, randomized, dose-finding study. Mov Disord 28(13):1838–1846. https://doi.org/10.1002/mds.25561

106. Tison F, Keywood C, Wakefield M, Durif F, Corvol JC, Eggert K, Lew M, Isaacson S, Bezard E, Poli SM, Goetz CG, Trenkwalder C, Rascol O (2016) A phase 2A trial of the novel mGluR5-negative allosteric modulator dipraglurant for levodopa-induced dyskinesia in Parkinson's disease. Mov Disord 31(9):1373–1380. https://doi.org/10.1002/mds.26659

107. Durif F, Debilly B, Galitzky M, Morand D, Viallet F, Borg M, Thobois S, Broussolle E, Rascol O (2004) Clozapine improves dyskinesias in Parkinson disease: a double-blind, placebo-controlled study. Neurology 62(3):381–388

108. Katzenschlager R, Manson AJ, Evans A, Watt H, Lees AJ (2004) Low dose quetiapine for drug induced dyskinesias in Parkinson's disease: a double blind cross over study. J Neurol Neurosurg Psychiatry 75(2):295–297

109. Kleiner-Fisman G, Herzog J, Fisman DN, Tamma F, Lyons KE, Pahwa R, Lang AE, Deuschl G (2006) Subthalamic nucleus deep brain stimulation: summary and meta-analysis of outcomes. Mov Disord 21 Suppl 14:S290–S304. https://doi.org/10.1002/mds.20962

110. Deuschl G, Schade-Brittinger C, Krack P, Volkmann J, Schäfer H, Bötzel K, Daniels C, Deutschländer A, Dillmann U, Eisner W, Gruber D, Hamel W, Herzog J, Hilker R, Klebe S, Kloss M, Koy J, Krause M, Kupsch A, Lorenz D, Lorenzl S, Mehdorn HM, Moringlane JR, Oertel W, Pinsker MO, Reichmann H, Reuss A, Schneider GH, Schnitzler A, Steude U, Sturm V, Timmermann L, Tronnier V, Trottenberg T, Wojtecki L, Wolf E, Poewe W, Voges J, German Parkinson Study Group NuS (2006) A randomized trial of deep-brain stimulation for Parkinson's disease. N Engl J Med 355(9):896–908. https://doi.org/10.1056/NEJMoa060281

111. Schuepbach WM, Rau J, Knudsen K, Volkmann J, Krack P, Timmermann L, Halbig TD, Hesekamp H, Navarro SM, Meier N, Falk D, Mehdorn M, Paschen S, Maarouf M, Barbe MT, Fink GR, Kupsch A, Gruber D, Schneider GH, Seigneuret E, Kistner A, Chaynes P, Ory-Magne F, Brefel Courbon C, Vesper J, Schnitzler A, Wojtecki L, Houeto JL, Bataille B, Maltete D, Damier P, Raoul S, Sixel-Doering F, Hellwig D, Gharabaghi A, Kruger R, Pinsker MO, Amtage F, Regis JM, Witjas T, Thobois S, Mertens P, Kloss M, Hartmann A, Oertel WH, Post B, Speelman H, Agid Y, Schade-Brittinger C, Deuschl G (2013) Neurostimulation for Parkinson's disease with early motor complications. N Engl J Med 368(7):610–622. https://doi.org/10.1056/NEJMoa1205158

112. Mestre TA, Espay AJ, Marras C, Eckman MH, Pollak P, Lang AE (2014) Subthalamic nucleus-deep brain stimulation for early motor complications in Parkinson's disease-the EARLYSTIM trial: early is not always better. Mov Disord 29(14):1751–1756. https://doi.org/10.1002/mds.26024

113. Schupbach WM, Rau J, Houeto JL, Krack P, Schnitzler A, Schade-Brittinger C, Timmermann L, Deuschl G (2014) Myths and facts about the EARLYSTIM study. Mov Disord 29(14):1742–1750. https://doi.org/10.1002/mds.26080

114. Odekerken VJ, van Laar T, Staal MJ, Mosch A, Hoffmann CF, Nijssen PC, Beute GN, van Vugt JP, Lenders MW, Contarino MF, Mink MS, Bour LJ, van den Munckhof P,

Schmand BA, de Haan RJ, Schuurman PR, de Bie RM (2013) Subthalamic nucleus versus globus pallidus bilateral deep brain stimulation for advanced Parkinson's disease (NSTAPS study): a randomised controlled trial. Lancet Neurol 12(1):37–44. https://doi.org/10.1016/s1474-4422(12)70264-8

115. Katzenschlager R, Hughes A, Evans A, Manson AJ, Hoffman M, Swinn L, Watt H, Bhatia K, Quinn N, Lees AJ (2005) Continuous subcutaneous apomorphine therapy improves dyskinesias in Parkinson's disease: a prospective study using single-dose challenges. Mov Disord 20(2):151–157. https://doi.org/10.1002/mds.20276

116. Martinez-Martin P, Reddy P, Katzenschlager R, Antonini A, Todorova A, Odin P, Henriksen T, Martin A, Calandrella D, Rizos A, Bryndum N, Glad A, Dafsari HS, Timmermann L, Ebersbach G, Kramberger MG, Samuel M, Wenzel K, Tomantschger V, Storch A, Reichmann H, Pirtosek Z, Trost M, Svenningsson P, Palhagen S, Volkmann J, Chaudhuri KR (2015) EuroInf: a multicenter comparative observational study of apomorphine and levodopa infusion in Parkinson's disease. Mov Disord 30(4):510–516. https://doi.org/10.1002/mds.26067

117. Holloway RG, Shoulson I, Fahn S, Kieburtz K, Lang A, Marek K, McDermott M, Seibyl J, Weiner W, Musch B, Kamp C, Welsh M, Shinaman A, Pahwa R, Barclay L, Hubble J, LeWitt P, Miyasaki J, Suchowersky O, Stacy M, Russell DS, Ford B, Hammerstad J, Riley D, Standaert D, Wooten F, Factor S, Jankovic J, Atassi F, Kurlan R, Panisset M, Rajput A, Rodnitzky R, Shults C, Petsinger G, Waters C, Pfeiffer R, Biglan K, Borchert L, Montgomery A, Sutherland L, Weeks C, DeAngelis M, Sime E, Wood S, Pantella C, Harrigan M, Fussell B, Dillon S, Alexander-Brown B, Rainey P, Tennis M, Rost-Ruffner E, Brown D, Evans S, Berry D, Hall J, Shirley T, Dobson J, Fontaine D, Pfeiffer B, Brocht A, Bennett S, Daigneault S, Hodgeman K, O'Connell C, Ross T, Richard K, Watts A, Group PS (2004) Pramipexole vs levodopa as initial treatment for Parkinson disease: a 4-year randomized controlled trial. Arch Neurol 61(7):1044–1053. https://doi.org/10.1001/archneur.61.7.1044

118. Bracco F, Battaglia A, Chouza C, Dupont E, Gershanik O, Marti Masso JF, Montastruc JL (2004) The long-acting dopamine receptor agonist cabergoline in early Parkinson's disease: final results of a 5-year, double-blind, levodopa-controlled study. CNS Drugs 18(11):733–746

119. Rascol O, Brooks DJ, Korczyn AD, De Deyn PP, Clarke CE, Lang AE (2000) A five-year study of the incidence of dyskinesia in patients with early Parkinson's disease who were treated with ropinirole or levodopa. 056 study group. N Engl J Med 342(20):1484–1491. https://doi.org/10.1056/nejm200005183422004

120. Fox SH, Katzenschlager R, Lim SY, Ravina B, Seppi K, Coelho M, Poewe W, Rascol O, Goetz CG, Sampaio C (2011) The movement disorder society evidence-based medicine review update: treatments for the motor symptoms of Parkinson's disease. Mov Disord 26 Suppl 3:S2–S41. https://doi.org/10.1002/mds.23829

121. Charles D, Konrad PE, Davis TL, Neimat JS, Hacker ML, Finder SG (2015) Deep brain stimulation in early stage Parkinson's disease. Parkinsonism Relat Disord 21(3):347–348. https://doi.org/10.1016/j.parkreldis.2014.10.032

122. Hacker ML, Tonascia J, Turchan M, Currie A, Heusinkveld L, Konrad PE, Davis TL, Neimat JS, Phibbs FT, Hedera P, Wang L, Shi Y, Shade DM, Sternberg AL, Drye LT, Charles D (2015) Deep brain stimulation may reduce the relative risk of clinically important worsening in early stage Parkinson's disease. Parkinsonism Relat Disord 21(10):1177–1183. https://doi.org/10.1016/j.parkreldis.2015.08.008

123. Goetz CG, Damier P, Hicking C, Laska E, Muller T, Olanow CW, Rascol O, Russ H (2007) Sarizotan as a treatment for dyskinesias in Parkinson's disease: a double-blind placebo-controlled trial. Mov Disord 22(2):179–186. https://doi.org/dx.doi.org/10.1002/mds.21226.

124. Bara-Jimenez W, Bibbiani F, Morris MJ, Dimitrova T, Sherzai A, Mouradian MM, Chase TN (2005) Effects of serotonin 5-HT1A agonist in advanced Parkinson's disease. Mov Disord 20(8):932–936. https://doi.org/10.1002/mds.20370.

125. Lewitt PA, Hauser RA, Lu M, Nicholas AP, Weiner W, Coppard N, Leinonen M, Savola JM (2012) Randomized clinical trial of fipamezole for dyskinesia in Parkinson disease (FJORD study). Neurology 79(2):163–169. https://doi.org/10.1212/WNL.0b013e31825f0451.

126. Manson AJ, Iakovidou E, Lees AJ (2000) Idazoxan is ineffective for levodopa-induced dyskinesias in Parkinson's disease. Mov Disord 15(2):336–337.

127. Rascol O, Arnulf I, Peyro-Saint Paul H, Brefel-Courbon C, Vidailhet M, Thalamas C, Bonnet AM, Descombes S, Bejjani B, Fabre N, Montastruc JL, Agid Y (2001) Idazoxan, an alpha-2 antagonist, and L-DOPA-induced dyskinesias in patients with Parkinson's disease. Mov Disord 16(4):708–713.

128. Svenningsson P, Rosenblad C, Af Edholm Arvidsson K, Wictorin K, Keywood C,

Shankar B, Lowe DA, Bjorklund A, Widner H (2015) Eltoprazine counteracts l-DOPA-induced dyskinesias in Parkinson's disease: a dose-finding study. Brain 138(Pt 4):963–973. https://doi.org/10.1093/brain/awu409

129. Bonifati V, Fabrizio E, Cipriani R, Vanacore N, Meco G (1994) Buspirone in levodopa-induced dyskinesias. Clin Neuropharmacol 17(1):73–82.

130. Mestre TA, Shah B, Connolly B, de Aquino C, Al Dhakeel A, Walsh R, Prashanth LK, Ghate T, Lui J, Fox SH (2014) A pilot study evaluating the histamine H2 antagonist, famotidine, for levodopa-induced dyskinesia in Parkinson's disease. Mov Disord 29:S250–S251.

131. Carroll CB, Bain PG, Teare L, Liu X, Joint C, Wroath C, Parkin SG, Fox P, Wright D, Hobart J, Zajicek JP (2004) Cannabis for dyskinesia in Parkinson disease: a randomized double-blind crossover study. Neurology 63(7):1245–1250. https://doi.org/10.1212/01. wnl.0000140288.48796.8e.

132. Sieradzan KA, Fox SH, Hill M, Dick JP, Crossman AR, Brotchie JM (2001) Cannabinoids reduce levodopa-induced dyskinesia in Parkinson's disease: a pilot study. Neurology 57(11):2108–2111.

133. Stathis P, Konitsiotis S, Tagaris G, Peterson D (2011) Levetiracetam for the management of levodopa-induced dyskinesias in Parkinson's disease. Mov Disord 26(2):264–270. https://doi.org/10.1002/mds.23355.

134. Wolz M, Lohle M, Strecker K, Schwanebeck U, Schneider C, Reichmann H, Grahlert X, Schwarz J, Storch A (2010) Levetiracetam for levodopa-induced dyskinesia in Parkinson's disease: a randomized, double-blind, placebo-controlled trial. J Neural Trans 117(11):1279–1286. https://doi. org/10.1007/s00702-010-0472-x.

135. Kobylecki C, Burn DJ, Kass-Iliyya L, Kellett MW, Crossman AR, Silverdale MA (2014) Randomized clinical trial of topiramate for levodopa-induced dyskinesia in Parkinson's disease. Parkinsonism Relat Disord 20(4):452–455. https://doi.org/10.1016/j. parkreldis.2014.01.016.

136. Price PA, Parkes JD, Marsden CD (1978) Sodium valproate in the treatment of levodopa-induced dyskinesia. JNNP 41(8):702–706. https://doi.org/10.1136/ jnnp.41.8.702.

137. Trenkwalder C, Berg D, Rascol O, Eggert K, Ceballos-Baumann A, Corvol JC, Storch A, Zhang L, Azulay JP, Broussolle E, Defebvre L, Geny C, Gostkowski M, Stocchi F, Tranchant C, Derkinderen P, Durif F, Espay AJ, Feigin A, Houeto JL, Schwarz J, Di Paolo T, Feuerbach D, Hockey HU, Jaeger J, Jakab A, Johns D, Linazasoro G, Maruff P, Rozenberg I, Sovago J, Weiss M, Gomez-Mancilla B (2016) A placebo-controlled trial of AQW051 in patients with moderate to severe levodopa-induced dyskinesia. Mov Disord 31(7):1049–1054. https://doi.org/10.1002/mds.26569.

138. Lindeboom SF, Lakke JP (1978) Deanol and physostigmine in the treatment of L-dopa-induced dyskinesias. Acta Neurol Scand 58(2):134–138.

139. Rascol O, Fabre N, Blin O, Poulik J, Sabatini U, Senard JM, Ane M, Montastruc JL, Rascol A (1994) Naltrexone, an opiate antagonist, fails to modify motor symptoms in patients with Parkinson's disease. Mov Disord 9(4):437–440. https://doi.org/10.1002/ mds.870090410.

140. Fox S, Silverdale M, Kellett M, Davies R, Steiger M, Fletcher N, Crossman A, Brotchie J (2004) Non-subtype-selective opioid receptor antagonism in treatment of levodopa-induced motor complications in Parkinson's disease. Mov Disord 19(5):554–560. https://doi.org/10.1002/mds.10693.

141. Rascol O, Ferreira J, Negre-Pages L, Perez-Lloret S, Lacomblez L, Galitzky M, Lemarie JC, Corvol JC, Brotchie JM, Bossi L (2012) A proof-of-concept, randomized, placebo-controlled, multiple cross-overs (n-of-1) study of naftazone in Parkinson's disease. Fund Clin Pharmacol 26(4):557–564. https://doi. org/10.1111/j.1472-8206.2011.00951.x.

142. Corvol JC, Durif F, Meissner WG, Azulay JP, Haddad R, Guimaraes-Costa R, Mariani LL, Cormier-Dequaire F, Thalamas C, Galitzky M, Boraud T, Debilly B, Eusebio A, Houot M, Dellapina E, Chaigneau V, Salis A, Lacomblez L, Benel L, Rascol O (2018) Naftazone in advanced Parkinson's disease: an acute L-DOPA challenge randomized controlled trial. Parkinsonism Relat Disord 60:51–56. https:// doi.org/10.1016/j.parkreldis.2018.10.005.

143. Group PS (2001) Evaluation of dyskinesias in a pilot, randomized, placebo-controlled trial of remacemide in advanced Parkinson disease. Arc Neurol 58(10):1660–1668.

144. Arnulf I, Vidailhet M, Bonnet AM, Descombes S, Jaillon C, Agid Y, Brefel C, Rascol O, Xie J, Pollak P, Cattelin F (2001) Blockade of cholecystokinin-A receptors has no effect on dyskinesias in Parkinson's disease. J Neurol Neurosurg Psychiatry 70(6):812–813.

145. Reardon KA, Mendelsohn FA, Chai SY, Horne MK (2000) The angiotensin converting enzyme (ACE) inhibitor, perindopril, modifies the clinical features of Parkinson's disease. Aust N Z J Med 30(1):48–53.

146. Blanchet PJ, Fang J, Hyland K, Arnold LA, Mouradian MM, Chase TN (1999) Short-term effects of high-dose 17beta-estradiol in postmenopausal PD patients: a crossover study. Neurology 53(1):91–95.

Chapter 6

Clinical Trials for Gait Disorders in Parkinson's Disease

Zuzana Kosutzka, Urban M. Fietzek, and Peter Valkovic

Abstract

Subtle gait disorders are an integral part of the clinical picture from the early stages of Parkinson's disease (PD), and these may worsen as the disease progresses. Disorders of gait have a substantial effect on patient quality of life. Despite immense research efforts, gait disorders in PD and related complications remain challenging, especially concerning therapy. This chapter summarizes the current knowledge about the clinical picture of gait disorders. Special focus is given to the different selection criteria for clinical trials with a recommendation of a five-step selection process. Furthermore, we describe the observational and interventional methods that can be used in the clinical trials focusing on PD gait disorders.

Key words Parkinson's disease, Gait, Freezing of gait, Clinical trials

1 Introduction

Beginning in the early stages of Parkinson's disease (PD), at least subtle gait disorders are a clinical manifestation that worsens with disease progression [1]. The gait of people with PD (PwP) is typically characterized by shortened steps, increased stride variability, increased cadence, and reduced speed [2]. Gait disorders substantially impact patient quality of life [3] and are among the most significant contributors to poor quality of life in nondemented PD patients [4]. Fear of falling plays a major role in this deterioration [5], and quality of life is also significantly diminished by freezing of gait (FoG) [6].

Falls are the most significant consequences of gait disorders [7]. Their etiology in PD is complex and multidimensional, and several risk factors have been identified [8]. A prior fall is the strongest predictor of future fall [9]. Although rates vary, on average 61% of people with PD fall each year compared with 33% of all older adults [10]. Among newly diagnosed PD patients, the combination of slow gait speed, decreased stance time, and Hoehn and Yahr III emerged as a significant baseline predictor with 92% sensi-

Santiago Perez-Lloret (ed.), *Clinical Trials In Parkinson's Disease*, Neuromethods, vol. 160,
https://doi.org/10.1007/978-1-0716-0912-5_6, © Springer Science+Business Media, LLC, part of Springer Nature 2021

tivity and 62% specificity [11]. Cognitive impairment is an independent contributing factor to falls [12]. FoG and falls are intrinsically interconnected [13]. Complications connected with related injuries increase morbidity and mortality, and direct and indirect costs increase health-care expenditures [14].

The pathophysiologies of PD gait disorders involve central and peripheral neuroanatomical structures, and affect multiple neurotransmitters [15]. Gait pathology is associated with other non-motor symptoms and may appear before motor cardinal signs [16].

1.1 Clinical Picture of Gait Disorders in PD

PD patients in early stages may present with reduced arm swing. In fact, asymmetrical arm swing may be one of the first detectable motor signs in PD [17, 18]. PwP is an example of a mixed type gait disorder with continuous and episodic pathology [19]. Affected patients present with bradyhypokinetic gait features including slow and small steps that may be more pronounced depending on the medication cycle in fluctuating patients.

Episodic gait disorders in PD occasionally occur during the gait. Probably the most impressive episodic gait disturbance in PD is FoG [20]. The main clinical feature is the inability to step forward. FoG can be divided into three subtypes: small steps forward, trembling in place, and complete akinesia [21]. Another type of disturbance is festination of gait, which is characterized by an unintentional increase in speed, usually with small steps. Festinating gait can be classified as gait with progressive step shortening and a compensatory increase of cadence and gait with forward trunk leaning with small balance-correcting steps [22].

2 Selection Criteria for Gait Trials in PD

Selection of an appropriate patient population is crucial for valid study outcomes. A five-step selection flowchart can be used to efficiently design clinical trials assessing gait (Fig. 1).

2.1 Premotor, Early Motor, and Advanced Stages

Patients may be selected for a trial depending on the stage of the disease. Due to the delicate motor–cognitive interplay that gait requires, it seems to be a sensitive biomarker of different neurodegenerative processes from very early stages [23]. Subtle changes are even detectable in the premotor stages of PD in patients without obvious clinical signs [24].

The main early stage gait abnormalities are reduced arm swing and subtle gait variability. Early and longitudinal gait assessment is important to understand how it is impaired in early PD and to identification features that are responsive and nonresponsive to dopaminergic treatment. Depending on disease phenotype, people with postural instability and gait difficulty (PIGD) naturally have more walking difficulties compared with a tremor-dominant phe-

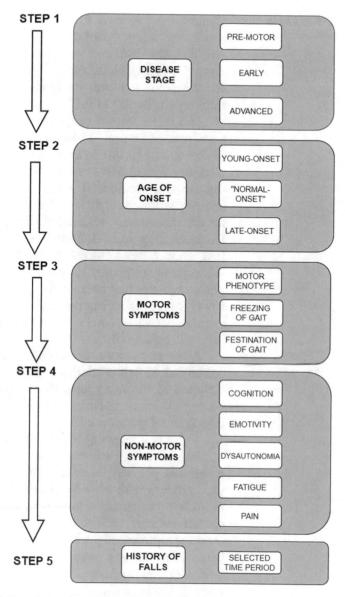

Fig. 1 Five-step approach to patient selection for gait trials in PD

notype, especially with regard to temporal characteristics. Interestingly, pace and variability may deteriorate faster in patients with the tremor-dominant phenotype [25]. Advanced stages of PD are characterized by motor fluctuations and dyskinesia that can lead to a very unusual dyskinetic gait pattern [26].

Hoehn and Yahr staging was introduced in 1967 and remains the most widely used system to characterize populations under clinical investigation [27]. In gait trials, patient groups are usually divided based on the absence/presence of postural instability.

2.2   PD Phenotype

Clinical subtyping of PD is important for prediction of disease course and subgroup stratification in clinical trials. PD is generally classified into different subtypes based on the presenting symptoms. Two main classification systems are used in clinical practice and research. Jankovic categorizes PD into three subtypes: (1) tremor-dominant, (2) PIGD, and (3) indeterminate type [28].

The other system proposed by Schiess et al. recognizes also three main subtypes: (1) tremor-dominant, (2) akinetic-rigid, and (3) mixed [29]. An interesting concept introduced by Kotagal considers that PIGD is not a legitimate subtype but rather a multidimensional continuum influenced by several overlapping age-related pathologies [30]. Patients are generally categorized based on the dominating selected features on the Unified Parkinson's Disease Rating Scale (UPDRS) or the Movement Disorder Society (MDS)-Sponsored Revision of the UPDRS (MDS-UPDRS). When comparing subjects with PIGD and tremor-dominant subtypes, objective gait parameters (gait speed, stride length, gait variability) may overlap, but the PIGD group shows worse performance in these parameters when measured under challenging conditions like dual-tasking [31].

A recent retrospective cohort study of subjects with autopsy-confirmed PD showed that age at the diagnosis should be also included in the subtyping, that is, patients with younger disease onset had faster disease progression with earlier recurrent falls and wheelchair dependence [32].

There is a growing body of evidence that phenotypes should not be rigorously dichotomized but rather perceived as a multidimensional phenotypic continuum. This is supported by the fact that patients with initially classified disease phenotype based on motor symptoms often convert to a different phenotype [33]. Gait characteristics, levodopa responsiveness, and the plethora of nonmotor symptoms might be important factors [1]. PD phenotyping based on nonmotor symptoms may better reflect neurotransmitter pathology with potentially good correlations in objective measures. It has been also suggested that this approach would be more stable over time compared with standard motor phenotyping [34]. Nevertheless, this concept must be verified in larger population studies with longer follow-up.

Overall, patient selection based on motor phenotyping should be complemented with nonmotor symptom assessment.

2.3   Episodic Gait Phenomena

2.3.1   Fallers and Nonfallers

Despite extensive research efforts underlying neural mechanisms associated with falls in PD, patients to study are scarce. One goal is to identify predictors of near falls and falls. Dyskinesia and the score in frontal assessment battery can predict future falls [35]. Paul et al. developed a simple clinical tool with sound predictive accuracy of future falls based on the history of falls in the previous year, FoG, and gait velocity [36, 37]. Some common inclusion

Table 1
Examples of inclusion criteria mentioning fall frequency

Study	Inclusion criterion of falls
Galperin et al., 2019 [65]	Two falls in the last 6 months
Jehu et al., 2018 [127]	Fall in the last 3 months
Chung et al., 2010 [77]	One to three falls in 6 weeks
Allcock et al., 2009 [128]	Two falls in 12 months
Wood et al., 2002 [129]	Fall in 1 year

criteria based on fall frequency are shown in Table 1. For further review of risk factors, assessment, and future research directions concerning falls, please *see* the publication by Fasano et al. [7].

2.3.2 Freezers and Nonfreezers

How to define a PwP as a freezer is a challenging task. The majority of studies define freezers according to the question, "Do you feel that your feet are glued to the floor?" [38]. In addition, some studies require that the patient is observed during challenging tasks and that freezing must be observed before the status is established [39].

It is important to clearly define the type of FoG when including a patient in a study. Patients can show FoG during ON and/or OFF medication states, so three types can be delineated. Patients with OFF-FoG are termed OFF-Freezer. Patients with ON (only) FoG are termed SUPRA-ON-Freezers, and patients with freezing in both conditions are (OFF+) ON-Freezer [40, 41]. A pseudo-ON label can be given if the patient's L-DOPA-response has not been firmly established, but this is not helpful for clinical research as it does not allow further interpretation of results in a definite fashion.

3 Methodology for the Study of Gait in PD

Standardized assessment using widely accepted instruments and adhering to strict protocols will contribute to a better understanding of gait pathophysiology.

3.1 Questionnaires

Questionnaires are a widely used method to study gait and balance issues in PD because they are simple and relatively quick to administer. Patient-reported data are still the main information reported in gait research. Based on the recent MDS Critique and Recommendations dedicated to Measurement instruments, four questionnaires have been recommended [42].

The Activities-specific Balance Confidence Scale (ABC Scale) assesses balance confidence during activities without losing balance [43]. The Falls Efficacy Scale reflects the fear of falling and self-efficacy to avoid falls during activities of daily living [44]. The Freezing of Gait Questionnaire consists of six multiple choice questions that enable the so-called freezers to quantify the severity of gait freezing [45]. The last recommended option is the Survey of Activities and Fear of Falling in the Elderly-Modified (SAFFE-m), which reflects patients' avoidance of certain activities in case of falls [46].

3.2 Scales

Even though the MDS published the recommendations of tests that should be used for the assessment of gait and balance there is no gold standard for clinical assessment [42].

The MDS-UPDRS includes five items assessing gait and balance (2.12, 2.13, 3.10–3.12). The sum of all items forms the PIGD Score [47]. Probably the most controversial item is the pull-test, and a good alternative might be the push and release test [48]. Another recommended clinical scale is the Berg Balance Scale that assesses mainly static balance requiring additional accessories [49]. A more complex test is the Mini-BESTest assessing gait stability and posture that can be completed in 10–15 min [50]. If more focus on gait abilities is required, the Dynamic Gait Index might be a valid option [51]. The Functional Gate Assessment is similar test with sound psychometric properties [52]. Kryszton et al. suggest that the BESTest with its shortened versions and the Fullerton Advanced Balance Scale be used for complex balance assessment of PD patients [53]. The Timed Up-and-Go clinical test includes verticalization, gait, and turns with the outcome of measured time [54].

One clinical rating instrument for FoG assessment has received a label of "suggested use" by the above-mentioned scale committee. Unlike other approaches, the FoG score is rated from the observation of patients' gait phenomenology during performance of parkour instead of counting or timing FoG episodes. The rater assigns a 1 for the observation of any shuffling steps of festination, a 2 for trembling-in-place or akinetic freezing, and a score of 3 for any aborted movement or a need for assistance during the 12 varied situations/tasks. Thus, patients' scores can range from 0 to 36 [55]. A minimal clinically relevant change was reported and allows interpretation of the magnitude of the effects [56]. Rapid documentable changes in FoG score (e.g., after the administration of pharmacological agents or from deep brain stimulation [DBS]) cannot be captured using patient questionnaires that are designed for weekly changes of gait status [57].

3.3   Imaging

Recent advances in imaging techniques have provided new insight into the underlying mechanisms of gait disorders in PwP. A seemingly dynamic process like walking can be investigated despite the static condition of various imaging techniques.

Functional MRI (fMRI) is the most widely used imaging technique to study FoG in PwP. It permits investigation of whole-brain networks during all parts of the gait cycle. fMRI was involved in identifying the cortical and subcortical neural correlates of FoG [58]. In a recent study combining MRI volumetry with fMRI, there was a clear association between cortical abnormalities and falls [59]. Patients with frequent falls had significantly lower grey matter volume in the right superior temporal gyrus and right inferior parietal lobule. Fallers also had lower connectivity in the right posterior perisylvian region. An interesting concept combining virtual reality (VR) and fMRI was introduced by Shine et al. to shed light on neural correlates of FoG [60]. Nuclear methods can elucidate metabolic processes associated with gait disturbance, while neurotransmitter imaging can be also useful for assessing the underlying pathology of gait disorders.

Thus, it is a good practice to exclude structural changes of the brain when recruiting patients for trials assessing gait.

3.4   Movement Analysis

Gait analysis can be done in many different ways, ranging from affordable sensors to expensive three-dimensional (3D) motion capture assessment. Technology-based systems with user friendly environment have attracted the research interest in the last years (for overview *see* [61]).

3D motion capture systems provide extremely accurate analyses and are often used as an objective standard [62]. However, these systems lack mobility and are often confined to laboratory research. Economically more feasible are RGB or infrared camera systems that also accurately measure spatiotemporal characteristics of gait with fair accuracy [25].

There are also sensor-based approaches using inertial measurement units (IMUs), that is, accelerometers, gyroscopes, or magnetometers. These systems have been developed to a high sophistication, recording up to nine-degrees-of-freedom-data in real-time [63]. Miniaturization of sensing technology and data storage allow for a multitude of body-worn devices. Using those sensors to monitor patients has allowed gait parameter assessment in flexible environments [64]. Thus, body-worn IMUs can potentially assess the everyday activities of daily living in natural conditions for up to weeks [65].

The selection of key outcomes for gait research is crucial with respect to purpose and pathology. Gait measurement can be broadly grouped within a structure that captures (1) spatiotemporal characteristics that reflect a typical gait cycle, expressed as the mean of multiple steps; and (2) dynamic features of gait, which represent the inconsistency of spatiotemporal measures across these steps [66]. Gait variability (stride-to-stride fluctuations) may discriminate important clinical features better than routine spatio-temporal measures such as mean gait speed or step time [67]. From a clinimetric point of view, at least 30 steps should be collected when using a continuous walking protocol [68].

In principal, less precise raw signals will require more elaborate data post-processing, often involving proprietary algorithms to interpret the data. Future research will have to answer questions opening up through the use of artificial intelligence to detect patterns and pathologies. Researchers should keep in mind their motivation to help people with disability and disease. They should recognize the privacy of their patients, as data increasingly cannot be kept anonymous due to the sheer power of calculation and connectivity. This is true for motion data, in addition to genetic data. Open platforms and transparent algorithms and methodologies should be mandated for publications.

4 Interventional Methods to Study Gait Disorders in PD

4.1 Medication

Dopaminergic medication is highly effective in targeting gait impairments in uncomplicated early-to-mid stage PD [69]. Most gait impairments are responsive to dopaminergic treatment, underscoring the importance of the dopaminergic contribution to gait.

Another line of evidence results from two studies with DBS in which the MED ON STIM OFF arm always demonstrated highly improved gait parameters compared to the MED OFF STIM OFF arm [70, 71]. This observation also pertains to FoG. Festination and FoG respond excellently to L-DOPA in early stage PD and can minimize the clinical observation of these symptoms for years. The good responsiveness of FoG to L-DOPA has been shown repeatedly [21, 41, 72].

In advanced PD stages with more widespread neuronal involvement, nondopaminergic mechanisms that regulate axial motor symptoms play an increasingly important role. This might be why dissociation of upper limb (appendicular) and lower limb (gait) levodopa responses can be observed in heterogeneous populations that span Hoehn and Yahr disease stages 0–4 [73]. Several large trials comparing L-DOPA and dopamine agonists have shown that gait performance is worse in the agonist treatment arm. This could be due to the relative undertreatment compared to L-DOPA when current comparative dosing calculations are applied [74]. Similarly, FoG has been observed more often in the pramipexole or ropinirole arm in the head-to-head comparison studies [75, 76].

Nondopaminergic agents such as central cholinesterase inhibitors (e.g., rivastigmine or donepezil) have been investigated for the management of gait disorders [77]. Although beneficial effects were observed, these agents might also introduce new problems such as syncope and falling. Methylphenidate was intensively researched, but the yielded incongruent results. Positive observations might have resulted from the intrinsic dopaminergic-enhancing effect of the drug [78–80].

In clinical trials studying gait, the patients' medication regimens should be marked including the total doses of dopaminergic and nondopaminergic medication per day. When assessing gait in the OFF stage, the last dose of dopaminergic medication should be mentioned.

4.2 Noninvasive Stimulation

Transcranial magnetic stimulation (TMS) has been used an elegant, safe, and noninvasive method to elucidate central mechanisms of gait physiology and pathophysiology [81, 82]. One of the initial findings in PwP was reported by Pascual-Leone, who found improved reaction times and changes of motor threshold after repetitive stimulation and speculated upon its use in therapeutic settings [83]. It was shown that high-frequency repetitive TMS (rTMS) (>1.0 Hz) can enhance cortical excitability, whereas low-frequency ($\leq$1.0 Hz) rTMS decreases it [84]. Theta burst stimulation (TBS) is a variation of rTMS with intermittent TBS having an enhancing character and continuous TBS decreasing excitability [85]. rTMS can also affect other brain regions, presumably via corticostriatothalamocortical circuits [86]. However, only a few studies reported effects on gait in PwP. In one protocol, the frontal regions received low-frequency stimulation (0.2 Hz) for 2 weeks, and 30 stimuli were applied every day for 10 min with the coil over the vertex [87]. The authors reported no changes in gait speed over a 10-min walking distance. Another protocol delivered rTMS to the M1 also at low frequency (0.5 Hz) and found improvements in gait speed [88].

In conclusion, rTMS has never been used to treat PwP apart from experimental settings, as a clear treatment effect could not be established, and the technical requirements are not easily minimized. Such a minimization might be achieved using a different technology that was initially developed at the University of Göttingen by Walter Paulus and his group [89]. Anodal transcranial direct current stimulation (tDCS) has a facilitating effect on cortical structures. It also modifies arterial cerebral blood flow so that both direct electrical and indirect perfusional mechanisms could explain the observed effects from tDCS [90]. Recently, tDCS has gained attention as several authors reported an alleviation of FoG. One crossover protocol in ten PwP stimulated the motor cortex during 5 days and found reduced episodes and episode duration compared to the control group that lasted for 4 weeks [91]. The same authors conducted a similar crossover study with 20 PwP who received either motor cortex, motor and premotor cortex, or sham stimulation [92]. Using the FoG score as their primary target, the authors described superior effects for the multitarget arm compared to the motor cortex alone and sham stimulation. This finding was further backed by the secondary analyses and by a solid demonstration that the blinding to treatment had been effective.

In conclusion, tDCS is a promising tool to modify cortical functioning and might be of therapeutic use for patients for whom we cannot establish good symptom control with traditional treatment approaches. However, far more research is needed before a general recommendation can be issued.

4.3 DBS

DBS of the subthalamic nucleus (STN) has been established as an effective therapeutic alternative for PwP with fluctuating PD [93, 94]. Other target structures besides the STN have been evaluated, but they have shown less overall benefit, especially in terms of gait improvement [95].

A number of trials were assessed the specific effect of DBS on gait. In an early study of nine patients, Stolze et al. demonstrated that STN-DBS had a similar effect on gait as levodopa therapy [71]. There was controversial discussion regarding whether DBS would increase the incidence of FoG. While DBS of the internal pallidum could directly induce freezing [96], STN-DBS was less prone to this side effect [97]. This is why current clinical suggestions prefer STN-DBS in patients with freezing in their medical history [98].

Nonetheless, STN-DBS may be complicated by a stimulation refractory freezing disorder that can pose a serious clinical problem. Several troubleshooting mechanisms were proposed Moreau et al. suggested stimulation with lower frequencies (e.g., 60 Hz instead of 130 Hz) [99]. Fasano et al. tried to stimulate with less intensity to make gait more symmetrical, as this was shown to be one of the potential drivers for freezing [70, 100]. Finally, Weiss, et al. used an interleaving approach and simultaneously costimulated the dorsal substantia nigra and saw improved gait in terms of reduced FoG and festination [101]. Neurostimulators with constant current steering and new programming options may bring new options for PwP and postoperative gait issues [102].

The effect of DBS on gait parameters remains controversial [103]. It remains unclear if the deterioration is DBS provoked or it is a natural disease progression. In the ideal setting, a control group of complementary PD patients without DBS should be included in the study designs. However, this might be unethical because most centers offer this advanced treatment option to patients experiencing motor fluctuations.

4.4 Physiotherapy

Targeted physiotherapy is crucial in the management of gait disorders in PD. The approach to the patient is usually multidimensional and includes learning compensatory strategies and strengthening the musculoskeletal and cardiovascular system. One of the main principles is to learn strategies that compensate for bradykinesia through bypassing defective basal ganglia with the frontal cortex [104]. Nevertheless, the lack of systematic randomized control trials complicates the systematic application of physiotherapy. Visual and tactile cueing strategies seem to be efficient nonpharmacological options to manage FoG, with new cueing possibilities afforded by wearable technologies [105]. Interestingly, internal cueing strategies (e.g., singing) seem to be more efficient than external cueing approaches [106].

Recent technological developments gave rise to exergames that include rehabilitation based on computer games (for review *see* Barry et al. [107]). It can be an affordable home-based tool to complement traditional rehabilitation strategies. Ethical constraints of the home-based exercising still need to addressed. Adding a VR component to conventional physiotherapy may provide more natural everyday training conditions with a good safety profile. Larger patient population studies are needed to confirm the real benefit of VR in gait physiotherapy for PD patients, and unified VR environments should be used across study designs [108, 109].

In conclusion, rigorous multicentric large-scale randomized control trials need to be implemented to specify the optimal durations and types of gait physiotherapy for PwP.

5 Particularities of Gait and Balance Research in PD

5.1 Motor Fluctuations

Motor complications including dyskinesias and motor fluctuations affect 75–80% of PwP with levodopa exposure for more than 5–10 years [110]. With regard to gait, this can span from dyskinesia that severely interferes with the gait cycle to a complete inability to walk [69]. Surprisingly, most falls happen in the ON state, but falls related to FoG happen in the OFF state [111]. Gait protocols should be done in both motor stages, but this can be complicated by involuntary movements that can interfere with the recordings. On the other hand, OFF stages also limit the gait, and PwP are either not able to walk or are very prone to the falls.

5.2 Polyneuropathy and Other Comorbidities

Peripheral polyneuropathy is not uncommon in PwP, especially in those with higher levodopa equivalent doses [112]. Generally, patients with polyneuropathy have lower walking speed and are more prone to falls [113]. Screening for this condition is recommended, and patients with signs of polyneuropathy should be excluded from the studies or taken into consideration when interpreting the results. Another confounding factor in gait studies could be primary and/or secondary orthopedic complications of PD like spine deformities and osteoarthritis. These pathologies can also lead to pathological gait patterns, mainly decreased walking velocity [114]. Cardiovascular system comorbidities should be also noted [31]. Even though orthostatic hypotension does not reduce the walking speed, PwP with orthostatic hypotension have higher postural sway and fall more often.

5.3   Neuropsychiatric Aspects

Nonmotor symptoms like cognition and emotional disturbance should be taken into account when assessing gait [115].

5.3.1   Cognitive Deficits

Cognitive deficits are strongly associated with gait disorders in PD [12, 116, 117]. Visuospatial abilities and executive functions are impaired in patients with gait disorders, even in the prediagnostic phase [118]. Patients with FoG are deficient mainly in response inhibition, attention, and problem-solving [119]. Motor–cognitive dual tasks (e.g., subtracting 7 s while walking) are used to explore the interplay between gait and cognition. Dual-tasking performance is also a good predictor of future falls in PwP [120].

5.3.2   Anxiety and Depression

Affective neuropsychiatric disturbances have a considerable impact on gait patterns in PwP. There is a growing body of evidence that anxiety is an independent risk factor for gait disorders, namely, FoG [121]. Anxiety can also contribute to deficits in set-shifting, and this may subsequently contribute to FoG [122]. Depression is also associated with slower gait and higher gait variability, even in the early stages of the disease [66].

In summary, we suggest screening all PwP with gait abnormalities for nonmotor symptoms (e.g., by use of the nonmotor symptom questionnaires) and with at least one screening neuropsychological tool.

6   Ten Simple Rules for Designing Clinical Trials for Gait Disorders in PD

1. Select the patient population by considering (a) age; (b) disease stage (e.g., Hoehn and Yahr or Schwab and England); (c) motor state: OFF vs. ON vs. dysON; (d) presence of episodic gait disorders (e.g., Freezing and Festination; (e) fallers versus nonfallers, adhere to standardized and objective criteria; (f) existing treatment approach (e.g., best medication, DBS); (g) intensity of any therapeutic interventions (e.g., exerciser vs. nonexerciser).

2. Select the study type: observational, interventional or combined.

3. Select the study design: longitudinal, cross-sectional, crossover; blinding issues; sites.

4. Prefer the use of patient-reported outcomes such as questionnaires with objective measurement and vice versa, and assess the validity and clinimetric features of the selected methods;

determine if there is a need for more than one repetition of the task/test.

5. Determine if the patient can complete the task; when examining gait, take into account the walking abilities during the worst stage.

6. Gait protocols ideally should include turns to both directions and dual-/multitasking.

7. The clinical protocol should include a neuropsychological screening or full assessment.

8. Complementary data (MRI, EMG of the lower limbs, somatosensory evoked potentials) should be assessed only if expected results clearly support the hypothesis under investigation.

9. If follow-up is planned, be realistic with respect to disease progression and dropout rate.

10. Follow your initial hypothesis, and report negative results when the initial hypothesis is falsified by the data; remember that a positive study might be true positive only in about two-thirds of cases, while a negative study is reliably true negative more than 95% of the time [123]

7   Examples of Own Clinical Research

7.1   Turning Is the Critical Situation for PwP with Freezing

The initial reporting on FoG, for example, by Jean-Martin Charcot or Ambani and Van Woert [124, 125] stressed the start hesitation aspect of this gait phenomenon, a view was perpetuated until this century. Schaafsma et al. were the first to experimentally show that turning most effectively provokes freezing [21]. In a group of 19 PwP, turning elicited freezing in 63% of turning events, but only in 23% while starting in the OFF condition, compared to 15% and 4% in the ON condition, respectively. This important observation explains why turns were repeatedly shown to be one of the most relevant triggers for falling [7, 13]. Consequently, turns have been put in the center of clinical evaluations [38, 55].

Still, several aspects of turning were not well examined. In 2016, we evaluated the impact of spatial constraint during turning in freezing and nonfreezing PwP [126]. To objectively evaluate the hypothesis, we decided to use sensor-based assessments, that is, number of steps and turn duration. The step number was assessed using the wavelet decomposition of the accelerometry signal (Fig. 2). We selected 20 PwP with FoG and 20 without according to established criteria and compared the results to 20 healthy controls. We were able to reproduce known effects of increased step count in the PD group vs. healthy controls. The new finding was a highly significant interaction of spatial constraint with the status of the freezer, which corresponds to the clinical observation why nar-

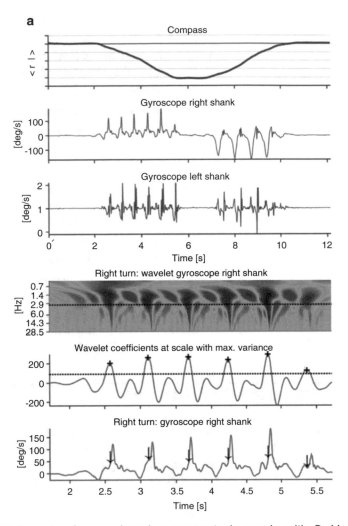

Fig. 2 Example of sensor-based assessments in people with Parkinson's disease

row spaces and turning synergistically deteriorate gait for PwP. This observation impacts clinical routine because it demonstrates that turning evaluations should be done in a similar spatial situation (e.g., a floor area of 40 × 40 cm).

7.2 Cueing Improves Continuous and Episodic Gait Disorders in PD

One of the most astonishing features of gait phenomenology in PwP is the dramatic switch from gait inability during a freeze to full functionality if the patient is using a different motor program, such as cycling [38]. Such observations harken back to Luria who described a PD patient who was able to walk in a situation of severe danger. Nonnekes et al. recently published a video series of these cases demonstrating an amazing variety of these motor program switches [22]. Besides this shift of motor programs, cue application has been shown to effectively ameliorate bradykinetic gait [2] and is very helpful for freezing situations.

A large European collaborative study evaluated the use of cueing on gait in PwP [39]. However, a dedicated research study that evaluated freezing as the primary target was an unmet need. There were multiple methodological difficulties involved in this study. Recent suggestions for freezing improving therapy stress the need for program personalization and matching of difficulty level to the individual need.

8 Conclusion

Future studies should focus on biomarkers for gait disorders in the premotor stages of PD. Gait is a very complex activity with unique profiles in different neurodegenerative entities, which could mediate the early application of disease-modifying treatment. Patients with implanted DBS systems should also be included in trials assessing potential benefits of interventions including medication, noninvasive stimulation, and tailored physiotherapy. There is a serious need of standardize gait assessment. Uniform questionnaires and scales with respect to clinimetric properties should be used across trials. Measurement instruments should include assessments performed in the patients' natural conditions. VR might be an option, but the selection of VR environments should be unified. Home-based systems could also be implemented. Ethical issues and safety measures should be addressed systematically. Accelerating gait disorder research requires strengthening cooperation between centers to prevent design duplication.

References

1. Mirelman A, Bonato P, Camicioli R, Ellis TD, Giladi N, Hamilton JL, Hass CJ, Hausdorff JM, Pelosin E, Almeida QJ (2019) Gait impairments in Parkinson's disease. Lancet Neurol 18(7):697–708. https://doi.org/10.1016/S1474-4422(19)30044-4

2. Morris M, Iansek R, McGinley J, Matyas T, Huxham F (2004) Three-dimensional gait biomechanics in Parkinson's disease: evidence for a centrally mediated amplitude regulation disorder. Mov Disord 20(1):40–50. https://doi.org/10.1002/mds.20278

3. Rahman S, Griffin HJ, Quinn NP, Jahanshahi M (2008) Quality of life in Parkinson's disease: the relative importance of the symptoms. Mov Disord 23(10):1428–1434. https://doi.org/10.1002/mds.21667

4. Santos Garcia D, de Deus FT, Suarez Castro E, Borrue C, Mata M, Solano Vila B, Cots Foraster A, Alvarez Sauco M, Rodriguez Perez AB, Vela L, Macias Y, Escalante S, Esteve P, Reverte Villarroya S, Cubo E, Casas E, Arnaiz S, Carrillo Padilla F, Pueyo Morlans M, Mir P, Martinez-Martin P (2019) Non-motor symptoms burden, mood, and gait problems are the most significant factors contributing to a poor quality of life in non-demented Parkinson's disease patients: results from the COPPADIS Study cohort. Parkinsonism Relat Disord 66:151–157. https://doi.org/10.1016/j.parkreldis.2019.07.031

5. Brozova H, Stochl J, Roth J, Ruzicka E (2009) Fear of falling has greater influence than other aspects of gait disorders on quality of life in patients with Parkinson's disease. Neuro Endocrinol Lett 30(4):453–457

6. Walton CC, Shine JM, Hall JM, O'Callaghan C, Mowszowski L, Gilat M, Szeto JYY,

Naismith SL, Lewis SJG (2014) The major impact of freezing of gait on quality of life in Parkinson's disease. J Neurol 262(1):108–115. https://doi.org/10.1007/s00415-014-7524-3

7. Fasano A, Canning CG, Hausdorff JM, Lord S, Rochester L (2017) Falls in Parkinson's disease: a complex and evolving picture. Mov Disord 32(11):1524–1536. https://doi.org/10.1002/mds.27195

8. Bloem BR, Boers I, Cramer M, Westendorp RG, Gerschlager W (2001) Falls in the elderly. I Identification of risk factors. Wiener Klin Wochensch 113(10):352–362

9. Pickering RM, Grimbergen YAM, Rigney U, Ashburn A, Mazibrada G, Wood B, Gray P, Kerr G, Bloem BR (2007) A meta-analysis of six prospective studies of falling in Parkinson's disease. Mov Disord 22(13):1892–1900. https://doi.org/10.1002/mds.21598

10. Allan LM, Ballard CG, Rowan EN, Kenny RA (2009) Incidence and prediction of falls in dementia: a prospective study in older people. PLoS One 4(5):e5521–e5521. https://doi.org/10.1371/journal.pone.0005521

11. Lord S, Galna B, Yarnall AJ, Morris R, Coleman S, Burn D, Rochester L (2017) Natural history of falls in an incident cohort of Parkinson's disease: early evolution, risk and protective features. J Neurol 264(11):2268–2276. https://doi.org/10.1007/s00415-017-8620-y

12. Amboni M, Barone P, Hausdorff JM (2013) Cognitive contributions to gait and falls: evidence and implications. Mov Disord 28(11):1520–1533. https://doi.org/10.1002/mds.25674

13. Bloem BR, Hausdorff JM, Visser JE, Giladi N (2004) Falls and freezing of gait in Parkinson's disease: a review of two interconnected, episodic phenomena. Mov Disord 19(8):871–884. https://doi.org/10.1002/mds.20115

14. Huse DM, Schulman K, Orsini L, Castelli-Haley J, Kennedy S, Lenhart G (2005) Burden of illness in Parkinson's disease. Mov Disord 20(11):1449–1454. https://doi.org/10.1002/mds.20609

15. Devos D, Defebvre L, Bordet R (2010) Dopaminergic and non-dopaminergic pharmacological hypotheses for gait disorders in Parkinson's disease. Fundam Clin Pharmacol 24(4):407–421. https://doi.org/10.1111/j.1472-8206.2009.00798.x

16. Del Din S, Elshehabi M, Galna B, Hobert MA, Warmerdam E, Suenkel U, Brockmann K, Metzger F, Hansen C, Berg D, Rochester L, Maetzler W (2019) Gait analysis with wearables predicts conversion to parkinson disease.

Ann Neurol 86(3):357–367. https://doi.org/10.1002/ana.25548

17. Mirelman A, Bernad-Elazari H, Thaler A, Giladi-Yacobi E, Gurevich T, Gana-Weisz M, Saunders-Pullman R, Raymond D, Doan N, Bressman SB, Marder KS, Alcalay RN, Rao AK, Berg D, Brockmann K, Aasly J, Waro BJ, Tolosa E, Vilas D, Pont-Sunyer C, Orr-Urtreger A, Hausdorff JM, Giladi N (2016) Arm swing as a potential new prodromal marker of Parkinson's disease. Mov Disord 31(10):1527–1534. https://doi.org/10.1002/mds.26720

18. Nürnberger L, Klein C, Baudrexel S, Roggendorf J, Hildner M, Chen S, Kang JS, Hilker R, Hagenah J (2014) Ultrasound-based motion analysis demonstrates bilateral arm hypokinesia during gait in heterozygous PINK1 mutation carriers. Mov Disord 30(3):386–392. https://doi.org/10.1002/mds.26127

19. Giladi N, Horak FB, Hausdorff JM (2013) Classification of gait disturbances: distinguishing between continuous and episodic changes. Mov Disord 28(11):1469–1473. https://doi.org/10.1002/mds.25672

20. Fahn S (1995) The freezing phenomenon in parkinsonism. Adv Neurol 67:53–63

21. Schaafsma JD, Balash Y, Gurevich T, Bartels AL, Hausdorff JM, Giladi N (2003) Characterization of freezing of gait subtypes and the response of each to levodopa in Parkinson's disease. Eur J Neurol 10(4):391–398. https://doi.org/10.1046/j.1468-1331.2003.00611.x

22. Nonnekes J, Giladi N, Guha A, Fietzek UM, Bloem BR, Růžička E (2019) Gait festination in parkinsonism: introduction of two phenotypes. J Neurol 266(2):426–430. https://doi.org/10.1007/s00415-018-9146-7

23. Montero-Odasso M, Pieruccini-Faria F, Bartha R, Black SE, Finger E, Freedman M, Greenberg B, Grimes DA, Hegele RA, Hudson C, Kleinstiver PW, Lang AE, Masellis M, McLaughlin PM, Munoz DP, Strother S, Swartz RH, Symons S, Tartaglia MC, Zinman L, Strong MJ, Investigators O, McIlroy W (2017) Motor phenotype in neurodegenerative disorders: gait and balance platform Study design protocol for the Ontario neurodegenerative research initiative (ONDRI). J Alzheimers Dis 59(2):707–721. https://doi.org/10.3233/JAD-170149

24. Mirelman A, Gurevich T, Giladi N, Bar-Shira A, Orr-Urtreger A, Hausdorff JM (2011) Gait alterations in healthy carriers of the LRRK2 G2019S mutation. Ann Neurol 69(1):193–197. https://doi.org/10.1002/ana.22165

25. Galna B, Lord S, Burn DJ, Rochester L (2014) Progression of gait dysfunction in incident Parkinson's disease: impact of medication and phenotype. Mov Disord 30(3):359–367. https://doi.org/10.1002/mds.26110

26. Ruzicka E, Zarubova K, Nutt JG, Bloem BR (2011) "Silly walks" in Parkinson's disease: unusual presentation of dopaminergic-induced dyskinesias. Mov Disord 26(9):1782–1784. https://doi.org/10.1002/mds.23667

27. Goetz CG, Poewe W, Rascol O, Sampaio C, Stebbins GT, Counsell C, Giladi N, Holloway RG, Moore CG, Wenning GK, Yahr MD, Seidl L (2004) Movement Disorder Society task force report on the Hoehn and Yahr staging scale: status and recommendations. Mov Disord 19(9):1020–1028. https://doi.org/10.1002/mds.20213

28. Thenganatt MA, Jankovic J (2014) Parkinson disease subtypes. JAMA Neurol 71(4):499–504. https://doi.org/10.1001/jamaneurol.2013.6233

29. Schiess MC, Suescun J (2015) Clinical determinants of progression of Parkinson disease: predicting prognosis by subtype. JAMA Neurol 72(8):859–860. https://doi.org/10.1001/jamaneurol.2015.1067

30. Kotagal V (2016) Is PIGD a legitimate motor subtype in Parkinson disease? Ann Clin Transl Neurol 3(6):473–477. https://doi.org/10.1002/acn3.312

31. Herman T, Rosenberg-Katz K, Jacob Y, Auriel E, Gurevich T, Giladi N, Hausdorff JM (2013) White matter hyperintensities in Parkinson's disease: do they explain the disparity between the postural instability gait difficulty and tremor dominant subtypes? PLoS One 8(1):e55193–e55193. https://doi.org/10.1371/journal.pone.0055193

32. De Pablo-Fernandez E, Lees AJ, Holton JL, Warner TT (2019) Prognosis and neuropathologic correlation of clinical subtypes of Parkinson disease. JAMA Neurol 76(4):470–479. https://doi.org/10.1001/jamaneurol.2018.4377

33. Erro R, Picillo M, Amboni M, Savastano R, Scannapieco S, Cuoco S, Santangelo G, Vitale C, Pellecchia MT, Barone P (2019) Comparing postural instability and gait disorder and akinetic-rigid subtyping of Parkinson disease and their stability over time. Eur J Neurol 26(9):1212–1218. https://doi.org/10.1111/ene.13968

34. Titova N, Qamar MA, Chaudhuri KR (2017) The nonmotor features of Parkinson's disease. Int Rev Neurobiol 132:33–54. https://doi.org/10.1016/bs.irn.2017.02.016

35. Lindholm B, Eek F, Skogar O, Hansson EE (2019) Dyskinesia and FAB score predict future falling in Parkinson's disease. Acta Neurol Scand 139(6):512–518. https://doi.org/10.1111/ane.13084

36. Duncan RP, Cavanaugh JT, Earhart GM, Ellis TD, Ford MP, Foreman KB, Leddy AL, Paul SS, Canning CG, Thackeray A, Dibble LE (2015) External validation of a simple clinical tool used to predict falls in people with Parkinson disease. Parkinsonism Relat Disord 21(8):960–963. https://doi.org/10.1016/j.parkreldis.2015.05.008

37. Paul SS, Canning CG, Sherrington C, Lord SR, Close JC, Fung VS (2013) Three simple clinical tests to accurately predict falls in people with Parkinson's disease. Mov Disord 28(5):655–662. https://doi.org/10.1002/mds.25404

38. Snijders AH, Haaxma CA, Hagen YJ, Munneke M, Bloem BR (2012) Freezer or non-freezer: clinical assessment of freezing of gait. Parkinsonism Relat Disord 18(2):149–154. https://doi.org/10.1016/j.parkreldis.2011.09.006

39. Nieuwboer A, Rochester L, Herman T, Vandenberghe W, Emil GE, Thomaes T, Giladi N (2009) Reliability of the new freezing of gait questionnaire: agreement between patients with Parkinson's disease and their carers. Gait Posture 30(4):459–463. https://doi.org/10.1016/j.gaitpost.2009.07.108

40. Espay AJ, Fasano A, van Nuenen BFL, Payne MM, Snijders AH, Bloem BR (2012) "on" state freezing of gait in Parkinson disease: a paradoxical levodopa-induced complication. Neurology 78(7):454–457. https://doi.org/10.1212/WNL.0b013e3182477ec0

41. Fietzek UM, Zwosta J, Schroeteler FE, Ziegler K, Ceballos-Baumann AO (2013) Levodopa changes the severity of freezing in Parkinson's disease. Parkinsonism Relat Disord 19(10):894–896. https://doi.org/10.1016/j.parkreldis.2013.04.004

42. Bloem BR, Marinus J, Almeida Q, Dibble L, Nieuwboer A, Post B, Ruzicka E, Goetz C, Stebbins G, Martinez-Martin P, Schrag A (2016) Measurement instruments to assess posture, gait, and balance in Parkinson's disease: critique and recommendations. Mov Disord 31(9):1342–1355. https://doi.org/10.1002/mds.26572

43. Franchignoni F, Giordano A, Ronconi G, Rabini A, Ferriero G (2014) Rasch validation of the activities-specific balance confidence scale and its short versions in patients with Parkinson's disease. J Rehabil Med 46(6):532–539. https://doi.org/10.2340/16501977-1808

44. Tinetti ME, Richman D, Powell L (1990) Falls efficacy as a measure of fear of falling.

J Gerontol 45(6):P239–P243. https://doi.org/10.1093/geronj/45.6.p239

45. Giladi N, Shabtai H, Simon ES, Biran S, Tal J, Korczyn AD (2000) Construction of freezing of gait questionnaire for patients with parkinsonism. Parkinsonism Relat Disord 6(3):165–170. https://doi.org/10.1016/s1353-8020(99)00062-0

46. Yardley L, Smith H (2002) A prospective study of the relationship between feared consequences of falling and avoidance of activity in community-living older people. The Gerontologist 42(1):17–23. https://doi.org/10.1093/geront/42.1.17

47. Goetz CG, Tilley BC, Shaftman SR, Stebbins GT, Fahn S, Martinez-Martin P, Poewe W, Sampaio C, Stern MB, Dodel R, Dubois B, Holloway R, Jankovic J, Kulisevsky J, Lang AE, Lees A, Leurgans S, LeWitt PA, Nyenhuis D, Olanow CW, Rascol O, Schrag A, Teresi JA, van Hilten JJ, LaPelle N, Movement Disorder Society URTF (2008) Movement Disorder Society-sponsored revision of the unified Parkinson's disease rating scale (MDS-UPDRS): scale presentation and clinimetric testing results. Mov Disord 23(15):2129–2170. https://doi.org/10.1002/mds.22340

48. Valkovič P, Brožová H, Bötzel K, Ee R, Benetin J (2008) Push and release test predicts better Parkinson fallers and nonfallers than the pull test: comparison in OFF and ON medication states. Mov Disord 23(10):1453–1457. https://doi.org/10.1002/mds.22131

49. Qutubuddin AA, Pegg PO, Cifu DX, Brown R, McNamee S, Carne W (2005) Validating the Berg balance scale for patients with Parkinson's disease: a key to rehabilitation evaluation. Arch Phys Med Rehabil 86(4):789–792. https://doi.org/10.1016/j.apmr.2004.11.005

50. Di Carlo S, Bravini E, Vercelli S, Massazza G, Ferriero G (2016) The mini-BESTest: a review of psychometric properties. Int J Rehabil Res 39(2):97–105. https://doi.org/10.1097/MRR.0000000000000153

51. Dye DC, Eakman AM, Bolton KM (2013) Assessing the validity of the dynamic gait index in a balance disorders clinic: an application of Rasch analysis. Phys Ther 93(6):809–818. https://doi.org/10.2522/ptj.20120163

52. Yang Y, Wang Y, Zhou Y, Chen C, Xing D, Wang C (2014) Validity of the functional gait assessment in patients with Parkinson disease: construct, concurrent, and predictive validity. Phys Ther 94(3):392–400. https://doi.org/10.2522/ptj.20130019

53. Krzyszton K, Stolarski J, Kochanowski J (2018) Evaluation of balance disorders in Parkinson's disease using simple diagnostic tests-not so simple to choose. Front Neurol 9:932. https://doi.org/10.3389/fneur.2018.00932

54. Barry E, Galvin R, Keogh C, Horgan F, Fahey T (2014) Is the timed up and go test a useful predictor of risk of falls in community dwelling older adults: a systematic review and meta-analysis. BMC Geriatr 14:14. https://doi.org/10.1186/1471-2318-14-14

55. Ziegler K, Schroeteler F, Ceballos-Baumann AO, Fietzek UM (2010) A new rating instrument to assess festination and freezing gait in parkinsonian patients. Mov Disord 25(8):1012–1018. https://doi.org/10.1002/mds.22993

56. Fietzek UM, Schulz SJ, Ziegler K, Ceballos-Baumann AO (2020) The minimal clinically relevant change of the FOG score. J Parkinsons Dis 10(1):325–332. https://doi.org/10.3233/jpd-191783

57. Giladi N, Tal J, Azulay T, Rascol O, Brooks DJ, Melamed E, Oertel W, Poewe WH, Stocchi F, Tolosa E (2009) Validation of the freezing of gait questionnaire in patients with Parkinson's disease. Mov Disord 24(5):655–661. https://doi.org/10.1002/mds.21745

58. Fasano A, Herman T, Tessitore A, Strafella AP, Bohnen NI (2015) Neuroimaging of freezing of gait. J Parkinsons Dis 5(2):241–254. https://doi.org/10.3233/jpd-150536

59. Otomune H, Mihara M, Hattori N, Fujimoto H, Kajiyama Y, Konaka K, Mitani Y, Watanabe Y, Mochizuki H (2019) Involvement of cortical dysfunction in frequent falls in patients with Parkinson's disease. Parkinsonism Relat Disord 64:169–174. https://doi.org/10.1016/j.parkreldis.2019.04.007

60. Matar E, Shine JM, Gilat M, Ehgoetz Martens KA, Ward PB, Frank MJ, Moustafa AA, Naismith SL, Lewis SJG (2019) Identifying the neural correlates of doorway freezing in Parkinson's disease. Hum Brain Mapp 40(7):2055–2064. https://doi.org/10.1002/hbm.24506

61. Maetzler W, Klucken J, Horne M (2016) A clinical view on the development of technology-based tools in managing Parkinson's disease. Mov Disord 31(9):1263–1271. https://doi.org/10.1002/mds.26673

62. Vilas-Boas MDC, Choupina HMP, Rocha AP, Fernandes JM, Cunha JPS (2019) Full-body motion assessment: concurrent validation of two body tracking depth sensors versus a gold standard system during gait. J Biomech 87:189–196. https://doi.org/10.1016/j.jbiomech.2019.03.008

63. Schlachetzki JCM, Barth J, Marxreiter F, Gossler J, Kohl Z, Reinfelder S, Gassner H, Aminian K, Eskofier BM, Winkler J, Klucken J (2017) Wearable sensors objectively measure gait parameters in Parkinson's disease. PLoS One 12(10):e0183989. https://doi.org/10.1371/journal.pone.0183989

64. Del Din S, Hickey A, Hurwitz N, Mathers JC, Rochester L, Godfrey A (2016) Measuring gait with an accelerometer-based wearable: influence of device location, testing protocol and age. Physiol Meas 37(10):1785–1797. https://doi.org/10.1088/0967-3334/37/10/1785

65. Galperin I, Hillel I, Del Din S, Bekkers EMJ, Nieuwboer A, Abbruzzese G, Avanzino L, Nieuwhof F, Bloem BR, Rochester L, Della Croce U, Cereatti A, Giladi N, Mirelman A, Hausdorff JM (2019) Associations between daily-living physical activity and laboratory-based assessments of motor severity in patients with falls and Parkinson's disease. Parkinsonism Relat Disord 62:85–90. https://doi.org/10.1016/j.parkreldis.2019.01.022

66. Lord S, Galna B, Verghese J, Coleman S, Burn D, Rochester L (2013) Independent domains of gait in older adults and associated motor and nonmotor attributes: validation of a factor analysis approach. J Gerontol A Biol Sci Med Sci 68(7):820–827. https://doi.org/10.1093/gerona/gls255

67. Hausdorff JM (2007) Gait dynamics, fractals and falls: finding meaning in the stride-to-stride fluctuations of human walking. Hum Mov Sci 26(4):555–589. https://doi.org/10.1016/j.humov.2007.05.003

68. Galna B, Lord S, Rochester L (2013) Is gait variability reliable in older adults and Parkinson's disease? Towards an optimal testing protocol. Gait Posture 37(4):580–585. https://doi.org/10.1016/j.gaitpost.2012.09.025

69. Curtze C, Nutt JG, Carlson-Kuhta P, Mancini M, Horak FB (2015) Levodopa is a double-edged sword for balance and gait in people with Parkinson's disease. Mov Disord 30(10):1361–1370. https://doi.org/10.1002/mds.26269

70. Fasano A, Herzog J, Seifert E, Stolze H, Falk D, Reese R, Volkmann J, Deuschl G (2011) Modulation of gait coordination by subthalamic stimulation improves freezing of gait. Mov Disord 26(5):844–851. https://doi.org/10.1002/mds.23583

71. Stolze H, Klebe S, Poepping M, Lorenz D, Herzog J, Hamel W, Schrader B, Raethjen J, Wenzelburger R, Mehdorn HM, Deuschl G, Krack P (2001) Effects of bilateral subthalamic nucleus stimulation on parkinsonian gait. Neurology 57(1):144–146. https://doi.org/10.1212/wnl.57.1.144

72. Bartels AL, Balash Y, Gurevich T, Schaafsma JD, Hausdorff JM, Giladi N (2003) Relationship between freezing of gait (FOG) and other features of Parkinson's: FOG is not correlated with bradykinesia. J Clin Neurosci 10(5):584–588. https://doi.org/10.1016/s0967-5868(03)00192-9

73. Vokaer M, Azar NA, de Beyl DZ (2003) Effects of levodopa on upper limb mobility and gait in Parkinson's disease. J Neurol Neurosurg Psychiatry 74(9):1304–1307. https://doi.org/10.1136/jnnp.74.9.1304

74. Tomlinson CL, Stowe R, Patel S, Rick C, Gray R, Clarke CE (2010) Systematic review of levodopa dose equivalency reporting in Parkinson's disease. Mov Disord 25(15):2649–2653. https://doi.org/10.1002/mds.23429

75. Holloway RG, Shoulson I, Fahn S, Kieburtz K, Lang A, Marek K, McDermott M, Seibyl J, Weiner W, Musch B, Kamp C, Welsh M, Shinaman A, Pahwa R, Barclay L, Hubble J, LeWitt P, Miyasaki J, Suchowersky O, Stacy M, Russell DS, Ford B, Hammerstad J, Riley D, Standaert D, Wooten F, Factor S, Jankovic J, Atassi F, Kurlan R, Panisset M, Rajput A, Rodnitzky R, Shults C, Petsinger G, Waters C, Pfeiffer R, Biglan K, Borchert L, Montgomery A, Sutherland L, Weeks C, DeAngelis M, Sime E, Wood S, Pantella C, Harrigan M, Fussell B, Dillon S, Alexander-Brown B, Rainey P, Tennis M, Rost-Ruffner E, Brown D, Evans S, Berry D, Hall J, Shirley T, Dobson J, Fontaine D, Pfeiffer B, Brocht A, Bennett S, Daigneault S, Hodgeman K, O'Connell C, Ross T, Richard K, Watts A, Parkinson Study G (2004) Pramipexole vs levodopa as initial treatment for Parkinson disease: a 4-year randomized controlled trial. Arch Neurol 61(7):1044–1053. https://doi.org/10.1001/archneur.61.7.1044

76. Rascol O, Brooks DJ, Korczyn AD, De Deyn PP, Clarke CE, Lang AE (2000) A five-year study of the incidence of dyskinesia in patients with early Parkinson's disease who were treated with ropinirole or levodopa. N Engl J Med 342(20):1484–1491. https://doi.org/10.1056/nejm200005183422004

77. Chung KA, Lobb BM, Nutt JG, Horak FB (2010) Effects of a central cholinesterase inhibitor on reducing falls in Parkinson disease. Neurology 75(14):1263–1269. https://doi.org/10.1212/WNL.0b013e3181f6128c

78. Espay AJ, Dwivedi AK, Payne M, Gaines L, Vaughan JE, Maddux BN, Slevin JT, Gartner M, Sahay A, Revilla FJ, Duker AP, Shukla R (2011) Methylphenidate for gait impairment in Parkinson disease: a randomized clinical trial. Neurology 76(14):1256–1262. https://doi.org/10.1212/WNL.0b013e3182143537

79. Foley PB, Espay AJ, Maddux BN, Revilla FJ, Duker AP, Slevin JT (2011) Methylphenidate for gait impairment in Parkinson disease: a randomized clinical trial. Neurology 77(23):e140–e140. https://doi.org/10.1212/wnl.0b013e318239c081

80. Moreau C, Delval A, Defebvre L, Dujardin K, Duhamel A, Petyt G, Vuillaume I, Corvol J-C, Brefel-Courbon C, Ory-Magne F, Guehl D, Eusebio A, Fraix V, Saulnier P-J, Lagha-Boukbiza O, Durif F, Faighel M, Giordana C, Drapier S, Maltête D, Tranchant C, Houeto J-L, Debû B, Sablonniere B, Azulay J-P, Tison F, Rascol O, Vidailhet M, Destée A, Bloem BR, Bordet R, Devos D (2012) Methylphenidate for gait hypokinesia and freezing in patients with Parkinson's disease undergoing subthalamic stimulation: a multicentre, parallel, randomised, placebo-controlled trial. Lancet Neurol 11(7):589–596. https://doi.org/10.1016/s1474-4422(12)70106-0

81. Barker AT, Jalinous R, Freeston IL (1985) Non-invasive magnetic stimulation of human motor cortex. Lancet 325(8437):1106–1107. https://doi.org/10.1016/s0140-6736(85)92413-4

82. Lomarev MP, Kanchana S, Bara-Jimenez W, Iyer M, Wassermann EM, Hallett M (2005) Placebo-controlled study of rTMS for the treatment of Parkinson's disease. Mov Disord 21(3):325–331. https://doi.org/10.1002/mds.20713

83. Pascual-Leone A, Valls-Sole J, Brasil-Neto JP, Cammarota A, Grafman J, Hallett M (1994) Akinesia in Parkinson's disease. II Effects of subthreshold repetitive transcranial motor cortex stimulation. Neurology 44(5):892–892. https://doi.org/10.1212/wnl.44.5.892

84. Maeda F, Keenan JP, Tormos JM, Topka H, Pascual-Leone A (2000) Modulation of corticospinal excitability by repetitive transcranial magnetic stimulation. Clin Neurophysiol 111(5):800–805. https://doi.org/10.1016/s1388-2457(99)00323-5

85. Huang YZ, Edwards MJ, Rounis E, Bhatia KP, Rothwell JC (2005) Theta burst stimulation of the human motor cortex. Neuron 45(2):201–206. https://doi.org/10.1016/j.neuron.2004.12.033

86. Strafella AP, Paus T, Barrett J, Dagher A (2001) Repetitive transcranial magnetic stimulation of the human prefrontal cortex induces dopamine release in the caudate nucleus. J Neurosci 21(15):RC157

87. Ikeguchi M, Touge T, Nishiyama Y, Takeuchi H, Kuriyama S, Ohkawa M (2003) Effects of successive repetitive transcranial magnetic stimulation on motor performances and brain perfusion in idiopathic Parkinson's disease. J Neurol Sci 209(1–2):41–46. https://doi.org/10.1016/s0022-510x(02)00459-8

88. Lefaucheur J-P, Drouot X, Von Raison F, Ménard-Lefaucheur I, Cesaro P, Nguyen J-P (2004) Improvement of motor performance and modulation of cortical excitability by repetitive transcranial magnetic stimulation of the motor cortex in Parkinson's disease. Clin Neurophysiol 115(11):2530–2541. https://doi.org/10.1016/j.clinph.2004.05.025

89. Nitsche MA, Paulus W (2000) Excitability changes induced in the human motor cortex by weak transcranial direct current stimulation. J Physiol 537(Pt 3):633–639. https://doi.org/10.1111/j.1469-7793.2000.t01-1-00633.x

90. Amadi U, Ilie A, Johansen-Berg H, Stagg CJ (2014) Polarity-specific effects of motor transcranial direct current stimulation on fMRI resting state networks. NeuroImage 88(100):155–161. https://doi.org/10.1016/j.neuroimage.2013.11.037

91. Valentino F, Cosentino G, Brighina F, Pozzi NG, Sandrini G, Fierro B, Savettieri G, D'Amelio M, Pacchetti C (2014) Transcranial direct current stimulation for treatment of freezing of gait: a cross-over study. Mov Disord 29(8):1064–1069. https://doi.org/10.1002/mds.25897

92. Dagan M, Herman T, Harrison R, Zhou J, Giladi N, Ruffini G, Manor B, Hausdorff JM (2018) Multitarget transcranial direct current stimulation for freezing of gait in Parkinson's disease. Mov Disord 33(4):642–646. https://doi.org/10.1002/mds.27300

93. Deuschl G, Schade-Brittinger C, Krack P, Volkmann J, Schäfer H, Bötzel K, Daniels C, Deutschländer A, Dillmann U, Eisner W, Gruber D, Hamel W, Herzog J, Hilker R, Klebe S, Kloß M, Koy J, Krause M, Kupsch A, Lorenz D, Lorenzl S, Mehdorn HM, Moringlane JR, Oertel W, Pinsker MO, Reichmann H, Reuß A, Schneider G-H, Schnitzler A, Steude U, Sturm V, Timmermann L, Tronnier V, Trottenberg T, Wojtecki L, Wolf E, Poewe W, Voges J (2006) A randomized trial of deep-brain stimulation for Parkinson's disease. N Engl J Med

355(9):896–908. https://doi.org/10.1056/nejmoa060281

94. Schuepbach WM, Rau J, Knudsen K, Volkmann J, Krack P, Timmermann L, Halbig TD, Hesekamp H, Navarro SM, Meier N, Falk D, Mehdorn M, Paschen S, Maarouf M, Barbe MT, Fink GR, Kupsch A, Gruber D, Schneider GH, Seigneuret E, Kistner A, Chaynes P, Ory-Magne F, Brefel Courbon C, Vesper J, Schnitzler A, Wojtecki L, Houeto JL, Bataille B, Maltete D, Damier P, Raoul S, Sixel-Doering F, Hellwig D, Gharabaghi A, Kruger R, Pinsker MO, Amtage F, Regis JM, Witjas T, Thobois S, Mertens P, Kloss M, Hartmann A, Oertel WH, Post B, Speelman H, Agid Y, Schade-Brittinger C, Deuschl G, Group ES (2013) Neurostimulation for Parkinson's disease with early motor complications. N Engl J Med 368(7):610–622. https://doi.org/10.1056/NEJMoa1205158

95. Bronstein JM, Tagliati M, Alterman RL, Lozano AM, Volkmann J, Stefani A, Horak FB, Okun MS, Foote KD, Krack P, Pahwa R, Henderson JM, Hariz MI, Bakay RA, Rezai A, Marks WJ Jr, Moro E, Vitek JL, Weaver FM, Gross RE, DeLong MR (2011) Deep brain stimulation for Parkinson disease: an expert consensus and review of key issues. Arch Neurol 68(2):165–165. https://doi.org/10.1001/archneurol.2010.260

96. Schrader C, Capelle HH, Kinfe TM, Blahak C, Bazner H, Lutjens G, Dressler D, Krauss JK (2011) GPi-DBS may induce a hypokinetic gait disorder with freezing of gait in patients with dystonia. Neurology 77(5):483–488. https://doi.org/10.1212/wnl.0b013e318227b19e

97. Ferraye MU, Debu B, Fraix V, Xie-Brustolin J, Chabardes S, Krack P, Benabid AL, Pollak P (2008) Effects of subthalamic nucleus stimulation and levodopa on freezing of gait in Parkinson disease. Neurology 70(16, Part 2):1431–1437. https://doi.org/10.1212/01.wnl.0000310416.90757.85

98. Gilat M, Lígia Silva de Lima A, Bloem BR, Shine JM, Nonnekes J, SJG L (2018) Freezing of gait: promising avenues for future treatment. Parkinsonism Relat Disord 52:7–16. https://doi.org/10.1016/j.parkreldis.2018.03.009

99. Moreau C, Defebvre L, Destee A, Bleuse S, Clement F, Blatt JL, Krystkowiak P, Devos D (2008) STN-DBS frequency effects on freezing of gait in advanced Parkinson disease. Neurology 71(2):80–84. https://doi.org/10.1212/01.wnl.0000303972.16279.46

100. Plotnik M, Bartsch RP, Zeev A, Giladi N, Hausdorff JM (2013) Effects of walking speed on asymmetry and bilateral coordination of gait. Gait Posture 38(4):864–869. https://doi.org/10.1016/j.gaitpost.2013.04.011

101. Weiss D, Walach M, Meisner C, Fritz M, Scholten M, Breit S, Plewnia C, Bender B, Gharabaghi A, Wächter T, Krüger R (2013) Nigral stimulation for resistant axial motor impairment in Parkinson's disease? A randomized controlled trial. Brain 136(Pt 7):2098–2108. https://doi.org/10.1093/brain/awt122

102. Dayal V, Grover T, Limousin P, Akram H, Cappon D, Candelario J, Salazar M, Tripoliti E, Zrinzo L, Hyam J, Jahanshahi M, Hariz M, Foltynie T (2018) The effect of short pulse width settings on the therapeutic window in subthalamic nucleus deep brain stimulation for Parkinson's disease. J Parkinsons Dis 8(2):273–279. https://doi.org/10.3233/jpd-171272

103. Fasano A, Aquino CC, Krauss JK, Honey CR, Bloem BR (2015) Axial disability and deep brain stimulation in patients with Parkinson disease. Nat Rev Neurol 11(2):98–110. https://doi.org/10.1038/nrneurol.2014.252

104. Morris ME, Martin CL, Schenkman ML (2010) Striding out with Parkinson disease: evidence-based physical therapy for gait disorders. Phys Ther 90(2):280–288. https://doi.org/10.2522/ptj.20090091

105. Ginis P, Nackaerts E, Nieuwboer A, Heremans E (2018) Cueing for people with Parkinson's disease with freezing of gait: a narrative review of the state-of-the-art and novel perspectives. Ann Phys Rehabil Med 61(6):407–413. https://doi.org/10.1016/j.rehab.2017.08.002

106. Harrison EC, Horin AP, Earhart GM (2018) Internal cueing improves gait more than external cueing in healthy adults and people with Parkinson disease. Sci Rep 8(1):15525. https://doi.org/10.1038/s41598-018-33942-6

107. Barry G, Galna B, Rochester L (2014) The role of exergaming in Parkinson's disease rehabilitation: a systematic review of the evidence. J Neuroeng Rehabil 11:33. https://doi.org/10.1186/1743-0003-11-33

108. Dockx K, Bekkers EM, Van den Bergh V, Ginis P, Rochester L, Hausdorff JM, Mirelman A, Nieuwboer A (2016) Virtual reality for rehabilitation in Parkinson's disease. Cochrane Database Syst Rev 12:CD010760. https://doi.org/10.1002/14651858.CD010760.pub2

109. Kosutzka Z, Kusnirova A, Hajduk M, Straka I, Minar M, Valkovic P (2019) Gait disorders questionnaire-promising tool for virtual reality designing in patients with Parkinson's disease. Front Neurol 10:1024. https://doi.org/10.3389/fneur.2019.01024

110. Bhidayasiri R, Truong DD (2008) Motor complications in Parkinson disease: clinical manifestations and management. J Neurol Sci 266(1–2):204–215. https://doi.org/10.1016/j.jns.2007.08.028

111. Okuma Y, Silva de Lima AL, Fukae J, Bloem BR, Snijders AH (2018) A prospective study of falls in relation to freezing of gait and response fluctuations in Parkinson's disease. Parkinsonism Relat Disord 46:30–35. https://doi.org/10.1016/j.parkreldis.2017.10.013

112. Mancini F, Comi C, Oggioni GD, Pacchetti C, Calandrella D, Coletti Moja M, Riboldazzi G, Tunesi S, Dal Fante M, Manfredi L, Lacerenza M, Cantello R, Antonini A (2014) Prevalence and features of peripheral neuropathy in Parkinson's disease patients under different therapeutic regimens. Parkinsonism Relat Disord 20(1):27–31. https://doi.org/10.1016/j.parkreldis.2013.09.007

113. Hanewinckel R, Drenthen J, Verlinden VJA, Darweesh SKL, van der Geest JN, Hofman A, van Doorn PA, Ikram MA (2017) Polyneuropathy relates to impairment in daily activities, worse gait, and fall-related injuries. Neurology 89(1):76–83. https://doi.org/10.1212/wnl.0000000000004067

114. Tramonti C, Di Martino S, Unti E, Frosini D, Bonuccelli U, Rossi B, Ceravolo R, Chisari C (2017) Gait dynamics in Pisa syndrome and Camptocormia: the role of stride length and hip kinematics. Gait Posture 57:130–135. https://doi.org/10.1016/j.gaitpost.2017.05.029

115. Avanzino L, Lagravinese G, Abbruzzese G, Pelosin E (2018) Relationships between gait and emotion in Parkinson's disease: a narrative review. Gait Posture 65:57–64. https://doi.org/10.1016/j.gaitpost.2018.06.171

116. Vandenbossche J, Deroost N, Soetens E, Coomans D, Spildooren J, Vercruysse S, Nieuwboer A, Kerckhofs E (2013) Freezing of gait in Parkinson's disease: disturbances in automaticity and control. Front Hum Neurosci 6:356–356. https://doi.org/10.3389/fnhum.2012.00356

117. Vandenbossche J, Deroost N, Soetens E, Spildooren J, Vercruysse S, Nieuwboer A, Kerckhofs E (2011) Freezing of gait in Parkinson disease is associated with impaired conflict resolution. Neurorehabil Neural Repair 25(8):765–773. https://doi.org/10.1177/1545968311403493

118. Chastan N, Bair W-N, Resnick SM, Studenski SA, Decker LM (2019) Prediagnostic markers of idiopathic Parkinson's disease: gait, visuospatial ability and executive function. Gait Posture 68:500–505. https://doi.org/10.1016/j.gaitpost.2018.12.039

119. Peterson DS, King LA, Cohen RG, Horak FB (2016) Cognitive contributions to freezing of gait in Parkinson disease: implications for physical rehabilitation. Phys Ther 96(5):659–670. https://doi.org/10.2522/ptj.20140603

120. Heinzel S, Maechtel M, Hasmann SE, Hobert MA, Heger T, Berg D, Maetzler W· (2016) Motor dual-tasking deficits predict falls in Parkinson's disease: a prospective study. Parkinsonism Relat Disord 26:73–77. https://doi.org/10.1016/j.parkreldis.2016.03.007

121. Ehgoetz Martens KA, Ellard CG, Almeida QJ (2015) Anxiety-provoked gait changes are selectively dopa-responsive in Parkinson's disease. Eur J Neurosci 42(4):2028–2035. https://doi.org/10.1111/ejn.12928

122. Martens KAE, Hall JM, Gilat M, Georgiades MJ, Walton CC, Lewis SJG (2016) Anxiety is associated with freezing of gait and attentional set-shifting in Parkinson's disease: a new perspective for early intervention. Gait Posture 49:431–436. https://doi.org/10.1016/j.gaitpost.2016.07.182

123. The Economist (2013) Unlikely results. The Economist. https://www.economist.com/graphic-detail/2013/10/21/unlikely-results. Accessed 26 Apr 2019

124. Ambani LM, Van Woert MH (1973) Start hesitation – a side effect of long-term levodopa therapy. N Engl J Med 288(21):1113–1115. https://doi.org/10.1056/nejm197305242882108

125. Charcot JM On paralysis agitans. Lectures on the diseases of the nervous system, delivered at La Salpêtrière, 2nd edn. Henry C Lea, Philadelphia, PA. https://doi.org/10.1037/12839-005

126. Fietzek UM, Stuhlinger L, Plate A, Ceballos-Baumann A, Bötzel K (2017) Spatial constraints evoke increased number of steps during turning in Parkinson's disease. Clin Neurophysiol 128(10):1954–1960. https://doi.org/10.1016/j.clinph.2017.07.399

127. Jehu D, Nantel J (2018) Fallers with Parkinson's disease exhibit restrictive trunk control during walking. Gait Posture 65:246–250. https://doi.org/10.1016/j.gaitpost.2018.07.181

128. Allcock LM, Rowan EN, Steen IN, Wesnes K, Kenny RA, Burn DJ (2009) Impaired attention predicts falling in Parkinson's disease. Parkinsonism Relat Disord 15(2):110–115. https://doi.org/10.1016/j.parkreldis.2008.03.010

129. Wood BH, Bilclough JA, Bowron A, Walker RW (2002) Incidence and prediction of falls in Parkinson's disease: a prospective multidisciplinary study. J Neurol Neurosurg Psychiatry 72(6):721–725. https://doi.org/10.1136/jnnp.72.6.721

Chapter 7

Clinical Trials for Disease-Modifying Agents in Parkinson's Disease

Orlando Artavia and Lana Chahine

Abstract

Clinically meaningful disease-modifying therapies for Parkinson's disease (PD) have yet to be identified. There are several factors that may account for this. PD is clinically and pathophysiologically heterogeneous and has relatively insidious and protracted progression. Symptomatic therapies confound measurement of the disease as well. With this complexity comes the challenge of identifying optimal study design and outcome measures for clinical trials for disease-modifying therapies. Here, these two main considerations in clinical trial design for demonstrating disease modification in PD are reviewed. Outcome measures discussed include patient-reported outcomes, clinical features, biofluid biomarkers, and structural/functional imaging. Examples of application of these biomarkers in ongoing clinical trials are presented. Study designs that mitigate effects of symptomatic therapies are discussed. Special emphasis is placed on considerations for clinical trials testing immunotherapies. Much progress has been made in identifying who is at risk of PD and characterizing the prodromal PD phase, and clinical trials to prevent PD are on the horizon. Important aspects of clinical trial design in individuals at risk for PD are also reviewed.

Key words Parkinson's disease, Neuroprotection, Clinical trials, Disease modification

1 Introduction

Parkinson's disease (PD) is a neurodegenerative disorder marked by a wide range of motor and nonmotor manifestations. It is inexorably progressive, and several clinical trials aimed at stopping or reducing progression have failed. The reasons for these failed trials are many [1]. Drug development in PD is made particularly challenging by the complex pathophysiology of PD and a lack of animal models that recapitulate the disease pathology. As a consequence, some trials may have failed because the tested agent lacks efficacy as a disease-modifying agent. However, there is concern in the field that there were lost opportunities, where the disease-modifying effect of the interventions was not demonstrated where in fact one may have existed. In turn, there are several reasons for this, but many are related to a single factor: the primary outcome measure of the trial. To date, there have as yet been no

Santiago Perez-Lloret (ed.), *Clinical Trials In Parkinson's Disease*, Neuromethods, vol. 160,
https://doi.org/10.1007/978-1-0716-0912-5_7, © Springer Science+Business Media, LLC, part of Springer Nature 2021

satisfying options for primary outcome measures that can be confidently considered a true reflection of disease modification. Related to the latter is the confounding effects of symptomatic treatments (whether by the agent being tested or by concomitant medications). Further compounding these issues is the natural history of PD, whereby key outcomes that would be prevented by disease-modifying interventions may have a prolonged latency of emergence, sometimes over several years. This chapter focuses on outcome measures in PD clinical trials for disease modification, and also includes an overview of study designs that maybe of utility in testing for disease modification.

2   Outcome Measures for PD Disease Modification Trials

Disease modification may be broadly defined as favorably changing the clinical course of PD, independent of a symptomatic benefit [2]. This is distinct from, though related to, neuroprotection, which is the prevention of neuronal loss. An agent could be neuroprotective, but not disease modifying. An agent could also be disease modifying, without being neuroprotective.

2.1 Clinical Measures and Patient Reported Outcomes

As mentioned, many of the key obstacles to demonstrating disease modification in PD center around the outcome of interest, and its measurement. Several manifestations of PD carry high morbidity and impaired quality of life. In a broad sense, preventing or delaying these, or reducing their severity, would be a modification of the disease course. Motor outcomes are often the focus of disease modification trials in PD, because many of the disabling manifestations of PD are motor. Commonly used outcome measures include motor rating scales (such as the Unified Parkinson's Disease Rating Scale—UPDRS—and the Movement Disorders Society-Unified Parkinson's Disease Rating Scale -MDS-UPDS-). The limitations of these scales as outcome measures include (but are not limited to) the rater dependency (making them subjective, even with high intra-/interrater reliability), and their fluctuation over time. Most importantly, once symptomatic therapy is started, extricating the effects of therapy from the effect of the agent being tested becomes difficult. Time to symptomatic therapy initiation is also often used as an outcome measure in early-stage PD disease modification trials, but the decision to initiate treatment is individualized and influenced by many factors so as to make this measure less favorable. Another approach to measuring disease modification has focused on the emergence of motor or nonmotor manifestations in PD such as falling, freezing of gait, motor complications (dyskinesias, fluctuations), dementia, or psychosis [3]. These outcomes carry high morbidity and are associated with reduced quality of life. Preventing them would certainly be invaluable. However, many of these outcomes have prolonged latencies, and time-to-event for these outcomes is therefore often not feasible for clinical trials that are of only a few months or at most a few years duration.

Another type of clinical outcome measure centers on patient-reported outcomes, such as quality-of-life scales. Indeed, an intervention that prevents a sustained reduction in quality of life would meet the definition of a disease modifying agent. However, here again extricating symptomatic therapy effects and the effect of various confounders is made challenging when using such measures.

These and other factors point to a critical need: objective measures of PD. The ideal objective outcome measure for PD disease modification trials would have the following characteristics. Logistically, in order to be used in multicenter studies, it would need to be affordable, simple to measure across centers, and with minimal training requirements. It would need to measure the disease "activity" and its progression over the trial period but perhaps more importantly, it would need to serve as a surrogate endpoint for PD manifestations/complications that have a prolonged latency (i.e., a surrogate of the delayed outcomes discussed above). Importantly, it would not be susceptible to the effect of symptomatic therapies.

Many objective measures of motor function in PD are in development [4]. These have several advantages including portability, and potential to capture motor function in the patient's natural environment. However, they will likely have many of the same limitations imposed by clinical rating scales, especially in terms of effect to symptomatic therapies. They will thus require extensive validation to ensure they are sound measures of disease modification. They will not be further reviewed here.

2.2 α-Synuclein and Other Biologic Biomarkers

Parkinson's disease is an α-synucleinopathy, which is a group of diseases characterized by the pathological accumulation of α-synuclein in different neuronal populations and in different regions of the brain [5, 6]. Evidence suggests that α-synuclein deposition is pathogenic [7] and accruing lines of evidence indicate that PD manifestations, severity, and progression are at least in part related to the extent of α-synuclein deposition [5, 6]. For example, α-synuclein deposition has been directly associated with nigral cell loss [5, 6]. Thus, α-synuclein is a key candidate biomarker in PD. Intensive efforts are underway to identify robust measures of α-synuclein in biologic fluids, tissue, and on imaging. This will be especially critical to explore as an outcome measure in clinical trials employing therapies targeting α-synuclein [8]. Measures of α-synuclein in that context are needed both to demonstrate target engagement as well as outcome measures. To that end, assays to detect and quantitate α-synuclein in several biologic fluids including cerebrospinal fluid (CSF), saliva, blood, and peripheral tissue are underway. A critical aspect of this biomarker development centers on the detection of abnormal, pathologic forms of α-synuclein.

The pathophysiology of PD is complex. A central aspect of it is the accumulation of α-synuclein, but this is likely the final common pathway for abnormalities in many different processes including lysosomal function, mitochondrial function), calcium homeostasis, and axonal transport [5, 6]. Protein misfolding, oxidative stress and neuroinflammation contribute to α-synuclein accumulation as well [5, 6]. There are also likely mechanisms leading to neuronal loss in PD that are entirely independent of α-synuclein. Disease modifying agents may target any number of these processes, and in turn, objective markers are needed to measure the effects of the agent.

2.3 Structural and Functional Imaging

Another potential objective outcome in PD disease modification trials is not related to PD pathology per se but rather its consequences, such as changes on imaging [9]. For example, cortical atrophy on MRI precedes clinical onset of dementia in PD by years [10]. With further development and validation, cortical volume on MRI could come to serve as an outcome measure for disease modification interventions, as exemplified in Huntington's disease clinical trials [11]. Ligand-based imaging can also be used to measure the consequences of underlying pathology. Dopamine transporter (DAT) SPECT is associated with clinically significant long-term outcomes in PD [12], and the rate of change in DAT binding in early PD is measurable over a 1-year period [13]. Fluorodopa positron emission tomography (PET) is also a promising imaging biomarker for monitoring disease-modifying response to therapy [3, 14]. However, further validation of dopaminergic imaging biomarkers, and specifically establishing if they are useful as surrogates of long-term outcomes, is needed.

A few examples of the use of objective measures of the processes that contribute to PD pathology in PD disease modification clinical are emerging (Table 1). However, again, these markers require further development and validation especially in terms of their ability to not only be an accurate reflection of the biologic process at hand but also, more importantly, serve as surrogates of a clinically meaningful outcome.

2.4 Potential Benefits and Pitfalls of Composite or Multiple Independent Endpoints

Few trials in PD have used composite endpoints as the primary endpoint for clinical trials. For example, for a clinical trial investigating an intervention to reduce disabling gait changes, a composite endpoint could be occurrence of freezing of gait, requirement for an assistive device, *or* a fall with significant injury, over the period of the trial. There would be certain advantages to doing so, namely, increasing improving statistical efficiency, decreasing the required sample size to adequate power trials, and possibly shortening trial duration. However, several limitations to composite endpoints exist; great consideration would need to be taken to

Table 1
Applications of putative PD biomarkers in PD disease modification trials

	Applications	Examples
Imaging		
Dopamine transporter SPECT	Diagnostic biomarker	Safety of Urate Elevation in Parkinson's Disease (SURE-PD) (NCT00833690)
Dopamine transporter SPECT	Trial secondary endpoint: Rate of change in DAT binding	Safety of Urate Elevation in Parkinson's Disease (SURE-PD) (CTI: NCT00833690)
		A Study to Evaluate the Efficacy of Prasinezumab (RO7046015/ PRX002) in Participants with Early Parkinson's Disease (PASADENA) (CTI: NCT03100149)
		Evaluating the Safety, Pharmacokinetics, and Pharmacodynamics of BIIB054 in Participants with Parkinson's Disease (SPARK) (CTI: NCT03318523)
[18F]-fluorodopa (F-Dopa)	Trial secondary endpoint: Percent change from baseline (as a measure of aromatic acid decarboxylase expression)	VY-AADC02 for Parkinson's Disease with Motor Fluctuations (CTI: NCT03562494)
Biofluid		
Cerebrospinal fluid	α-Synuclein	A Study to Evaluate the Efficacy of Prasinezumab (RO7046015/ PRX002) in Participants with Early Parkinson's Disease (PASADENA) (CTI: NCT03100149)
Cerebrospinal fluid	Homovanillic acid: Surrogate of dopaminergic neuron function and/or levels of CNS dopamine (exploratory)	Impact of Nilotinib on Safety, Tolerability, Pharmacokinetics and Biomarkers in Parkinson's Disease (PD Nilotinib) (CTI: NCT02954978)
Cerebrospinal fluid glucocerebrosidase activity	Trial secondary endpoint: Measure of glucocerebrosidase (GCase) activity	Ambroxol in Disease Modification in Parkinson Disease (AiM-PD) (CTI: NCT02941822)
Tissue		
Skin α-synuclein		A Study to Evaluate the Efficacy of Prasinezumab (RO7046015/ PRX002) in Participants with Early Parkinson's Disease (PASADENA) (CTI: NCT03100149)

CTI ClinicalTrials.gov Identifier

ensure that all contributors to the composite, and the composite itself, is clinically meaningful, and that a result of the trial would be interpretable.

3   Study Designs Considerations

3.1   Sample Selection and Enrichment

As therapies emerge that target specific molecular pathways in the pathogenesis of PD, identification of the appropriate samples to test these agents on will be key. Therapies targeting specific genetic pathways will logically first be tested on individuals with known abnormalities in these pathways. An example of this are therapies targeting function of the enzyme leucine rich repeat kinase 2 (LRRK2) [15]. LRRK2 G2019S mutations occur in over one-third of PD patients that are of North African Berber origin [16], and at least 15% of PD patients of Ashkenazi Jewish origin [17], reaching 30% in familial PD cases. Thus, carriers of G2019S mutations would logically form the initial samples LRRK2-therapies may be targeted at. Similarly, mutations in the glucocerebrosidase gene (GBA) account for a large proportion of familial PD in Ashkenazi Jewish populations, and around 6% of cases in the USA [18].

But gene-targeted therapies may also be of use at least in some patients without known pathogenic mutations in the target genes. For example, while LRRK2 mutations occur in only about 1–2% of sporadic PD cases, LRRK2 dysfunction is also seen in sporadic PD [19]. Likewise, glucocerebrosidase activity is abnormal even in some PD patients without known pathogenic mutations in the GBA gene [20]. Thus, it is possible that at least some patients without pathogenetic mutations may still benefit from gene-targeted therapies. Better understanding of specific genetic polymorphisms that may also predict response to such therapies, as well as assays to measure gene product/function, will be needed to help identify such patients for clinical trials.

3.2   Addressing the Confounding Effect of Symptomatic Therapy

As mentioned earlier, one of the issues related to the interpretation of clinical trials for disease modification is the confounding effect of symptomatic therapy. A proposed solution for this issue is called "washout" [2]. In placebo-controlled studies using the washout design, symptomatic therapies are discontinued after the prespecified exposure period to the agent under investigation has been passed, and the intervention arm is compared to the placebo arm. However, the symptomatic effects of PD medications can last weeks, which poses several challenges for washout including the ethical considerations of withholding symptomatic therapies as well as risk of nocebo effect (a deterioration of symptoms due to the expectation of symptomatic therapy withdrawal) [2].

Another clinical trial design proposed to overcome the confounding effects of symptomatic therapy is the "delayed start" design. In such studies, the initial phase is a conventional study with a placebo arm, but in the second stage, the placebo arm also receives the intervention. This allows for examination of whether a *delay* in treatment is related to different (worse) outcomes. Determination the optimal duration of delay requires careful consideration. In the application of the delayed start design in testing the disease-modifying potential of rasagiline [21], concern has been raised that the study was powered to detect differences in motor scale scores that were not clinically meaningful [2].

Another way to determine neuro protection and disease modification is using "staggered studies" in which data from different cohort studies can be pooled, comparing outcomes in the treatment group and placebo groups with different initial date of intervention.

4 Specific Considerations for Studies Investigating Immunotherapies as Disease Modifying Therapies

Given the role of α-synuclein in PD pathophysiology, α-synuclein has become a prime target for disease modification. One approach involves immunotherapies. Much progress has been made in passive immunization via administration of antibodies against α-synuclein [22]; active immunization approaches are in development as well. Much remains to be learned about the optimal epitope target for such trials, as trials in Alzheimer's disease have taught us [23, 24]. In addition to therapeutic development, several clinical trial considerations are important as these agents are investigated. Obviously, the sample targeted by these therapies is key, and whether agents targeting α-synuclein will have demonstrable disease-modifying effects in manifest vs. prodromal or at-risk PD remains to be seen. As with other types of investigational interventions, primary outcome measures continue to be a key consideration in demonstrating disease modifying effects of immunotherapies. Studies in Alzheimer's disease have indicated that immunotherapies targeting the underlying proteinopathy may engage and remove the target without having a clinical benefit [25–27]. Thus, while markers of target engagement are essential in such trials, demonstrating clinically meaningful disease modification will be critical. Another important consideration in immunotherapy trials variability is the antibody response across treated subjects, whether due to a variable age-related reduction in the immune competency or development of other antibodies [23, 28].

5 Specific Considerations for Clinical Trials for PD Prevention in Prodromal Populations

There is increasing recognition that in order to modify the course of PD pathology, intervention must occur at the earliest stages, possibly before the full clinical syndrome of motor PD is manifest [3].

Fortunately, many prodromal features of PD have been described, including constipation, neuropsychiatric symptoms, subtle motor abnormalities, retinal changes, vision changes, olfactory loss and REM sleep behavior disorder (Table 2) [29–33]. Carriers of pathogenic mutations associated with PD who do not have PD are another at-risk group of great interest. It is now possible to accurately identify individuals who go on to develop PD, especially when clinical prodromal features are combined with imaging biomarkers [34–39]. Indeed, 2-stage screening system for PD risk has been proposed, whereby low-cost, noninvasive measures such as olfactory testing are administered, and in those screening positive, more costly, specific testing such as dopaminergic imaging using PET or SPECT are then performed. Studies to date have demonstrated proof of this concept in research cohorts (though this has yet to be extended to the general population). In a study of 361 asymptomatic first-degree relatives of PD patients screened for olfactory deficits [40], 40 hyposmics and 38 normosmics were identified. They then underwent dopamine transporter SPECT. Four of the 40 hyposmic relatives and none of the normosmic relatives developed PD within 2 years of initial assessment; all those that developed PD also had associated abnormal SPECT scan (reduced striatal binding ratios) at baseline [40], indicative of subclinical dopaminergic system degeneration. An abnormality of odor discrimination was the best predictor of future risk of PD [41]. The 5-year risk of developing PD among first-degree relatives of PD patients with unexplained olfactory deficits was estimated at 12.5% [42]. These results have also been replicated in the Parkinson At Risk Syndrome cohort (PARS) [43]. PARS screened over 10,000 individuals with smell testing and identified 303 individuals with olfactory deficit. These individuals in turn underwent dopamine transporter (DAT) SPECT scanning and longitudinal extensive clinical characterization. Of the 280 individuals with at least 1 year of follow-up, 16 went on to develop PD [38, 39]. There was a relative risk of 17.5 of developing PD among those with a combination of olfactory deficit and DAT binding. In individuals with both olfactory dysfunction and DAT binding deficit, risk of PD was 12% per year.

Several questions remain to be answered as we prepare for trials in at-risk samples. We do not know how early intervention needs to occur to meaningfully modify long-term outcome. Aside from what the target sample for PD prevention trials would be, the other main

Table 2
Prodromal PD features and considerations related to clinical trials in at-risk and prodromal populations

Category	Marker	Clinical trial considerations
Motor abnormalities	Objective and subjective measures of motor function	Short- and long-term longitudinal changes in at-risk groups and utility as predictive marker and/or outcome measure remains to be defined
Sensory	Olfactory dysfunction	Strongly associated with future risk of PD, especially when combined with dopaminergic imaging or substantia nigra hypoechogenicity. Likely useful as sample selection criterion. Unlikely to be useful as primary outcome measure due to low rate of change once abnormal (few data on rate of change in transition from normal to abnormal olfaction; additional studies required)
	Retinal and vision changes (such as reduced color discrimination)	Short- and long-term longitudinal changes in at-risk groups and utility as predictive marker for PD remain to be defined
Sleep/sleepiness	REM sleep behavior disorder	Highly predictive of future risk of synucleinopathy; likely useful as sample selection criterion. Further data are needed to understand if incident RBD could be useful as an outcome measure in, for example, asymptomatic carriers of genetic mutations
	Excessive daytime sleepiness	Possibly useful as sample selection criterion (when combined with other measures). Short- and long-term longitudinal changes in at-risk groups and utility as predictive marker and outcome measure for PD remain to be defined
	Restless legs syndrome	Short- and long-term longitudinal changes in at-risk groups remain to be defined
Autonomic	Constipation	Nonspecific (common in older adults) but strongly associated with future risk of PD when combined with other prodromal features. Likely useful as sample selection criterion. Unlikely to be useful as outcome measure
	Orthostatic hypotension	Short- and long-term longitudinal changes in at-risk groups and utility as predictive marker and/or outcome measure remains to be defined
	Reduced heart rate variability	Short- and long-term longitudinal changes in at-risk groups and utility as predictive marker and/or outcome measure remains to be defined
Neuropsychiatric	Cognitive abnormalities	Detectable changes preceding incident PD have been reported, but additional data are needed in order to determine utility as predictive marker and/or outcome measure

(continued)

Table 2
(continued)

Category	Marker	Clinical trial considerations
	Specific personality traits	Possibly useful as sample selection criterion (when combined with other measures). Unlikely to be useful as primary outcome measure as unlikely to have detectable rate of change, but deserves further study
	Depression	Possibly useful as sample selection criterion (when combined with other measures). Short- and long-term longitudinal changes in at-risk groups and utility as predictive marker and outcome measure for PD remain to be defined
	Anxiety	Possibly useful as sample selection criterion (when combined with other measures). Short- and long-term longitudinal changes in at-risk groups and utility as predictive marker and outcome measure for PD remain to be defined
Imaging	Striatal dopamine deficiency (SPECT)	When abnormal, strongly associated with future risk of PD. likely useful as sample selection criterion. Strong candidate as outcome measure in clinical trials, but requires further study
	PD-related covariance pattern (PET and SPECT)	Short- and long-term longitudinal changes in at-risk groups and utility as predictive marker and/or outcome measure remains to be defined
	Substantia nigra hyperechogenicity transcranial doppler sonography	Strongly associated with future risk of PD. Likely useful as a sample selection criterion. Unlikely to be useful as a primary outcome measure due to low rate of change once abnormal. Rate of change in transition from normal to abnormal requires further study
	Hippocampal hyperperfusion on SPECT	Short- and long-term longitudinal changes in at-risk groups and utility as predictive marker and/or outcome measure remains to be defined
Electrophysiologic	Severity of REM atonia loss	Short- and long-term longitudinal changes in at-risk groups remain to be defined
	EEG slowing	Short- and long-term longitudinal changes in at-risk groups remain to be defined

consideration is the primary outcome measure. Obviously, incident PD would be the "gold-standard" outcome measure, but the time and cost required for such trials would not be feasible. Instead, surrogate measures of "phenoconversion" to PD are needed. Here, special attention needs to be given to which measures have predictable change in the short-term. Also, importantly, attention needs to be given to the finding that many measures of prodromal features have ceiling effects or other patterns of change (or lack thereof) that would not make them good candidates as outcome measures [44, 45]. However, some measures may be useful in the earliest stages of transition from normal to abnormal; data are limited on this phase for most measures and studies will be needed to address this critical gab in knowledge (Table 2).

6 Conclusion

Clinical trials for disease-modifying agents in Parkinson's disease need primary outcome measures that would show a true reflection of disease modification and would be a long-term predictor of the disease course. Clinical outcomes like motor scales and time to initiate symptomatic therapy have the limitation of being subjective, while the emergence of motor/nonmotor symptoms takes a long time (years) in the course of the disease to show up, making them not feasible for clinical trials. Adding to that, there is also the confounding caused by symptomatic treatment, and although there are study techniques designed to remediate the confounding problem (washout, delayed start, staggered studies), they also present with ethical concerns and uncertainty regarding length of suspension of treatment.

Imaging markers like cortical volume on MRI, DAT-SPECT, and fluorodopa-PET are promising markers and may offer the advantage of not being confounded by symptomatic therapy. However, they need further validation both in terms of their ability to serve as surrogate markers of long-term outcomes and, more importantly, that they serve as surrogates of clinically meaningful outcomes.

Since PD is an α-synucleinopathy, there are efforts underway to detect α-synuclein in CSF, saliva, blood, and other biofluids. It is hoped that with the right assays, α-synuclein could serve as both a diagnostic biomarker and a surrogate outcome in clinical trials. These would be especially important for current studies of immunotherapies that target alpha-synuclein.

Another important consideration for clinical trials is that PD is pathophysiology complex. There are several drug development programs targeting the many possible pathophysiologic pathways in PD. As these targeted molecular therapies emerge, understanding which PD patients would benefit from which therapies will be essential. Subtyping based on clinical, imaging, biomarker, and genetic characteristics may be useful to that end. In addition, and particularly in relation to assays targeting genetic abnormalities that may contribute to PD pathogenesis, assays to measure gene product/function may help in the future both for sample selection as well as possibly as outcome measures.

Finally, a critical aspect of clinical trials for disease modifying agents is whether the appropriate target for specific therapies are at-risk or prodromal individuals rather than those with fully manifest PD. It is likely that when intervention is done at the earliest stages of the disease, it is more likely to cause disease modification and, following that trend, maybe even prevent the disease. This poses the challenge of accurately identifying people at the prodromal phase and people at risk for PD. Fortunately, many prodromal features have already been described, and some studies have already demonstrated the feasibility and utility of combining prodromal markers to improve specificity of detection. In the case of prodromal clinical trials, the outcome measure to be employed will be critical, and surrogate measures of future PD risk will be essential to allow for relatively short trials with realistic sample sizes.

References

1. Lang AE, Espay AJ (2018) Disease modification in Parkinson's disease: current approaches, challenges, and future considerations. Mov Disord 33(5):660–677. https://doi.org/10.1002/mds.27360

2. Rascol O (2009) "Disease-modification" trials in Parkinson disease: target populations, endpoints and study design. Neurology 72(7 Suppl):S51–S58. https://doi.org/10.1212/WNL.0b013e318199049e

3. Lang AE (2010) Clinical trials of disease-modifying therapies for neurodegenerative diseases: the challenges and the future. Nat Med 16(11):1223–1226. https://doi.org/10.1038/nm.2220

4. Espay AJ, Bonato P, Nahab FB, Maetzler W, Dean JM, Klucken J, Eskofier BM, Merola A, Horak F, Lang AE, Reilmann R, Giuffrida J, Nieuwboer A, Horne M, Little MA, Litvan I, Simuni T, Dorsey ER, Burack MA, Kubota K, Kamondi A, Godinho C, Daneault JF, Mitsi G, Krinke L, Hausdorff JM, Bloem BR, Papapetropoulos S, Movement Disorders Society Task Force on T (2016) Technology in Parkinson's disease: challenges and opportunities. Mov Disord 31(9):1272–1282. https://doi.org/10.1002/mds.26642

5. Rocha EM, De Miranda B, Sanders LH (2018) Alpha-synuclein: pathology, mitochondrial dysfunction and neuroinflammation in Parkinson's disease. Neurobiol Dis 109(Pt B):249–257. https://doi.org/10.1016/j.nbd.2017.04.004

6. Poewe W, Seppi K, Tanner CM, Halliday GM, Brundin P, Volkmann J, Schrag AE, Lang AE (2017) Parkinson disease. Nat Rev Dis Primers 3:17013. https://doi.org/10.1038/nrdp.2017.13

7. Mahlknecht P, Seppi K, Poewe W (2015) The concept of prodromal Parkinson's disease. J Parkinsons Dis 5(4):681–697. https://doi.org/10.3233/JPD-150685

8. Zheng B, Liao Z, Locascio JJ, Lesniak KA, Roderick SS, Watt ML, Eklund AC, Zhang-James Y, Kim PD, Hauser MA, Grunblatt E, Moran LB, Mandel SA, Riederer P, Miller RM, Federoff HJ, Wullner U, Papapetropoulos S, Youdim MB, Cantuti-Castelvetri I, Young AB, Vance JM, Davis RL, Hedreen JC, Adler CH, Beach TG, Graeber MB, Middleton FA, Rochet JC, Scherzer CR, Global PDGEC (2010) PGC-1alpha, a potential therapeutic target for early intervention in Parkinson's disease. Sci Transl Med 2(52):52ra73. https://doi.org/10.1126/scitranslmed.3001059

9. Strafella AP, Bohnen NI, Pavese N, Vaillancourt DE, van Eimeren T, Politis M, Tessitore A, Ghadery C, Lewis S, Group IP-NS (2018) Imaging markers of progression in Parkinson's disease. Mov Disord Clin Pract 5(6):586–596. https://doi.org/10.1002/mdc3.12673

10. Weintraub D, Dietz N, Duda JE, Wolk DA, Doshi J, Xie SX, Davatzikos C, Clark CM, Siderowf A (2012) Alzheimer's disease pattern of brain atrophy predicts cognitive decline in Parkinson's disease. Brain 135(Pt 1):170–180. https://doi.org/10.1093/brain/awr277

11. Tabrizi SJ, Scahill RI, Durr A, Roos RA, Leavitt BR, Jones R, Landwehrmeyer GB, Fox NC, Johnson H, Hicks SL, Kennard C, Craufurd D, Frost C, Langbehn DR, Reilmann R, Stout JC, Investigators T-H (2011) Biological and clinical changes in premanifest and early stage Huntington's disease in the TRACK-HD study: the 12-month longitudinal analysis.

Lancet Neurol 10(1):31–42. https://doi.org/10.1016/S1474-4422(10)70276-3

12. Ravina B, Marek K, Eberly S, Oakes D, Kurlan R, Ascherio A, Beal F, Beck J, Flagg E, Galpern WR, Harman J, Lang AE, Schwarzschild M, Tanner C, Shoulson I (2012) Dopamine transporter imaging is associated with long-term outcomes in Parkinson's disease. Mov Disord 27(11):1392–1397. https://doi.org/10.1002/mds.25157

13. Marek K, Chowdhury S, Siderowf A, Lasch S, Coffey CS, Caspell-Garcia C, Simuni T, Jennings D, Tanner CM, Trojanowski JQ, Shaw LM, Seibyl J, Schuff N, Singleton A, Kieburtz K, Toga AW, Mollenhauer B, Galasko D, Chahine LM, Weintraub D, Foroud T, Tosun-Turgut D, Poston K, Arnedo V, Frasier M, Sherer T, Parkinson's Progression Markers I (2018) The Parkinson's progression markers initiative (PPMI) – establishing a PD biomarker cohort. Ann Clin Transl Neurol 5(12):1460–1477. https://doi.org/10.1002/acn3.644

14. Feigin A, Kaplitt MG, Tang C, Lin T, Mattis P, Dhawan V, During MJ, Eidelberg D (2007) Modulation of metabolic brain networks after subthalamic gene therapy for Parkinson's disease. Proc Natl Acad Sci U S A 104(49):19559–19564. https://doi.org/10.1073/pnas.0706006104

15. West AB (2017) Achieving neuroprotection with LRRK2 kinase inhibitors in Parkinson disease. Exp Neurol 298(Pt B):236–245. https://doi.org/10.1016/j.expneurol.2017.07.019

16. Hulihan MM, Ishihara-Paul L, Kachergus J, Warren L, Amouri R, Elango R, Prinjha RK, Upmanyu R, Kefi M, Zouari M, Sassi SB, Yahmed SB, El Euch-Fayeche G, Matthews PM, Middleton LT, Gibson RA, Hentati F, Farrer MJ (2008) LRRK2 Gly2019Ser penetrance in Arab-Berber patients from Tunisia: a case-control genetic study. Lancet Neurol 7(7):591–594. https://doi.org/10.1016/S1474-4422(08)70116-9

17. Ozelius LJ, Senthil G, Saunders-Pullman R, Ohmann E, Deligtisch A, Tagliati M, Hunt AL, Klein C, Henick B, Hailpern SM, Lipton RB, Soto-Valencia J, Risch N, Bressman SB (2006) LRRK2 G2019S as a cause of Parkinson's disease in Ashkenazi Jews. N Engl J Med 354(4):424–425. https://doi.org/10.1056/NEJMc055509

18. Sidransky E, Nalls MA, Aasly JO, Aharon-Peretz J, Annesi G, Barbosa ER, Bar-Shira A, Berg D, Bras J, Brice A, Chen CM, Clark LN, Condroyer C, De Marco EV, Durr A, Eblan MJ, Fahn S, Farrer MJ, Fung HC, Gan-Or Z, Gasser T, Gershoni-Baruch R, Giladi N, Griffith A, Gurevich T, Januario C, Kropp P, Lang AE, Lee-Chen GJ, Lesage S, Marder K, Mata IF, Mirelman A, Mitsui J, Mizuta I, Nicoletti G, Oliveira C, Ottman R, Orr-Urtreger A, Pereira LV, Quattrone A, Rogaeva E, Rolfs A, Rosenbaum H, Rozenberg R, Samii A, Samaddar T, Schulte C, Sharma M, Singleton A, Spitz M, Tan EK, Tayebi N, Toda T, Troiano AR, Tsuji S, Wittstock M, Wolfsberg TG, Wu YR, Zabetian CP, Zhao Y, Ziegler SG (2009) Multicenter analysis of glucocerebrosidase mutations in Parkinson's disease. N Engl J Med 361(17):1651–1661. https://doi.org/10.1056/NEJMoa0901281

19. Di Maio R, Hoffman EK, Rocha EM, Keeney MT, Sanders LH, De Miranda BR, Zharikov A, Van Laar A, Stepan AF, Lanz TA, Kofler JK, Burton EA, Alessi DR, Hastings TG, Greenamyre JT (2018) LRRK2 activation in idiopathic Parkinson's disease. Sci Transl Med 10(451):eaar5429. https://doi.org/10.1126/scitranslmed.aar5429

20. Alcalay RN, Levy OA, Waters CC, Fahn S, Ford B, Kuo SH, Mazzoni P, Pauciulo MW, Nichols WC, Gan-Or Z, Rouleau GA, Chung WK, Wolf P, Oliva P, Keutzer J, Marder K, Zhang X (2015) Glucocerebrosidase activity in Parkinson's disease with and without GBA mutations. Brain 138(Pt 9):2648–2658. https://doi.org/10.1093/brain/awv179

21. Olanow CW, Rascol O, Hauser R, Feigin PD, Jankovic J, Lang A, Langston W, Melamed E, Poewe W, Stocchi F, Tolosa E, Investigators AS (2009) A double-blind, delayed-start trial of rasagiline in Parkinson's disease. N Engl J Med 361(13):1268–1278. https://doi.org/10.1056/NEJMoa0809335

22. Merchant KM, Cedarbaum JM, Brundin P, Dave KD, Eberling J, Espay AJ, Hutten SJ, Javidnia M, Luthman J, Maetzler W, Menalled L, Reimer AN, Stoessl AJ, Weiner DM, The Michael JFFASCPWG (2019) A proposed roadmap for Parkinson's disease proof of concept clinical trials investigating compounds targeting alpha-Synuclein. J Parkinsons Dis 9(1):31–61. https://doi.org/10.3233/JPD-181471

23. George S, Brundin P (2015) Immunotherapy in Parkinson's disease: micromanaging alpha-Synuclein aggregation. J Parkinsons Dis 5(3):413–424. https://doi.org/10.3233/JPD-150630

24. Lemere CA, Masliah E (2010) Can Alzheimer disease be prevented by amyloid-beta immunotherapy? Nat Rev Neurol 6(2):108–119. https://doi.org/10.1038/nrneurol.2009.219

25. Holmes C, Boche D, Wilkinson D, Yadegarfar G, Hopkins V, Bayer A, Jones RW, Bullock R, Love S, Neal JW, Zotova E, Nicoll JA (2008) Long-term effects of Abeta42 immunisation in Alzheimer's disease: follow-up of a randomised, placebo-controlled phase I trial. Lancet 372(9634):216–223. https://doi.org/10.1016/S0140-6736(08)61075-2

26. Davis DG, Schmitt FA, Wekstein DR, Markesbery WR (1999) Alzheimer neuropathologic alterations in aged cognitively normal subjects. J Neuropathol Exp Neurol 58(4):376–388

27. Neuropathology Group of the Medical Research Council Cognitive Function and Ageing Study (MRC CFAS) (2001) Pathological correlates of late-onset dementia in a multicentre, community-based population in England and Wales. Lancet 357(9251):169–175

28. Lannfelt L, Relkin NR, Siemers ER (2014) Amyloid-ss-directed immunotherapy for Alzheimer's disease. J Intern Med 275(3):284–295. https://doi.org/10.1111/joim.12168

29. Berg D, Postuma RB, Adler CH, Bloem BR, Chan P, Dubois B, Gasser T, Goetz CG, Halliday G, Joseph L, Lang AE, Liepelt-Scarfone I, Litvan I, Marek K, Obeso J, Oertel W, Olanow CW, Poewe W, Stern M, Deuschl G (2015) MDS research criteria for prodromal Parkinson's disease. Mov Disord 30(12):1600–1611. https://doi.org/10.1002/mds.26431

30. Guo L, Normando EM, Shah PA, De Groef L, Cordeiro MF (2018) Oculo-visual abnormalities in Parkinson's disease: possible value as biomarkers. Mov Disord 33(9):1390–1406. https://doi.org/10.1002/mds.27454

31. Mirelman A, Bernad-Elazari H, Thaler A, Giladi-Yacobi E, Gurevich T, Gana-Weisz M, Saunders-Pullman R, Raymond D, Doan N, Bressman SB, Marder KS, Alcalay RN, Rao AK, Berg D, Brockmann K, Aasly J, Waro BJ, Tolosa E, Vilas D, Pont-Sunyer C, Orr-Urtreger A, Hausdorff JM, Giladi N (2016) Arm swing as a potential new prodromal marker of Parkinson's disease. Mov Disord 31(10):1527–1534. https://doi.org/10.1002/mds.26720

32. Postuma RB, Gagnon JF, Vendette M, Montplaisir JY (2009) Markers of neurodegeneration in idiopathic rapid eye movement sleep behaviour disorder and Parkinson's disease. Brain 132(Pt 12):3298–3307. https://doi.org/10.1093/brain/awp244

33. Postuma RB, Gagnon JF, Vendette M, Desjardins C, Montplaisir JY (2011) Olfaction and color vision identify impending neurodegeneration in rapid eye movement sleep behavior disorder. Ann Neurol 69(5):811–818. https://doi.org/10.1002/ana.22282

34. Berg D, Seppi K, Behnke S, Liepelt I, Schweitzer K, Stockner H, Wollenweber F, Gaenslen A, Mahlknecht P, Spiegel J, Godau J, Huber H, Srulijes K, Kiechl S, Bentele M, Gasperi A, Schubert T, Hiry T, Probst M, Schneider V, Klenk J, Sawires M, Willeit J, Maetzler W, Fassbender K, Gasser T, Poewe W (2011) Enlarged substantia nigra hyperechogenicity and risk for Parkinson disease: a 37-month 3-center study of 1847 older persons. Arch Neurol 68(7):932–937. https://doi.org/10.1001/archneurol.2011.141

35. Heldmann M, Heeren J, Klein C, Rauch L, Hagenah J, Munte TF, Kasten M, Bruggemann N (2018) Neuroimaging abnormalities in individuals exhibiting Parkinson's disease risk markers. Mov Disord 33(9):1412–1422. https://doi.org/10.1002/mds.27313

36. Iranzo A, Valldeoriola F, Lomena F, Molinuevo JL, Serradell M, Salamero M, Cot A, Ros D, Pavia J, Santamaria J, Tolosa E (2011) Serial dopamine transporter imaging of nigrostriatal function in patients with idiopathic rapid-eye-movement sleep behaviour disorder: a prospective study. Lancet Neurol 10(9):797–805. https://doi.org/10.1016/S1474-4422(11)70152-1

37. Iranzo A, Santamaria J, Valldeoriola F, Serradell M, Salamero M, Gaig C, Ninerola-Baizan A, Sanchez-Valle R, Llado A, De Marzi R, Stefani A, Seppi K, Pavia J, Hogl B, Poewe W, Tolosa E, Lomena F (2017) Dopamine transporter imaging deficit predicts early transition to synucleinopathy in idiopathic rapid eye movement sleep behavior disorder. Ann Neurol 82(3):419–428. https://doi.org/10.1002/ana.25026

38. Jennings D, Siderowf A, Stern M, Seibyl J, Eberly S, Oakes D, Marek K, Investigators P (2014) Imaging prodromal Parkinson disease: the Parkinson associated risk syndrome study. Neurology 83(19):1739–1746. https://doi.org/10.1212/WNL.0000000000000960

39. Jennings D, Siderowf A, Stern M, Seibyl J, Eberly S, Oakes D, Marek K, Investigators P (2017) Conversion to Parkinson disease in the PARS Hyposmic and dopamine transporter-deficit prodromal cohort. JAMA Neurol 74(8):933–940. https://doi.org/10.1001/jamaneurol.2017.0985

40. Ponsen MM, Stoffers D, Booij J, van Eck-Smit BL, Wolters E, Berendse HW (2004) Idiopathic hyposmia as a preclinical sign of Parkinson's disease. Ann Neurol 56(2):173–181. https://doi.org/10.1002/ana.20160

41. Ponsen MM, Stoffers D, Twisk JW, Wolters E, Berendse HW (2009) Hyposmia and executive dysfunction as predictors of future

Parkinson's disease: a prospective study. Mov Disord 24(7):1060–1065. https://doi.org/10.1002/mds.22534

42. Berendse HW, Ponsen MM (2009) Diagnosing premotor Parkinson's disease using a two-step approach combining olfactory testing and DAT SPECT imaging. Parkinsonism Relat Disord 15(Suppl 3):S26–S30. https://doi.org/10.1016/S1353-8020(09)70774-6

43. Siderowf A, Jennings D, Eberly S, Oakes D, Hawkins KA, Ascherio A, Stern MB, Marek K, Investigators P (2012) Impaired olfaction and other prodromal features in the Parkinson at-risk syndrome study. Mov Disord 27(3):406–412. https://doi.org/10.1002/mds.24892

44. Berg D, Merz B, Reiners K, Naumann M, Becker G (2005) Five-year follow-up study of hyperechogenicity of the substantia nigra in Parkinson's disease. Mov Disord 20(3):383–385. https://doi.org/10.1002/mds.20311

45. Behnke S, Runkel A, Kassar HA, Ortmann M, Guidez D, Dillmann U, Fassbender K, Spiegel J (2013) Long-term course of substantia nigra hyperechogenicity in Parkinson's disease. Mov Disord 28(4):455–459. https://doi.org/10.1002/mds.25193

Chapter 8

Value and Methods of Pharmacovigilance in the Monitoring of Drug Safety in Parkinson's Disease

Santiago Perez-Lloret, James A. G. Crispo, Maria Veronica Rey, Donald Mattison, and Daniel Krewski

Abstract

Pharmacovigilance encompasses the detection, assessment, understanding and prevention of drug-related adverse outcomes. This important discipline plays a key role in evaluating drug safety and efficacy under real-world conditions of use following market authorization. Pharmacovigilance focuses mainly on adverse drug reactions (ADRs), which are unintended adverse medical events caused by an approved drug used at normal human doses for the prophylaxis, diagnosis, or therapy of disease, or for modification of physiological function. Spontaneous reporting of ADRs is a key source of pharmacovigilance data on marketed drugs. Analytic pharmacoepidemiologic studies can also be used to study real world safety and effectiveness (RWSE) of marketed drugs. Evaluation of spontaneous reporting data is often referred to as passive pharmacovigilance, with analytic epidemiological studies based on electronic health records or claims data representing active surveillance.

Pharmacovigilance has contributed substantially to our knowledge of the safety of antiparkinsonian drugs. Though patients on dopamine agonists and/or L-DOPA usually develop diurnal somnolence or impulse-control disorders, because they only occur infrequently, they were only identified in the marketing phase by spontaneous reporting of the adverse events. Tolcapone has been shown to produce fulminant hepatitis with a very low frequency, an adverse effect that was unnoticed during premarketing clinical development. Heart valvular disorders in patients treated with ergolinic dopamine agonists were also identified using spontaneous reporting data and later confirmed by large-scale pharmacoepidemiologic studies. Heart failure has been related to pramipexole use by several pharmacoepidemiological studies, after a signal was detected in phase III clinical trials. In this chapter, the main concepts underlying modern pharmacovigilance are outlined, along with the specific role of pharmacovigilance in assessing the safety of antiparkinsonian drugs.

Key words Parkinson's disease, Drug safety, Pharmacovigilance, Adverse drug reactions, Clinical trials

---

1 Introduction

As with any other substance, drug products can produce noxious effects. Adverse drug reactions (ADRs) are defined by World Health Organization as noxious medical events that are unintended and have a causal relationship with a medical intervention used at doses normally used in man for the prophylaxis, diagnosis, or therapy of disease, or for the modifications of physiological function [1] (*see* Appendix). Importantly, ADRs represent between 11% and 16% of hospital admissions [1], and increase death rates by 20%, extend hospitalization and augment medical cost by 20% [2]. As will be discussed later, the study of drug safety begins in the preclinical phases of drug development and continues after marketing authorization, when large numbers of patients are treated with the drug. Pharmacovigilance can be defined as "the science and activities relating to the detection, assessment, understanding and prevention of adverse effects or any other drug-related problem" [3]. While it encompasses drug safety assessment before and after commercialization, the bulk of pharmacovigilance activities take place after marketing approval is granted. In this chapter, Pharmacovigilance value and its role in the assessment of drug safety will be reviewed. Examples relevant to Parkinson's Disease (PD) will be given. We begin with a discussion of different types of drug effects.

2   Classification of Drug Effects

Medications contain active moieties, also called drugs, and excipients (Fig. 1) [4]. Drugs act on one or more biological targets to produce one or more responses, called *pharmacodynamic effects* [5, 6] (Fig. 1). They can be of interest for the treatment of human diseases through biological mechanisms such as dopamine receptor activation by dopaminergic agonists or the blockage of Catechol-O-methyl transferase by entacapone [7]. These are called *therapeutic effects* and are studied in detail during Phases II and III of the premarketing clinical development program of the drug to define the indications of the drug product [8]. On the other hand, *side effects* may occur as a result of the action of the drug on other biological targets, unrelated to those causing the therapeutic effect (*see* Appendix) [5, 6]. These effects can be desirable, undesirable, or inconsequential. An example of an undesirable side effect is mouth dryness caused by the anticholinergic effects of tricyclic antidepressants, unrelated to the blockage of serotonin uptake responsible of the antidepressant effect [9]. Some drugs also have *nonpharmacodynamic effects*, including immunological or idiosyncratic reactions, which are not therapeutic [5, 6]. Finally, *toxic effects* can occur with excessive drug doses and lack of efficacy may result from underdosage (Fig. 1) [5, 6].

Excipients, which are essential for drug stability and affect some pharmacokinetic processes, are believed to be biologically inactive. Although excipients are generally inert, this is not always the case, as shown by the cases of asthma worsened by benzoate contained in some anti-asthmatic drug formulations or skin pertur-

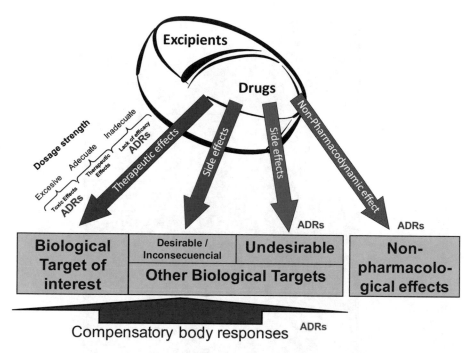

Fig. 1 Medications' constituents and drugs effects classification. The label "ADRs" indicates the mechanisms underlying the generation of Adverse Drug Reaction

bations after transdermal drug delivery systems [10, 11]. Excipients are part of the formulation of the drug are thus subsumed within the exposure profile for the drug of interest.

3 Definition and Classification of Adverse Drug Reactions

During the course of drug therapy, patients may experience a number of clinical responses, many of which are not related to the drug itself. All untoward medical occurrences that may present during treatment are called "adverse events (AEs)" [1] (*see* Appendix). If an AE has a causal link with the drug, or at least causality can be reasonably suspected, then it may be called an ADR. They can be expected on the basis of the "Summary of Product Characteristics" (or Package Label) or unexpected [6].

Adverse drug reactions can be produced by several mechanisms and can be related or unrelated to medication dose. As shown in Table 1, they can be grouped into distinct categories [6, 12]. Dose-related type A reactions ("augmented") are related to toxic or side effects. Type B ("bizarre") reactions are non–dose-related and are caused by nonpharmacodynamic mechanisms such as immunotoxicity. "Chronic" type C reactions are related to compensatory body-responses, such as the alteration of endocrine axes produced by hormone administration. Type D reactions

Table 1
Classification of adverse drug reactions

	A	B	C	D	E	F
Type	Dose-related	Non–dose-related	Dose- and time-related	Time-related	End of use	Therapy failure
Mnemonic	Augmented	Bizarre	Chronic	Delayed	End of use	Failure
Mechanism	Toxic effects Side effects	Immunological reactions Idiosyncratic reactions	Compensatory body responses	Mixed	Compensatory body responses	Inadequate dose strength
Principal features	The most common. Predictable. Related to drugs' pharmacology	Uncommon Unpredictable Not related to drugs' pharmacology	Uncommon. Seen after long-term exposure. Related to drug's pharmacology	Uncommon. Dose-related Appears after drug discontinuation	Relatively common. Usually dose-related. Occurs after drug withdrawal	Common. Usually related to drug interactions, but not always
Example	L-DOPA-induced dyskinesias	Aspirin hypersensitivity	Endocrine axis suppression by hormones	Teratogenesis Tardive dyskinesias	Insomnia after BZD withdrawal	CBZ-failure with CYP3A4 inductors

BZD benzodiazepines, *CBZ* carbamazepine

("delayed") are of mixed origin and may be related or unrelated to medication dose. Type E reactions ("end of use") are also related to compensatory body-responses and reflect withdrawal syndromes. Finally type F ("therapy failure") can be related to underdosage resulting from drug interactions, among other causes.

As described earlier, pharmacovigilance is the complete set of activities relating to the detection, assessment, and prevention of adverse effects or any other drug-related problems (modified from World Health Organization) [3], with a focus on ADRs. The history of this discipline goes back to the 1960s, when the Australian physician WG McBride published a letter describing his observation that babies whose mothers had used thalidomide during pregnancy were born with congenital abnormalities more often than babies who had not been exposed to thalidomide in utero [13]. Between 1957 and 1962, thalidomide caused severe birth defects in more than 10,000 children [14]. These observations prompted the cancelation of the marketing authorization of the drug and stimulated the worldwide discussion about drug safety [1, 3]. This subsequently led to the creation of the WHO Program for International Drug Monitoring, which is coordinated by the Uppsala Monitoring Centre in Uppsala, Sweden [1, 3]. It also produced changes in the US legislation to strengthen the Food Drug and Cosmetic Act [15].

4 Methods Used in Pharmacovigilance

The main way of gathering safety data of a drug in the premarketing phase are double-blind, randomized, controlled clinical trials (RCT) [16, 17]. Moreover, they are the most valid tool for assessing the causality relationship between an adverse event and the drug administered [5]. Nonetheless they suffer from considerable shortcomings, which often include a small number of participants and limited external validity.

The number of subjects exposed to the drug in RCTs is limited, which impairs the detection of rare ADRs (1/1000 subjects or less) [5]. Exposure time is usually insufficient to detect ADRs with long latency times, as was the case with sclerosing peritonitis occurring up to 4 years exposure to the beta-adrenergic receptor blocker practolol [18]. The patient population in clinical trials is highly selected most of the times, excluding patients with high risk of suffering ADRs. Indeed, polymedicated patients or those with most severe disease are often excluded, which reduces the risk of experiencing an ADR, and thus of the chance of observing them. These limitations makes difficult to extrapolate safety result to wider populations [5, 19].

The evaluation of drug safety in the postmarket approval phase under real-world conditions of use is therefore of vital importance

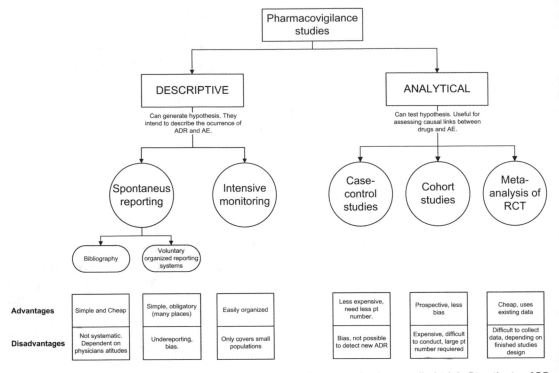

Fig. 2 Approaches for studying drugs' safety in humans. *RCT* randomized, controlled trial, *Pt* patients, *ADR* adverse drug reactions

[5, 19]. Pharmacovigilance should be a collaborative effort between academic researchers, regulatory authorities and the pharmaceutical industry [19]. Several strategies besides the RCT are available for studying drug safety in the humans (Fig. 2) [5, 19]. Spontaneous reporting is the basic method for ADR detection. Adverse drug reactions reporting to local Regulatory Authorities is mandatory for physicians in many countries [20]. Pharmaceutical manufacturers have to ensure that suspected adverse reactions to their products are reported to the relevant authority [20]. Local regulatory authorities can then report the ADR to the WHO Uppsala Monitoring Centre, which has accrued more than 16 million reports since its creation in the 1960s [1, 3, 21]. Spontaneous reporting systems (SRS) are best for early detection of signals (empirical evidence of a possible causal relationship between an adverse event and a drug, which should further be explored) of new, rare, and serious ADRs [19]. Interestingly, spontaneous reporting produced the initial evidence that led to the withdrawal of drugs from the French market in 90% of cases over the period 1998–2004 [22]. Similar results were observed for the European Union during the period 2002–2011 [23].

Underreporting, which may be as high as 95% [24], and selective reporting are the greatest limitations of SRS [19]. Underreporting can lead to the false conclusion that a real risk is absent, while selected reporting of suspected risks may introduce bias in the safety profile inferred from gathered reports [19]. Additionally, since the size of the exposed population to the drug is unknown, ADR rates cannot be precisely calculated. Nonetheless, it should be kept in mind that of 11 products withdrawn from US and UK markets, the decision was made on the basis of SRS results in eight of them (73%) [19]. Spontaneous reporting system database explorations may be enhanced through the use of data-mining processes [25, 26]. Active surveillance may be performed by implementing dedicated reporting structures or by conducting well-designed observational studies, including case-control or cohort studies (Fig. 2) [5].

Observational studies can be conducted with reasonable efficiency by undertaking the analysis of large databases [27]. Examples of databases are the Taiwan National Health Insurance Research Database, UK Clinical Practice Research Datalink (CPRD), US Medicaid database, Kaiser Permanente California, and the Cerner HealthFacts® data warehouse. All of these databases include detailed health information for large numbers of patients, which may be used to assess drug utilization trends and associations between drug exposure and adverse health outcomes.

An interesting case of active surveillance is the Food and Drug Administration (FDA)-sponsored "mini-Sentinel Initiative" [28], launched in 2008. This initiative seeks to reuse existing health care information to enable FDA to conduct active postmarketing safety surveillance. By using data from administrative and claims databases, electronic health record systems, and registries, FDA will monitor regulated medical products in near real-time, to better understand product safety. The main object of the initiative is to refine safety *signals*. *Signal generation* refers to the use of statistical methods to identify adverse outcome associated with the use of medical products, while *refinement* can be defined as the process by which an identified potential safety signal is further investigated to determine whether evidence exists to support a relationship between the medical product exposure and the outcome. *Signals* can come from premarketing experience or by observations during the marketing period (as for example, by SRS). If during refinement a signal is confirmed, then it will often be necessary to conduct additional analyses to validate it. Currently, the "Mini-Sentinel" safety pilot program is up and running (https://www.sentinelinitiative.org/). It includes about 99 million individuals, with 2.9 billion prescription drug dispensings and 2.4 billion unique medical encounters. Between 2012 and 2017, it served as the basis for the issue of six new safety communications [28].

5   Assessment of ADR

During the course of a drug treatment, events may occur which do not necessarily relate to exposure to active moieties or excipients. Nonetheless, medicines should be considered potential causes of all events occurring in patients who are being treated by such medicines [12]. It is important to keep in mind that subjects may not remember all their medications, especially if they are over-the-counter formulations, herbal or traditional medicines, recreational or abuse drugs, and oral contraceptives. Therefore, an exhaustive medical history needs to be taken in these cases.

Causality assessment is indispensable but often difficult aspect of pharmacovigilance. The following aspects of the event should be taken into account [6]:

– Appropriate temporal relationship. When symptoms start soon after initiation of treatment initiation, diagnosis can be easily achieved. However, connecting symptoms with the long-term use of a drug may be more difficult.

– Plausibility of the event. If the event result from the pharmacodynamic action of the drug or is a side effect, it may be easier to attribute the ADR to the drug. Nonetheless, many ADRs result from nonpharmacodynamic mechanisms (Fig. 1) and are thus not expected and difficult to predict.

– Exclusion of other causes. Adverse drug reactions may be diagnosed after exclusion of every possible cause of the observed event.

– Laboratory data. Identification of above-normal drug levels may be of value in the diagnosis of ADRs, when they are dose-related.

– Nature of the event. Some reactions are usually observed with drugs, such as the occurrence of Stevens–Johnson syndrome with some antiepileptic drugs [29].

– Dechallenge and rechallenge. In many cases, nonessential drugs may be discontinued or changed. Event disappearance is indicative of a link with the drug. Rechallenge may be ethically difficult, but of great value in causality assessment.

Methods such as the Naranjo scale [30], Kramer's algorithm [31], Karsh and Lasagna scale [32], WHO-Uppsala Monitoring Centre causality assessment criteria [6], or Begaud's algorithm [33] can be used to assess causality of AEs observed in patients. Based on different probabilistic approaches, they are able to classify causality from certain to unlikely. A simplified algorithm based in Edwards and colleagues [12] is provided in Table 2.

Table 2
Imputation of adverse drug reactions

	Time relationship	Alternative explanations	Dechallenge	Rechallenge
Certain	Plausible	Absent	Positive	Positive
Probable/likely	Reasonable	Unlikely	Plausible	Not required
Possible	Reasonable	Present	Not done	Not done
Unlikely	Improbable	Present	Not done or negative	Not done or negative

6 Pharmacovigilance of Antiparkinsonian Drugs

Parkinson's disease is a condition that affects all populations and cultures. Approximately 90% of people with PD are diagnosed after 50 years of age, with the condition having a worldwide prevalence of 1.6% among individuals 65 years of and older [34]. Its symptoms are most frequently treated by drugs including anticholinergics (trihexiphenidyl), monoamine oxidase B (MAO-B) inhibitors (selegiline, rasagiline), catechol-O-methyltransferase (COMT) inhibitors (tolcapone, entacapone), amantadine, dopamine agonists (bromocriptine, ropinirole, pramipexole, and apomorphine, among others), or L-3,4-dihydroxyphenylalanine (L-DOPA) [7]. Although these drugs are commonly used as antiparkinsonian agents, their beneficial therapeutic effects are sometimes counterbalanced by the occurrence of ADRs. Examples of such adverse reactions include diurnal somnolence or impulse-control disorders by dopamine agonists or fulminant hepatitis by tolcapone. Pharmacovigilance has been of great importance in the characterization of these reactions, which passed unnoticed in the premarketing studies. Some interesting examples will be reviewed in the next paragraphs.

It has long been recognized that dopaminergic medications can cause sleepiness [35]. Nonetheless, it was only in 1999, more than 10 years after the approval of the first dopamine agonist for the treatment of PD, that Frutch and colleagues reported the occurrence of "sleep attacks" in patients on ropinirole and pramipexole [36]. Shortly after, reports of sleep attacks on bromocriptine, lisuride, piribedil [37], pergolide and levodopa alone [35, 38] followed. Sleep attacks were initially suggested to be a class effect of dopaminergic medications [39], but results published later on challenged this view [40]. Their physiopathology appears to be related to the somnogenic effect of low doses of dopamine [35], so they can be classified as undesirable side effects.

Impulse-control disorders (ICDs) also went unnoticed in the premarketing phase, and were reported for the first time in 2000 by Giovanni and colleagues [41]. ICDs includes pathological gambling, hypersexuality, binge eating, or compulsive buying, and affects more than 10% of patients [42, 43]. They severely affect the psychosocial wellbeing of the patients and their caregivers [43, 44]. It is possible that these disorders had passed unnoticed because patients usually do not commonly disclose such conditions to the physician overseeing their pharmacological treatment [45].

Another example is tolcapone-induced hepatitis. Assal and colleagues described the first case of fulminant hepatitis related to tolcapone in 1998 [46]. Two months later, the European Medicines Agency (EMA) withdrew the marketing application of Tasmar® (tolcapone) based on three cases of fulminant hepatitis [47]. The mechanism of this event is unknown and it appears to be idiosyncratic, unrelated to the drug's inhibitory action on the catechol-O-methyl transferase and thus related to a nonpharmacodynamic effect, which was not observed during the preclinical and clinical studies [47, 48]. Nowadays, tolcapone regained marketing authorization in Europe, but is indicated only as a second-line treatment for motor fluctuations and with strict monitoring of liver function [49].

Another interesting example is the association between heart valve disorders and exposure to ergolinic dopamine agonists. The first reports of an association between the development of valvular dysfunction and dopamine agonists were published in 2002 [50, 51], long after commercialization of the drugs. Initially, pergolide treatment was implicated, followed by cabergoline [52]. The echocardiographic and histological features were very similar to those of fenfluramine-induced or ergotamine-induced disease and carcinoid heart disease, thus suggesting a possible involvement of serotoninergic receptor activation [53].

Subsequent large nested case–control studies confirmed these results [54]. As a result of these studies, the manufacturer voluntarily removed pergolide from the US market in 2007, while EMA issued a recommendation to limit the maximum dose of both pergolide and cabergoline to 3 mg and to monitor patients with regular echocardiograms [53]. This is an example of an undesirable side effect, as opposed to sleep attacks and ICDs that are related to dopamine agonists toxic effects (i.e., an abnormal stimulation of dopamine receptors).

Finally, heart failure has been recently associated with exposure to pramipexole in PD [55]. Originally, FDA conducted a review of phase two and three clinical trial data of pramipexole and found that, compared to placebo, heart failure was more frequently diagnosed in pramipexole users, both in parkinsonian and restless-leg syndrome patients [56]. Five pharmacoepidemiological studies later demonstrated an elevated risk of heart failure following use of pramipexole or cabergoline for the treatment of PD [57–61].

Although the effects of cabergoline may be explained by the induction of cardiac valve fibrosis, the basis for the significantly increased risk associated with pramipexole is unclear, thus precluding classification of this ADR. It is also unclear if the observed effects are related to medication dose, time course, or the population at risk.

7 Conclusion

Pharmacovigilance is essential to the study of and safety of medications in humans under real world conditions of use. Several approaches are available, ranging from passive approaches based on SRS data, to complex and expensive studies, which may include analytic epidemiological studies of large patient populations exposed to the drug of interest. When available, administrative health records provide a potentially rich source of data to support pharmacoepidemiology. Causality assessment is of vital importance for the diagnosis and management of ADRs. It can be achieved by several probabilistic approaches, but can sometimes be challenging. While more complex and expensive, active surveillance is probably the most effective approach to r the assessment of drug safety.

Pharmacovigilance has enabled the identification of many important safety issues associated with the use of antiparkinsonian medications, such as the possibility of experiencing sleep attacks or impulse-control disorders with dopamine agonists, heart valve disease with ergolinic dopamine agonists, fulminant hepatitis with tolcapone, and heart failure after pramipexole. Both passive and active pharmacovigilance remain important tools for the evaluate of the real-world safety and effectiveness of drugs used to treat Parkinson's and other diseases.

Appendix

Adverse event or experience: any untoward medical occurrence that may present during treatment with a medicine but which does not necessarily have a causal relationship with this treatment. The basic point here is the coincidence in time without any suspicion of a causal relationship.

Serious adverse event: any event that: is fatal; is life-threatening; is permanently/significantly disabling; requires or prolongs hospitalization; causes a congenital anomaly; or requires intervention to prevent permanent impairment or damage.

Adverse drug reaction (ADR): a response to a medicine which is noxious and unintended, and which occurs at doses normally used in man for the prophylaxis, diagnosis, or therapy of disease, or for the modifications of physiological function. In this description it is of importance that it concerns the response of a patient, in which individual factors may play an important role, and that the phenomenon is noxious (an unexpected therapeutic response, for example, may be a side effect but not an adverse reaction).

Unexpected adverse reaction: an adverse reaction, the nature or severity of which is not consistent with domestic labeling or market authorization, or expected from characteristics of the drug.

Side effect: an unintended pharmacodynamic effect of a pharmaceutical product occurring at doses normally used by a patient which is related to the pharmacological properties of the drug.

Modified from: Safety of Medicines, A guide to detecting and reporting adverse drug reactions. WHO, Geneva, 2002 [1].

References

1. World Health Organization Collaborating Centre for International Drug Monitoring (2002) Safety of Medicines. A guide to detecting and reporting adverse drug reactions. World Health Organization, Geneva

2. Bord CA, Rach CL (2006) Adverse drug reactions in United States hospitals. Pharmacotherapy 26:601–608

3. World Health Organization Collaborating Centre for International Drug Monitoring (2002) The importance of pharmacovigilance. Safety monitoring of medicinal products. World Health Organization, Geneva

4. The Council of the European Economic Community (1906) Directive 65/65/EEC of the 26 Jan 1965 on the approximation of provisions laid down by law, regulation or administrative action relating to medicinal products. The Council of the European Economic Community, Brussels

5. Montastruc JL, Sommet A, Lacroix I et al (2006) Pharmacovigilance for evaluating adverse drug reactions: value, organization, and methods. Joint Bone Spine 73:629–632

6. Rehan HS, Chopra D, Kakkar AK (2009) Physician's guide to pharmacovigilance: terminology and causality assessment. Eur J Intern Med 20:3–8

7. Olanow CW, Stern MB, Sethi K (2009) The scientific and clinical basis for the treatment of Parkinson disease. Neurology 72:S1–S136

8. Strom BL (2000) What is pharmacoepidemiology? In: Strom BL (ed) Pharmacoepidemiology. Wiley & Sons, New York, NY, pp 3–15

9. Uher R, Farmer A, Henigsberg N et al (2009) Adverse reactions to antidepressants. Br J Psychiatry 195:202–210

10. Balatsinou L, Di GG, Sabatino G et al (2004) Asthma worsened by benzoate contained in some antiasthmatic drugs. Int J Immunopathol Pharmacol 17:225–226

11. Brown MB, Traynor MJ, Martin GP et al (2008) Transdermal drug delivery systems: skin perturbation devices. Methods Mol Biol 437:119–139

12. Edwards IR, Aronson JK (2000) Adverse drug reactions: definitions, diagnosis, and management. Lancet 356:1255–1259

13. Mcbride WG (1961) Thalidomide and congenital malformations. Lancet 2:1358–1358

14. Vargesson N (2015) Thalidomide-induced teratogenesis: history and mechanisms. Birth Defects Res C Embryo Today 105:140–156

15. Greene JA, Podolsky SH (2012) Reform, regulation, and pharmaceuticals—the Kefauver-Harris amendments at 50. N Engl J Med 367:1481–1483

16. Grimes DA, Schulz KF (2002) Bias and causal associations in observational research. Lancet 359:248–252

17. Grimes DA, Schulz KF (2002) An overview of clinical research: the lay of the land. Lancet 359:57–61

18. Mann RD (2007) An instructive example of a long-latency adverse drug reaction—sclerosing peritonitis due to practolol. Pharmacoepidemiol Drug Saf 16:1211–1216

19. Harmark L, Van Grootheest AC (2008) Pharmacovigilance: methods, recent developments and future perspectives. Eur J Clin Pharmacol 64:743–752

20. World Health Organization Collaborating Centre for International Drug Monitoring (2000) Safety monitoring of medicinal products: guidelines for setting up and running a Pharmacovigilance Centre. World Health Organization, Geneva

21. World Health Organization Collaborating Centre for International Drug Monitoring

(2019) Uppsala Monitoring Centre: VigiBase. World Health Organization, Geneva

22. Olivier P, Montastruc JL (2006) The nature of the scientific evidence leading to drug withdrawals for pharmacovigilance reasons in France. Pharmacoepidemiol Drug Saf 15:808–812

23. Mcnaughton R, Huet G, Shakir S (2014) An investigation into drug products withdrawn from the EU market between 2002 and 2011 for safety reasons and the evidence used to support the decision-making. BMJ Open 4:e004221

24. Hazell L, Shakir SA (2006) Under-reporting of adverse drug reactions: a systematic review. Drug Saf 29:385–396

25. Stephenson WP, Hauben M (2007) Data mining for signals in spontaneous reporting databases: proceed with caution. Pharmacoepidemiol Drug Saf 16:359–365

26. Weaver J, Willy M, Avigan M (2008) Informatic tools and approaches in postmarketing pharmacovigilance used by FDA. AAPS J 10:35–41

27. Chan EW, Liu KQ, Chui CS et al (2015) Adverse drug reactions – examples of detection of rare events using databases. Br J Clin Pharmacol 80:855–861

28. Robb MA, Racoosin JA, Sherman RE et al (2012) The US Food and Drug Administration's sentinel initiative: expanding the horizons of medical product safety. Pharmacoepidemiol Drug Saf 21(Suppl 1):9–11

29. Ordonez L, Salgueiro E, Jimeno FJ et al (2015) Spontaneous reporting of Stevens-Johnson syndrome and toxic epidermal necrolysis associated with antiepileptic drugs. Eur Rev Med Pharmacol Sci 19:2732–2737

30. Naranjo CA, Busto U, Sellers EM et al (1981) A method for estimating the probability of adverse drug reactions. Clin Pharmacol Ther 30:239–245

31. Kramer MS, Leventhal JM, Hutchinson TA et al (1979) An algorithm for the operational assessment of adverse drug reactions. I. Background, description, and instructions for use. JAMA 242:623–632

32. Karch FE, Lasagna L (1975) Adverse drug reactions. A critical review. JAMA 234:1236–1241

33. Begaud B, Evreux JC, Jouglard J et al (1985) Imputation of the unexpected or toxic effects of drugs. Actualization of the method used in France. Therapie 40:111–118

34. Dorsey ER, Constantinescu R, Thompson JP et al (2007) Projected number of people with Parkinson disease in the most populous nations, 2005 through 2030. Neurology 68:384–386

35. Cantor CR, Stern MB (2002) Dopamine agonists and sleep in Parkinson's disease. Neurology 58:S71–S78

36. Frucht S, Rogers JD, Greene PE et al (1999) Falling asleep at the wheel: motor vehicle mishaps in persons taking pramipexole and ropinirole. Neurology 52:1908–1910

37. Ferreira JJ, Galitzky M, Montastruc JL et al (2000) Sleep attacks and Parkinson's disease treatment. Lancet 355:1333–1334

38. Ferreira JJ, Thalamas C, Montastruc JL et al (2001) Levodopa monotherapy can induce "sleep attacks" in Parkinson's disease patients. J Neurol 248:426–427

39. Homann CN, Wenzel K, Suppan K et al (2002) Sleep attacks in patients taking dopamine agonists: review. BMJ 324:1483–1487

40. Micallef J, Rey M, Eusebio A et al (2009) Antiparkinsonian drug-induced sleepiness: a double-blind placebo-controlled study of L-dopa, bromocriptine and pramipexole in healthy subjects. Br J Clin Pharmacol 67:333–340

41. Giovannoni G, O'sullivan JD, Turner K et al (2000) Hedonistic homeostatic dysregulation in patients with Parkinson's disease on dopamine replacement therapies. J Neurol Neurosurg Psychiatry 68:423–428

42. Antonini A, Cilia R (2009) Behavioural adverse effects of dopaminergic treatments in Parkinson's disease: incidence, neurobiological basis, management and prevention. Drug Saf 32:475–488

43. Voon V, Fox SH (2007) Medication-related impulse control and repetitive behaviors in Parkinson disease. Arch Neurol 64:1089–1096

44. Voon V, Thomsen T, Miyasaki JM et al (2007) Factors associated with dopaminergic drug-related pathological gambling in Parkinson disease. Arch Neurol 64:212–216

45. Perez-Lloret S, Rey MV, Fabre N et al (2012) Do Parkinson's disease patients disclose their adverse events spontaneously? Eur J Clin Pharmacol 68:857–865

46. Assal F, Spahr L, Hadengue A et al (1998) Tolcapone and fulminant hepatitis. Lancet 352:958

47. Colosimo C (1999) The rise and fall of tolcapone. J Neurol 246:880–882

48. Benabou R, Waters C (2003) Hepatotoxic profile of catechol-O-methyltransferase inhibitors in Parkinson's disease. Expert Opin Drug Saf 2:263–267

49. Borges N (2005) Tolcapone in Parkinson's disease: liver toxicity and clinical efficacy. Expert Opin Drug Saf 4:69–73

50. Pritchett AM, Morrison JF, Edwards WD et al (2002) Valvular heart disease in patients taking pergolide. Mayo Clin Proc 77:1280–1286

51. Antonini A, Poewe W (2007) Fibrotic heart-valve reactions to dopamine-agonist treatment in Parkinson's disease. Lancet Neurol 6:826–829

52. Steiger M, Jost W, Grandas F et al (2009) Risk of valvular heart disease associated with the use of dopamine agonists in Parkinson's disease: a systematic review. J Neural Transm 116:179–191

53. Bhattacharyya S, Schapira AH, Mikhailidis DP et al (2009) Drug-induced fibrotic valvular heart disease. Lancet 374:577–585

54. Schade R, Andersohn F, Suissa S et al (2007) Dopamine agonists and the risk of cardiac-valve regurgitation. N Engl J Med 356:29–38

55. Perez-Lloret S, Rey MV, Crispo J et al (2014) Risk of heart failure following treatment with dopamine agonists in Parkinson's disease patients. Expert Opin Drug Saf 13:351–360

56. Administration USFDA (2012) FDA drug safety communication: ongoing safety review of Parkinson's drug Mirapex (pramipexole) and possible risk of heart failure. FDA, Silver Spring, MD

57. Arbouw ME, Movig KL, Guchelaar HJ et al (2012) Dopamine agonists and ischemic complications in Parkinson's disease: a nested case-control study. Eur J Clin Pharmacol 68:83–88

58. Mokhles MM, Trifiro G, Dieleman JP et al (2012) The risk of new onset heart failure associated with dopamine agonist use in Parkinson's disease. Pharmacol Res 65:358–364

59. Renoux C, Dell'aniello S, Brophy JM et al (2012) Dopamine agonist use and the risk of heart failure. Pharmacoepidemiol Drug Saf 21:34–41

60. Hsieh PH, Hsiao FY (2013) Risk of heart failure associated with dopamine agonists: a nested case-control study. Drugs Aging 30:739–745

61. Crispo J, Willis AW, Thibault DP et al (2016) Associations between cardiovascular events and nonergot dopamine agonists in Parkinson's disease. Mov Disord Clin Pract 2016:1–11

Part III

Clinical Trials for Nonmotor Symptoms in Parkinson's Disease

Chapter 9

Clinical Trials for Cognitive Impairment and Dementia in Parkinson's Disease

Brandon Barton

Abstract

Parkinson's disease (PD)-related cognitive impairment is one of the most disabling and refractory components of the disease, with the majority of patients developing dementia after several years of disease progression. The degree of impairment can progress from mild complaints without clear deficits to mild cognitive impairment (PD-MCI) to dementia (PDD). Specific definitions of PDD and PD-MCI are relatively new in differentiation from other types of dementia. Cognitive dysfunction in PD is not completely understood, only modestly responds to pharmaceutical interventions, fluctuates significantly, is affected by multiple other PD-related disease factors, and is highly heterogeneous by nature, all of which make trial design and interpretation highly challenging. A small number of large drug trials have given insights into important issues of study design, choosing outcome measures, and adverse events that complicate the ability to study this class of patients. A major obstacle is the lack of biomarkers or universally agreed-up outcome measures, and while many PD-specific cognitive scales are now recommended for use, most have not been used in large trials. Drug trials may be limited and a number of nonpharmaceutical approaches are being explored.

Key words Parkinson's disease, Cognition, Cognitive dysfunction, Treatment, Clinical trials

1 Introduction

1.1 Cognitive Impairment in Parkinson's Disease

Though in his early descriptions of "The Shaking Palsy," James Parkinson did not describe a cognitive component to the disease that would later bear his eponym, we now know that cognitive impairment is common in Parkinson's disease (PD) [1]. The understanding of how PD causes cognitive impairment has evolved as patients live longer with modern PD therapies. A longer life span due to treatment of motor symptoms allows the disease pathology to progress further through the brain, subsequently affecting the structures and networks associated with cognition [2]. Cognitive impairment in PD contributes heavily to reduced qualify of life and morbidity in PD [3], and for patients and families remains one of the most feared of all disease complications. Unfortunately, while

Santiago Perez-Lloret (ed.), *Clinical Trials In Parkinson's Disease*, Neuromethods, vol. 160,
https://doi.org/10.1007/978-1-0716-0912-5_9, © Springer Science+Business Media, LLC, part of Springer Nature 2021

the amount of publications about cognitive impairment in PD have vastly increased in the last two decades, the number of quality interventional clinical trials remains relatively few, given the complexity of the problem. Current treatment for cognitive symptoms is much less effective than treatment of motor symptoms and sometimes comes at the cost of worsening motor or autonomic symptoms.

1.2 Epidemiology of PD-Related Cognitive Impairment

With increasing life expectancy, Parkinson's disease dementia (PDD) is set to become more prevalent. Dementia may occur in up to 75% of patients with PD at 10 years after diagnosis and up to 83% at 20 years, as seen in long-duration observational studies [4, 5]. The incidence rate of dementia in PD is increased 5–6 times compared to that of the general population. The prevalence rate of dementia in the general PD population is around 40% in cross-sectional studies [6]. Risk factors for Parkinson's disease dementia include advanced age, treatment-induced visual hallucinations, and more severe motor symptoms [7]. Additionally, the prevalence of mild cognitive impairment in PD (PD-MCI), increasingly recognized as a potential precursor stage to dementia, has an estimated prevalence of 27% (range 19–38%) [8].

1.3 Pathophysiology of PD Cognitive Impairment

There are multiple pathological correlates of PDD, both anatomical and neurochemical. Widespread Lewy bodies gradually accumulate in cortical and subcortical structures, with subsequent neuronal dysfunction and death. In a substantial portion of patients, Alzheimer's disease (AD) pathological changes are also seen with deposition of neurofibrillary tangles and amyloid plaques [9–11]. Cholinergic deficits are consistently noted, with greater deficits in the cortex than are seen in Alzheimer's disease (AD) which correlate to cognitive as well as psychiatric symptoms such as hallucinations [12–14]. Imbalance in dopaminergic pathways, though largely causing motor dysfunction, are also contributory to cognitive dysfunction in PD. However, dopaminergic deficit only partially explains the pathophysiological mechanism underlying cognitive impairment. There is often improvement of cognitive symptoms, especially in functions of attentional control, early in the disease course through the administration of levodopa [15]. Serotonergic, noradrenergic, and glutamatergic systems are all involved in cognitive symptoms to varying degrees [16].

Ongoing prospective studies, such as in the Parkinson's Progression Markers Initiative (PPMI) [17] are following a cohort of several hundred PD patients to ascertain risk factors and biomarkers for development of cognitive decline. While there is no definitive biomarker for cognitive function, researchers found that a combination of five variables allowed a prediction of cognitive impairment if present after 2 years of diagnosis, including age, smell abnormalities, presence of REM behavioral sleep disorder (RBD), CSF AB_{42} levels, and reduced caudate update on dopamine transporter (DAT) imaging, with some of these factors being confirmed as possible cognitive impairment risks in other studies [18]. MRI findings may also predict PD cognitive impairment including AD-like atrophy on MRI scans [19]. Multiple individual risk factors including genetic markers (e.g., GBA, APOE, MAPT) and cognitive reserve may also influence rates of decline and neuropsychiatric comorbidities [16]. Cerebrovascular disease, often clinically silent, is often discovered concurrently with PD cognitive impairment and contributes to cognitive burden proportionate to quantity seen on neuroimaging [20].

Dementia in PD can occur due to pathological Lewy body spread to the alone or to the interaction of comorbid pathologies. A significant number of patients with dementia have mixed pathology, even early in the course of the cognitive dysfunction [21] which complicates premorbid diagnosis and classification of patients. In particular, amyloid deposition similar to AD can occur in a large portion of PD patients (ref/clarify). In addition, the general aging of society increases the burden not only for PD but for multifactorial dementias, and many patients who develop dementia in the course of PD may have not only evidence of PD but also strokes, AD, tau, and so on, and the clinical features of all these disorders may overlap [22]. Pathophysiology: mostly LB but other degenerative pathways and multiple neurotransmitters.

1.4 Cognitive Profiles in PD

Cognitive symptoms in PD range a wide gamut, from cognitive complaints without clear deficits, through mild dysfunction, to marked dementia. Early cognitive deficits predominantly affect executive function and attention, which are resultant from dopamine depletion of the dorsolateral striatofrontal circuitry and the mesocortical pathways [23, 24]. The phenotype is variable between patients but generally shows cognitive slowing (bradyphrenia), and executive, attentional, visuospatial dysfunction, with memory impairments [25]. Compared to Alzheimer's disease, PDD has larger fluctuations of attention, more frequent visual hallucinations, less severe memory problems (at least in the early phases), and more severe visuospatial deficits [26].

Cognitive impairment is often associated with other neuropsychiatric disorders, and the medications used to treat these disorders could influence cognitive performance. Patients with cognitive issues have a number of other comorbid features that affect cognition directly or indirectly including sleep, apathy, depression, anxiety, psychosis, hypotension, and dysarthria/communication difficulties. Inclusion and exclusion criteria often need to consider these items as they may interfere with cognitive assessments.

1.5 Burden and Cost of PD Cognitive Impairment

Parkinson disease dementia results in markedly reduced quality of life for both patients and caregivers [27], particularly with the progressive functional and motor decline in the more advanced stages of the disease, resulting in higher costs, twice the risk of death and increased risk of nursing home placement [28–30]. Dementia complicates optimal treatment of motor symptoms since most antiparkinsonian agents can worsen neuropsychiatric features and eventually need to be minimized, which leads to worsening mobility.

1.6 Definitions in PD Cognitive Impairment for Research

Until recently, PD cognitive impairment did not have specific developed criteria separate from those of general dementia or the more common diagnosis of AD. The International Parkinson's Disease and Movement Disorder Society (MDS) has formally defined PDD and provided diagnostic guidelines for clinical guidance [31]. After extensive literature review, expert committees suggested diagnostic criteria specifically for dementia associated with PD developed based on characteristic clinical features identified on a comprehensive review of the existing literature [32] (Table 1). Key differences from previously used Diagnostic and Statistical Manual of Mental Disorders Version 4 (DSM-IV) criteria for diagnosis of dementia are that memory does not necessarily need to be impaired in PDD and impairment in other cognitive domains can define dementia (e.g., visuospatial or executive dysfunction, which are often more impaired in PD). PDD patients must perform poorly in more than one cognitive domain to a degree severe enough to interfere with activities of daily living. Additionally, with the greater observance of co-occurrence of dementia with PD, and based on prevalence of cognitive impairment even in early PD, recent revisions of diagnostic criteria for PD suggest that dementia is not an excluding condition for PD diagnosis as it has been previously [33].

As cognition has been studied in more intense detail with neuropsychological testing at all stages of PD, it is now clear that subtle cognitive deficits are nearly universally identified, even in very early PD [34, 35]. Cognitive dysfunction in PD without dementia has been labeled "mild cognitive impairment" (PD-MCI) to parallel definitions in the AD field, although the utility and validity of the term PD-MCI is subject to debate [36, 37]. PD-MCI is presumably a state of pre-dementia related to future PDD in which cognitive testing is abnormal but activities of daily living (ADLs) are not impaired significantly. The fine line between PD-MCI and PDD depends on the severity of how ADLs are affected, which can sometimes be subjective or lead to underdiagnosis based on patient under-reporting of symptoms. In 2012, the parallel concept of MCI was formally defined by an MDS taskforce [8], which has gradually gained traction in the PD cognitive impairment research community (Table 2). Up to one third of PD patients have at least mild cognitive impairment at baseline upon diagnosis, with varying rates reported between different studies. Like PDD, PD-MCI is a heterogeneous entity caused by several brain pathologies [16]. PD-MCI can be amnestic, nonamnestic, or multidomain [38]. By

Table 1
Clinical features of Parkinson's disease dementia

1. Core features:

   (a) Diagnosis of PD by Queen Square Brain Bank Criteria

   (b) A slowly progressive dementia syndrome in the context of established PD, clinically diagnosed with the following features:

   • Impairment in more than one cognitive domain

   • Decline from premorbid levels of function

   • Demonstration of impairment in daily activities

2. Associated clinical features:

   (a) Cognitive features: Deficits in

   • Attention

   • Executive function

   • Memory

   • Visuospatial function

   • Language

   (b) Behavioral features:

   • Apathy

   • Hallucinations

   • Delusions

   • Excessive daytime sleepiness

   • Personality and mood changes

3. Features that make the diagnosis uncertain but do not exclude PD cognitive impairment

   • Time interval of development of dementia unknown, unclear baseline, comorbidities that may cause cognitive impairment but not thought to be cause of dementia

4. Features suggesting other conditions as the cause of cognitive impairment that make it impossible to diagnose PD cognitive impairment

   • Example: acute delirium, major depression, cerebrovascular disease, other probable causes of dementia

definition, PD-MCI patients perform 1–2 standard deviations below the age-matched mean on two or more neurocognitive tests. Longitudinal studies suggest that nearly all patients with Parkinson's disease who have MCI will eventually fulfil criteria for dementia [39].

1.7 Diagnosis of PD Cognitive Impairment

The diagnostic process for PD related cognitive impairment is not highly standardized and may vary by type of center, specialization

Table 2
Parkinson's disease-mild cognitive impairment criteria

1. Inclusion criteria
(a) Diagnosis of PD by Queen Square Brain Bank Criteria
(b) Slowly progressive cognitive decline in the context of established PD
• Cognitive deficits on scales of global cognitive abilities or neuropsychological testing
• No significant impairment in daily activities, though complex functional tasks may encounter more subtle difficulties
2. Exclusion criteria
(a) Diagnosis of PD dementia
(b) Other explanations for cognitive impairment (e.g., strokes, depression, medications, delirium)
(c) Other PD comorbidities (e.g., mood disorders, psychosis, motor impairment, sleepiness) that impair cognitive testing
3. Fulfillment of specific cognitive criteria
(a) Abbreviated assessments: Impairment on validated scale for global cognitive activities used in PD or impairment on at least two tests in a limited neuropsychological battery (covering attention, visuospatial, memory, executive, language domains)
(b) Comprehensive assessments: Impairment in at least two neuropsychological tests performed in a more comprehensive battery
4. (Optional) subtype classification:
(a) PD-MCI single-Domain: Only one domain impaired
(b) PD-MCI multiple domain: Impaired in two or more cognitive domains

of evaluating physicians, and available resources. The gold standard is considered to be full neuropsychological testing, which in itself can be a heterogeneous mix of different tests, looking at function in all major cognitive domains. With lack of neuropsychologist access or limited time, some patients are diagnosed on the basis of short screening tests, which have varying degrees of sensitivity and specificity. Cognition testing can assess for global function or for tests targeting individual cognitive domains (i.e., working memory processing speed, orientation, attention, executive function, abstract reasoning, visuospatial abilities, praxis) [40]. For example, stimulants such as atomoxetine have been studied to look at cognitive batteries specifically constructed to evaluate executive function [41].

There is currently no consensus as to the best battery of specific set of tests, with different expert centers performing different batteries. Consultation with a neuropsychologist is required to choose the tests best related to the study question at hand. Specific diagnostic tests have been proposed for PDD and PD-MCI by MDS taskforces based on the new definitions proposed by MDS (Tables 1 and 2), which can be referenced in the respective articles for PDD [31, 42] and PD-MCI [40]. These test recommendations have undergone validation in several studies since and are available for future clinical trials, though not yet used as the basis of major trials to date.

A condition related to PDD should be mentioned, called dementia with Lewy bodies (DLB). PDD and DLB are both are related to alpha-synuclein deposition in the form of Lewy bodies and represent part of a disease spectrum that could be joined under the general term "Lewy body diseases." They share the same pathological substrate as well as clinical phenomenology but have a different temporal presentation. Patients who develop criteria for dementia and motor symptoms of parkinsonism within the same year concurrently are labeled with a DLB diagnosis Complicating DLB diagnosis is the fact that PD-MCI can occur in the first year in up to 42%, necessitating establishment of a careful history in order to arrive at the most likely diagnosis [43]. DLB exists on a spectrum with PDD and share the same pathology, although evolving on a different timescale. Drugs or trials that apply to PDD should therefore be theoretically interchangeable between those with DLB, and often the two conditions are lumped together given they pathological and phenotypical similarities. PDD and DLB, taken together, are the second most common causes of dementia after AD, accounting for up to 15–20% of the global incidences of dementia [44]. PDD exists on a spectrum with DLB, with many patients falling in a spectrum between classic PD and DLB. DLB also has more pronounced fluctuations of attention, RBD, and hallucinations, similar to advanced PDD. The definition and diagnosis of DLB are constantly being revised, with the fourth consensus report published in 2017 [45].

2 Preclinical Models

Preclinical models of Parkinson's are largely based on toxic induction of dopamine loss through agents such as paraquat or 6-hydroxydopamine(6-OHDA), causing an acute motor syndrome [46]. These models are mainly used to study the motor effects of dopamine cell loss and the effects of possible disease modifying agents. More recently, animal models have been used to explore models of nonmotor symptoms (i.e., sleep disturbances, neuropsychiatric and cognitive deficits), though there is a lack of good animal models for nonmotor symptoms in PD [47]. Rats injected with 6-OHDA bilaterally have shown cognitive deficits as well as anxiety and depressive-like states, with alterations in cognitive performance tasks testing working memory, spatial reference memory, or reversal learning [48, 49]. Behavioral symptoms have been explored in animal models more extensively than cognitive symptoms and there is little data about cognitive outcomes.

3   Summary of Clinical Tools and Outcome Measures

As there are no accepted biomarkers for cognition or progressive disease in PD, previous trials have relied on a variety of endpoints, only some of which have been used in large clinical trials. Many of these were created for AD or dementia in general and are not specific to PD, but the development of PD-specific tools has been more recently attempted. Tools designed for AD (a predominantly cortical dementia) may not fully capture some of the unique features of PD-related cognitive impairment, particularly when testing patients early in the disease course (where frontal and subcortical deficits predominate). Regardless, AD-based scales have been used in some large pivotal PD trials, such as the ADAS-Cog [50], which is a standard neuropsychological test consisting of 11 items assessing orientation, memory, language, reasoning, and praxis, with scores ranging from 0 (least impaired) to 70 (most impaired). Although the ADAS-cog has been developed for use in AD patients, its use in PDD is supported by results of studies of donepezil and rivastigmine described in the trial section. However, the ADAS-cog may underestimate the clinical improvement in patients with PDD because of a lack of sensitivity to the impairments seen in these patients, particularly in those with strong reserve and higher levels of education.

There are no universally agreed upon clinical scales specifically for PD cognitive impairment, but a number have been used and investigated, including the following from the larger drug trials in PD (Table 3). The scale chosen for a given study should reflect the type of study and the purpose of the cognitive testing (e.g., some scales are quick and good for screening or long-term repeat assessments, while others are more sensitive to change and can take longer to perform and should be done less frequently).

In an attempt to help researchers choose from the large number of available scales for PD cognitive impairment, the MDS performed a systematic literature review to identify scales used to assess global cognitive function in PD [51]. Of 12 scales included in the review, only three were classified as "recommended" and two were "recommended with caveats." Recommended scales had been shown to apply to PD patients with data on its use beyond the developers of the scales, *and* found to be reliable, sensitive to change, and valid, based on opinion of an expert panel and literature review. Scales "with caveats" were found to have generally adequate properties but lack evaluation of some measurement properties or not being evaluated specifically in PD patients. The

Table 3
Commonly used outcome measures

Primary outcome measures	Secondary outcome measures	Study primary author
MMSE	–	Aarsland [76]
CIBIC+	–	Aarsland [76]
ADAS-cog	–	Emre [66], Ravina [77], Dubois [79]
ADCS-CGIC	–	Emre [66], Emre [88]
DRS	–	Leroi [86] [75]
CGIC	–	Aarsland [87]
–	NPI	Aarsland [76], Emre [66], Leroi [75], Aarsland [87], Leroi [86], Emre [88], Dubois [79]
–	MMSE	Emre [66], Ravina [77], Aarsland [87], Dubois [79]
–	ADCS-ADL	Emre [66], Emre [88]
–	CDR	Emre [66]
–	D-KEFS	Emre [66]
–	BTA	Leroi [75], Dubois [79]
–	UPDRS-ADL	Leroid [75]
–	MDRS	Ravina [77]
–	CGI	Ravina [77]
–	BPRS	Ravina [77]
–	DAD	Aarsland [87], Dubois [79]
–	AQT	Aarsland [87]
–	CIBIC+	Leroi [86]
–	ZBI	Emre [88]

MMSE Mini–Mental State Examination, *CIBIC+* clinician's interview-based impression of change with caregiver's input, *ADAS-cog* Alzheimer's Disease Assessment Scale, cognitive subscale, *ADCS-CGIC* Alzheimer's Disease Cooperative Study—Clinician's Global Impression of Change, *DRS* Dementia Rating Scale, *CGIC* clinical global impression of change, *NPI* Neuropsychiatric Inventory, *ADCS-ADL* Alzheimer's Disease Cooperative Study—Activities of Daily Living Inventory, *CDR* Clinical Dementia Rating scale, *D-KEFS BTA* Brief Test of Attention, *UPDRS-ADL* Unified Parkinson's Disease Rating Scale—Activities of Daily Living Subscale (Part II), *MDRS* Mattis Dementia Rating Scale, *BPRS* Brief Psychotic Rating Scale, *DAD* disability assessment in dementia, *AQT* A Quick Test of Cognitive Speed, *ZBI* Zarit Burden Interview

recommended scales were the Montreal Cognitive Assessment (MoCA) [52], Mattis Dementia Rating Scale (MDRS) [53], and Parkinson's Disease-Cognitive Rating Scale (PD-CDR) [54]. The scales **recommended with caveats** were the Mini–Mental Parkinson [55] (MMP, which has limited coverage of executive function) and the Scales for Outcomes in Parkinson's Disease Cognition [56] (SCOPA-COG, which has limited data on sensitivity to changes). Six scales were listed as **suggested** (i.e. applied to PD populations, data on use beyond developers *or* reliable, valid, and sensitive to change).

Neuropsychological testing batteries can be designed and adapted to the cognitive aspects in question, with a variety of scales to choose from, as long as one evaluates the required domains of interest. There is flexibility depending on the patient population, desired cognitive outcome (global vs. subscale for attention, memory, executive function, etc.), and the involved neuropsychologist's preferences.

Neuroimaging outcomes for PD-related cognitive impairment have not been used for major trials so far, but may have a growing use as methods for predicting cognitive impairment and measuring longitudinal anatomical and functional changes, and are being refined with further studies [57]. Unlike AD, there are no current direct neuroimaging biomarkers specific for PD pathology.

Safety and tolerability are typically assessed by comparing treatment-emergent adverse events (AE) frequency between treatment groups. Other safety assessments include vital signs, ECG, clinical laboratory testing, and physical and neurological exams.

4   Design of Clinical Trials Used for PD Cognitive Impairment

Cognitive studies in neurogenerative diseases are challenging for a number of reasons. Owing to the markedly high placebo (or "nocebo") rate seen in most trials for PD [60], *open-label studies* have little value in recommending definitive treatment for serious conditions such as dementia. The most impactful trials involve *randomized controlled trials (RCT)* with double-blind placebo, but these are difficult to perform in populations with dementia, for reasons explored later in this chapter. RCTs allow for minimization of bias in allocation to treatments groups, produce comparable groups in each treatment arm, and control for the many variables occurring in patients with dementia, and assure optimal validity of statistical comparisons [61]. There are no currently referenced *historical datasets* for PD cognition trials as the patient populations, dementia definitions, and stage of disease studied have varied in different trials, though the building of large PD cohorts with regular measurements may develop into historical controls/databases for future use [17]. *Cross-over designs* have been used in some studies of PD cognitive impairment, allowing each patient to serve as their own control as they are randomized to the treatment or control therapy sequentially. This helps reduce the number of study participants needed and reduces variability between groups, but can be a problem when there are many dropouts due to drug side effects or fluctuating disease condition, or when a washout period in-between study arms is difficult to achieve. *Withdrawal studies* have been a useful tool when it is difficult to find enough patients that are not already using treatments that are widely prescribed, which includes the limited number of medications used for cognitive function. Patients can be randomized to the study drug continuation or to withdrawal, with observation of whether the withdrawal groups do worse. *Large simple clinical trials* might be ideal for chronic conditions like PD with cognitive impairment, but have not been carried out in a structured way to specifically assess cognitive function in PD, which is one of many complications in advanced PD over longer periods [4]. Power calculations can be based on the outcomes of change seen in the major clinical trials, though the newer PD-specific scales have not been used in large randomized clinical trials to date.

5 Summary of Major Clinical Trials

A review of the major clinical trials for PD cognitive impairment is revealing as to what the most common trial methods have been in used to justify treatment and government approvals of these medications.

5.1 Typical Inclusion Criteria

- Younger cutoff age: typically 40–50 years old.
- Older cutoff age: typically not given, since dementia is more common with age, and many PDD cases may be excluded as a result. Instead of age, features of other possible types of dementia (such as AD) are excluded.
- Clinical diagnosis of PD established by Queen Square Brain Bank criteria [62] which assures a probable diagnosis of PD based on clinicopathologic correlations.
- Documented occurrence of onset of dementia at least 1 year after the diagnosis of PD, avoiding confusion with other primary dementia syndromes with parkinsonism (e.g. DLB, AD with parkinsonism).
- A prespecified H&Y stage (typically II–IV or $\leq$IV in the "off" state).
- Cognitive decline meeting DSM-IV criteria for dementia due to PD; the newer MDS-based criteria could also be used, but they have not been frequently used in clinical trials.
- Scoring in a dementia range on a screening scale other than the measured outcomes. This is often the Mini–Mental State Exam (MMSE) [63] score $\geq$10 and $\leq$26 prior to randomization.

- Patients are typically required to have a reliable caregiver given the potential unreliability of taking drugs as prescribed, accurately reporting adverse events, and for consent purposes.
- Reasonable workup to exclude alternative causes of cognitive impairment.

5.2 Typical Exclusion Criteria

- Clinical diagnosis of dementia with Lewy bodies (DLB), excluded according to the Consensus Guideline criteria [31].
- Neuropsychological testing profile consistent with concomitant possible or probable AD, such as Alzheimer's disease and Related Disorders Association (NINCDS-ADRDA) criteria [64].
- Other causes of dementia reasonably ruled out, typically by examining the following:
  - Serum vitamin B12/folate levels, thyroid function testing, basic electrolytes.
  - Structural brain imaging (CT/MRI) within a specified time frame before baseline to rule out multiple strokes, subdural hemorrhage, hydrocephalus, and so on.
- Absence of history of recurrent major psychiatric or sleep issues.
  - Screening scales may include the following:
    - Patient interview or psychiatric evaluation.
    - Montgomery–Åsberg Depression Rating Scale [65] if depressed.
    - Sleep study if excessive sleepiness.
- Good general health (often defined based on risk factors of studied drug, e.g., history of bradycardia excluding use of cholinesterase inhibitor).
- Not previously treated with drugs related to the study drug within a specified time frame, or ever.
- Concomitant drugs that would interfere with cognitive function, such as anticholinergics and cholinergic agents.
- Concomitant drugs that would interfere with measurement of other variables, unless at a predicted stable dose and tolerated for a prespecified interval before the study. Examples include atypical antipsychotics (e.g. quetiapine, clozapine) and antidepressants, typically allowed if used at a stable dose prior to the study with expectations that the dose would not change during the study.
- Stable motor parkinsonian symptoms that would not be anticipated to change for the duration of the trial, with last change at a specified interval before the trial (typically 1–3 months).
- Possible hypersensitivity to drug class.

5.3 Cognitive Impairment Medication Trials

The majority of large trials in PD cognitive impairment have been with cholinesterase inhibitors. There have been two published large, randomized, controlled cholinesterase inhibitor studies in Parkinson's disease dementia (PDD). The first was a positive study for rivastigmine, which led to its approval by the US FDA for an indication of PDD, the second was a trial for donepezil, which was negative. Details of the larger, higher quality trials are discussed below and summarized in Table 4.

5.3.1 Rivastigmine

In the pivotal study for rivastigmine [66], a multicenter, parallel-group, double-blind, placebo-controlled trial was employed. PD patients were diagnosed by the United Kingdom Parkinson's Disease Society Brain Bank Criteria [62], were at least 50 years old and carried a diagnosis of mild-moderate dementia due to Parkinson's disease according to the fourth edition of the Diagnostic and Statistical Manual of Mental Disorders (code 294.1) [67], with onset of dementia at least 2 years after PD diagnosis. Enrolled patients had mild-to-moderately severe dementia, as defined by a Mini–Mental State Examination (MMSE) [63] score of 10–24.

Randomization of patients was in a 2:1 ratio to receive either rivastigmine (n = 362) or placebo (n = 172). During the 16-week

Table 4
Summary of major clinical trials

Drug	Dose/day	Trial primary author	Number of subjects enrolled	Duration in weeks	Study design
Donepezil	5 and 10 mg	Aarsland [76]	14	10	Crossover trial, double-blind, placebo-controlled
Donepezil	10 mg	Leroi [75]	16	18	RCT, double-blind, placebo-controlled
Donepezil	10 mg	Ravina [77]	28	10	Crossover trial, double-blind, placebo-controlled
Donepezil	5 and 10 mg	Dubois [79]	550	24	RCT, double-blind, placebo-controlled
Memantine	20 mg	Aarsland [87]	75	24	RCT, double-blind, placebo-controlled
Memantine	20 mg	Leroi [86]	25	16	RCT, double-blind, placebo-controlled
Memantine	20 mg	Emre [88]	199	24	RCT, double-blind, placebo-controlled
Rivastigmine	12 mg	Emre [66]	541	24	RCT, double-blind, placebo-controlled

drug escalation period, study drug doses were increased by 3 mg/day at intervals of at least 4 weeks. The highest well-tolerated dose for each patient was maintained for the maintenance phase, with a mean dose of 8.6 mg/day (highest possible dose 12 mg), with more than half of the patients ($n = 201$, 55.5%) receiving 9–12 mg/day.

Primary efficacy variables included the Alzheimer's Disease Cooperative Study—Clinician's Global Impression of Change (ADCS-CGIC) [68] and cognitive subscale of the ADAS-cog. Secondary outcomes were the Alzheimer's Disease Cooperative Study—Activities of Daily Living (ADCS-ADL) [69], the ten-item NPI [70], the MMSE, the Cognitive Drug Research (CDR) Computerized Assessment System power of attention tests [71], the Delis-Kaplan Executive Function System (D-KEFS) Verbal Fluency test, and the ten-point Clock Drawing Test [72]. Motor function was assessed with the UPDRS Part III. Statistical analysis included intention to treat and per protocol populations (both results were consistent).

Efficacy assessments were made at baseline, week 16, and week 24. Data was analyzed of patients who were assessed for one of the primary efficacy variables at least once after baseline and who received at least one dose of study medication. Dropout rates were significant: 131 patients (24.2%) prematurely discontinued the study, mainly due to AEs (27.3% in the treatment group, 17.1% in the placebo group). After 24 weeks, the rivastigmine group showed mean (±SD) improvement on the ADAS-cog of 2.1 (±8.2) points (an 8.8% improvement from baseline). The placebo group instead had a worsening of 0.7 (±7.5) points (a 2.9% deterioration) from baseline. The overall absolute difference between groups was 11.7%.

The mean (±SD) scores for the ADCS-CGIC at week 24 were 3.8 (±1.4) with rivastigmine and 4.3 (±1.5) with placebo group. Of the seven possible responses, the rivastigmine group had significantly more clinically meaningful responses than placebo ($p = 0.007$). Secondary outcome measures also were all significantly better with rivastigmine, including significantly more improvement of at least 30% on the NPI (42.5% vs. 33.7%). Adverse events were more frequent with rivastigmine: nausea (in 29% vs. 11.2% placebo), vomiting, leading to some study withdrawals. While there were no statistically significant changes in UPDRS, there was more tremor in the rivastigmine group (10.2% vs. 3.9%, $p = 0.01$, some causing study withdrawal) and more parkinsonian symptoms (27.3% vs. 15.6%, $p = 0.002$). AE in general were more common on rivastigmine (83.7% vs. 70.9%, $p < 0.001$), though serious adverse events were equal between groups.

The trial and approval of the drug was not accepted without some controversy [73], including criticisms that there was a small size of the effect, not necessarily outweighing the burden of cholinergic AEs, higher dropout rate in rivastigmine group, and the burden side effects (all of which are already common in PD, such as tremor, sialorrhea, and urinary frequency, which potentially "unblinded" some patients). Regardless it still stands to date as the only pivotal trial to achieve US FDA approval for the indication of PD cognitive impairment.

One large (n = 583) open-label, randomized study evaluated the long-term safety of patients randomized to rivastigmine capsules versus patches in PD dementia [74]. Though open-label, this was considered a high-quality safety with longer-term observation over 76 weeks. Efficacy outcomes included Alzheimer's Disease Cooperative Study—Activities of Daily Living (ADCS-ADL), Neuropsychiatric Inventory (NPI-10), and Mattis Dementia Rating Scale (MDRS). The results indicated no new safety concerns, with the patch having overall less AEs (other than application site reactions) but less efficacy than capsules on outcome measures at different time points. Though open-label studies cannot prove efficacy over placebo, they may provide comparative information between different types of treatment delivery methods and provide more information on long-term safety, which is critical for patients with advanced illness.

5.3.2 Donepezil

There have been two crossover and two parallel-group placebo-controlled RCT examining the effects of donepezil in PDD. These mostly have included small numbers of patients. Despite promise from an earlier open-label study, a large double-blind trial was negative.

In the 18-week double-blind, placebo-controlled, parallel-group RCT study, Leroi et al. randomized seven patients to donepezil (2.5–10 mg/day, mean dose 6.4 mg/day) and nine patients to placebo [75]. Inclusion criteria included diagnosis of PD and DSM-IV criteria for dementia, with MMSE ≥ 10. The primary efficacy outcomes included a neuropsychological battery of several tests that assessed global cognition, memory, visuospatial and executive functions, attention, and psychomotor speed. Secondary outcome measures included the UPDRS for motor function, psychiatric scales, and ADL rating scales. There was no statistically significant difference in any outcome variable except for a slight improvement in the dementia rating scale (DRS). Only ten patients (62.5%) completed, with six withdrawing due to AEs, including five from the donepezil group (nausea, vomiting, sialorrhea, urinary frequency, worsening PD symptoms, falls, lightheadedness, diplopia, rhinorrhea). One withdrawal from the placebo group had visual hallucinations, diarrhea, and disorientation. There were no differences on the UPDRS scale despite complaints of motor worsening.

In the study by Aarsland et al., which was the first placebo-controlled trial for a cholinesterase inhibitor in PDD, a crossover double-blind, placebo-controlled methodology was employed to study donepezil for two sequential 10-week treatment period without any time allocated for drug washout [76]. The study enrolled PD patients who met DSM-IV criteria for dementia at least 1 year after onset of motor symptoms, with MMSE score ranging from 16 to 26. The drug was dosed at 5 mg daily and increased to 10 mg daily as tolerated. There were eight patients in the donepezil group and six in the placebo group to start. Outcome measures included MMSE, clinician interview-based impression of change (CIBIC+), iwhere a lower score signifies improvement. The motor section of the UPDRS was used to monitor parkinsonism. Secondary outcome measures included a battery of neuropsychological tests and the Neuropsychiatric Inventory (NPI). Efficacy analysis was done on patients with at least 1 score after baseline. Two patients, both on donepezil, withdrew due to AD (typical cholinergic effects of dizziness, nausea, diarrhea) in the first study period but the rest finished the study. After the first 10 weeks, patients on donepezil had improvement of MMSE scores (increased by 2.1 (2.7) points compared to placebo (increased by 0.3 ± 3.2 points), $p = 0.013$. The mean CIBC+ score was 3.3 (0.9) on donepezil and 4.1 (0.8) on placebo, $p = 0.034$. On the basis of the CIBIC+ score, five (42%) patients on donepezil and two (17%) on placebo were rated as improved. There were no carryover or residual effects. Parkinsonism didn't get worse during donepezil treatment. The power of the small study was too low to make conclusive statements on the prevalence of side effects, but AEs were reported in 71% of donepezil patients and 75% of placebo. After excluding patients with missing data for CIBIC+ outcomes, the comparisons of CIBIC+ between donepezil and placebo was not significant ($p =0.79$).

In the study by Ravina et al., a double-blind, placebo-controlled, crossover trial was employed for a trial of donepezil in 22 patients with PD and dementia [77]. Similar to the Aarsland trial, there were two sequential 10-week treatment periods but this time with a 6 week washout period in between. Inclusion criteria included DSM-IV criteria for dementia with onset 2 years or more after onset of motor symptoms, and MMSE scores ranging from 17 to 26. The drugs were dosed at 5 mg daily and increased to 10 mg after 4 weeks if tolerated. The primary outcome measure was the ADAS-Cog, with secondary measures of MMSE, MDRS (Mattis Dementia Rating scale [78], clinical global impression of change (CGI-C), brief psychiatric rating scale (BPRS), and motor outcomes were monitored with the UPDRS. There were no carryover effects between the periods of treatment. Results showed a nonsignificant improvement of ADAS-Cog of 1.9 (1.4) on donepezil, with scores of 22.5 (6.9) on donepezil vs. 24.4 (9.4) on placebo, $p = 0.18$. The MDRS scores showed no significant treatment effects but the MMSE was significantly higher ($p = 0.004$) on donepezil with a difference of 2 (0.61) points. Three subjects withdrew from the first period (two on donepezil for exacerbation of psychosis and arrhythmia, one on placebo for exacerbation of psychosis) and three from the second period (two on donepezil, one on placebo). AE rates (mainly worsening psychosis or agitation) were 52% for donepezil and 45% for placebo.

In the Dubois et al. study a randomized, placebo-controlled, double-blind study with a large number of patients ($N = 550$ from 108 sites in 13 countries) was reported [79]. PD patients H&Y stage IV or less in off-state meeting DSM-IV criteria for dementia and MMSE between 10 and 26 were included, excluding DLB by Consensus criteria. In this study, Donepezil was randomized 1:1:1 with donepezil 5 mg, donepezil 10 mg, or placebo for 24 weeks. Two coprimary endpoints were chosen: the ADAS-cog and the CIBIC+. Secondary endpoints included neurocognitive tests measuring executive function, attention, ADLs, and behavior (including Brief Test of Attention, MMSE, Delis-Kaplan Executive Functions System (D-KEFS), DAD, and NPI [70]. The mean change from baseline of the ADAC-cog mean was −0.3, −2.45, and −3.72 for the placebo, 5 mg, and 10 mg treatment groups, respectively. After statistical adjustment for treatment-by-country interaction (an unforeseen event), the comparisons between 5 and 10 mg with placebo were not significant for primary endpoints ($p = 0.5$ and $p = 076$, respectively). The CIBIC+ mean changes were 3.7, 3.6, and 3.9, for the placebo, 5 mg, and 10 mg treatment groups, respectively; while the 10 mg donepezil group was significantly better, the 5 mg group was not. Secondary outcomes showed significant benefits on MMSE, D-KEFS, and BTA but not ADLs. There were more cholinergic side effects and higher discontinuation rates in the donepezil groups (i.e., nausea, vomiting, and diarrhea). Secondary outcomes with the D-KEFS and BTA, relevant cognitive tests for PDD, showed significant superiority compared to placebo. AEs were cholinergic in nature consistent with other cholinesterase inhibitor trials and were more frequent on donepezil. There were slightly more reports of parkinsonian symptoms on donepezil than on placebo, but this did not correlate to major overall changes on the PD motor scales (CGIC-PD or the UPDRS motor scale). The authors of the report felt that cognitive function and global status was improved though the planned primary goals were not achieved. The trial patients were not as cognitively impaired at baseline as compared to those in the rivastigmine trial, and thus may not have had as much room to improve. Interestingly the magnitude of the change manifested by donepezil treatment on the ADAS-cog was greater than the change seen in donepezil trials in AD [80]. For the CIBIC+, a treatment effect was observed for donepezil, with 10 mg showing the best effect, confirming some clinical relevance of the drug despite study outcomes. There was no significant benefit in ADLs.

In one study, galantamine was given to 41 patients with PDD [81]. The study design was a 24-week, parallel-group, open-label RCT, randomized to either drug or preexisting therapy. Inclusion criteria was ICD-10 criteria for dementia, onset 2 years or more after onset of motor symptoms, and MMSE less than 25. The drug was titrated from 4 mg twice a day to 8 mg twice a day after 4 weeks. There was not a prespecified primary outcome variable, with various outcome measures including MMSE, ADAS-Cog, frontal assessment battery (FAB), clock-drawing test, the HPI, assessment of distress of relatively, UPDRS for motor score stability, and the disability assessment for dementia (DAD) [82]. Two patients in the control group dropped out due to inability to complete the testing battery. Galantamine appeared to have statistically significant benefit on all measures of severity of cognitive impairment with no significant worsening of motor function. AEs occurred in 30% of galantamine treated patients including drooling, orthostatic hypotension, more tremor, nausea, and increased urinary frequency. Unfortunately, the lower quality of design and open label nature of the study lead to a rating of insufficient evidence for use of galantamine for the treatment of dementia in PD.

In summary for cholinesterase inhibitors: the initial MDS Evidence Based Medicine (EBM) review in 2011 [83] concluded that there is insufficient evidence for donepezil to be recommended for the treatment of dementia in PD. On the updated review in 2019 [84], while there was still "insufficient evidence for galantamine and donepezil for treatment of PDD, the MDS EBM review decided to upgrade the clinical implications of these drugs to "possibly useful" after noting that some benefits were found in some of the outcome measures. Additionally, a meta-analysis of treatments in PDD demonstrated that cholinesterase inhibitors overall modestly improves cognitive function [85]. In current clinical practice, most physicians interchangeably use different cholinesterase inhibitor, with availability of different drugs subject to review and restrictions by availability and insurance approval plans. Unlike AD, there isn't evidence to show any slowing of cognitive progression with this drug class.

5.3.4 NMDA Receptor Antagonists: Memantine

Three randomized placebo-controlled, parallel-group trials have been performed using memantine for the treatment of PDD; of these, two were completed in patients with a diagnosis either of PDD or of DLB.

In the trial by Leroi et al. [86], a placebo-controlled, parallel-group RCT was performed with memantine over a 22-week period, including discontinuation of study drug at week 16. Inclusion criteria were PDD by DSM-IV criteria at least 1 year after onset of motor symptoms and MMSE score from 10 to 27. Cholinesterase inhibitors were allowed if stable for 6 months previous to entry into the study and if there was no change in behavior or cognition for at least 1 month prior to entry. Twenty-five patients were randomized to a fixed dose of 20 mg/day memantine or placebo. Primary outcome measures were improvement on the Dementia Rating Scale (DRS)[53]. Secondary outcomes included MMSE, NPI, and CIBIC+. At 16 weeks, there was no difference on any outcome measure between memantine and placebo. Global function improvement as measured by CIBIC+ scores improved with memantine vs. placebo (60% vs. 43% respectively, $p = 0.07$). After the 6-week withdrawal from the study drug at week 22, there was more deterioration in the group that was on memantine vs. placebo (70% vs. 29% respectively, $p = 0.04$); mean CIBIC+ ± SD score: 5.4 ± 1.2 vs. 4.4 ± 0.5, respectively. No significant changes were seen on other outcome variables, and motor symptoms based on UPDRS were normal. AE rates were similar between the treatment groups.

In the trial by Aarsland et al. [87], 72 patients were enrolled who had either PDD (by DSM-IV definition, or at least 1 year after onset of motor symptoms) or DLB (by revised consensus operational criterial for DLB [45], and randomly assigned memantine or placebo in a placebo-controlled, parallel group RCT for 24 weeks. Patients could have mild or moderate dementia with MMSE ≥ 12). The primary outcome measure was the CGI-C and secondary outcome measures included the MMSE, NPI, a quick test of cognitive speed (AQT), DAD, and a UPDRS motor subscale to assess for motor stability. Only 56 patients completed the study (78%, 27 on memantine and 29 on placebo). At 24 weeks, the memantine group had better CGI scores than placebo: 3.5 (1.5) vs. 4.2 (1.2), $p = 0.03$. A subgroup analysis showed that the change in CGI-C for PDD (4.3 in placebo, 2.9 in memantine group) versus no differences in CGI-C for DLB; this may suggest that there is a more global and robust response in PDD patients, though the study was not powered to detect differences between these subgroups. A moderate clinical improvement was found in 27% patients on memantine versus none on placebo; however, 17% on memantine reported moderate worsening versus 18% on placebo. On secondary measures, there was improved speed on attentional tasks in the memantine group (AQT difference 12.4; $p = 0.004$), otherwise there were no significant differences between the groups in secondary outcome measures. 47% (20 on placebo and 15 on memantine) reported AEs; 16 patients (21%; nine on placebo and seven on memantine) dropped out due to AEs with 11 (seven on placebo and four on memantine) of them due to worsening disease.

In another study, Emre et al. used memantine 20 mg/day in a randomized, double-blind, placebo-controlled study for treatment of PDD or DLB for 24 weeks [88]. Inclusion criteria were similar to that of the Aarsland trial except that MMSE range was 10–24 and H&Y stage of parkinsonism was required to be ≤3. There was no predefined primary endpoint, with efficacy variables including the ADCS-CGIC scale, ADCS-ADL, ACDS-ADL23, Zarit burden interview (for caregiver burden), NPI, UPDRS motor scores, and several cognitive tests over the five core domains of attention, executive function, language, memory, and visuospatial function. A larger number of patients were recruited: 199 with either DLB ($n = 78$) or PDD ($n = 121$) being randomly assigned to memantine or placebo. Patients who had at least one valid post-baseline assessment were included in analysis. The full analysis set included 93 patients given memantine (33 with DLB, and 60 with PDD) and 97 given placebo (41 with DLB, and 56 with PD dementia). While the ADCS-CGIC score had significant improvement in the memantine group at week 12 ($p = 0.14$), both in the total population and those with PDD, this was not sustained at week 24. Primary endpoint values at week 24 for the whole population (3.5 with memantine vs. 3.8 with placebo, treatment difference −0.3]; $p = 0.120$) and for the patients with PDD (3.6 with memantine vs. 3.8 with placebo, treatment difference −0.1; $p = 0.576$) were not significantly different. Interestingly at 24 weeks, patients with DLB on memantine had greater improvement based on subanalysis (mean change from baseline 3.3 vs. 3.9, respectively, difference −0.6, with NPI scores improving more significantly in the memantine group than placebo (−4.3 vs. 1.7, respectively, difference −5.9; $p = 0.041$) but not in those with PDD ($p = 0.522$) or in the total patient population ($p = 0.092$). In most individual cognitive tests scores, Zarit caregiver burden, and ADCS-ADL23 there were no significant changes, and UPDRS did not differ between groups. AE incidence was similar between both groups, most commonly falls (8% in each arm) and somnolence (5% of memantine, 3% of placebo), with proportion of AEs leading to withdrawal being the same in both groups.

Overall efficacy conclusions for memantine, despite the number of studies, was considered insufficient evidence for treatment of dementia in PD by the MDS EBM review. The drug is considered acceptably safe without need for specialized monitoring.

5.3.5 Dopaminergic medications

In the early days of dopamine therapy, the "awakening" effect of levodopa on cognitive functions had already been recognized [89]. Dopaminergic medications have a variable effect on cognition (e.g., better working memory or worse reversal learning) [90]. Effects of dopaminergic treatment for cognition varies by several factors, including the specific task demand, the duration of disease (early vs. late), existence of more fixed cognitive decline, and genetic risk factors. Clinically, effects vary greatly: while some improve, many have no improvement and some even worsen [91]. In some cases, cognitive function is worse or neutral in effect.

Table 5
Evidence-based review of drugs for PD cognitive impairment

Drug	Class drug	Efficacy	Safety rating	Implications for practice
Rivastigmine	Cholinesterase inhibitor	Efficacious	Acceptable risk	Clinically useful
Donepezil	Cholinesterase inhibitor	Insufficient evidence	Acceptable risk	Possibly useful
Galantamine	Cholinesterase inhibitor	Insufficient evidence	Acceptable risk	Possibly useful
Memantine	NMDA antagonist	Insufficient evidence	Acceptable risk	Investigational

NMDA N-methyl-D-aspartate (NMDA) antagonist

5.3.6 Summary of Drug Treatments in PD

The practice implications for the treatment dementia in PD are summarized in Table 5. None of the studies exceeded 24 weeks; therefore, all recommendations given here are for the short-term treatment of dementia in PD.

5.4 Non-pharmacological trials for cognitive impairment

Nonpharmacological trials have also been used in a growing number of trials, which is important since medication options are limited, drug trials are difficult and expensive to carry out, and medications can have associated side effects and are not effective at slowing or stopping progression of cognitive decline. Such novel treatment options are needed and currently underexplored.

Noninvasive neurostimulation: Trials for active transcranial Direct Current Stimulation (t-DCS) over the left dorsolateral prefrontal cortex versus sham t-DCS have been examined for improvement of mild cognitive impairment in PD patients receiving computer-based cognitive training [92]. Despite some positive effects on some cognitive scales, and despite the relative safety of the intervention, studies are small and with unclear generalization to the PD population, and continue to be investigated.

Invasive neurostimulation: A new field of intervention has recently opened after the discovery that deep brain stimulation (DBS) may influence activity in cognitive circuits [93, 94]. DBS may be to directly modulate neural networks to relieve cognitive symptoms, by means that are not fully elucidated [95]. A recent randomized trial of DBS in the nucleus basalis of Meynert for PD showed negative results [96], though in a small number of patients ($n = 6$) with relatively modest cognitive impairment (MMSE 21–26). Additionally, larger RCT of DBS for motor symptoms of PD have incorporated more cognitive outcome measurements and have increased understanding of how DBS might affect cognitive function, though by definition these studies only include patients with normal cognition or PD-MCI, since demented patients are excluded from receiving surgical treatment for motor symptoms. Compared to GPi stimulation, STN stimulation had subtle decline in several cognitive measures including attention, phenomic fluency, and memory; this has influenced some surgery teams to opt for GPi in patients with more baseline cognitive issues, though the effect on global outcomes and quality of life is not clear [97].

Physical exercise: Exercise trials have expanded exponentially in number and scope in recent years given the observations that various types of exercises can have a markedly positive effect on the motor and nonmotor symptoms of PD. Few are designed for cognitive improvement alone, but many have used cognitive scales or correlates as outcome measures. Gait and cognitive function are highly correlated in complex ways, and gait may serve as one surrogate marker for cognitive function, with rehabilitation of one factor being important for improving the other [98]. Some trials demonstrate improvement of gait in PD after treatment with cholinesterase inhibitors, suggesting common neurochemical deficits for both cognitive and gait systems [99]. Exercise trials have the disadvantage of being nearly impossible to blind (therefore not accounting for placebo effect), having a difficult-to-construct control group, since randomizing to no activity at all is becoming unethical, and the effects of socialization and group activity of any degree may improve both cognitive or physical outcomes [100]). Many studies show positive outcomes on chosen cognitive outcome scales [101], though no large studies are primarily centered around cognitive function. Research on the effects of exercise on gait during dual tasks are sensitive to assess the relationship between gait and cognition, but there are few such studies completed. Exercise may influence functional activity in the basal ganglia and cortical areas known to be involved in gait and cognition, and may alleviate PD symptoms and also counteract the effects of PD pathology in the brain, with some arguments for a neuroprotective effect [102].

There is currently insufficient evidence to identify if one modality of exercise is superior for improving cognition in PD patients. A systematic review of RCTs found that physical exercise programs generally promote positive and significant effects on global cognitive function, processing speed, sustained attention, and mental flexibility in mild-to-moderate stage PD participants, including interventions such as tango, cognitive training alongside motor training, and treadmill. Treadmill training with higher-intensity three times a week for about 60 min over 24 weeks produced larger improvements in cognition compared to other modalities [103].

Cognitive exercises/cognitive rehabilitation: Cognitive rehabilitation is based on the hypothesis that increased mental effort can lead to protection of current cognition and possibly even improvement, though strengthening cognitive functional networks or anatomical correlates. One study evaluated cognitive rehabilitation for improving cognitive impairment in PD patients receiving computer-based cognitive training. While some significant effects were reported, the study was exploratory with a small sample size, and effects cannot be generalized [104]. However, cognitive rehabilitation studies may have some effects based on a metanalysis [105] and can be a popular alternative for patients wishing to get actively engaged in research without risks of side effects or burden of medical surveillance.

Diet: at present, there is limited data connecting the role of diet to cognitive impairment, though some newer data suggest that adherence to a Mediterranean diet might reduce the probability of prodromal PD (i.e., symptoms and signs of PD are present, but are yet insufficient to define disease). This raises the possibility of studying the role of diet in the progression of disease, which may ultimately impact cognition later in the disease course [106].

Computer based testing: Since in-person testing may take a longer amount of time, some studies have employed computer-based or technology-enhanced outcome measures, which allow the patient to pace themselves with a computer task-based measuring tool and engage in cognitively unique way, such as exergaming [107].

5.5 Treatment of Nondementia Cognitive Impairment

In real-world practice, cognitive medications are often used to treat patients who have cognitive complaints but are not felt to have PDD based on objective testing. However, there is currently no evidence-based support for this practice. In the MDS evidence-based review [84] there were five high quality studies that investigated the effect of medications on nondementia cognitive impairment. Rivastigmine [108] and rasagiline [109] were negative studies for PD-MCI outcomes.

6 Challenges and Current Deficiencies in PD-Related Cognitive Impairment Trials

A detailed review of the trials previously performed and experience with conceptual difficulties of designing trials for degenerative cognitive disorders reveals a large number of challenges, The following points should be reviewed in consideration of trial design and interpretation:

1. Differences in definitions of dementia between trials historically hampers the ability to synthesize data from previous trials and provide metanalysis or power calculations.

2. There is a general lack of use of the newer and more specific and recommended outcome measures for PD cognitive impairment in past trials, again hampering the ability to syn-

thesize data from previous trials and provide metanalysis or power calculations. Continued development and validation of PDD and PD-MCI scales is essential.

3. The inherent pathological and clinical heterogeneity of cognitive problems in PD make generalization of results difficult and uniform recruitment challenging. This is probably why single ligand-targeted drug therapy or single mechanism-modulating treatments have been limited in effect. Drug "cocktails" or multimechanism approaches may be a more optimal approach in the future. Additionally, alpha synuclein may be a driver of other comorbid pathologies including AD-related pathology and TDP-43 [110] which threatens accurate diagnosis and stratification of patients for clinical trials and warrants better protein-specific markers for neurodegenerative conditions. Therefore, the clinical response to the PD-pathology related treatment may be drowned out in the influence of the other pathologies.

4. Multiple other nonmotor factors influence cognitive outcomes in PD,: including influence from anxiety, depression, apathy, fatigue and hypersomnolence, hypotension, psychosis, and sleep disorders, all of which may be present and hard to control for in all patients.

5. Lack of reliable biomarkers specific to cognitive function.

6. Higher rate of adverse events leading to higher dropout rates, likely because patients are more advanced with high burden of symptoms that could be interpreted as adverse events, many of which may cause dropout despite being in the placebo group. Higher rates of adverse events with more dropouts may therefore exaggerate the treatment benefits, when using no change as the imputed value in a last-observation-carried-forward analysis for a condition in which deterioration is expected.

7. The inherently fluctuating nature of the cognition symptoms in PD: patients may enroll when getting worse, with deviation back to mean after enrollment in the trial, independent of the effects of the planned interventions.

8. Inherent difficulties in determining the difference between dementia and MCI, which relies on adequate assessment of activities of daily living, which is particularly difficult in patients who have high baseline function (and who tend to be over-represented in trials at major academic centers).

9. Challenges inherent to studying cognition outcomes: Patients have limited tolerance for outcome testing due to fatigue, anxiety, or frustration. There is a need to balance the requirement a focused primary outcome measure with the need for assessing more global or secondary outcomes. This is challenging since the better cognitive batteries tend to be longer and mul-

tiple tests may be needed to fully explore all cognitive domains of interest. Cognitive function is complex and studying one cognitive subdomain may not be enough to show global change, but too much testing can reduce the reliability of the outcome measurements secondary to reduced patient tolerance.

10. Recruitment strategies are challenging: more advanced patients have difficulties with transport, may cancel or reschedule more frequently, may forget appointments, may comply less with study tasks, and have impaired ability to consent for studies, sometimes requiring more complicated consent processes.

11. Unblinding is possible due to higher rates of side effects, such as the higher incidence of nausea, tremor, and vomiting among patients taking cholinesterase inhibitors.

12. As seen in past trials, the degree of change of cognition in a short period is modest at best, requiring large numbers of patients to detect small differences, which may lead to statistical challenges with variations between recruitment at multiple centers or in multiple countries.

13. Cognitive testing results may vary depending on the motor state of the patient and fluctuations in alertness that occur with PD-related motor and cognitive impairment, therefore cognitive scales may not reflect average valuates and may vary significantly at different times with each patient. Tests should be done in the best "on" state as a general goal.

14. The definition of a successful trial (i.e., incremental changes in points on a cognitive scale) does not always correlate to meaningful change for the patient unless the scale used incorporates meaningful changes. The effects of the cognitive drugs are subtle and a minimal meaningfully clinical difference of change is not well defined.

15. There are challenges inherent to studying cognition in neurodegenerative disease in general. There are so few drugs for even marginally influencing the symptoms that it is hard to recruit patients who are naïve to the effects of these drugs, or who may be willing to get off of available treatments for weeks or months to try a new and unproven agent.

16. Often caregivers need to be involved for consent and help carrying through study, they are not always screened specifically for their qualifications and how their decisions and care affect trial outcomes.

17. Age, sex, culture, language, and education can impact the outcomes of cognitive tests and the results need to be adjusted for different populations. Scoring of scales and cutoffs for normality may need to be adjusted, some scales already have norma-

tive values but many do not. Translation and cultural adaptations of scales need to be considered. The MDS recommends somes scales that have validated language translations.

18. Patients with higher premorbid intelligence may need more extensive testing and investigations to determine if they meet cutoffs for PD-MCI or PDD ratings, with change from baseline being more important than deviations from age-related norms.

19. Determination of ADL impairment is difficult and may require multiple independent sources of information.

20. If scales are administered longitudinally in a trial the learning effect through repeat testing may alter results, and scales with alternate versions/editions may be favorable (such as MoCA and PD-CRS).

21. Cognitive impairment takes years to develop, yet trials are only up to 24 weeks in duration. The long-term effects and impact of interventions, in a trial or tracked prospectively, is largely unknown.

7 Conclusions, Future Challenges, and Suggestions

Cognitive interventions and studies in PD are still in relative infancy and the newest tools to define and measure PDD and PD-MCI are only beginning to be used in this field of research. The coming decades are anticipated to have higher-impact studies as the pathophysiology of the disorder and modifiable risks for cognitive impairment are better realized. Incorporation of putative biomarkers along with more refined clinical scales will reveal new information on how these markers correlate with real-world patients individually and not just as a large population. Until there are better markers for cognitive outcomes, it will be hard to prove clincally relevent endpoints. Neuroimaging markers may also have promise for clinical trials.

Ideally, researchers will identify more homogeneous populations through more unified inclusion/exclusion criteria and move toward more consistent definition of dementia and meaningful interventions, updated periodically based on review of new technology and biomarkers. Using multifactorial interventions and outcome measures (e.g., cognitive rehabilitation, exercise, and medication in combination) may be more effective than the current strategy of univariable interventions. Comparative treatments between different interventions has yet to be explored.

Since cognitive issues are typically a late complication related to spread of the disease from subcortical to cortical structures, trials focusing on reducing the rate of disease progression may ultimately be the best way to treat the cognitive effects of the disease;

once the damage is done to the brain structures, there may be no reasonably effective way to restore function in these networks. Cognitive dysfunction may be better prevented rather than resolved. Ultimately an intervention that is applied before cognitive dysfunction starts or at the very beginning of cognitive symptoms would be most ideal to interrupt or reduce the progression of change, but earlier markers of cognitive function are lacking and such trials would take many years to carry out due to the slow progression of the problem. Focus on practical outcomes meaningful to patients and meeting criteria for minimally significant cognitive changes will be welcome, as well as trials for investigating best practical and palliative care for patients and caregivers affected by cognitive issues.

Barriers to trial participation are many for these frailer patients. Bringing the trials closer to home or in the home, to the communities where patients live, will help to enrich sample size and expand the numbers of patients that can be involved. Larger simple trials can gather more patients, computer-based techniques can allow patients to do the testing in their normal environment where they are not stressed or distracted by a different setting. Telemedicine may be one way to bring trial assessments to the home setting, though limited by economic and technical issues this practice is rapidly expanding in some countries. Patients in a home environment may be more engaged and willing to participate, or may show a baseline true to their state (patients are anecdotally often more alert and engaged in the medical office). Patient registries for PDD or DLB, social media, and patient organizations can come together to get larger numbers of interested participants in trials that would otherwise be missed.

References

1. Parkinson J (2002) An essay on the shaking palsy. J Neuropsychiatry Clin Neurosci 14(2):223–236. https://doi.org/10.1176/jnp.14.2.223
2. Braak H, Tredici KD, Rüb U, de Vos RAI, Jansen Steur ENH, Braak E (2003) Staging of brain pathology related to sporadic Parkinson's disease. Neurobiol Aging 24(2):197–211. https://doi.org/10.1016/s0197-4580(02)00065-9
3. Aarsland D, Kramberger MG (2015) Neuropsychiatric symptoms in Parkinson's disease. J Parkinsons Dis 5(3):659–667. https://doi.org/10.3233/jpd-150604
4. Reid WGJ, Hely MA, Morris JGL, Loy C, Halliday GM (2011) Dementia in Parkinson's disease: a 20-year neuropsychological study (Sydney multicentre study). J Neurol Neurosurg Psychiatry 82(9):1033–1037. https://doi.org/10.1136/jnnp.2010.232678
5. Aarsland D, Andersen K, Larsen JP, Lolk A (2003) Prevalence and characteristics of dementia in Parkinson disease. Arch Neurol 60(3):387. https://doi.org/10.1001/archneur.60.3.387
6. Cummings JL (1988) Intellectual impairment in Parkinson's disease: clinical, pathologic, and biochemical correlates. Top Geriatr 1(1):24–36. https://doi.org/10.1177/089198878800100106
7. Hobson P, Meara J (2004) Risk and incidence of dementia in a cohort of older subjects with Parkinson's disease in the United Kingdom. Mov Disord 19(9):1043–1049. https://doi.org/10.1002/mds.20216
8. Litvan I, Aarsland D, Adler CH, Goldman JG, Kulisevsky J, Mollenhauer B, Rodriguez-

Oroz MC, Tröster AI, Weintraub D (2011) MDS task force on mild cognitive impairment in Parkinson's disease: critical review of PD-MCI. Mov Disord 26(10):1814–1824. https://doi.org/10.1002/mds.23823

9. Leech RW, Brumback RA, Poduslo SE, Schiffer R, Adesina A (2001) Dementia: the University of Oklahoma autopsy experience. J Okla State Med Assoc 94:507–511. https://doi.org/10.1007/978-1-4612-0811-2_6

10. Apaydin H, Ahlskog JE, Parisi JE, Boeve BF, Dickson DW (2002) Parkinson disease neuropathology. Arch Neurol 59(1):102. https://doi.org/10.1001/archneur.59.1.102

11. Compta Y, Parkkinen L, Kempster P, Selikhova M, Lashley T, Holton JL, Lees AJ, Revesz T (2013) The significance of α-Synuclein, amyloid-β and tau pathologies in Parkinson's disease progression and related dementia. Neurodegener Dis 13(2–3):154–156. https://doi.org/10.1159/000354670

12. Perry EK, Curtis M, Dick DJ, Candy JM, Atack JR, Bloxham CA, Blessed G, Fairbairn A, Tomlinson BE, Perry RH (1985) Cholinergic correlates of cognitive impairment in Parkinson's disease: comparisons with Alzheimer's disease. J Neurol Neurosurg Psychiatry 48(5):413–421. https://doi.org/10.1136/jnnp.48.5.413

13. Tiraboschi P, Hansen LA, Alford M, Sabbagh MN, Schoos B, Masliah E, Thal LJ, Corey-Bloom J (2000) Cholinergic dysfunction in diseases with Lewy bodies. Neurology 54(2):407–407. https://doi.org/10.1212/wnl.54.2.407

14. Bohnen NI, Kaufer DI, Ivanco LS, Lopresti B, Koeppe RA, Davis JG, Mathis CA, Moore RY, DeKosky ST (2003) Cortical cholinergic function is more severely affected in parkinsonian dementia than in Alzheimer disease. Arch Neurol 60(12):1745. https://doi.org/10.1001/archneur.60.12.1745

15. Rowe JB, Hughes L, Ghosh BCP, Eckstein D, Williams-Gray CH, Fallon S, Barker RA, Owen AM (2008) Parkinson's disease and dopaminergic therapy—differential effects on movement, reward and cognition. Brain 131(8):2094–2105. https://doi.org/10.1093/brain/awn112

16. Halliday GM, Leverenz JB, Schneider JS, Adler CH (2014) The neurobiological basis of cognitive impairment in Parkinson's disease. Mov Disord 29(5):634–650. https://doi.org/10.1002/mds.25857

17. Initiative PPM (2011) The Parkinson progression marker Initiative (PPMI). Prog Neurobiol 95:629–635

18. Schrag A, Siddiqui UF, Anastasiou Z, Weintraub D, Schott JM (2017) Clinical variables and biomarkers in prediction of cognitive impairment in patients with newly diagnosed Parkinson's disease: a cohort study. Lancet Neurol 16(1):66–75. https://doi.org/10.1016/s1474-4422(16)30328-3

19. Weintraub D, Dietz N, Duda JE, Wolk DA, Doshi J, Xie SX, Davatzikos C, Clark CM, Siderowf A (2011) Alzheimer's disease pattern of brain atrophy predicts cognitive decline in Parkinson's disease. Brain 135(1):170–180. https://doi.org/10.1093/brain/awr277

20. Veselý B, Rektor I (2016) The contribution of white matter lesions (WML) to Parkinson's disease cognitive impairment symptoms: a critical review of the literature. Parkinsonism Relat Disord 22:S166–S170. https://doi.org/10.1016/j.parkreldis.2015.09.019

21. Fiorenzato E, Biundo R, Cecchin D, Frigo AC, Kim J, Weis L, Strafella AP, Antonini A (2018) Brain amyloid contribution to cognitive dysfunction in early-stage Parkinson's disease: the PPMI dataset. J Alzheimers Dis 66(1):229–237. https://doi.org/10.3233/jad-180390

22. Foguem C, Manckoundia P (2018) Lewy body disease: clinical and pathological "overlap syndrome" between Synucleinopathies (Parkinson disease) and Tauopathies (Alzheimer disease). Curr Neurol Neurosci Rep 18(5):24. https://doi.org/10.1007/s11910-018-0835-5

23. Lawson RA, Yarnall AJ, Johnston F, Duncan GW, Khoo TK, Collerton D, Taylor JP, Burn DJ (2016) Cognitive impairment in Parkinson's disease: impact on quality of life of carers. Int J Geriatr Psychiatry 32(12):1362–1370. https://doi.org/10.1002/gps.4623

24. Anderkova L, Barton M, Rektorova I (2017) Striato-cortical connections in Parkinson's and Alzheimer's diseases: relation to cognition. Mov Disord 32(6):917–922. https://doi.org/10.1002/mds.26956

25. Emre M (2003) Dementia associated with Parkinson's disease. Lancet Neurol 2(4):229–237. https://doi.org/10.1016/s1474-4422(03)00351-x

26. Calderon J (2001) Perception, attention, and working memory are disproportionately impaired in dementia with Lewy bodies compared with Alzheimer's disease. J Neurol Neurosurg Psychiatry 70(2):157–164. https://doi.org/10.1136/jnnp.70.2.157

27. Leroi I, McDonald K, Pantula H, Harbishettar V (2012) Cognitive impairment in Parkinson disease. J Geriatr Psychiatry Neurol 25(4):208–214. https://doi.org/10.1177/0891988712464823

28. Schrag A (2000) What contributes to quality of life in patients with Parkinson's disease? J Neurol Neurosurg Psychiatry 69(3):308–312. https://doi.org/10.1136/jnnp.69.3.308

29. Aarsland D, Larsen JP, Karlsen K, Lim NG, Tandberg E (1999) Mental symptoms in Parkinson's disease are important contributors to caregiver distress. Int J Geriatr Psychiatry 14(10):866–874. https://doi.org/10.1002/(sici)1099-1166(199910)14:10<866::aid-gps38>3.0.co;2-z

30. Huber SJ, Paulson GW, Shuttleworth EC (1988) Relationship of motor symptoms, intellectual impairment, and depression in Parkinson's disease. J Neurol Neurosurg Psychiatry 51(6):855–858. https://doi.org/10.1136/jnnp.51.6.855

31. Dubois B, Burn D, Goetz C, Aarsland D, Brown RG, Broe GA, Dickson D, Duyckaerts C, Cummings J, Gauthier S, Korczyn A, Lees A, Levy R, Litvan I, Mizuno Y, McKeith IG, Olanow CW, Poewe W, Sampaio C, Tolosa E, Emre M (2007) Diagnostic procedures for Parkinson's disease dementia: recommendations from the movement disorder society task force. Mov Disord 22(16):2314–2324. https://doi.org/10.1002/mds.21844

32. Goetz CG, Emre M, Dubois B (2009) Parkinson's disease dementia: definitions, guidelines, and research perspectives in diagnosis. Ann Neurol 64(S2):S81–S92. https://doi.org/10.1002/ana.21455

33. Postuma RB, Berg D, Stern M, Poewe W, Olanow CW, Oertel W, Obeso J, Marek K, Litvan I, Lang AE, Halliday G, Goetz CG, Gasser T, Dubois B, Chan P, Bloem BR, Adler CH, Deuschl G (2015) MDS clinical diagnostic criteria for Parkinson's disease. Mov Disord 30(12):1591–1601. https://doi.org/10.1002/mds.26424

34. Lees AJ, Smith E (1983) Cognitive deficits in the early stages of Parkinson's disease. Brain 106(2):257–270. https://doi.org/10.1093/brain/106.2.257

35. Getz SJ, Levin B (2017) Cognitive and neuropsychiatric features of early Parkinson's disease. Arch Clin Neuropsychol 32(7):769–785. https://doi.org/10.1093/arclin/acx091

36. Dubois B (2007) Is PD-MCI a useful concept? Mov Disord 22(9):1215–1216. https://doi.org/10.1002/mds.21566

37. Tröster AI (2008) Neuropsychological characteristics of dementia with Lewy bodies and Parkinson's disease with dementia: differentiation, early detection, and implications for "mild cognitive impairment" and biomarkers. Neuropsychol Rev 18(1):103–119. https://doi.org/10.1007/s11065-008-9055-0

38. Janvin CC, Larsen JP, Aarsland D, Hugdahl K (2006) Subtypes of mild cognitive impairment in Parkinson's disease: progression to dementia. Mov Disord 21(9):1343–1349. https://doi.org/10.1002/mds.20974

39. Pigott K, Rick J, Xie SX, Hurtig H, Chen-Plotkin A, Duda JE, Morley JF, Chahine LM, Dahodwala N, Akhtar RS, Siderowf A, Trojanowski JQ, Weintraub D (2015) Longitudinal study of normal cognition in Parkinson disease. Neurology 85(15):1276–1282. https://doi.org/10.1212/wnl.0000000000002001

40. Litvan I, Goldman JG, Tröster AI, Schmand BA, Weintraub D, Petersen RC, Mollenhauer B, Adler CH, Marder K, Williams-Gray CH, Aarsland D, Kulisevsky J, Rodriguez-Oroz MC, Burn DJ, Barker RA, Emre M (2012) Diagnostic criteria for mild cognitive impairment in Parkinson's disease: Movement Disorder Society task force guidelines. Mov Disord 27(3):349–356

41. Warner CB, Ottman AA, Brown JN (2018) The role of atomoxetine for Parkinson disease-related executive dysfunction. J Clin Psychopharmacol 38(6):627–631. https://doi.org/10.1097/jcp.0000000000000963

42. Emre M, Aarsland D, Brown R, Burn DJ, Duyckaerts C, Mizuno Y, Broe GA, Cummings J, Dickson DW, Gauthier S, Goldman J, Goetz C, Korczyn A, Lees A, Levy R, Litvan I, McKeith I, Olanow W, Poewe W, Quinn N, Sampaio C, Tolosa E, Dubois B (2007) Clinical diagnostic criteria for dementia associated with Parkinson's disease. Mov Disord 22(12):1689–1707. https://doi.org/10.1002/mds.21507

43. Yarnall AJ, Breen DP, Duncan GW, Khoo TK, Coleman SY, Firbank MJ, Nombela C, Winder-Rhodes S, Evans JR, Rowe JB, Mollenhauer B, Kruse N, Hudson G, Chinnery PF, O'Brien JT, Robbins TW, Wesnes K, Brooks DJ, Barker RA, Burn DJ (2013) Characterizing mild cognitive impairment in incident Parkinson disease: the ICICLE-PD study. Neurology 82(4):308–316. https://doi.org/10.1212/wnl.0000000000000066

44. Jellinger KA, Attems J (2011) Prevalence and pathology of dementia with Lewy bodies in the oldest old: a comparison with other dementing disorders. Dement Geriatr Cogn Disord 31(4):309–316. https://doi.org/10.1159/000327360

45. McKeith IG, Boeve BF, Dickson DW, Halliday G, Taylor JP, Weintraub D, Aarsland D (2017) Diagnosis and management of dementia with Lewy bodies: fourth consensus report of the DLB consortium. Neurology 89(1):88–100

46. Blesa J, Przedborski S (2014) Parkinson's disease: animal models and dopaminergic cell vulnerability. Front Neuroanat 8. https://doi.org/10.3389/fnana.2014.00155

47. Konnova EA, Swanberg M (2018) Animal models of Parkinson's disease. Parkinson's

disease: pathogenesis and clinical aspects. Codon Publ. https://doi.org/10.15586/codonpublications.parkinsonsdisease.2018.ch5

48. Campos FL, Carvalho MM, Cristovão AC, Je G, Baltazar G, Salgado AJ, Kim Y-S, Sousa N (2013) Rodent models of Parkinson's disease: beyond the motor symptomatology. Front Behav Neurosci 7. https://doi.org/10.3389/fnbeh.2013.00175

49. Drui G, Carnicella S, Carcenac C, Favier M, Bertrand A, Boulet S, Savasta M (2013) Loss of dopaminergic nigrostriatal neurons accounts for the motivational and affective deficits in Parkinson's disease. Mol Psychiatry 19(3):358–367. https://doi.org/10.1038/mp.2013.3

50. Rosen WG, Mohs RC, Davis KL (1984) A new rating scale for Alzheimer's disease. Am J Psychiatr 141(11):1356–1364. https://doi.org/10.1176/ajp.141.11.1356

51. Skorvanek M, Goldman JG, Jahanshahi M, Marras C, Rektorova I, Schmand B, van Duijn E, Goetz CG, Weintraub D, Stebbins GT, Martinez-Martin P (2017) Global scales for cognitive screening in Parkinson's disease: critique and recommendations. Mov Disord 33(2):208–218. https://doi.org/10.1002/mds.27233

52. Nasreddine ZS, Phillips NA, Bedirian V, Charbonneau S, Whitehead V, Collin I, Cummings JL, Chertkow H (2005) The Montreal cognitive assessment, MoCA: a brief screening tool for mild cognitive impairment. J Am Geriatr Soc 53(4):695–699. https://doi.org/10.1111/j.1532-5415.2005.53221.x

53. Johnson-Greene D (2004) Dementia Rating Scale-2 (DRS-2). PJ Jurica, CL Leitten, S Mattis: Psychological assessment resources, 2001. Arch Clin Neuropsychol 19 (1):145–147. doi:https://doi.org/10.1016/j.acn.2003.07.003

54. Pagonabarraga J, Kulisevsky J, Llebaria G, García-Sánchez C, Pascual-Sedano B, Gironell A (2008) Parkinson's disease-cognitive rating scale: a new cognitive scale specific for Parkinson's disease. Mov Disord 23(7):998–1005. https://doi.org/10.1002/mds.22007

55. Mahieux F, Michelet D, Manifacier MJ, Boller F, Fermanian J, Guillard A (1995) Mini-mental Parkinson: first validation study of a new bedside test constructed for Parkinson's disease. Behav Neurol 8(1):15–22. https://doi.org/10.1155/1995/304876

56. Marinus J, Visser M, Verwey NA, Verhey FRJ, Middelkoop HAM, Stiggelbout AM, van Hilten JJ (2003) Assessment of cognition in Parkinson's disease. Neurology 61(9):1222–1228. https://doi.org/10.1212/01.wnl.0000091864.39702.1c

57. Lanskey JH, McColgan P, Schrag AE, Acosta-Cabronero J, Rees G, Morris HR, Weil RS (2018) Can neuroimaging predict dementia in Parkinson's disease? Brain. https://doi.org/10.1093/brain/awy211

58. Fahn S, Elton R, Members of the UPDRS Development Committee (1987) Unified Parkinson's disease rating scale. In: Fahn S, Marsden CD, Calne DB, Goldstein M (eds) Recent developments in Parkinson's disease, vol 2. Macmillan Health Care Information, Florham Park, NJ, pp 293–304. https://doi.org/10.1016/b978-0-12-374028-1.00001-4

59. Schneider LS, Olin JT, Doody RS, Clark CM, Morris JC, Reisberg B, Schmitt FA, Grundman M, Thomas RG, Ferris SH (1997) Validity and reliability of the Alzheimer's disease cooperative study-clinical global impression of change. Alzheimer Dis Assoc Disord 11:22–32. https://doi.org/10.1097/00002093-199700112-00004

60. Witek N, Stebbins GT, Goetz CG (2018) What influences placebo and nocebo responses in Parkinson's disease? Mov Disord 33(8):1204–1212. https://doi.org/10.1002/mds.27416

61. Friedman LM, Furberg CD, DeMets DL (1998) Basic study design. Fundamentals of clinical trials. Springer, New York, NY. https://doi.org/10.1007/978-1-4757-2915-3_4

62. Hughes AJ, Daniel SE, Kilford L, Lees AJ (1992) Accuracy of clinical diagnosis of idiopathic Parkinson's disease: a clinico-pathological study of 100 cases. J Neurol Neurosurg Psychiatry 55(3):181–184. https://doi.org/10.1136/jnnp.55.3.181

63. Folstein M, Folstein S, McHugh PR (1975) "Mini-mental status". A practical method for grading the cognitive state of patients for the clinician. J Psychiatric Res 12(3):189–198

64. McKhann G, Drachman D, Folstein M, Katzman R, Price D, Stadlan EM (1984) Clinical diagnosis of Alzheimer's disease: report of the NINCDS-ADRDA work group* under the auspices of Department of Health and Human Services Task Force on Alzheimer's disease. Neurology 34(7):939–939. https://doi.org/10.1212/wnl.34.7.939

65. Montgomery SA, Åsberg M (1979) A new depression scale designed to be sensitive to change. Br J Psychiatry 134(4):382–389. https://doi.org/10.1192/bjp.134.4.382

66. Emre N, Aarsland D, Albanese A (2004) Rivastigmine for dementia associated with Parkinson's disease. New Engl J Med 351:2509–2518

67. American Psychiatric Association (2000) Diagnostic and statistical manual of mental disorders, fourth edition, text revision (DSM-IV-TR). American Psychiatric Association, Washington, DC. https://doi.org/10.1176/appi.books.9780890423349

68. Schneider LS, Olin JT (1996) Clinical global impressions in Alzheimer's clinical trials. Int Psychogeriatr 8(2):277–290. https://doi.org/10.1017/s1041610296002645

69. Galasko D, Bennett D, Sano M, Ernesto C, Thomas R, Grundman M, Ferris S (1997) An inventory to assess activities of daily living for clinical trials in Alzheimer's disease. Alzheimer Dis Assoc Disord 11:33–39. https://doi.org/10.1097/00002093-199700112-00005

70. Cummings JL, Mega M, Gray K, Rosenberg-Thompson S, Carusi DA, Gornbein J (1994) The neuropsychiatric inventory: comprehensive assessment of psychopathology in dementia. Neurology 44(12):2308–2308. https://doi.org/10.1212/wnl.44.12.2308

71. Simpson PM, Surmon DJ, Wesnes KA, Wilcock GK (1991) The cognitive drug research computerized assessment system for demented patients: a validation study. Int J Geriatr Psychiatry 6(2):95–102. https://doi.org/10.1002/gps.930060208

72. Manos PJ, Wu R (1994) The ten point clock test: a quick screen and grading method for cognitive impairment in medical and surgical patients. Int J Psychiatry Med 24(3):229–244. https://doi.org/10.2190/5a0f-936p-vg8n-0f5r

73. Press DZ (2004) Parkinson's disease dementia – a first step? N Engl J Med 351(24):2547–2549. https://doi.org/10.1056/nejme048305

74. Emre M, Poewe W, De Deyn PP, Barone P, Kulisevsky J, Pourcher E, van Laar T, Storch A, Micheli F, Burn D, Durif F, Pahwa R, Callegari F, Tenenbaum N, Strohmaier C (2014) Long-term safety of Rivastigmine in Parkinson disease dementia. Clin Neuropharmacol 1. https://doi.org/10.1097/wnf.0000000000000010

75. Leroi I, Brandt J, Reich SG, Lyketsos CG, Grill S, Thompson R, Marsh L (2004) Randomized placebo-controlled trial of donepezil in cognitive impairment in Parkinson's disease. Int J Geriatr Psychiatry 19(1):1–8. https://doi.org/10.1002/gps.993

76. Aarsland D (2002) Donepezil for cognitive impairment in Parkinson's disease: a randomised controlled study. J Neurol Neurosurg Psychiatry 72(6):708–712. https://doi.org/10.1136/jnnp.72.6.708

77. Ravina B (2005) Donepezil for dementia in Parkinson's disease: a randomised, double blind, placebo controlled, crossover study. J Neurol Neurosurg Psychiatry 76(7):934–939. https://doi.org/10.1136/jnnp.2004.050682

78. Mattis S (2011) Mattis dementia rating scale. Encyclopedia of clinical neuropsychology. Springer, New York, NY. https://doi.org/10.1007/978-0-387-79948-3_3657

79. Dubois B, Tolosa E, Katzenschlager R, Emre M, Lees AJ, Schumann G, Pourcher E, Gray J, Thomas G, Swartz J, Hsu T, Moline ML (2012) Donepezil in Parkinson's disease dementia: a randomized, double-blind efficacy and safety study. Mov Disord 27(10):1230–1238. https://doi.org/10.1002/mds.25098

80. Whitehead A, Perdomo C, Pratt RD, Birks J, Wilcock GK, Evans JG (2004) Donepezil for the symptomatic treatment of patients with mild to moderate Alzheimer's disease: a meta-analysis of individual patient data from randomised controlled trials. Int J Geriatr Psychiatry 19(7):624–633. https://doi.org/10.1002/gps.1133

81. Litvinenko IV, Odinak MM, Mogil'naya VI, Emelin AY (2008) Efficacy and safety of galantamine (reminyl) for dementia in patients with Parkinson's disease (an open controlled trial). Neurosci Behav Physiol 38(9):937–945. https://doi.org/10.1007/s11055-008-9077-3

82. Gelinas I, Gauthier L, McIntyre M, Gauthier S (1999) Development of a functional measure for persons with Alzheimer's disease: the disability assessment for dementia. Am J Occup Ther 53(5):471–481. https://doi.org/10.5014/ajot.53.5.471

83. Seppi K, Weintraub D, Coelho M, Perez-Lloret S, Fox SH, Katzenschlager R, Hametner E-M, Poewe W, Rascol O, Goetz CG, Sampaio C (2011) The Movement Disorder Society evidence-based medicine review update: treatments for the non-motor symptoms of Parkinson's disease. Mov Disord 26(S3):S42–S80. https://doi.org/10.1002/mds.23884

84. Seppi K, Ray Chaudhuri K, Coelho M, Fox SH, Katzenschlager R, Perez Lloret S, Weintraub D, Sampaio C, Chahine L, Hametner EM, Heim B, Lim SY, Poewe W, Djamshidian-Tehrani A (2019) Update on treatments for nonmotor symptoms of Parkinson's disease—an evidence-based medicine review. Mov Disord 34(2):180–198. https://doi.org/10.1002/mds.27602

85. Wang HF, Yu JT, Tang SW, Jiang T, Tan CC, Meng XF, Wang C, Tan MS, Tan L (2014) Efficacy and safety of cholinesterase inhibi-

tors and memantine in cognitive impairment in Parkinson's disease, Parkinson's disease dementia, and dementia with Lewy bodies: systematic review with meta-analysis and trial sequential analysis. J Neurol Neurosurg Psychiatry 86(2):135–143. https://doi.org/10.1136/jnnp-2014-307659

86. Leroi I, Overshott R, Byrne EJ, Daniel E, Burns A (2009) Randomized controlled trial of memantine in dementia associated with Parkinson's disease. Mov Disord 24(8):1217–1221. https://doi.org/10.1002/mds.22495

87. Aarsland D, Ballard C, Walker Z, Bostrom F, Alves G, Kossakowski K, Leroi I, Pozo-Rodriguez F, Minthon L, Londos E (2009) Memantine in patients with Parkinson's disease dementia or dementia with Lewy bodies: a double-blind, placebo-controlled, multicentre trial. Lancet Neurol 2009(8):613–618

88. Emre M, Tsolaki M, Bonuccelli U, Destée A, Tolosa E, Kutzelnigg A, Ceballos-Baumann A, Zdravkovic S, Bladström A, Jones R (2010) Memantine for patients with Parkinson's disease dementia or dementia with Lewy bodies: a randomised, double-blind, placebo-controlled trial. Lancet Neurol 9(10):969–977. https://doi.org/10.1016/s1474-4422(10)70194-0

89. Marsh GG, Markham CM, Ansel R (1971) Levodopa's awakening effect on patients with parkinsonism. J Neurol Neurosurg Psychiatry 34(3):209–218. https://doi.org/10.1136/jnnp.34.3.209

90. Cools R (2001) Enhanced or impaired cognitive function in Parkinson's disease as a function of dopaminergic medication and task demands. Cereb Cortex 11(12):1136–1143. https://doi.org/10.1093/cercor/11.12.1136

91. Rektorová I (2010) Effects of dopamine agonists on neuropsychiatric symptoms of Parkinson's disease. Neurodegener Dis 7(1–3):206–209. https://doi.org/10.1159/000295665

92. Manenti R, Cotelli MS, Cobelli C, Gobbi E, Brambilla M, Rusich D, Alberici A, Padovani A, Borroni B, Cotelli M (2018) Transcranial direct current stimulation combined with cognitive training for the treatment of Parkinson disease: a randomized, placebo-controlled study. Brain Stimul 11(6):1251–1262. https://doi.org/10.1016/j.brs.2018.07.046

93. Freund H-J, Kuhn J, Lenartz D, Mai JK, Schnell T, Klosterkoetter J, Sturm V (2009) Cognitive functions in a patient with Parkinson-dementia syndrome undergoing deep brain stimulation. Arch Neurol 66(6):781. https://doi.org/10.1001/archneurol.2009.102

94. Kuhn J, Hardenacke K, Lenartz D, Gruendler T, Ullsperger M, Bartsch C, Mai JK, Zilles K, Bauer A, Matusch A, Schulz RJ, Noreik M, Bührle CP, Maintz D, Woopen C, Häussermann P, Hellmich M, Klosterkötter J, Wiltfang J, Maarouf M, Freund HJ, Sturm V (2014) Deep brain stimulation of the nucleus basalis of Meynert in Alzheimer's dementia. Mol Psychiatry 20(3):353–360. https://doi.org/10.1038/mp.2014.32

95. Lv Q, Du A, Wei W, Li Y, Liu G, Wang XP (2018) Deep brain stimulation: a potential treatment for dementia in Alzheimer's disease (AD) and Parkinson's disease dementia (PDD). Front Neurosci 12. https://doi.org/10.3389/fnins.2018.00360

96. Gratwicke J, Zrinzo L, Kahan J, Peters A, Beigi M, Akram H, Hyam J, Oswal A, Day B, Mancini L, Thornton J, Yousry T, Limousin P, Hariz M, Jahanshahi M, Foltynie T (2018) Bilateral deep brain stimulation of the nucleus basalis of Meynert for Parkinson disease dementia. JAMA Neurol 75(2):169. https://doi.org/10.1001/jamaneurol.2017.3762

97. Wang J-W, Zhang Y-Q, Zhang X-H, Wang Y-P, Li J-P, Li Y-J (2016) Cognitive and psychiatric effects of STN versus GPi deep brain stimulation in Parkinson's disease: a meta-analysis of randomized controlled trials. PLoS One 11(6):e0156721. https://doi.org/10.1371/journal.pone.0156721

98. Intzandt B, Beck EN, Silveira CRA (2018) The effects of exercise on cognition and gait in Parkinson's disease: a scoping review. Neurosci Biobehav Rev 95:136–169. https://doi.org/10.1016/j.neubiorev.2018.09.018

99. Perez-Lloret S, Peralta MC, Barrantes FJ (2016) Pharmacotherapies for Parkinson's disease symptoms related to cholinergic degeneration. Expert Opin Pharmacother 17(18):2405–2415. https://doi.org/10.1080/14656566.2016.1254189

100. David FJ, Robichaud JA, Leurgans SE, Poon C, Kohrt WM, Goldman JG, Comella CL, Vaillancourt DE, Corcos DM (2015) Exercise improves cognition in Parkinson's disease: the PRET-PD randomized, clinical trial. Mov Disord 30(12):1657–1663. https://doi.org/10.1002/mds.26291

101. Amara AW, Memon AA (2018) Effects of exercise on non-motor symptoms in Parkinson's disease. Clin Ther 40(1):8–15. https://doi.org/10.1016/j.clinthera.2017.11.004

102. Hindle JV, Petrelli A, Clare L, Kalbe E (2013) Nonpharmacological enhancement of cognitive function in Parkinson's disease: a systematic review. Mov Disord 28(8):1034–1049. https://doi.org/10.1002/mds.25377

103. da Silva FC, Iop RR, de Oliveira LC, Boll AM, de Alvarenga JGS, Gutierres Filho PJB, de Melo LMAB, Xavier AJ, da Silva R (2018) Effects of physical exercise programs on cognitive function in Parkinson's disease patients: a systematic review of randomized controlled trials of the last 10 years. PLoS One 13(2):e0193113. https://doi.org/10.1371/journal.pone.0193113

104. Bernini S, Alloni A, Panzarasa S, Picascia M, Quaglini S, Tassorelli C, Sinforiani E (2019) A computer-based cognitive training in mild cognitive impairment in Parkinson's disease. NeuroRehabilitation 44(4):555–567. https://doi.org/10.3233/nre-192714

105. Alzahrani H, Venneri A (2018) Cognitive rehabilitation in Parkinson's disease: a systematic review. J Parkinsons Dis 8(2):233–245. https://doi.org/10.3233/jpd-171250

106. Maraki MI, Yannakoulia M, Stamelou M, Stefanis L, Xiromerisiou G, Kosmidis MH, Dardiotis E, Hadjigeorgiou GM, Sakka P, Anastasiou CA, Simopoulou E, Scarmeas N (2018) Mediterranean diet adherence is related to reduced probability of prodromal Parkinson's disease. Mov Disord 34(1):48–57. https://doi.org/10.1002/mds.27489

107. Schaeffer E, Busch J-H, Roeben B, Otterbein S, Saraykin P, Leks E, Liepelt-Scarfone I, Synofzik M, Elshehabi M, Maetzler W, Hansen C, Andris S, Berg D (2019) Effects of exergaming on attentional deficits and dual-tasking in Parkinson's disease. Front Neurol 10. https://doi.org/10.3389/fneur.2019.00646

108. Mamikonyan E, Xie SX, Melvin E, Weintraub D (2015) Rivastigmine for mild cognitive impairment in Parkinson disease: a placebo-controlled study. Mov Disord 30(7):912–918. https://doi.org/10.1002/mds.26236

109. Weintraub D, Hauser RA, Elm JJ, Pagan F, Davis MD, Choudhry A (2016) Rasagiline for mild cognitive impairment in Parkinson's disease: a placebo-controlled trial. Mov Disord 31(5):709–714. https://doi.org/10.1002/mds.26617

110. Visanji NP, Lang AE, Kovacs GG (2019) Beyond the synucleinopathies: alpha synuclein as a driving force in neurodegenerative comorbidities. Transl Neurodegener 8(1):28. https://doi.org/10.1186/s40035-019-0172-x

Clinical Trials for Depression, Anxiety, Fatigue, and Apathy in Parkinson's Disease

Matej Skorvanek and Marek Balaz

Abstract

Depression, anxiety, apathy, and fatigue are common manifestations of Parkinson's disease with a significant impact on quality of life and everyday functioning of patients. Despite their clinical relevance, only limited evidence is available regarding treatment interventions for these disorders, and properly designed trials with depression, anxiety, apathy, and fatigue as primary outcome measures have been scarce to this date. All of these disorders present multidimensional constructs with differing underlying neurobiological basis, which will likely necessitate development of distinct therapeutic approaches. This chapter will discuss the terminology and definition of depression, anxiety, apathy, and fatigue in PD, their differentiation from other PD- and non–PD-related symptoms, subtyping, physiological markers, as well as considerations related to the optimal outcome measures and design of clinical trials.

Key words Parkinson's disease, Nonmotor, Depression, Anxiety, Fatigue, Apathy, Therapy, Rating scale, Trial design

1 Introduction

Nonmotor symptoms (NMS) are integral part of Parkinson's disease (PD) and may be present from early prodromal to advanced stages of the disease with increasing prevalence and severity [1–3]. Depression, anxiety, apathy, as well as fatigue, have a high cumulative prevalence and a significant impact on everyday functioning of patients, including activities of daily living, caregiver burden and care dependency [4–9]. In fact, these symptoms were repeatedly identified among the most important individual determinants of worse quality of life (QoL) in PD [10, 11]. Despite their clinical relevance, these symptoms are often underreported and undertreated [12]. The pathophysiology of depression, anxiety, apathy, and fatigue is complex, reflecting the widespread brainstem and cortical pathology in PD, with involvement of several neurotransmitters, including dopaminergic, serotonergic, noradrenergic and cholinergic systems [13]. Management strategies for depression,

Santiago Perez-Lloret (ed.), *Clinical Trials In Parkinson's Disease*, Neuromethods, vol. 160,
https://doi.org/10.1007/978-1-0716-0912-5_10, © Springer Science+Business Media, LLC, part of Springer Nature 2021

anxiety, apathy, and fatigue have limited evidence, are mostly empirical and often based on approaches applied in other disorders due to general lack of properly designed clinical trials in PD, with these symptoms targeted as primary outcomes. In fact, no specific treatment is established for PD-related anxiety and fatigue and evidence is available only for a limited number of interventions for depression and apathy [14–16].

Several challenges are related to clinical trials in depression, anxiety, apathy, and fatigue, including the following:

- Variable terminology used across studies and nonuniform definition of syndromes.

- Differentiation of depression, anxiety, apathy, and fatigue may be challenging, as these symptoms often share common features and have a high coincidence.

- Necessity to differentiate PD-related symptoms from symptoms etiologically nonrelated to PD, such as fatigue caused by anemia or congestive heart disease.

- Depression, anxiety, apathy, and fatigue present multidimensional constructs with heterogeneous neurobiological basis. It is therefore likely that several therapeutic approaches are necessary to target different subtypes of these disorders and their underlying etiologies.

- The pathogenesis of depression, anxiety, apathy, and fatigue in PD is likely to differ considerably from non-PD patients, and treatments used in general psychiatry services may not be as effective in PD and will require clearer clarification in well-designed clinical studies.

- There is lack of gold standard assessment tools and absence of responsiveness parameters like minimal clinically important difference (MCID) for majority of the rating scales available in PD.

- Most of the treatment trials for these disorders, especially in anxiety, apathy, and fatigue were underpowered and/or were performed with poor methodological standards in general.

This chapter will discuss the terminology and definition of depression, anxiety, apathy, and fatigue in PD, their differentiation from other PD- and non–PD-related symptoms, subtyping, physiological markers, as well as considerations related to the optimal outcome measures and design of clinical trials.

2 Depression

Depression is one of the most common of neuropsychiatric symptoms in PD [17].

In patients with Parkinson's disease (PD), clinically relevant symptoms of depression occur in up to 35%, with overall prevalence ranging from 2.7% to 90% [18]. Approximately 17% of patients fulfil criteria for major depressive disorder (MDD) [4]. Depression has a large impact on overall functioning. Depressed PD patients score lower on scales assessing activities of daily living, show a more rapid deterioration of motor and cognitive functioning, and have a higher mortality [19–21]. Depression carries not only the burden of a symptom itself but can also limit patient's access to advanced interventional treatments for PD, such as deep brain stimulation. It also has an important significance for caregivers, as it is associated with distress above the stress associated with functional impairment related to PD itself [21]. Depression is frequently underdiagnosed and undertreated in PD, despite being one of the most studied and most common psychiatric manifestations of PD.

2.1 Definition of PD-Related Depression and Selection of Study Subjects

2.1.1 Terminology and Diagnostic Criteria for Depression in PD

Diagnosis of depressive disorders is based on standard criteria, including ICD-10 diagnostic criteria and the definition of depressive disorders in the DSM Diagnostic and Statistical Manual of Mental Disorders (DSM) V criteria by the American Psychiatric Association [22].

The types of depressive disorders seen in PD include major depression, minor depression, and dysthymia. Key psychiatric symptoms of depression include lowered mood, anhedonia, feelings of guilt and worthlessness, irritability, anxiety, and difficulties in concentration. Somatic symptoms of depression may partially overlap with and are intrinsic part of PD symptomatology (hypomimia, sleep disturbances, loss of appetite)—see also below. In the setting of PD, depression requires special criteria for its definition. Both anxiety and depression have particular phenotypes in PD that at least statistically distinguish them from the same disorders in the general population [23, 24].

Depression can occur at any time during the course of the disease, while younger age at onset, lower education, increased motor complications, higher dopaminergic medication, and the postural instability and/or gait dysfunction have been identified as risk factors [25]. The diagnosis of depression is confounded also by masking of neurovegetative symptoms of depression, such as fatigue, insomnia, and sleep disturbances by intrinsic PD symptoms [26], and this generates high variability in the clinical interpretation and brings uncertainty into diagnosis of depression among PD patients. A cross-sectional study from the Netherlands and Norway, involving depressed patients with PD and depressed patients without PD, showed that while the cognitive level of patients was similar in both groups, patients with PD had a different depressive profile. PD patients showed less endogenous symptoms such as feelings of guilt and sadness, but more somatic symptoms such as concentration problems [27].

*2.1.2 Distinguishing
PD-Related Depression
from Other PD-Related
Constructs*

Recognizing depressive symptoms in Parkinson's disease may be challenging because the psychomotor slowing and emotional bluntness commonly seen in depression can resemble the bradykinesia and masked facial expression of PD. Depression and PD share overlapping symptoms such as reduced facial expression, problems with sleeping, fatigue, psychomotor retardation, and reduced appetite. Overlap in symptoms is likely to cause difficulties in the accurate identification and diagnosis of depression in PD and may contribute to the underdiagnosis of depression in patients with PD [25, 28]. Furthermore, depression must be differentiated from apathy, which commonly occurs in Parkinson's disease, is characterized by diminished motivation, and has a significant overlap with depressive symptoms. A potentially useful discriminating feature is mood, which is negative in depression and neutral in apathy [29]. However, according to recently published data, apathy and fatigue may be associated with depression based on dopaminergic depletion in the mesocorticolimbic structures, and disruption of the prefrontal cortex–basal ganglia axis [30]. Thus, exact and definite separation of these symptoms is far from easy.

2.2 Outcome Measures for Depression

2.2.1 Rating Scales for Depression

Clinical interviews using formal diagnostic criteria are the gold standard for establishing psychiatric diagnoses in research studies. A number of clinician-driven and self-report scales and questionnaires are available. Scales used to assess depression in PD were reviewed by a Movement Disorder Society (MDS) Task Force in 2007 [31]. For the purpose of screening Hamilton Depression Rating Scale (HAM-D), Beck Depression Inventory (BDI), Hospital Anxiety and Depression Scale (HADS), Montgomery–Åsberg Depression Rating Scale (MADRS), and Geriatric Depression Scale (GDS) were recommended [31]. HAM-D, MADRS, BDI, and Zung Self-Rating Depression were recommended for measuring the severity of depressive symptoms. In order to adjust for confounding symptoms, assessments of motor dysfunction (Unified Parkinson's Disease Rating Scale or Movement Disorder Society-Unified Parkinson's Disease Rating Scale) is recommended as well [17, 31]. The scales recommended for both screening and rating of severity of depression in PD are summarized below.

The BDI-II [32] is a generic, self-reported scale. Administration time is 5–10 min and the scale is of public domain. It consists of 21 items, scored 0 (not present or least severe) to 3 (most severe); total scores range from 0 to 63 points, and the cutoff used in PD is ≥14 points. The scale has been shown to be reliable, valid, and sensitive to change [33].

Advantages and disadvantages: The scale is commonly used and well researched. However, it contains numerous somatic items and most validation studies were performed for the original version BDI-I.

The HAM-D [34] is a generic, clinician-rated scale. Administration time is 15 min and the scale is of public domain. It consists of 21 items, ten items scored 0–4, nine items scored 0–2, and two items scored 0–3. Higher scores point to more severe symptoms, while frequency is not rated. Cutoff score of ≥10 points is suggested for screening of depression in PD and ≥14 points is diagnostic of major depressive disorder. The scale has been shown to be reliable, valid, and sensitive to change [33].

Advantages and disadvantages. The HAM-D is the most widely used interviewer-rated outcome scale, it assesses frequent comorbid symptoms and is more sensitive to detect depression than MADRS. On the other hand, appropriate training is required to obtain reliable scores, the scale does not fit well with DSM-IV criteria, somatic symptoms of depression are overrepresented, it lacks a consistent rating metric and differential item weighing. Some items include multiple content constructs.

The MADRS [35] is a generic, clinician-rated scale, which covers all DSM-IV criteria of major depressive disorder except psychomotor retardation and agitation. Administration time is 15 min and the scale is of public domain. It consists of ten items, all scored 0 (not present or normal) to six (most severe symptoms), total scores range from 0 to 60 points. Cutoff score of ≥15 points has been validated in PD for screening of depression. The scale has been shown to be reliable, valid, and sensitive to change [33].

Advantages and disadvantages. The scale has good concurrent validity with DSM criteria, can be used for screening with appropriate cutoff and sensitive to change of depression severity. On the other hand, only sparse data from patients with severe cognitive impairment are available and administration requires some clinical experience with depression.

2.2.2   Neuroimaging

The exact pathophysiology of depression in PD is unclear, but alterations in the serotonin system with greater cell loss in the dorsal raphe nucleus in depressed vs. nondepressed patients were reported [36]. This led to the hypothesis that depression in PD may be caused by serotonin deficiency. However, positron emission tomography (PET) studies have not shown an association between depression and serotonin dysfunction [37]. Several studies and clinical experience have suggested that depression in PD could be regarded as a hypodopaminergic state in the ventral striatum [24]. Lower ventral striatal and mesolimbic dopamine levels may result in a lower stimulation of cortical brain areas, which are associated with motivation and reward processing [38]. In line with this, a positron emission tomography study (PET) study found lower availability of dopamine and noradrenaline transporter binding in the ventral striatum, which correlated with the severity of apathy, whereas lower binding in the limbic system correlated with severity of anxiety [39]. Depression has also been identified as a risk factor for impulse control disorders, particularly in patients with motor fluctuations and dyskinesias [40].

PET or SPECT studies of depression in PD examined neural metabolic activity in the resting state using various radiotracers [41]. Majority of these studies reported reduced neural metabolic activity in PD patients with depression (dPD) compared to PD alone or healthy controls (HCs) and some have found inverse correlations between depression and neural metabolism. The abnormal regions were predominantly in the frontal lobe and striatum, as well as the subcortical or limbic regions including thalamus, amygdala, hippocampus, anterior cingulate cortex and insula.

Of the seven studies using T1-weighted imaging, only one reported bilaterally increased thalamic grey matter (GM) volume in dPD patients compared to PD patients without depression [42]. Studies showed that dPD patients had decreased GM volumes in the prefrontal, parietal and insular regions as well as the limbic system (anterior cingulate cortices and amygdala) [43, 44] compared to HCs or nondepressed PD patients. Involvement of the limbic circuit in PD-related depressive symptoms has been found in a VBM MRI study [45].

Using DTI techniques, three out of four studies reported compromised white matter connectivity indexed by decreased fractional anisotropy (FA) in various tracts, including the bilateral anterior cingulate cortex and thalamus and multiple tracts connecting to the left frontal and deep temporal lobes [46–48].

Compared to DTI findings, resting state functional MRI studies of depression in PD revealed both increased and decreased resting state neural activities in depressed PD patients, compared to nondepressed PD patients and healthy controls. In depressed PD patients, increased functional connectivity between spatially discrete regions in the resting state was observed in the subcortical areas [41].

The implication of an extra-nigrostriatal pathway in the prefrontal, temporal and limbic cortices implies that other neurotransmitters, such as serotonin and noradrenaline, may also be involved in PD-depression [21]. This view is supported by reported changes of serotonin or noradrenaline in PD patients with depression, compared to healthy controls or PD patients without depression [39, 49, 50] in addition to dopaminergic alterations [51, 52]. Although it is challenging to identify a single model of neuropathology of depression in PD, these observations suggest that depression is unlikely to be the result of a single brain region or neurotransmitter system, but involves the dysregulation of cortical-limbic networks in addition to that of the nigrostriatal pathway.

A growing body of evidence suggests the existence of a link between inflammatory-immune responses and the occurrence of depression and cognitive impairment in PD patients [53].

2.2.3   Genetics

High percentage of individuals carrying mutations related to PD suffer mood disturbances [26]. For example, asymptomatic carriers of the glucocerebrosidase (GBA) mutation showed higher depression scores compared to controls, based on the Beck Depression Inventory (BDI) [54]. Similarly, mutations in Parkin appear to predispose to the occurrence of depression [55], and healthy carriers of the G2019S mutation in the LRRK2 gene were shown to develop depression during the prodromal stage [56].

2.3   Treatment Trials

2.3.1   Pharmacological Treatments

Several meta-analyses demonstrated that pharmacologic treatment with selective serotonin reuptake inhibitors (SSRI) [57–59] and behavioral interventions, including cognitive behavioral therapy (CBT), significantly improved depression symptoms among PD patients [60]. However, two other meta-analyses reported that antidepressants are in general not more efficacious than placebo despite their wide use in PD [61, 62]. It is, however, possible there were some negative effects from small sample sizes.

A randomized controlled "Study of Antidepressants in PD" (SAD-PD) demonstrated the effectiveness of SSRI paroxetine and selective serotonin and norepinephrine reuptake inhibitor (SNRI) venlafaxine extended release (XR) in the active treatment of depressive syndromes in PD patients as compared to a placebo treatment [63].

An 8-week randomized placebo-controlled study reported a significant improvement of the tricyclic antidepressant nortriptyline but no effect of the SSRI paroxetine compared to placebo [64]. In contrast, a large 12-week placebo-controlled randomized controlled study found that paroxetine and the SNRI venlafaxine were both efficacious and well tolerated [65]. However, no changes in anxiety or quality of life scores could be found.

Although some reports suggested that the dopamine agonist pramipexole, presumably independent of its motor benefits, is efficacious in treating depression in PD [66, 67], these studies did not require patients to have a diagnosis of depression according to DSM or ICD criteria, and the actual improvement of depression was small. A small study suggested that the transdermal dopamine agonist rotigotine may be useful in treating apathy and depression in PD [68]. Importance of dopaminergic agents in the occurrence of mood disorders of PD is reflected also in the context of dopamine withdrawal syndrome and increased rates of apathy and depression following rapid medication reduction after deep brain stimulation [69]. Dopamine agonists are not licensed for the treatment of depression and bear the risk of undesirable side effects, including daytime sleepiness and impulsive or compulsive behaviors.

2.3.2   Non-pharmacological Treatments

A randomized controlled trial of CBT versus placebo has shown a significant reduction of depression measured by the HAM-D compared with the clinically monitored patient group (56% vs. 8%). Patients in the CBT arm also coped better with their PD symptoms and reported a higher quality of life [70]. CBT was also helpful in reducing addictive behaviors in PD in a small randomized controlled trial, where significant reduction of anxiety and depression was found in patients receiving CBT [71].

A double-blind randomized controlled trial showed that 10 days of repetitive transcranial magnetic stimulation, but not sham stimulation over the left dorsolateral prefrontal cortex improved mild to moderate depressive symptoms in 22 PD patients [72]. Positive effects were observed up to 30 days after stimulation. However, in this study, patients with major depression were not included, and it is unclear whether the beneficial effects of repetitive transcranial magnetic stimulation would last longer than 30 days after cessation of stimulation. Electroconvulsive treatment has been shown to be useful in PD patients with therapy refractory major depression in open label reports [73]. However, because of the potential risks and insufficient data it should only be considered in selected patients.

2.3.3 Selection of Study Participants

- Entry criteria for a trial should include patients with clinically confirmed diagnosis of PD based on accepted criteria—for example, UK PD Brain bank criteria [74] or the more recent MDS clinical diagnostic criteria for PD [75].

- Diagnosis of depression should be further ascertained based on one of the validated diagnostic criteria for depression (DSM-V) or validated rating scales with appropriate clinimetric properties, such as specifically in PD specifically—for example, BDI-II, HAM-D, GDS.

- Inclusion/exclusion of subjects with other comorbid conditions (e.g., anxiety or dementia) depends on the study objectives—these may be ascertained either by formal diagnostic criteria, or by rating scales with cutoffs validated in PD specifically. In case patients with these conditions are included in the trial, outcomes should be ideally stratified for dementia status (e.g., normal cognition, PD-MCI, PD-dementia) and subtype or severity of anxiety.

- Inclusion or exclusion of patients with dementia should be carefully considered based on the study objectives. None of the available rating scales has been validated for PD patients with dementia. Cognitive impairment can be ascertained using either complex neuropsychological batteries or validated global cognitive scales, such as the DRS-2 [76], MoCA [77], or PD-CRS [78]. The MMSE is not preferred in PD specifically due to multiple shortages of the scale [79].

- Dopaminergic medication including levodopa and dopamine agonists is generally necessary for achievement of sufficient compensation of motor symptoms and thus should be allowed in case the dose is stable prior to study entry and during the study course (e.g., for ≥ 2 or ≥ 3 months).

- Motor state and autonomic functions should be monitored before the study and during the study course in order to rule out or describe possible influence of these symptoms on the depression symptomatology.

- Allowing comedication with cognitives should be considered, since these may influence treatment outcomes. In case these medications are allowed a stable use for a period of ≥ 2 or ≥ 3 months should be typically required [80].

- Other pharmacological and surgical treatments including anxiolytics, monoamine oxidase inhibitors, anticholinergics, stimulants, such as modafinil or methylphenidate, antipsychotics, or deep brain stimulation may affect mood or cognition and thus should be excluded or based on study objectives in certain cases may be kept if on a stable dose [80].

- Separate evaluation of inpatients and outpatients should be considered, as behavioral problems are more frequent in patients in hospital or residential care [81].

2.3.4 Selection of Outcome Measures

Primary outcome measure should be able to reflect symptomatic effect and its selection depends on the study design and subtype of depression targeted. Recommended scales include for purposes of screening—HAM-D, BDI-II, HADS, MADRS, and GDS [31]. HAM-D, MADRS, BDI, and Zung Self-Rating Depression are recommended for measuring the severity of depressive symptoms.

Secondary outcome measures should be selected based on relevance for the study objectives and should evaluate either potential study confounders or demonstrate the clinical relevance of treatment effect. For suggested secondary outcome measures *see* Table 1.

2.3.5 Required Study Duration

Period of 12–26 weeks should be sufficient for proving of substantial and lasting immediate effect of depression treatment in pharmacological study. Double-blind continuation of 6–12 months is suggested to capture also long-term effects of therapy and to observe additional changes, such as in ADLs and caregiver burden and offer significant time for buildup of antidepressant treatment effect. Also PD and depression are chronic conditions, which makes long term effect of treatment a key factor in their efficacy.

3   Anxiety

Anxiety belongs to the most important nonmotor symptoms in Parkinson's disease (PD) with wide impact on quality of life and PD management [10]. Anxiety may be defined as a future-oriented

Table 1
Suggested secondary outcome measures for clinical trials for depression in PD

Activities of daily living	UPDRS Part II [82], MDS-UPDRS Part II [83]
Anxiety	PAS [84], GAI [85]
Apathy	AS [86], LARS [87]
Autonomic dysfunction	SCOPA-AUT [88], NMSS [89], COMPASS [90]
Caregiver burden	ZBI [91]
Cognitive impairment	DRS-2 [76], MoCA [77], PD-CRS [78]
Depression	BDI-II [32], HAM-D [34], MADRS [35], GDS [92]
Fatigue	FSS [93], MFI [94], MFIS [95], PFS [96], FACIT-F [97]
Global	GCI (clinician- and/or patient-based) [98]
Motor examination	UPDRS Part III [82], MDS-UPDRS Part III [83]
Motor fluctuations and wearing off	MDS-UPDRS Part IV [83], UDysRS [99], WOQ9 [100]
Pain	King's PD Pain Scale [101]
Quality of life	PDQ8, PDQ39 [102]
Sleep problems	ESS [103], PDSS-2 [104]

MDS-UPDRS Movement Disorder Society-Unified Parkinson's Disease Rating Scale, PAS Parkinson Anxiety Scale, *GAI* Geriatric Anxiety Inventory, *AS* Apathy Scale, *LARS* Lille Apathy Rating Scale, *SCOPA-AUT* SCales for Outcomes in PArkinson's Disease—AUTomonic, *NMSS* Nonmotor Symptoms Scale, *COMPASS* COMPosite Autonomic Symptom Scale, *ZBI* Zarit Burden Interview, *DRS-2* Mattis Dementia Rating Scale, second version, *MoCA* Montreal Cognitive Assessment, *PD-CRS* Parkinson's Disease Cognitive Rating Scale, *BDI-II* Beck Depression Inventory-II, *HAM-D* Hamilton Depression Rating Scale, *MADRS* Montgomery–Åsberg Depression Rating Scale, *GDS* Geriatric Depression Scale, *FSS* Fatigue Severity Scale, *MFI* Multidimensional Fatigue Inventory, *MFIS* Modified Fatigue Impact Scale, *PFS* Parkinson Fatigue Scale, *FACIT-F* Functional Assessment of Chronic Illness Therapy—Fatigue Scale, *GCI* Global Clinical Impression, *UDysRS* Unified Dyskinesia Rating Scale, *WOQ9* 9-item Wearing off questionnaire, *PD* Parkinson's Disease, *PDQ* Parkinson's Disease Quality of Life Scale, *ESS* Epworth Sleepiness Scale, *PDSS-2* Parkinson's Disease Sleep Scale, second version

mood state that prepares a person for possible negative outcomes [105]. It is recognized as a common nonmotor, psychiatric symptom of PD that leads to high levels of care dependency, increased caregiver distress and may limit patient access to various advanced medical treatments [9, 106–112]. It also contributes to cognitive impairment, increased severity of motor symptoms such as on/off fluctuations or freezing episodes [5].

However, anxiety is seemingly underdiagnosed and under-treated in patients with PD due to diagnostic imprecision, symptom overlap with motor and cognitive features of PD, complexity of diagnosis, health care access and resources, as well as under-recognition of symptoms by patients and their caregivers [9, 113, 114]. The exact data on anxiety profile remain relatively scarce [115] and there are no solid treatment recommendations available [116]. Research on mood disorders in PD has predominantly focused on depressive disorders [117]. Additional studies on anxiety are warranted.

Prevalence estimates of anxiety disorders in PD range from 13% to 43%, and this variance has been largely ascribed to differences in instrumental measures and diagnostic criteria [5].

3.1 Definition of PD-Related Anxiety and Selection of Study Subjects

3.1.1 Terminology

Anxiety disorders include generalized anxiety disorder (GAD), panic disorder, agoraphobia, obsessive-compulsive disorder (OCD), social or specific phobia, and anxiety not otherwise specified (NOS). Anxiety is characterized by physiological responses, behaviors (such as avoidance behavior) and cognitive aspects (apprehensive thoughts). The symptoms can be episodic (such as during panic attacks) or continuous (as with GAD). Classification of anxiety subtypes is important due to treatment differences [118].

3.1.2 Distinguishing PD-Related Anxiety from Other PD-Related Constructs/Signs

Anxiety is closely related to depression. Patients with motor and nonmotor fluctuations related to PD might experience apathy, depression, and anxiety in the off periods, with euphoria, well-being, self-confidence, and hypomania characterizing the on-drug condition. Many symptoms of anxiety overlap with autonomic and other symptoms of PD—such as flushing, urinary frequency and urges, and dizziness. Distinction between PD-related and anxiety-related symptoms, such as concentration problems, executive dysfunction, muscle tension, dyspnea, restlessness, and dizziness, may be problematic. Patients with anxiety may describe an unsettling sensation of internal tremor that cannot be seen, that may or may not be improved by antiparkinsonian medications [119]. Inability to be reassured, somatic preoccupations, and preoccupation with details may limit benefits that patients can potentially draw from host of therapies available for PD itself.

3.1.3 Diagnostic Criteria for PD-Related Anxiety

PD-related anxiety is diagnosed based on *Diagnostic and Statistical Manual of Mental Disorders Fifth Edition* (DSM-V) criteria for symptomatic anxiety [22]. A careful history is vital for recognition of anxiety disturbances and establishing the diagnosis of anxiety disorder can be challenging in patients with PD. Diagnostic imprecision may occur, since several symptoms of anxiety overlap with mental and somatic symptoms commonly associated with PD, anxiety symptoms may be attributed to depression or psychotic anxiety. In a recent paper [120], anxiety disorder was diagnosed in 41.1% of 190 PD patients based on the DSM-V criteria. In addition, many patients may have clinically significant anxiety that does not correspond directly to the DSM-V criteria [22]. DSM-V residual category "Unspecified Anxiety Disorder" (DSM-IV category "Anxiety disorder not otherwise specified") is used to describe anxiety symptoms that do not meet DSM diagnostic criteria, but have a significant impact on patient's daily life. This category of sub-syndromal anxiety has been estimated to appear in 2–25% of PD patients when applying the previous DSM-IV criteria [121]. Issues in relation to diagnostic factors are confounded by the fact that DSM-V criteria for generalized anxiety disorder include fatigue, sleep disturbances, and impaired concentration, which are common PD symptoms. According to a recent paper, patients with PD show different syndromic profiles of anxiety that do not align with the symptom profiles represented by DSM-IV anxiety disorders and major depression [122]. Thus, DSM-IV criteria for anxiety disorders may not be clinically useful in PD.

3.2 Outcome Measures for Anxiety

3.2.1 Rating Scales for Anxiety

Shulman and colleagues reported that recognition of anxiety more than doubled (from 19% to 39%) when patients were screened with the Beck Anxiety Inventory (BAI) [123] in comparison to diagnosis based on clinical interview. Standardized scales provide systematic approaches to screening and evaluation of severity of anxiety disorders [118] and appear to be more suitable for clinical trials than general interviews. The BAI, the Hamilton Anxiety Rating Scale (HAM-A), and the Hospital Anxiety and Depression Scale—anxiety subscale (HADS-A) have been examined clinimetrically but demonstrate variable performance as predictors of anxiety disorders in PD population [124]. This may be a function of the heterogeneity of anxiety in general, as well as the presence of anxiety disturbances specific for PD (e.g., on–off fluctuations with associated anxiety).

Scales used to assess anxiety in PD were reviewed by an MDS Task Force in 2008 [125]. None of the reviewed scales reached the level of recommended, and thus members of the same committee subsequently performed a validation study that included the three most commonly used instruments: the Beck Anxiety Inventory, the Hospital Anxiety and Depression Scale, and the Hamilton Anxiety Rating Scale. However, none of these instruments showed satisfactory clinimetric properties [124]. As a result of these findings, a new PD-specific instrument, the Parkinson Anxiety Scale (PAS), was developed. Currently, based on the MDS review criteria, the PAS meets the level of recommended with caveats, since data on responsiveness of the scale are lacking, while the Geriatric Anxiety Inventory (GAI) also meets criteria for recommended with caveats.

The PAS [84] is a multidimensional PD-specific scale, which includes subscales for persistent anxiety, episodic anxiety, and avoidance behavior. It exists in both patient and clinician rated forms. Administration time is 10–15 min, and the scale is in public domain. It consists of 12 items scored 0 (not or never) to 4 (severe or almost always), total score range is 0–48 points, and the suggested cutoff for anxiety is 13/14 points for both the observer and the self-rated version. The scale has been shown to be reliable and valid; however, information on sensitivity to change is not available [33].

Advantages and disadvantages: It covers the most frequent forms of PD-related anxiety and enables a brief and easy administration. On the other hand, it should not be assessed in PD patients with dementia, and its sensitivity to change remains to be determined.

The GAI [85] is a generic self-rated scale, which provides an indication of the severity of anxiety symptoms experienced by elderly patients. Administration time is 10 min and the scale is of public domain. It consists of 20 items, rated as 0 (disagree) or 1 (agree), with total score ranging from 0 to 20 points, and the suggested cutoff in PD is 6/7 points, according to DSM-IV criteria. The scale has been shown to be reliable, valid, and sensitive to change; however, it has been validated in PD only partially [33, 126].

Advantages and disadvantages: The scale is brief and easy to administer with limited amount of somatic symptoms of anxiety, which may overlap with other PD symptoms. It does not cover all anxiety symptoms in PD and lacks more data from patients with dementia.

3.2.2 Physiological Measures of Anxiety in PD

In a PET study of PD patients with motor and mood fluctuations (some of them with anxiety), decreased blood flow after levodopa challenge in several brain regions, such as posterior cingulate cortex and medial frontal gyrus, was found [127]. In another PET study, Kaasinen et al. [128] have shown that PD patients with increased harm-avoidance personality score (associated with anxiety and depression) had increased F-dopa uptake in the right caudate nucleus. Functional imaging studies revealed an inverse correlation between dopaminergic density in the caudate and putamen and the severity of anxiety in PD [41]. In a resting-state connectivity study, presence of anxiety in PD predicted three distinct types of functional connectivity not described before: (1) increased limbic-orbitofrontal cortex, (2) decreased limbic-dorsolateral prefrontal cortex and orbitofrontal-dorsolateral prefrontal cortices, and (3) decreased sensorimotor-orbitofrontal cortices [129]. Another recent study has shown that decreased putamen functional connectivity with the orbitofrontal gyrus and cerebellum also correlated with increased anxiety in Parkinson's disease [130]. A combination of physiological measures could provide robust, continuous indicators of anxiety level. Main neurotransmitters involved in the pathogenesis of anxiety in psychiatric patients involve norepinephrine, serotonin, and GABA [131]. Cholecystokinin may be involved in pathogenesis of panic disorder and dopamine may have a role in social phobia [105]. Research on relationship between anxiety and the use of dopaminergic medications in PD has been limited [121].

3.3 Treatment Trials

To date, no specific pharmacological treatment for anxiety disorders in PD patients is established, and most of the therapies are based on observational studies, expert opinion, or clinical guidelines for anxiety disorders in patients without PD [132, 133]. The only controlled trials have examined psychotherapeutic approaches [118], and to date, there is only one randomized, controlled study focused on pharmacotherapy for anxiety in PD using a long-acting benzodiazepine bromazepam [134]. A double-blind crossover study has shown an association of anxiety score improvement with dosage of levodopa [135]. There is not enough solid evidence to provide recommendations on pharmacotherapy for anxiety in PD.

3.3.1 Selection of Study Participants

- Entry criteria for a trial should include patients with clinically confirmed diagnosis of PD based on accepted criteria—for example, UK PD Brain bank criteria [74] or the more recent MDS clinical diagnostic criteria for PD [75].

- Diagnosis of anxiety should be ascertained based on one of the validated diagnostic criteria for anxiety (DSM-IV) or validated rating scales with appropriate clinimetric properties—for example, the PAS [84] or GAI [85], meeting the MDS criteria for recommended with caveats. Also, patients identified as having significant anxiety symptoms, but not meeting the DSM-IV or V diagnostic criteria, and those who had PD-related anxiety and reported significant distress as a result of their anxiety that was caused by motor symptoms, should be included in the subsyndromal anxiety category. Anxiety related to motor symptoms of PD should be determined.

- Inclusion/exclusion of subjects with other comorbid conditions (e.g., depression or dementia) depends on the study objectives—these may be ascertained either by formal diagnostic criteria, or by rating scales with cutoffs validated in PD specifically. In case patients with these conditions are included in the trial, outcomes should be ideally stratified for dementia status (e.g., normal cognition, PD-MCI, PD-dementia) and/or severity of depressive symptomatology.

- In case patients with concomitant depression are to be excluded/included, several options are available.

  - Exclusion based on cutoff scores of a specific depression scale.
  - Exclusion based on presence of depression defined by diagnostic criteria.

- Assigning a categorical diagnosis of depression and subsequent prospective challenge with SSRI for 2 months. Treatment responders would be excluded, while nonresponders and those whose depression improves but anxiety persists would be included [80].

- Including patients having depression with dysphoric symptoms on a stable dose of medication for 3 months prior to randomization who were not expected to require a change in medication during the study [80].

- Depression rating scales should be included as secondary outcomes even in case depressed patients are excluded from the trial.

- Inclusion or exclusion of patients with dementia should be carefully considered based on the study objectives. None of the available rating scales has been validated for PD patients with dementia. Cognitive impairment can be ascertained using either complex neuropsychological batteries or validated global cognitive scales, such as the DRS-2 [76], MoCA [77], or PD-CRS [78]. The MMSE is not preferred in PD specifically due to multiple shortages of the scale [79].

- Dopaminergic medication including levodopa and dopamine agonists is generally necessary for achievement of sufficient compensation of motor symptoms and thus should be allowed in case the dose is stable prior to study entry (e.g., for ≥ 2 or ≥ 3 months).

- Presence of iatrogenic complications related to PD treatment should be noted (e.g., presence of motor and nonmotor fluctuations and impulse control disorders).

- Allowing comedication with antidepressants, anxiolytics and/ or cognitive enhancers should be considered, since these may influence treatment outcomes. In case these medications are allowed, a stable use for a period of ≥ 2 or ≥ 3 months should be typically required [80].

- Other pharmacological and surgical treatments including monoamine oxidase inhibitors, anticholinergics, antipsychotics, or deep brain stimulation may affect mood and thus should be excluded or based on study objectives in certain cases may be kept if on a stable dose [80].

- Presence of stressful life events should be noted [116].

- Anxiety is associated with severity of motor symptoms in PD, antiparkinsonian medication and stage of PD [116], therefore, optimization of dopaminergic medication with subsequent stable dose (e.g., for ≥ 2 or ≥ 3 months) and reassessment of presence and severity of anxiety before study entry may be considered.

- Separate evaluation of inpatients and outpatients should be considered, as behavioral problems are more frequent in patients in hospital or residential care [81]

3.3.2 Selection of Outcome Measures

Primary outcome measure should be able to reflect symptomatic effect of treatment on anxiety itself. Several studies include various outcome measure scales, such as PAS and HAM-A. Valid and reliable measures with excellent psychometric properties should be used. It should be, however, noted that information on sensitivity to change has not been properly determined for the PAS to this date and information on MCID is not available for any of the anxiety scales in PD specifically.

Secondary outcome measures should be selected based on relevance for the study objectives and should evaluate either potential study confounders or demonstrate the clinical relevance of treatment effect. For suggested secondary outcome measures *see* Table 2.

3.3.3 Required Study Duration

Annual psychiatric screening for anxiety (and other psychiatric symptoms) is a quality measure approved by American Academy of Neurology in the care of PD. However, most studies are focused on interventions with effects that may be apparent after 8–12 weeks. Follow-up for longer periods of time may be advisable.

4 Apathy

Apathy is a common NMS of PD with a cumulative prevalence of nearly 40% and present from early or even prodromal phases of the disease [2, 6]. Apathy significantly contributes to worse quality of life and is one of the leading factors of increased caregiver burden. Although apathy commonly coincides with other disorders, such as depression or cognitive impairment, it can be present as an isolated syndrome. Several attempts were made to define and classify apathetic syndromes in PD, which presents a multidimensional construct with behavioral, affective and cognitive features [136, 137]. Although pharmacological interventions with dopaminergic drugs, cholinesterase inhibitors, stimulants and antidepressants have been investigated in PD-related apathy, there is only a limited number of properly designed randomized trials with apathy as the primary outcome.

4.1 Definition of PD-Related Fatigue and Selection of Study Subjects

4.1.1 Terminology

Apathy is defined as a lack of motivation, characterized by reduced goal-directed behavior, goal-directed cognitive activity, and emotional responsivity [138]. Although different terms have been used to describe the features of apathy in the past, decrease in goal-directed behaviors is currently considered the most important feature of apathy in neurodegenerative disorders [137]. Several

Table 2
Suggested secondary outcome measures for clinical trials for anxiety in PD

Activities of daily living	UPDRS Part II [82], MDS-UPDRS Part II [83]
Apathy	AS [86], LARS [87]
Autonomic dysfunction	SCOPA-AUT [88], NMSS [89], COMPASS [90]
Caregiver burden	ZBI [91]
Cognitive impairment	DRS-2 [76], MoCA [77], PD-CRS [78]
Depression	BDI-II [32], HAM-D [34], MADRS [35], GDS [92]
Fatigue	FSS [93], MFI [94], MFIS [95], PFS [96], FACIT-F [97]
Global	GCI (clinician- and/or patient-based) [98]
Motor examination	UPDRS Part III [82], MDS-UPDRS Part III [83]
Motor fluctuations and wearing off	MDS-UPDRS Part IV [83], UDysRS [99], WOQ9 [100]
Pain	King's PD Pain Scale [101]
Quality of life	PDQ8, PDQ39 [102]
Sleep problems	ESS [103], PDSS-2 [104]

MDS-UPDRS Movement Disorder Society-Unified Parkinson's Disease Rating Scale, *AS* Apathy Scale, *LARS* Lille Apathy Rating Scale, *SCOPA-AUT* SCales for Outcomes in PArkinson's Disease—AUTomonic, *NMSS* Nonmotor Symptoms Scale, *COMPASS* COMPosite Autonomic Symptom Scale, *ZBI* Zarit Burden Interview, *DRS-2* Mattis Dementia Rating Scale, second version, *MoCA* Montreal Cognitive Assessment, *PD-CRS* Parkinson's Disease Cognitive Rating Scale, *BDI-II* Beck Depression Inventory-II, *HAM-D* Hamilton Depression Rating Scale, *MADRS* Montgomery–Åsberg Depression Rating Scale, *GDS* Geriatric Depression Scale, *FSS* Fatigue Severity Scale, *MFI* Multidimensional Fatigue Inventory, *MFIS* Modified Fatigue Impact Scale, *PFS* Parkinson Fatigue Scale, *FACIT-F* Functional Assessment of Chronic Illness Therapy—Fatigue Scale, *GCI* Global Clinical Impression, *UdysRS* Unified Dyskinesia Rating Scale, *PD* Parkinson's Disease, *PDQ* Parkinson's Disease Quality of Life Scale, *ESS* Epworth Sleepiness Scale, *PDSS-2* Parkinson's Disease Sleep Scale, second version, *WOQ9* 9-item Wearing Off Questionnaire

subdomains of goal-directed behavior have been associated with apathy: (a) reward deficiency syndrome (refers to state of emotional blunting or absence of emotional resonance, which prevents the patient from attaching motivational values and pleasure to external or internal stimuli) [139]; (b) autoactivation deficits (inability to spontaneously activate mental processing without external stimuli) [140]; (c) depression and emotional distress (associated with decrease in effortful and positively motivated behaviors) [141]; (d) executive dysfunction (cognitive inertia, difficulty to redirect attention to novel stimuli, work with information and generate plans for the future) [142]. These apathetic syndromes may be present individually or more typically in combination and they should be distinguished, as they have different underlying pathophysiological mechanisms and may require different therapeutic approach.

4.1.2 Distinguishing PD-Related Apathy from Other PD-Related Syndromes

Apathy may be commonly present with other features of PD, especially depression, dementia or fatigue [30, 141, 142]. It is, however, considered as an independent syndrome and can be found isolated in approximately 20% of patients [6]. Apathy may often be part of depression and in cases where these symptoms coincide distinguishing whether apathy is independent or results from depression may be more problematic. Pagonabarraga et al. [137] summarized specific features which may help distinguish these two conditions:

- Apathy (typical symptoms)—reduced initiative, decreased participation in external activities unless engaged by other person, loss of social interest in starting new activities, decreased interest in surrounding environment, emotional indifference, diminished emotional reactivity, less affection than usual, and lack of concern for others.

- Depression (exclusive symptoms for apathy)—sadness, feeling of guilt, negative thoughts and feelings, helplessness, hopelessness, pessimism, self-criticism, anxiety, and suicidal ideation.

- Overlapping symptoms—psychomotor retardation, anhedonia, anergia, less physical activity than usual, and decreased enthusiasm about usual interests.

Clinical diagnosis of apathy in these patients should be made based on symptoms causing decreased goal-directed behavior. Anhedonia is present in both syndromes and is less useful for their differentiation. Also blunted affect in apathy should be distinguished from negative affect commonly present in depression.

Apathy is present in nearly 60% of PD patients with dementia [143] and there are several reports showing that presence of apathy may herald onset of dementia [144]. This is not surprising, as executive dysfunction is a common correlate of apathy in neurodegenerative disorders [142], which commonly leads to difficulties in redirecting attention to novel stimuli and restricts search for new interests. Moreover, apathetic patients without dementia perform worse on global cognitive tests [144, 145]. Cognitive testing using either complex neuropsychological batteries or global cognitive scales with adequate coverage of different cognitive domains [79] should be a standard part of protocol in studies related to apathy.

Some patients may refer to apathy as to "fatigue," although these NMS present separate constructs [30]. While patients with apathy usually have decreased need or motivation to perform activities, patients with fatigue describe their problems more as lack of energy or a need for increased effort to attempt and perform daily activities (they want to do activities but do not have energy) [146].

4.1.3 Distinguishing PD-Related Apathy from Other Non–PD-Related Causes of Apathy

Apathy should be differentiated from other conditions, where symptoms of apathy may be present, such as change in level of consciousness, drug or medication abuse, or other physical and motor disabilities [137]. Apathy associated with other well-defined neuropsychological disorders should be distinguished as well.

4.1.4 Diagnostic Criteria for PD-Related Apathy

Three different diagnostic criteria for apathy applicable in PD populations were proposed [137, 147, 148]. However, only two of them were validated in PD [136, 148]. The diagnostic criteria proposed by Robert et al. [147] include four criteria which must be met: (a) diminished motivation in comparison to patients previous level of functioning; (b) presence of at least one symptom in at least two of the three following domains—goal-directed behavior, goal-directed cognitive activity, and emotions; (c) significant impact on patients' functioning; and (d) exclusion of other potential causes of the symptoms [147]. Prevalence of apathy in the respective validation study in 122 PD patients, based on these criteria, was 17.2% [136].

Diagnostic criteria proposed and validated by Starkstein et al. [148] require presence of: (a) lack of motivation relative to the previous level of patients' functioning; and (b) at least one symptom in each of the following domains—diminished goal-directed behavior, goal-directed cognition and concomitants of goal-directed behavior. These criteria have yielded a diagnosis of apathy in 32% of the 164 evaluated patients with PD [148].

None of these diagnostic criteria allows differentiation with other constructs including depression and cognitive impairment. This issue was targeted in the proposed clinical diagnostic criteria by Pagonabarraga et al. [137], which include criteria for (a) apathy, (b) depression, (c) cognitive impairment, and (d) exclusion criteria. Patients are classified as having isolated apathy if they meet only criteria (a), apathy with depression if they meet criteria (a) + (b), and finally apathy with cognitive impairment if the meet criteria (a) + (c). These criteria, however, still need formal validation.

4.2 Outcome Measures for Apathy

4.2.1 Rating Scales for Apathy in PD

Prevalence of apathy has been reported in the range of 17–70% and this is in large part based on the method of apathy assessment. Different uni- and multidimensional rating scales, which are rater-, caregiver-, and clinician-based, are available for assessment of apathy. A recent meta-analysis on PD-related apathy reported that Lille Apathy Rating Scale (LARS) and Neuropsychiatric Inventory

(NPI) had a nearly 10% lower rate of apathy compared to Apathy Scale (AS) [6]. Scales used to assess apathy and anhedonia in PD were reviewed by an MDS Task Force in 2008 [149]. Two scales were recommended: the AS and the Unified Parkinson's Disease Rating Scale (UPDRS) item 4 [82], although the authors state that the UPDRS item should be used strictly for screening because of the obvious limitations of a single-item construct. The LARS reached the suggested level, but since this original review, further clinimetric data have been published, and LARS currently meets the formal MDS criteria for recommended scales.

The AS [86] is a modified version of the Apathy Evaluation Scale [138] developed specifically for PD patients. Administration time is 5 min and the scale is of public domain. It consists of 14 items. Items 1–8 are scored 0 (a lot) to 3 (not at all) and items 9–14 are scored inversely 0 (not at all) to 3 (a lot), total scores range from 0 to 42 points, and the cutoff for presence of apathy is ≥14 points. The scale has been shown to be reliable, valid, and sensitive to change [33].

Advantages and disadvantages: The scale is brief, highly specific, has a defined cutoff, is sensitive to change and suitable for screening, assessing severity and treatment trials. On the other hand, it is a self-assessment scale, which is not adapted to patients with moderate/severe dementia. MCID is not available.

The LARS [87] is a multidimensional PD-specific scale covering issues like productivity, interests, initiative, novelty seeking, motivation, emotional response, and social life. Administration time is 10 min, and the scale is of public domain. It consists of 33 items. The first three items are scored on a five-point Likert scale (−2 to +2), while the remaining items are scored 1 (response indicative of apathy) or −1, total scores range from −36 to +36 points, and cutoff for apathy is ≥ −16 points. A version for the caregiver has also been validated [150]. The scale has been shown to be reliable, valid, and sensitive to change [33].

Advantages and disadvantages: The scale enables comprehensive assessment of apathy symptoms and domains, is validated in PD and in patients with mild to moderate dementia. On the other hand, it may potentially overlap with depression. MCID is not available.

4.2.2 Physiological Markers of Apathy

In studies assessing the neurobiological basis of apathy in PD, it is necessary to (a) distinguishing primary PD-related apathy from other causes (PD- or non–PD-related) as outlined above, (b) define and differentiate the subtypes and domains of apathy evaluated; and (c) control for potential confounders, such as depression, cognitive impairment and fatigue. Brain imaging studies, including MRI morphological imaging, functional and metabolic imaging techniques and neurotransmitter evaluation, represent the major tool for elucidation of neural basis of apathy so far. Compared to some other NMS, like fatigue, several well-defined subtypes of apathy with their respective neural substrates have been defined to this date [137], including the following:

- Reward deficiency syndrome associated with dysfunctions in orbitofrontal or ventromedial prefrontal cortex (PFC), anterior cingulate cortex (ACC), amygdala and ventral striatum, and substantia nigra pars compacta (SNpc) and ventral tegmental area (VTA).

- Emotional distress associated with increased activity in subgenual cingulate cortex and decreased activity in dorsolateral PFC and dorsal ACC.

- Executive dysfunction related to dysfunction in lateral PFC, lateral caudate nucleus and putamen and ACC.

- Autoactivation deficit involving SNpc, VTA, globus pallidus pars interna, mediodorsal and anterior thalamus, and dorsal-medial PFC (supplementary motor area and ACC).

In relation to neurotransmitter abnormalities, the role of mesolimbic, mesostriatal, and mesocortical dopaminergic denervation has been demonstrated in multiple studies from early untreated patients to late stages of disease with severe motor fluctuations [151, 152]. The role of dopamine deficiency in etiology of PD-related apathy was strongly supported also by clinical observations and therapeutic trials of patients after STN DBS [151] and treated with dopamine agonists [15, 153]. Involvement of other neurotransmitter systems including noradrenergic [39], serotonergic [154], and possibly also cholinergic [16] was described, with some of these abnormalities having potentially more prominent relationship to apathy than dopaminergic deficits, especially for serotonergic system in de novo untreated patients [154, 155].

4.3 Treatment Trials

Treatment trials in PD with apathy as primary outcome have been scarce. Two placebo-controlled, randomized double-blind trials have been performed with rivastigmine (31 patients) [16] and D2/D3 dopamine agonist piribedil (37 patients) [15], both showing significant improvement after 6 months and 12 weeks respectively. Other dopamine agonists including ropinirole [156], pramipexole [157], and rotigotine [68] have been shown effective in open-label trials as well. Effect of methylphenidate was reported in limited number of PD patients [158]. Studies using antidepressant yielded conflicting results with bupropion (norepinephrine–dopamine reuptake inhibitor) showing improvement in motivation [159], while SSRIs worsening apathy in PD [160].

4.3.1 Selection of Study Participants

- Entry criteria for a trial should include patients with clinically confirmed diagnosis of PD based on accepted criteria—for example, UK PD Brain bank criteria [74] or the more recent MDS clinical diagnostic criteria for PD [75].

- Diagnosis of apathy should be further ascertained based on one of the validated diagnostic criteria for apathy [147, 148] or validated rating scales with appropriate clinimetric properties, assessing the targeted subtype of apathy and having a well-defined and validated cutoff for apathy in PD specifically—for example, the AS [86] or LARS [87].

- Inclusion/exclusion of subjects with other comorbid conditions (e.g., depression or dementia) depends on the study objectives—these may be ascertained either by formal diagnostic criteria, or by rating scales with cutoffs validated in PD specifically. In case patients with these conditions are included in the trial, outcomes should be ideally stratified for dementia status (e.g., normal cognition, PD-MCI, PD-dementia) and/ or severity of depressive symptomatology.

- In case patients with concomitant depression are to be excluded/included, several options are available.

  – Exclusion based on cutoff scores of a specific depression scale.

  – Exclusion based on presence of depression defined by diagnostic criteria.

  – Assigning a categorical diagnosis of depression and subsequent prospective challenge with SSRI for 2 months. Treatment responders would be excluded, while nonresponders and those whose depression improves but apathy persists would be included [80]

  – Including patients having depression with dysphoric symptoms on a stable dose of medication for 3 months prior to randomization who were not expected to require a change in medication during the study [80]

  – Depression rating scales should be included as secondary outcomes even in case depressed patients are excluded from the trial.

- In case global cognitive scales are used for defining presence/ severity of cognitive impairment, the MMSE is not the preferred instrument in PD, despite its common use in clinical trials, based on the recent MDS recommendations for use of global cognitive scales in PD [79]. The generally suggested cutoffs for MMSE may not be appropriate for use in PD patients, because the scale contains a high proportion of items relating to cortical cognitive aspects which may be relatively preserved in PD, and it does not adequately assess executive and visuospatial functions, which are typically affected in PD. Moreover, it has low sensitivity to capture mild degree of deterioration and several recommended scales, including the DRS-2, MoCA, and PD-CRS have shown better capacity for detection of PD-MCI and PD-dementia [79].

- Dopaminergic medication including levodopa and dopamine agonists is generally necessary for achievement of sufficient compensation of motor symptoms and thus should be allowed in case the dose is stable prior to study entry (e.g., for ≥2 or ≥3 months).

- Allowing comedication with antidepressants and/or cognitive enhancers should be considered, since these may influence treatment outcomes. In case these medications are allowed a stable use for a period of ≥2 or ≥3 months should be typically required [80].

- Other pharmacological and surgical treatments including monoamine oxidase inhibitors, anticholinergics, stimulants, such as modafinil or methylphenidate, antipsychotics, or deep brain stimulation may affect mood, cognition, or apathy and thus should be excluded or based on study objectives in certain cases may be kept if on a stable dose [80].

- Apathy is associated with worse severity of motor symptoms in PD [161] and hypodopaminergic states [15]; therefore, optimization of dopaminergic medication with subsequent stable dose (e.g., for ≥2 or ≥3 months) and reassessment of presence and severity of apathy before study entry should be considered.

- Separate evaluation of inpatients and outpatients should be considered, as behavioral problems are more frequent in patients in hospital or residential care [80, 81].

4.3.2 Selection of Outcome Measures

Primary outcome measure should be able to reflect symptomatic effect and its selection depends on the study design and subtype of apathy targeted. Recommended scales for assessment of PD-related apathy include especially the AS with a generally accepted cutoff ≥14 points and the LARS with a cutoff for total scores ≥ −16 points, although MCIDs are not available for any available apathy rating scale.

Secondary outcome measures should be selected based on relevance for the study objectives and should evaluate either potential study confounders or demonstrate the clinical relevance of treatment effect. For suggested secondary outcome measures *see* Table 3.

4.3.3 Required Study Duration

Cummings et al. [80] suggest that a period of 6–8 weeks is sufficient to measure effect of intervention on apathy without risk of confounding from deterioration of underlying condition. Double-blind continuation of 3–6 months is suggested to capture also long-term effects of therapy and to observe additional changes, such as in ADLs and caregiver burden.

Table 3
Suggested secondary outcome measures for clinical trials for apathy in PD

Activities of daily living	UPDRS Part II [82], MDS-UPDRS Part II [83]
Anxiety	PAS [84], GAI [85]
Caregiver burden	ZBI [91]
Cognitive impairment	DRS-2 [76], MoCA [77], PD-CRS [78]
Depression	BDI-II [32], HAM-D [34], MADRS [35], GDS [92]
Fatigue	FSS [93], MFI [94], MFIS [95], PFS [96], FACIT-F [97]
Global	GCI (clinician- and/or patient-based) [98]
Motor examination	UPDRS Part III [82], MDS-UPDRS Part III [83]
Motor fluctuations and wearing off	MDS-UPDRS Part IV [83], UDysRS [99], WOQ9 [100]
Quality of life	PDQ8, PDQ39 [102]
Sleep problems	ESS [103], PDSS-2 [104]

MDS-UPDRS Movement Disorder Society-Unified Parkinson's Disease Rating Scale, *PAS* Parkinson Anxiety Scale, *GAI* Geriatric Anxiety Inventory, *ZBI* Zarit Burden Interview, *DRS-2* Mattis Dementia Rating Scale, second version, *MoCA* Montreal Cognitive Assessment, *PD-CRS* Parkinson's Disease Cognitive Rating Scale, *BDI-II* Beck Depression Inventory-II, *HAM-D* Hamilton Depression Rating Scale, *MADRS* Montgomery–Åsberg Depression Rating Scale, *GDS* Geriatric Depression Scale, *FSS* Fatigue Severity Scale, *MFI* Multidimensional Fatigue Inventory, *MFIS* Modified Fatigue Impact Scale, *PFS* Parkinson Fatigue Scale, *FACIT-F* Functional Assessment of Chronic Illness Therapy—Fatigue Scale, *GCI* Global Clinical Impression, *UDysRS* Unified Dyskinesia Rating Scale, *WOQ9* 9-item Wearing Off Questionnaire, *PDQ* Parkinson's Disease Quality of Life Scale, *ESS* Epworth Sleepiness Scale, *PDSS-2* Parkinson's Disease Sleep Scale, second version

5   Fatigue

Fatigue is one of the most common nonmotor symptoms (NMS) in Parkinson's disease (PD) [30]; it may have a significant impact on quality of life (QoL) [11], and a significant proportion of PD patients identify fatigue as one of their most disabling or even the most disabling symptom [162]. The first report on fatigue in PD was published nearly 25 years ago [162]; however, only limited progress in our understanding of its pathophysiology has been made since then, and there is no evidence-based treatment available to this date. This may be due to several key issues, which include (a) variable use of terminology and lack of definition of PD-related fatigue in the past; (b) not distinguishing fatigue from other similar PD-related constructs, such as depression, apathy, or somnolence; (c) not distinguishing PD-related fatigue from fatigue related to other disorders, such as anemia, congestive heart failure, and medication use; (d) lack of researchers and funding bodies with significant interest in fatigue research; and (e) poor methodological standards in this field in general. These and other issues will be covered in this part of the chapter.

5.1   Definition of PD-Related Fatigue

5.1.1   Terminology

Fatigue has been historically viewed as a vague subjective symptom that may overlap with better defined conditions, such as depression, apathy or excessive sleepiness. Although frequently viewed as a nonspecific symptom, fatigue can be clearly identified by patients and can be well-defined in both clinical and research contexts as a unique symptom worthy of attention for directing care and research efforts [163]. In fact, PD patients often characterize their fatigue as qualitatively different from normal physiological fatigue [96]. They report a sense of increased effort to do all or most activities and fatigue often leads to reduction in activities performed, which is different from excessive sleepiness, apathy (lack of motivation), or depression. Several isolated or combined subtypes of fatigue may be present in PD and their differentiation and definition is crucial for outcome of any fatigue-related trial in PD. Historically terms such as central and peripheral fatigue were used with certain degree of variability and often meaning different constructs [7]. Central fatigue may have referred to subjective feeling of fatigue, performance on a cognitive task, CNS causes of fatigue, and many other constructs, and so these terms should be used and interpreted with great caution. A more appropriate approach would be to distinguish subjective fatigue complaints from objective fatigability on performance and to specify what performance domains are affected by fatigue (motor, cognitive, etc.) [163].

5.1.2   Distinguishing PD-Related Fatigue from Other PD-Related Constructs

Fatigue may overlap with several other PD-related symptoms, especially depression, apathy, and excessive sleepiness [30, 164], but also orthostatic hypotension or pain [163]. Some of the NMS in PD have a tendency to cluster and recently distinct nonmotor subtypes of PD have been described [165], what makes the evaluation and management of fatigue even more complex. Nevertheless, it was shown that fatigue presents a separate construct, can be isolated and distinguished from these symptoms [30]. It is thus necessary to distinguish between the primary PD-related fatigue and fatigue associated with or resulting from other NMS. Therefore, if the study objective is to evaluate PD-related fatigue as such, based on study design and goals, patients with other comorbid and overlapping symptoms should be either excluded (e.g., based on presence of the confounding symptom evaluated by DSM criteria or other accepted measures), or should be properly assessed and included as a covariate to allow appropriate interpretation of results.

5.1.3 Distinguishing PD-Related Fatigue from Other Non–PD-related Causes of Fatigue

While fatigue is a common manifestation of PD, it is also commonly present in other medical conditions, which may have a causal treatment and inclusion of patients with these comorbidities may significantly bias the trial results. Therefore, (a) other medical conditions (e.g., anemia, congestive heart disease, symptomatic hypotension, respiratory problems, and vitamin D deficiency), (b) other psychiatric disorders, (e.g., major depression), and (c) selected concomitant medications, such as beta-blockers, need to be taken into consideration and either excluded from the study or considered as covariates.

5.1.4 Diagnostic Criteria for PD-Related Fatigue

The DSM-style diagnostic criteria for PD-related fatigue [166] were proposed recently by an international working group. Although they have not undergone formal validation yet, they present a major advance in definition of PD-related fatigue. These criteria focus on differentiating patients with clinically relevant fatigue, from those experiencing normal physiological fatigue and differentiating patients with fatigue from causes other than PD. This approach enables reduction of participant heterogeneity in trials aiming at understanding the pathophysiology and developing new therapeutic targets for fatigue in PD. Moreover, they provide a diagnostic category for fatigue-related medication coverage and disability claims.

5.2 Outcome Measures for Fatigue

Studies should choose specific outcome measures based on their research question and fatigue concept evaluated. For assessment of presence of fatigue, the abovementioned diagnostic criteria or other rating scales with a well-defined cutoff for fatigue may be utilized. In addition to other validated rating scales, visual analog scales may be utilized to measure the state (momentary) perceptions of fatigue. Outcome measures used for treatment trials should demonstrate adequate sensitivity to change. It is important to note that minimal clinically important difference (MCID) was determined only for the Modified Fatigue Impact Scale (MFIS) [163, 167] and thus clinical relevance of findings in fatigue treatment trials using other rating scales should be interpreted with caution, until their MCID is determined. Fatigue measures may be categorized into three general domains: (1) measures of perceptions of fatigue and subjective fatigue complaints; (2) measures of performance fatigability; and (3) physiologic factors associated with fatigue or fatigability [163].

5.2.1 Measures of Perceptions of Fatigue and Subjective Fatigue Complaints

Scales used to assess fatigue in PD were reviewed by an International Parkinson and Movement Disorder Society (MDS) Task Force in 2010 [168]. Four scales were recommended: Fatigue Severity Scale (FSS) [93] for severity and screening, Functional Assessment of Chronic Illness Therapy—Fatigue Scale (FACIT-F) [97] for screening, Multidimensional Fatigue Inventory (MFI) [94] for severity, and Parkinson Fatigue Scale (PFS) [96] for screening. The Modified Fatigue Impact Scale (MFIS) [95] was not included in this MDS review, but since then was validated in PD populations and can be recommended for use as well.

The FSS [93] is a unidimensional generic fatigue rating scale which emphasizes functional impact of fatigue and contains items on physical and mental fatigue and social aspects, although these are not divided in explicit domains. Administration time is 5 min, and the scale is copyrighted. It consists of nine items scored 1 (completely disagree) to 7 (completely agree); total score is calculated as mean score of the nine items (range 1–9). The scale has been shown to be reliable and valid in PD, while sensitivity to change has been demonstrated only in non-PD populations [33].

Advantages and disadvantages. The scale is brief, easy to use, is translated in multiple languages and validated in PD and other disorders. On the other hand, it is not validated in cognitively impaired patients and does not distinguish the types of fatigue.

The FACIT-F [97] is a generic scale, which covers the experience as well as impact of fatigue. Administration time is 5 min, scale is copyrighted. It consists of 13 items scored 0 (not at all) to 4 (very much), range 0–52 (no cutoff for PD, cutoff ≥34 for cancer-related fatigue, and ≥30 for significant fatigue). The scale has been shown to be reliable and valid in PD, while sensitivity to change has been demonstrated only in non-PD populations [33].

Advantages and disadvantages. The scale is brief, easy to use, and validated in different conditions. On the other hand, the scale lacks definition of the underlying variable and sensitivity to change was not demonstrated.

The MFI [94] is a generic multidimensional scale assessing five dimensions—general fatigue, physical fatigue, mental fatigue, reduced activity, and reduced motivation. Administration time is 10–20 min, scale is copyrighted. It consists of 20 items, scored 1 (that is true) to 5 (that is not true), with four items per each subscale—two items positive (need to be recoded) and two items negative. Scores range from 4 to 20 per subscale, higher indicating worse fatigue. The scale has been shown to be reliable, valid, and sensitive to change in PD populations [33].

Advantages and disadvantages. The scale has been recommended for measuring severity of fatigue in all settings and is able to distinguish different aspects of fatigue. On the other hand, it may overlap with NMS like apathy and depression, is not validated in cognitively impaired patients and is not suitable for screening and prevalence studies, as no cutoff was formally validated.

The PFS [96] assesses physical, but not cognitive and emotional aspects of fatigue. The scale is PD specific and copyrighted, administration time is 15 min. It consists of 16 items, rated 1 (strongly disagree) to 5 (strongly agree). Final scores can be calculated as (a) mean response across items (range 0–5, cutoff ≥2.95),

(b) binary approach, where options agree and strongly agree are scored as 1 and all others 0, scores are subsequently summed (range 0–16, cutoff ≥7), (c) scores can be also summed (range 16–80). The scale has been shown to be reliable, valid, and sensitive to change in PD populations [33].

Advantages and disadvantages. The scale is brief with easy scoring (in case of binary) and the construct is independent of mood, cognitive and sleep-related factors. On the other hand, it does not evaluate nonphysical aspects of fatigue, is not suitable for evaluating fatigue in relationship to fluctuations, and was not validated in conditions other than PD.

The MFIS [95] is a 21-item self-report generic scale that rates the impact of fatigue on various functions and includes a total score (primary outcome) as well as cognitive, physical, and psychosocial subscales. Each item is rated based on frequency of symptoms from 0 (never) to 4 (almost always) with a range of 0–84 for the total score, 0–36 for the physical subscore, and 0–40 for the cognitive subscore. The scale has been shown to be reliable, valid, and sensitive to change in PD populations and is the only fatigue rating scale with a defined minimal clinically important difference (MCID) [95, 167].

Advantages and disadvantages. The scale captures different aspects of fatigue, is validated in multiple conditions and is the only fatigue scale with a defined MCID in PD. On the other hand, it was not properly validated in patients with dementia.

5.2.2 Measures of Performance (Objective) Fatigability

There are several approaches to measure performance fatigability: (a) on a selected task before and after performing a different fatiguing task, and (b) at baseline before and during continuous performance of a task [163]. Performance fatigue may be measured during motor and cognitive tasks. In the motor domain tasks such as repetitive movements (e.g., finger tapping) or sustained force generation with various protocols including intermittent submaximal force and maximal sustained force exercise protocols were used [169, 170]. Fatigability in cognitive domains may be evaluated by examining changes in accuracy and response time, including intra-individual variability of response times [171, 172].

5.2.3 Physiological Markers of Fatigue or Fatigability

In studies elucidating the neurobiological basis of fatigue in PD, it is crucial to (a) distinguishing primary PD-related fatigue from other causes (PD- or non–PD-related) as outlined above, (b) define the fatigue constructs/domains evaluated and (c) define whether subjective trait fatigue, momentary state fatigue or other factors driving fatigability are being evaluated [163]. Moreover, mechanistic studies should be careful when attributing a causal relationship between physiologic markers and fatigue. In studies measuring performance fatigability over time, physiologic changes may reflect: (a) functional deterioration of systems associated with performance; (b) engagement of systems associated with perceptions or monitoring of effort and fatigue; (c) learning or other time-dependent processes not associated with fatigue; or (d) engagement of compensatory processes to maintain performance [173–180]. Studies should pay special attention to identify the neuroanatomic location and function of physiological factors as well as the domains of function affected; they should also specify how these factors may relate to fatigue (compensatory, contributory, etc.) [166].

Physiological measures with reasonable potential to elucidate the pathophysiological basis of PD-related fatigue include especially abnormalities in different neurotransmitter systems; metabolic changes; MRI morphologic, volumetric, and functional connectivity studies; autonomic tests and laboratory investigations including inflammatory markers (e.g., proinflammatory cytokines); hormonal changes; and others [181–188]. There are no data to this date correlating PD-related fatigue with any specific pathological changes or their distribution and predisposing genetic factors.

Several well-designed imaging studies have shown inconsistent relationship between dopaminergic system and fatigue in PD. In the ELLDOPA trial [123I]-b-CIT SPECT striatal dopamine transporter binding was not related to fatigue [182]. On the other hand, a previous PET study found a reduced ^{18}F-dopa uptake in the caudate and insula in patients with fatigue [181]. These inconsistent findings are supported also by a general lack of fatigue association with disease duration, PD stage and motor severity, daily levodopa equivalent dosages (LEDD), or specific dopaminergic medications [7, 189]. A link between PD-related fatigue and reduced serotonergic function in the limbic system and basal ganglia has been demonstrated in a study of Pavese et al. [181], who found reduced serotonin transporter binding in the caudate, putamen, ventral striatum, thalamus, cingulate, and amygdala. Other neurotransmitter systems, including cholinergic, noradrenergic, GABA, glutamate, or orexin, have not been studied in relation to PD fatigue so far. A recent FDG-PET study showed anticorrelated metabolic changes in cortical regions associated with the salience (i.e., right insular region) and default (i.e., bilateral posterior cingulate cortex) networks. The metabolic abnormalities detected in these brain regions displayed a significant correlation with level of fatigue and were associated with a disruption of the functional correlations with different cortical areas [184]. This is in line with a MRI study, which showed that distressing PD-related fatigue was associated with decreased connectivity in the supplementary motor area within the sensorimotor network and an increased connectivity in the prefrontal cortex and posterior cingulate cortices within the default mode network. Fatigue severity was correlated with both sensorimotor and default mode networks connectivity changes. Voxel-based morphometry analysis did not reveal any significant volume differences between all patients with PD and healthy controls and between patients with PD with and without fatigue [183]. In terms of laboratory investigations, fatigue in PD was associated with increased levels of inflammatory markers in serum and CSF, with high CRP levels significantly associated with more severe symptoms of fatigue [187, 190]. Also, serum interleukin-6 but not soluble tumor necrosis factor receptors were associated with fatigue in PD [191] and free testosterone levels were not associated with fatigue in a small study of male PD patients [192].

5.3 Treatment Trials

Treatment trials for fatigue have been relatively scarce until recently and no effective evidence-based treatment is available for PD-related fatigue to this date. Seventeen randomized-controlled trials for subjective fatigue have been identified in the literature. Ten were pharmacological studies investigating effects of levodopa–carbidopa [182], memantine [193], rasagiline [194, 195], caffeine [196], methylphenidate [197], and modafinil [198–200], and seven were nonpharmacological studies evaluating effects of physiotherapy [201–203], occupational therapy combined with transcranial direct current stimulation [204], dance [205], acupuncture [206], and deep brain stimulation (DBS) [207]. Only five of these studies [182, 196, 197, 200, 206] excluded subjects with potential secondary causes of fatigue, such as depressive symptoms or sleepiness and thus the remaining studies targeted fatigue with a heterogeneous range of pathophysiological backgrounds and not necessarily PD-related. Also, many of these treatment trials were underpowered and their results should thus be interpreted as inconclusive rather than negative.

Several studies have shown that dopaminergic medication including monoamine oxidase inhibitors, levodopa, and DBS may improve fatigue in some patients [182, 195, 208]. In fact, fatigue was found the most common manifestations of nonmotor fluctuations in PD [209], but unfortunately in many patients it remains clinically significant even after optimization of dopaminergic medication [210]. Therefore, optimization of dopaminergic therapy should be performed before enrollment of patients into other trials with nondopaminergic or nonpharmacological therapies. Other nondopaminergic approaches, such as treatment with methylphenidate [197], or nonpharmacological approaches, including physically engaging video games [203], noninvasive brain stimulation [204], exercise [211], and dance [205] have also a tendency to improve symptoms of fatigue, although these studies need confirmation in properly powered and designed trials. There is no evidence to support the use of any supplements, diets, or vitamins for fatigue in PD, including caffeine, although patients may occasionally report some benefits [163].

5.3.1 Selection of Study Participants

- Entry criteria for a trial should include patients with clinically confirmed diagnosis of PD based on accepted criteria—for example, UK PD Brain bank criteria [74] or the more recent MDS clinical diagnostic criteria for PD [75].

- Proposed diagnostic criteria for fatigue in PD [166] were not formally validated. Until their clinimetric properties, accuracy, and clinical usefulness are determined, fatigue should be diagnosed based on validated rating scales or objective performance measures with appropriate clinimetric properties, assessing the targeted subtype of fatigue and having a well-defined and validated cutoff for fatigue in PD specifically—for example, the FSS [93], PFS [96], or MFIS [95]. The MFI [94] and FACIT-F [97] do not have a well-defined cutoff in general or in PD specifically and thus their role is limited.

- Inclusion/exclusion of subjects with other comorbid conditions, especially depression, excessive sleepiness, or apathy, depends on the study objectives—these may be ascertained either by formal diagnostic criteria, or by rating scales with cutoffs validated in PD specifically. In case patients with these conditions are included in the trial, outcomes should be ideally stratified for severity/presence of depressive symptomatology, presence of excessive sleepiness and/or presence of apathy.

- In case patients with concomitant depression are to be excluded/included, several options are available [80].

  - Exclusion based on cutoff scores of a specific depression scale.

  - Exclusion based on presence of depression defined by diagnostic criteria.

  - Assigning a categorical diagnosis of depression and subsequent prospective challenge with SSRI for 2 month. Treatment responders would be excluded, while nonresponders and those whose depression improves, but apathy persists, would be included.

  - Including patients having depression with dysphoric symptoms on a stable dose of medication for 3 months prior to randomization who were not expected to require a change in medication during the study.

  - Depression rating scales should be included as secondary outcomes even in case depressed patients are excluded from the trial.

- Presence of excessive sleepiness for purposes of inclusion/exclusion of subjects in a fatigue trial can be ascertained, for example, by the Epworth Sleepiness Scale (ESS) using the standard cutoff >10 points [33, 103].

- Presence of apathy for the purpose of inclusion/exclusion of subjects can be diagnosed either based on diagnostic criteria [147, 148] or validated rating scale for apathy (e.g., the AS [86] or LARS [87]) as discussed in the previous section of this chapter.

- Inclusion or exclusion of patients with dementia should be carefully considered based on the study objectives. None of the available rating scales has been validated for PD patients with dementia. Cognitive impairment can be ascertained using either complex neuropsychological batteries or validated global cognitive scales, such as the DRS-2 [76], MoCA [77], or PD-CRS [78]. The MMSE is not preferred in PD specifically due to multiple shortages of the scale [79].

- Exclusion of patients with secondary non–PD-related causes of fatigue, such as anemia, congestive heart disease, oncological disorders, muscle disorders, symptomatic hypotension, respiratory problems, vitamin D deficiency, etc. should be strongly considered to minimize potential result bias.

- Dopaminergic medication including levodopa and dopamine agonists is generally necessary for achievement of sufficient compensation of motor symptoms and thus should be allowed in case the dose is stable prior to study entry (e.g., for ≥ 2 or ≥ 3 months).

- Allowing comedication with antidepressants and/or cognitives should be considered, since these may influence treatment outcomes. In case these medications are allowed a stable use for a period of ≥ 2 or ≥ 3 months should be typically required [80].

- Other pharmacological and surgical treatments including monoamine oxidase inhibitors, amantadine, anticholinergics, modafinil or methylphenidate, antipsychotics, or deep brain stimulation may affect mood, cognition, or apathy and thus should be excluded or exceptionally, based on study objectives, may be kept if on a stable dose [80].

- Several other treatments for conditions not related to PD, such as beta-blockers or other medications potentially causing symptomatic hypotension, etc., should be excluded, since these may worsen the symptoms of fatigue.

- Relationship of fatigue with motor symptoms and severity of disease is inconsistent [7, 212, 213]. However, fatigue may improve with dopaminergic medication in some cases [182], and it is the most common manifestation of nonmotor fluctuations [209]. Therefore, optimization of dopaminergic medication with subsequent stable dose (e.g., for ≥ 2 or ≥ 3 months) and reassessment of presence and severity of fatigue before study entry may be considered.

5.3.2 Selection of Outcome Measures

Primary outcome measures should be able to reflect symptomatic effect on fatigue (subjective or performance) and their selection depends on the study design and type of fatigue targeted—subjective (physical, cognitive, etc.) or performance (objective) fatigue. Several scales may be recommended for rating fatigue severity [168], especially the FSS [93], MFI [94], PFS [96], and MFIS [95]. Multidimensional fatigue rating scales like the MFI or MFIS could be advantageous, since they may capture treatment effects in different aspects of fatigue. The MFIS is the only fatigue scale available for now, which is able to detect clinical relevance of the study outcomes, since it has MCID determined. Scales used as primary outcomes must not necessarily be the same as rating scales used for making the diagnosis of fatigue at inclusion of subjects into the study. For example, MFI does not have a well-defined cutoff for diagnosing fatigue, but it has been recommended for the rating of severity of fatigue, and thus may reflect well the symptomatic treatment effects, especially if the MCID for this scale would be determined.

Secondary outcome measures should be selected based on relevance for the study objectives and should evaluate either potential study confounders or demonstrate the clinical relevance of treatment effect. For suggested secondary outcome measures *see* Table 4.

5.3.3 Required Study Duration

Majority of fatigue interventions have focused on short-term effects of therapy (typically 12 weeks) [195, 211]. Initial evaluation of treatment effects after a period of 6–8 weeks should be sufficient to measure effect of intervention on fatigue without risk of confounding from deterioration of underlying condition. Double-blind continuation and follow-up evaluations after 3, 6 and eventually 12 months are suggested to capture also long-term effects of therapy and to observe additional changes, such as in ADLs and caregiver burden.

6 Conclusions and Future Directions

Mood disorders, apathy, and fatigue have a significant clinical impact on everyday functioning of patients with PD and their effective management remains one of the major unmet needs. Well-designed and properly powered clinical trials have been very scarce and there is insufficient evidence to support any specific intervention for treatment of anxiety and fatigue in PD to this date. Several key issues need to be addressed in order to promote research and improve design of clinical trials for these disorders.

Table 4
Suggested secondary outcome measures for clinical trials for fatigue in PD

Activities of daily living	UPDRS Part II [82], MDS-UPDRS Part II [83]
Anxiety	PAS [84], GAI [85]
Apathy	AS [86], LARS [87]
Autonomic dysfunction	SCOPA-AUT [88], NMSS [89], COMPASS [90]
Caregiver burden	ZBI [91]
Cognitive impairment	DRS-2 [76], MoCA [77], PD-CRS [78]
Depression	BDI-II [32], HAM-D [34], MADRS [35], GDS [92]
Global	GCI (clinician- and/or patient-based) [98]
Motor examination	UPDRS Part III [82], MDS-UPDRS Part III [83]
Motor fluctuations and wearing off	MDS-UPDRS Part IV [83], UDysRS [99], WOQ9 [100]
Pain	King's PD Pain Scale [101]
Quality of life	PDQ8, PDQ39 [102]
Sleep problems	ESS [103], PDSS-2 [104]

MDS-UPDRS Movement Disorder Society-Unified Parkinson's Disease Rating Scale, *AS* Apathy Scale, *LARS* Lille Apathy Rating Scale, *SCOPA-AUT* SCales for Outcomes in PArkinson's Disease—AUTomonic, *NMSS* Nonmotor Symptoms Scale, *COMPASS* COMPosite Autonomic Symptom Scale, *ZBI* Zarit Burden Interview, *DRS-2* Mattis Dementia Rating Scale, second version, *MoCA* Montreal Cognitive Assessment, *PD-CRS* Parkinson's Disease Cognitive Rating Scale, *BDI-II* Beck Depression Inventory-II, *HAM-D* Hamilton Depression Rating Scale, *MADRS* Montgomery–Åsberg Depression Rating Scale, *GDS* Geriatric Depression Scale, *FSS* Fatigue Severity Scale, *MFI* Multidimensional Fatigue Inventory, *MFIS* Modified Fatigue Impact Scale, *PFS* Parkinson Fatigue Scale, *FACIT-F* Functional Assessment of Chronic Illness Therapy—Fatigue Scale, *GCI* Global Clinical Impression, *UDysRS* Unified Dyskinesia Rating Scale, *WOQ9* 9-item Wearing Off Questionnaire, *PD* Parkinson's Disease, *PDQ* Parkinson's Disease Quality of Life Scale, *ESS* Epworth Sleepiness Scale, *PDSS-2* Parkinson's Disease Sleep Scale, second version

- Inconsistent terminology and classification of subtypes of depression, anxiety, apathy, and fatigue should be unified and validated in future phenomenological, epidemiological, and treatment trials regarding their therapeutic utility and clinical meaningfulness.

- Depression, anxiety, apathy, and fatigue have often overlapping clinical manifestations and high coincidence, and were not adequately differentiated in a large proportion of previous clinical trials. Distinguishing these symptoms is crucial in order to increase homogeneity of enrolled subjects and decrease result bias.

- Clinical diagnostic criteria have been proposed for all discussed disorders. Nevertheless, the DSM-IV criteria for anxiety may not be clinically useful in PD specifically [122]) and the new DSM-V criteria [22] need more validation studies in PD. Also, both validated diagnostic criteria for apathy do not distinguish related constructs, such as depression and dementia [147, 148] and the proposed diagnostic criteria for fatigue have not been validated to this date [166].

- Statistical significance in clinical trials is not necessarily equal to clinical relevance. In relation to clinical trials, information on responsiveness parameters are rather limited for many of the instruments and the only rating scale with a defined MCID is the MFIS [95] for assessment of fatigue. This limits the interpretation of clinical relevance, as well as power calculations for future trials.

- Advances in understanding of the underlying neurobiological basis of mood disorders, apathy, and fatigue should help identify novel therapeutic targets, as well as physiological markers for monitoring of disease progression and treatment effects.

References

1. Martinez-Martin P, Schapira AH, Stocchi F et al (2007) Prevalence of nonmotor symptoms in Parkinson's disease in an international setting; study using nonmotor symptoms questionnaire in 545 patients. Mov Disord 22:1623–1629

2. Pont-Sunyer C, Hotter A, Gaig C et al (2015) The onset of nonmotor symptoms in Parkinson's disease (the ONSET PD study). Mov Disord 30:229–237

3. Skorvanek M, Martinez-Martin P, Kovacs N et al (2017) Differences in MDS-UPDRS scores based on Hoehn and Yahr stage and disease duration. Mov Disord Clin Pract 4:536–544

4. Reijnders JS, Ehrt U, Weber WE et al (2008) A systematic review of prevalence studies of depression in Parkinson's disease. Mov Disord 23:183–189

5. Dissanayaka NNNW, White E, O'sullivan JD et al (2014) The clinical spectrum of anxiety in Parkinson's disease. Mov Disord 29:967–975

6. Den Brok MGHE, van Dalen JW, van Gool WA et al (2015) Apathy in Parkinson's disease: a systematic review and meta-analysis. Mov Disord 30:759–769

7. Friedman JH, Brown RG, Comella C et al (2007) Fatigue in Parkinson's disease: a review. Mov Disord 22:297–308

8. Martinez-Martin P, Rodriguez-Blazquez C, Forjaz MJ et al (2015) Neuropsychiatric symptoms and caregiver's burden in Parkinson's disease. Parkinsonism Relat Disord 21:629–634

9. Riedel O, Dodel R, Deuschl G et al (2012) Depression and care-dependency in Parkinson's disease: results from a nationwide study of 1449 outpatients. Parkinsonism Relat Disord 18:598–601

10. Gallagher DA, Lees AJ, Schrag A (2010) What are the most important nonmotor symptoms in patients with Parkinson's disease and are we missing them? Mov Disord 25:2493–2500

11. Skorvanek M, Rosenberger J, Minar M et al (2015) Relationship between the non-motor items of the MDS–UPDRS and quality of life in patients with Parkinson's disease. J Neurol Sci 353:87–91

12. Chaudhuri KR, Prieto-Jurcynska C, Naidu Y et al (2010) The nondeclaration of nonmotor symptoms of Parkinson's disease to health care professionals. Mov Disord 25:704–709

13. Gallagher DA, Schrag A (2012) Psychosis, apathy, depression and anxiety in Parkinson's disease. Neurobiol Dis 46:581–589

14. Seppi K, Weintraub D, Coelho M et al (2011) The Movement Disorder Society evidence-based medicine review update: treatments for the non-motor symptoms of Parkinson's disease. Mov Disord 26:S42–S80

15. Thobois S, Lhommee E, Klinger H et al (2013) Parkinsonian apathy responds to dopaminergic stimulation of D2/D3 receptors with piribedil. Brain 136:1568–1577

16. Devos D, Moreau C, Maltete D et al (2014) Rivastigmine in apathetic but dementia and depression-free patients with Parkinson's disease: a double-blind, placebo controlled, randomised clinical trial. J Neurol Neurosurg Psychiatry 85:668–674

17. Svenningson P, Aarsland D (2013) Depression. In: Arsland D, Cummings J, Weintraub D, Chaudhuri KR (eds) Neuropsychiatric and cognitive changes in Parkinson's disease and related movement disorders. Cambridge University Press, Cambridge

18. Timmer MHM, van Beek MHCT, Bloem BR et al (2017) What a neurologist should know about depression in Parkinson's disease. Pract Neurol 17:359–368

19. Karlsen KH, Larsen JP, Tandberg E, Maeland JG (1999) Influence of clinical and demographic variables on quality of life in patients with Parkinson's disease. J Neurol Neurosurg Psychiatry 66:431–435

20. Schrag A, Jahanshahi M, Quinn N (2000) What contributes to quality of life in patients with Parkinson's disease? J Neurol Neurosurg Psychiatry 69:308–312

21. Aarsland D, Pahlhagen S, Ballard CG, Ehrt U, Svenningsson P (2012) Depression in Parkinson disease–epidemiology, mechanisms and management. Nat Rev Neurol 8:35–47

22. American Psychiatric Association (2013) Diagnostic and statistical manual of mental disorders, 5th edn. American Psychiatric Publishing, Arlington, VA

23. Ravina B, Camicioli R, Como PG et al (2007) The impact of depressive symptoms in early Parkinson disease. Neurology 69:342–347

24. Djamshidian A, Friedman JH (2014) Anxiety and depression in Parkinson's disease. Curr Treat Option Neurol 16:285

25. Dissanayaka NN, Sellbach A, Silburn PA, O'sullivan JD, Marsh R, Mellick GD (2011) Factors associated with depression in Parkinson's disease. J Affect Disord 132:82–88

26. Borgonovo J, Allende-Castro C, Almudena Laliena A, Guerrero N, Silva H, Concha ML (2017) Changes in neural circuitry associated with depression at pre-clinical, pre-motor and early motor phases of Parkinson's disease. Parkinsonism Relat Disord 35:17–24

27. Ehrt U, Bronnick K, Leentjens AF, Larsen JP, Aarsland D Depressive symptom profile in Parkinson's disease: a comparison with depression in elderly patients without Parkinson's disease. Int J Geriatr Psychiatry 21:252–258

28. Pachana NA, Egan SJ, Laidlaw K et al (2013) Clinical issues in the treatment of anxiety and depression in older adults with Parkinson's disease. Mov Disord 28:1930–1934

29. Richard IH (2006) Apathy does not equal depression in Parkinson disease: why we should care? Neurology 67:10–11

30. Skorvanek M, Gdovinova Z, Rosenberger J et al (2015) The associations between fatigue, apathy, and depression in Parkinson's disease. Acta Neurol Scand 131:80–87

31. Schrag A, Barone P, Brown RG et al (2007) Depression rating scales in Parkinson's disease: critique and recommendations. Mov Disord 22:1077–1092

32. Beck AT, Steer RA, Ball R, Ranieri WF (1996) Comparison of Beck depression inventories-IA and -II in psychiatric patient. J Pers Assess 67:588–597

33. Martinez-Martin P, Rodriguez-Blazquez C, Forjaz MJ, Kurtis MM, Skorvanek M (2017) Measurement of nonmotor symptoms in clinical practice. Int Rev Neurobiol 133:291–345

34. Hamilton M (1962) A rating scale for depression. J Neurol Neurosurg Psychiatry 23:56–62

35. Montgomery SA, Asberg M (1979) A new depression scale designed to be sensitive to change. Brit J Psychiatry 134:382–389

36. Paulus W, Jellinger K (1991) The neuropathologic basis of different clinical subgroups of Parkinson's disease. J Neuropathol Exp Neurol 50:743–755

37. Kim SE, Choi JY, Choe YS et al (2003) Serotonin transporters in the midbrain of Parkinson's disease patients: a study with 123I-beta-CIT SPECT. J Nucl Med 44:870–876

38. Vriend C, Pattij T, van der Werf YD et al (2013) Depression and impulse control disorders in Parkinson's disease: two sides of the same coin? Neurosci Biobehav Rev 38C:60–71

39. Remy P, Doder M, Lees A et al (2005) Depression in Parkinson's disease: loss of dopamine and noradrenaline innervation in the limbic system. Brain 128:1314–1322

40. Voon V, Sohr M, Lang AE et al (2011) Impulse control disorders in Parkinson disease: a multicenter case—control study. Ann Neurol 69:986–996

41. Wen MC, Chan LL, Tan LCS, Tan EK (2016) Depression, anxiety, and apathy in Parkinson's disease: insights from neuroimaging studies. Eur J Neurol 23:1001–1019

42. Huang P, Xuan M, Gu Q et al (2015) Abnormal amygdala function in Parkinson's disease patients and its relationship to depression. J Affect Disord 183:263–268

43. Feldmann A, Illes Z, Kosztolanyi P et al (2008) Morphometric changes of gray matter in Parkinson's disease with depression: a voxel-based morphometry study. Mov Disord 23:42–46

44. Kostić V, Agosta F, Petrović I et al (2010) Regional patterns of brain tissue loss associated with depression in Parkinson disease. Neurology 75:857–863

45. van Mierlo TJ, Chung C, Foncke EM, Berendse HW, van den Heuvel OA (2015) Depressive symptoms in Parkinson's disease are related to decreased hippocampus and amygdala volume. Mov Disord 30:245–252

46. Li W, Liu J, Skidmore F, Liu Y, Tian J, Li K (2010) White matter microstructure changes in the thalamus in Parkinson disease with depression: a diffusion tensor MR imaging study. AJNR Am J Neuroradiol 31:1861–1866

47. Huang P, Xu X, Gu Q et al (2014) Disrupted white matter integrity in depressed versus non-depressed Parkinson's disease patients: a tract-based spatial statistics study. J Neurol Sci 346:145–148

48. Matsui H, Nishinaka K, Oda M et al (2007) Depression in Parkinson's disease diffusion tensor imaging study. J Neurol 254:1170–1173

49. Mentis MJ, McIntosh AR, Perrine K et al (2002) Relationships among the metabolic patterns that correlate with mnemonic, visuospatial, and mood symptoms in Parkinson's disease. Am J Psychiatry 159:746–754

50. Politis M, Wu K, Loane C et al (2010) Depressive symptoms in PD correlate with higher 5-HTT binding in raphe and limbic structures. Neurology 75:1920–1927

51. Rektorova I, Srovnalova H, Kubikova R, Prasek J (2008) Striatal dopamine transporter imaging correlates with depressive symptoms and tower of London task performance in Parkinson's disease. Mov Disord 23:1580–1587

52. Vriend C, Raijmakers P, Veltman DJ et al (2014) Depressive symptoms in Parkinson's disease are related to reduced [^{123}I]FP-CIT binding in the caudate nucleus. J Neurol Neurosurg Psychiatry 85:159–164

53. Pessoa Rocha N, Reis HJ, Vanden Berghe P, Cirillo C (2014) Depression and cognitive impairment in Parkinson's disease: a role

for inflammation and immunomodulation. Neuroimmunomodulation 21:88–94

54. Beavan M, McNeill A, Proukakis C, Hughes DA, Mehta A, Schapira AH (2015) Evolution of prodromal clinical markers of Parkinson disease in a GBA mutation-positive cohort. JAMA Neurol 72:201–208

55. Srivastava A, Tang MX, Mejia-Santana H et al (2011) The relation between depression and parkin genotype: the CORE-PD study. Parkinsonism Relat Disord 17:740–744

56. Gaig C, Vilas D, Infante J et al (2014) Nonmotor symptoms in LRRK2 G2019S associated Parkinson's disease. PLoS One 9:e108982

57. Leentjens AF (2011) The role of dopamine agonists in the treatment of depression in patients with Parkinson's disease: a systematic review. Drugs 71:273–286

58. Marsh L (2013) Depression and Parkinson's disease: current knowledge. Curr Neurol Neurosci Rep 13:409

59. Cooney JW, Stacy M (2016) Neuropsychiatric issues in Parkinson's disease. Curr Neurol Neurosci Rep 16:49

60. Bomasang-Layno E, Fadlon I, Murray AN, Himelhoch S (2015) Antidepressive treatments for Parkinson's disease: a systematic review and meta-analysis. Parkinsonism Relat Disord 21:833–842

61. Rocha FL, Murad MG, Stumpf BP et al (2013) Antidepressants for depression in Parkinson's disease: systematic review and meta-analysis. J Psychopharmacol 27:417–423

62. Skapinakis P, Bakola E, Salanti G et al (2010) Efficacy and acceptability of selective serotonin reuptake inhibitors for the treatment of depression in Parkinson's disease: a systematic review and meta-analysis of randomized controlled trials. BMC Neurol 10:49

63. Friedman JH, Weintraub D (2012) Glad about SAD (PD). Neurology 78:1198–1199

64. Menza M, Dobkin RD, Marin H et al (2009) A controlled trial of antidepressants in patients with Parkinson disease and depression. Neurology 72:886–892

65. Richard IH, McDermott M, Kurlan R et al (2012) A randomized, double-blind, placebo-controlled trial of antidepressants in Parkinson disease. Neurology 78:1129–1136

66. Barone P, Scarzella L, Marconi R et al (2006) Pramipexole versus sertraline in the treatment of depression in Parkinson's disease: a national multicenter parallel-group randomized study. J Neurol 253:601–607

67. Barone P, Poewe W, Albrecht S et al (2010) Pramipexole for the treatment of depressive symptoms in patients with Parkinson's disease: a randomised, double-blind, placebo-controlled trial. Lancet Neurol 9:573–580

68. Chaudhuri KR, Martinez-Martin P, Antonini A et al (2013) Rotigotine and specific non-motor symptoms of Parkinson's disease: post hoc analysis of RECOVER. Parkinsonism Relat Disord 19:660–665

69. Williams NR, Okun MS (2013) Deep brain stimulation (DBS) at the interface of neurology and psychiatry. J Clin Invest 123:4546–4556

70. Dobkin RD, Menza M, Allen LA et al (2011) Cognitive-behavioral therapy for depression in Parkinson's disease: a randomized, controlled trial. Am J Psychiatry 168:1066–1074

71. Okai D, Askey-Jones S, Samuel M et al (2013) Trial of CBT for impulse control behaviors affecting Parkinson patients and their caregivers. Neurology 80:792–799

72. Pal E, Nagy F, Aschermann Z et al (2010) The impact of left prefrontal repetitive transcranial magnetic stimulation on depression in Parkinson's disease: a randomized, double-blind, placebo-controlled study. Mov Disord 25:2311–2317

73. Moellentine C, Rummans T, Ahlskog JE et al (1998) Effectiveness of ECT in patients with parkinsonism. J Neuropsychiatry Clin Neurosci 10:187–193

74. Hughes AJ, Daniel SE, Kilford L, Lees AJ (1992) Accuracy of clinical diagnosis of idiopathic Parkinson's disease: a clinicopathological study of 100 cases. J Neurol Neurosurg Psychiatry 55:181–184

75. Postuma R, Berg D, Stern M et al (2015) MDS clinical diagnostic criteria for Parkinson's disease. Mov Disord 30:1591–1599

76. Jurica PJ, Leitten CL, Mattis S (2004) Psychological assessment resources, 2001. Dementia rating Scale-2 (DRS-2). Arch Clin Neuropsychol 19:145–147

77. Nasreddine ZS, Phillips NA, Bédirian V et al (2005) The Montreal cognitive assessment, MoCA: a brief screening tool for mild cognitive impairment. J Am Geriatr Soc 53:695–699

78. Pagonabarraga J, Kulisevsky J, Llebaria G, Garcia-Sanchez C, Pascual-Sedano B, Gironell A (2008) Parkinson's disease-cognitive rating scale: a new cognitive scale specific for Parkinson's disease. Mov Disord 23:998–1005

79. Skorvanek M, Goldman J, Jahanshahi M et al (2017) Global scales for cognitive screening in Parkinson's disease: critique and recommendations. Mov Disord 33:208. https://doi.org/10.1002/mds.27233

80. Cummings J, Friedman JH, Garibaldi G et al (2015) Apathy in neurodegenerative diseases: recommendations on the design of clinical trials. J Geriatr Psychiatr Neurol 28:159–173

81. Seitz D, Purandare N, Conn D (2010) Prevalence of psychiatric disorders among older adults in long-term care homes: a systematic review. Int Psychogeriatr 22:1025–1039

82. Fahn S, Elton RL (1987) Members of the UPDRS development committee. Unified Parkinson's disease rating scale. In: Fahn S, Marsden CD, Calne DB, Lieberman A (eds) Recent developments in Parkinson's disease, vol 2. MacMillan Healthcare Information, Florham Park, NJ, pp 153–163

83. Goetz CG, Tilley BC, Schaftman SR et al (2008) Movement disorder Scoiety-sponsored revision of the unified Parkinson's disease rating scale (MDS-UPDRS): scale presentation and clinimetric testing results. Mov Disord 23:2129–2170

84. Leentjens AF, Dujardin K, Pontone GM, Starkstein SE, Weintraub D, Martinez-Martin P (2014) The Parkinson anxiety scale (PAS): development and validation of a new anxiety scale. Mov Disord 29:1035–1043

85. Pachana N, Byrne GJ, Siddle H, Koloski N, Harley E, Arnold E (2007) Development and validation of the geriatric anxiety inventory. Int Psychogeriatr 19:103–114

86. Starkstein SE, Mayberg HS, Preziosi TJ, Andrezejewski P, Leiguarda R, Robinson RG (1992) Reliability, validity and clinical correlates of apathy in Parkinson's disease. J Neuropsychiatry Clin Neurosci 4:134–139

87. Sockeel P, Dujardin K, Devos D, Deneve C, Destee A, Defebvre L (2006) The Lille apathy rating scale (LARS), a new instrument for detecting and quantifying apathy: validation in Parkinson's disease. J Neurol Neurosurg Psychiatry 77:579–584

88. Visser M, Marinus J, Stiggelbout AM, Van Hilten JJ (2004) Assessment of autonomic dysfunction in Parkinson's disease: the SCOPA-AUT. Mov Disord 19:1306–1312

89. Chaudhuri KR, Martinez-Martin P, Brown RG et al (2007) The metric properties of a novel non-motor symptoms scale for Parkinson's disease: results from an international pilot study. Mov Disord 22:1901–1911

90. Low PA (1993) Composite autonomic scoring scale for laboratory quantification of generalized autonomic failure. Mayo Clin Proc 68:748–752

91. Stern Y, Albert SM, Sano M et al (1994) Assessing patient dependence in Alzheimer's disease. J Gerontol 49:M216–M222

92. Yesavage JA, Brink TL, Rose TL et al (1983) Development and validation of a geriatric depression screening scale – a preliminary report. J Psychiatr Res 17:37–49

93. Krupp LB, LaRocca NG, Muir-Nash J, Steinberg AD (1989) The fatigue severity scale. Application to patients with multiple

sclerosis and systemic lupus erythematosus. Arch Neurol 46:1121–1123

94. Smets EMA, Garssen B, Bonke B, Deahes JCJM (1995) The multidimensional fatigue inventory (MFI) psychometric qualities of instrument to assess fatigue. J Psychosom Res 39:315–325

95. Schiehser DM, Ayers CR, Liu L, Lessig S, Song DS, Filoteo JV (2013) Validation of the modified fatigue impact scale in Parkinson's disease. Parkinsonism Relat Disord 19:335e338

96. Brown RG, Dittner A, Findley L, Wessely SC (2005) The Parkinson fatigue scale. Parkinsonism Relat Disord 11:49–55

97. Yellen SB, Cella DF, Webster K, Blendowski C, Kaplan E (1997) Measuring fatigue and other anemia-related symptoms with the functional assessment of cancer therapy (FACT) measurement system. J Pain Symptom Manag 13:63–74

98. Zarit SH, Zarit JM. (1987) Instructions for the burden interview. Technical document. University Park, PA: Pennsylvania State University

99. Goetz CG, Nutt JG, Stebbins GT (2008) The unified dyskinesia rating scale: presentation and clinimetric profile. Mov Disord 15:2398–2403

100. Stacy MA, Murphy JM, Greeley DR et al (2007) The sensitivity and specificity of the 9-item wearing off questionnaire. Parkinsonism Relat Disord 14:205–212

101. Chaudhuri KR, Rizos A, Trenkwalder C et al (2015) King's Parkinson's disease pain scale, the first scale for pain in PD: an international validation. Mov Disord 30:1623–1631

102. Peto V, Jenkinson C, Fitzpatrick R, Greenhall R (1995) The development and validation of a short measure of functioning and well-being for individuals with Parkinson's disease. Qual Life Res 4:241–248

103. Johns MW (1991) A new method for measuring daytime sleepiness: the Epworth sleepiness scale. Sleep 14:540–545

104. Trenkwalder C, Kohnen R, Hoegl B et al (2011) Parkinson's disease sleep scale-validation of the revised version PDSS-2. Mov Disord 26:644–652

105. Barlow D (2001) Anxiety and its disorders: the nature and treatment of anxiety and panic, 2nd edn. Guildford, New York, NY

106. Hanna K, Cronin-Golomb A (2012) Impact of anxiety on quality of life in Parkinson's disease. Parkinson Dis 2012:640707

107. Pontone G, Williams J, Anderson K et al (2011) Anxiety and self-perceived health status in Parkinson's disease. Parkinsonism Relat Disord 17:249–254

108. Weintraub D, Moberg P, Duda J, Katz I, Stern M (2004) Effect of psychiatric and other non-motor symptoms on disability in Parkinson's disease. J Am Geriatr Soc 52:784–788

109. Mathias J (2003) Neurobehavioral functioning of persons with Parkinson's disease. Appl Neuropsychol 10:57–68

110. Global Parkinson's Disease Survey (GPDS) Steering Committee (2002) Factors impacting on quality of life in Parkinson's disease: results from an international survey. Mov Disord 17:60–67

111. Cubo E, Bernard B, Leurgans S, Raman R (2000) Cognitive and motor function in patients with Parkinson's disease with and without depression. Clin Neuropharmacol 23:331–334

112. Richard I, Schiffer R, Kurlan R (1996) Anxiety and Parkinson's disease. J Neuropsychiatry Clin Neurosci 8:383–392

113. Marsh L, McDonald W, Cummings J, Ravina B, for the NINDS/NIMH Work Group on Depression and Parkinson's Disease (2006) Provisional diagnostic criteria for depression in Parkinson's disease: report of an NINDS/NIMH work group. Mov Disord 21:148–158

114. Weintraub D, Moberg P, Duda J, Katz I, Stern M (2003) Recognition and treatment of depression in Parkinson's disease. J Geriatr Psychiatr Neurol 16:178–183

115. Chen JJ, Marsh L (2014) Anxiety in Parkinson's disease: identification and management. Ther Adv Neurol Dis 7:52–59

116. Dissanayaka NNW, Torbey E, Pachana NA (2015) Anxiety rating scales in Parkinson's disease: a critical review updating recent literature. Int Psychogeriatr 27:1777

117. Broen MPG, Narayen NE, Kuijf ML, Dissanayaka NNW, Leentjens AFG (2016) Prevalence of anxiety in Parkinson's disease: a systematic review and meta-analysis. Mov Disord 31:1125

118. Marsh L, Calleo J (2013) Anxiety. In: Arsland D, Cummings J, Weintraub D, Chaudhuri KR (eds) Neuropsychiatric and cognitive changes in Parkinson's disease and related movement disorders. Cambridge University Press, Cambridge

119. Shulman L, Singer C, Bean J, Weiner W (1996) Internal tremor in patients with Parkinson's disease. Mov Disord 11:3–7

120. Kovacs M, Makkos A, Weintraut R, Karadi K, Kovacs N (2017) Prevalence of anxiety among Hungarian subjects with Parkinson's disease. Behav Neurol 2017:1470149

121. Weintraub D, Hoops S (2011) Anxiety syndromes and panic attacks. In: Olanow CW, Stocchi F, Lang AE (eds) Parkinson's disease: non-motor and non-dopaminergic features,

1st edn. Blackwell Publishing Ltd, Hoboken, NJ

122. Starkstein SE, Dragovic M, Dujardin K et al (2014) Anxiety has specific syndromal profiles in Parkinson disease: a data-driven approach. Am J Geriatr Psychiatr 22:1410–1417

123. Shulman L, Taback R, Rabinstein A, Weiner W (2002) Non-recognition of depression and other non-motor symptoms in Parkinson's disease. Parkinsonism Relat Disord 8:193–197

124. Leentjens A, Dujardin K, Marsh L, Richard I, Starkstein S, Martinez-Martin P (2011) Anxiety rating scales in Parkinson's disease: a validation study of the Hamilton anxiety rating scale, the Beck anxiety inventory, and the hospital anxiety and depression scale. Mov Disord 26:407–415

125. Leentjens AF, Dujardin K, Marsh L et al (2008) Anxiety rating scales in Parkinson's disease: critique and recommendations. Mov Disord 23:2015–2025

126. Matheson SF, Byrne GJ, Dissanayaka NN et al (2012) Validity and reliability of the geriatric anxiety inventory in Parkinson's disease. Australas J Ageing 31:13–16

127. Black KJ, Hershey T, Hartlein JM (2005) Levodopa challenge neuroimaging of levodopa-related mood fluctuations in Parkinson's disease. Neuropsychopharmacology 30:590–601

128. Kaasinen V, Nurmi E, Brück A et al (2001) Increased frontal [(18)F]fluorodopa uptake in early Parkinson's disease: sex differences in the prefrontal cortex. Brain 124:1125–1130

129. Dan R, Růžička F, Bezdicek O et al (2017) Separate neural representations of depression, anxiety and apathy in Parkinson's disease. Sci Rep 7:12164

130. Wang X, Li J, Yuan Y et al (2017) Altered putamen functional connectivity is associated with anxiety disorder in Parkinson's disease. Oncotarget 8:81377–81386

131. Nuss P (2015) Anxiety disorders and GABA neurotransmission: a disturbance of modulation. Neuropsychiatr Dis Treat 11:165–175

132. Walsh K, Bennett G (2001) Parkinson's disease and anxiety. Postgrad Med J 77:89–93

133. Prediger RD, Matheus FC, Schwarzbold ML, Lima MM, Vital MA (2012) Anxiety in Parkinson's disease: a critical review of experimental and clinical studies. Neuropharmacology 62:115–124

134. Casacchia M, Zamponi A, Squitieri G, Meco G (1975) Treatment of anxiety in Parkinson's disease with bromazepam. Riv Neurol 45:326–338

135. Kulisevsky J, Pascual-Sedano B, Barbanoj M et al (2007) Acute effects of immediate and controlled-release levodopa on mood in Parkinson's disease: a double-blind study. Mov Disord 22:62–67

136. Drijgers RL, Dujardin K, Reijnders JSAM, Defebvre L, Leentjens AFG (2010) Validation of diagnostic criteria for apathy in Parkinson's disease. Parkinsonism Relat Disord 16:656–660

137. Pagonabarraga J, Kulisevsky J, Strafella AP, Krack P (2015) Apathy in Parkinson's disease: clinical features, neural substrates, diagnosis, and treatment. Lancet Neurol 14:518–531

138. Marin RS (1991) Apathy: a neuropsychiatric syndrome. J Neuropsychiatry Clin Neurosci 3:243–254

139. Kuhl J, Atkinson JW (1986) Motivation, thought, and action. Praeger Publishers, New York, NY

140. Laplane D, Dubois B (2001) Auto-activation deficit: a basal ganglia related syndrome. Mov Disord 16:810–814

141. Liao C, Feng Z, Zhou D et al (2012) Dysfunction of fronto-limbic brain circuitry in depression. Neuroscience 201:231–238

142. Jahanshahi M (1998) Willed action and its impairments. Cogn Neuropsychol 15:483–533

143. Aarsland D, Brønnick K, Ehrt U et al (2007) Neuropsychiatric symptoms in patients with Parkinson's disease and dementia: frequency, profile and associated care giver stress. J Neurol Neurosurg Psychiatry 78:36–42

144. Dujardin K, Sockeel P, Delliaux M, Destee A, Defebvre L (2009) Apathy may herald cognitive decline and dementia in Parkinson's disease. Mov Disord 24:2391–2397

145. Martínez-Horta S, Pagonabarraga J, Fernández de Bobadilla R, García-Sanchez C, Kulisevsky J (2013) Apathy in Parkinson's disease: more than just executive dysfunction. J Int Neuropsychol Soc 19:571–582

146. Friedman JH, Beck JC, Chou KL et al (2016) Fatigue in Parkinson's disease: report from a multidisciplinary symposium. NPJ Parkinson Dis 2:15025

147. Robert P, Onyike CU, Leentjens AF et al (2009) Proposed diagnostic criteria for apathy in Alzheimer's disease and other neuropsychiatric disorders. Eur Psychiatry 24:98–104

148. Starkstein SE, Merello M, Jorge R, Brockman S, Bruce D, Power B (2009) The syndromal validity and nosological position of apathy in Parkinson's disease. Mov Disord 24:1211–1216

149. Leentjens AFG, Dujardin K, Marsh L et al (2008) Apathy and anhedonia rating scales in Parkinson's disease: critique and recommendations. Mov Disord 23:2004–2014

150. Dujardin K, Sockeel P, Delliaux M, Destée A, Defebvre L (2008) The Lille apathy rating scale: validation of a caregiver-based version. Mov Disord 23:845–849

151. Thobois S, Ardouin C, Lhommee E et al (2010) Non-motor dopamine withdrawal syndrome after surgery for Parkinson's disease: predictors and underlying mesolimbic denervation. Brain 133:1111–1127

152. Santangelo G, Vitale C, Picillo M et al (2015) Apathy and striatal dopamine transporter levels in de-novo, untreated Parkinson's disease patients. Parkinsonism Relat Disord 21:489–493

153. Rabinak CA, Nirenberg MJ (2010) Dopamine agonist withdrawal syndrome in Parkinson disease. Arch Neurol 67:58–63

154. Maillet A, Krack P, Lhommée E et al (2016) The prominent role of serotonergic degeneration in apathy, anxiety and depression in de novo Parkinson's disease. Brain 139:2486–2502

155. Thobois S, Prange S, Sgambato-Faure V, Tremblay L, Broussolle E (2017) Imaging the etiology of apathy, anxiety and depression in Parkinson's disease: implication for treatment. Curr Neurol Neurosci Rep 17:76

156. Czernecki V, Schupbach M, Yaici S et al (2008) Apathy following subthalamic stimulation in Parkinson disease: a dopamine responsive symptom. Mov Disord 23:964–969

157. Leentjens AF, Koester J, Fruh B, Shephard DT, Barone P, Houben JJ (2009) The effect of pramipexole on mood and motivational symptoms in Parkinson's disease: a meta-analysis of placebo-controlled studies. Clin Ther 31:89–98

158. Chatterjee A, Fahn S (2002) Methylphenidate treats apathy in Parkinson's disease. J Neuropsychiatr Clin Neurosci 14:461–462

159. Corcoran C, Wong ML, O'Keane V (2004) Bupropion in the management of apathy. J Psychopharmacol 18:133–135

160. Zahodne LB, Marsiske M, Okun MS, Bowers D (2012) Components of depression in Parkinson disease. J Geriatr Psychiatr Neurol 25:131–137

161. Pedersen KF, Alves G, Aarsland D, Larsen JP (2009) Occurrence and risk factors for apathy in Parkinson disease: a 4-year prospective longitudinal study. J Neurol Neurosurg Psychiatry 80:1279–1282

162. Friedman J, Friedman H (1993) Fatigue in Parkinson's disease. Neurology 43:2016–2018

163. Kluger BM (2017) Fatigue in Parkinson's disease. Int Rev Neurobiol 133:743–768

164. Havlikova E, van Dijk JP, Rosenberger J et al (2008) Fatigue in Parkinson's disease is not related to excessive sleepiness or quality of sleep. J Neurol Sci 270:107–113

165. Marras C, Chaudhuri KR (2016) Nonmotor features of Parkinson's disease subtypes. Mov Disord 31:1095–1102

166. Kluger BM, Herlofson K, Chou KL et al (2016) Parkinson's disease-related fatigue: a case definition and recommendations for clinical research. Mov Disord 31:625–631

167. Kluger BM, Garimella S, Garvan C (2017) Minimal clinically important difference of the modified fatigue impact scale in Parkinson's disease. Parkinsonism Relat Disord 43:101–104

168. Friedman JH, Alves G, Hagell P et al (2010) Fatigue rating scales critique and recommendations by the movement disorders society task force on rating scales for Parkinson's disease. Mov Disord 25:805–822

169. Hwang SS, Chang VT, Fairclough DL et al (2003) Longitudinal quality of life in advanced cancer patients: pilot study results from a VA medical cancer center. J Pain Symptom Manag 25:225–235

170. Lou JS (2012) Techniques in assessing fatigue in neuromuscular diseases. Phys Med Rehabil Clin N Am 23:11–22

171. Lou JS (2009) Physical and mental fatigue in Parkinson's disease: epidemiology, pathophysiology and treatment. Drugs Aging 26:195–208

172. Wang C, Ding M, Kluger BM (2014) Change in intraindividual variability over time as a key metric for defining performance-based cognitive fatigability. Brain Cognit 85:251–258

173. Gandevia SC (2001) Spinal and supraspinal factors in human muscle fatigue. Physiol Rev 81:1725–1789

174. Bruce JM, Bruce AS, Arnett PA (2010) Response variability is associated with self-reported cognitive fatigue in multiple sclerosis. Neuropsycholgy 24:77–83

175. Holtzer R, Foley F (2009) The relationship between subjective reports of fatigue and executive control in multiple sclerosis. J Neurol Sci 281:46–50

176. Kluger BM, Palmer C, Shattuck JT et al (2012) Motor evoked potential depression following repetitive central motor initiation. Exp Brain Res 216:585–590

177. Esposito F, Otto T, Zijlstra FR, Goebel R (2014) Spatially distributed effects of mental exhaustion on resting-state FMRI networks. PLoS One 9:e94222

178. Gevins AS, Morghan NH, Bressler SL et al (1987) Human neuroelectric patterns predict performance accuracy. Science 235:580–585

179. Ishii A, Tanaka M, Shigiharo Y et al (2013) Neural effects of prolonged mental fatigue: a magnetoencephalography study. Brain Res 1529:105–112

180. Liu JZ, Dai TH, Sahgal V et al (2002) Nonlinear cortical modulation of muscle fatigue: a functional MRI study. Brain Res 957:320–329

181. Pavese N, Metta V, Bose SK et al (2010) Fatigue in Parkinson's disease is linked to striatal and limbic serotonergic dysfunction. Brain 133:3434–3443

182. Schifitto G, Friedman JH, Oakes D et al (2008) Fatigue in levodopa-naive subjects with Parkinson disease. Neurology 71:481–485

183. Tessitore A, Giordano A, De Micco R et al (2016) Functional connectivity underpinnings of fatigue in "drug-naïve" patients with Parkinson's disease. Mov Disord 31:1497–1505

184. Cho SS, Aminian K, Li C et al (2017) Fatigue in Parkinson's disease: the contribution of cerebral metabolic changes. Hum Brain Mapp 38:283–292

185. Abe K, Takanashi M, Yanagihara T (2000) Fatigue in patients with Parkinson's disease. Behav Neurol 12:103–106

186. Fabbrini G, Latorre A, Suppa A et al (2013) Fatigue in Parkinson's disease: motor or non-motor symptom? Parkinsonism Relat Disord 19:148–152

187. Lindqvist D, Kaufman E, Brindin L et al (2012) Nonmotor symptoms in patients with Parkinson's disease—correlations with inflammatory cytokines in serum. PLoS One 7:e47387

188. Nakamura T et al (2011) Does cardiovascular autonomic dysfunction contribute to fatigue in Parkinson's disease? Mov Disord 26:1869–1874

189. Skorvanek M, Nagyova I, Rosenberger J et al (2013) Clinical determinants of primary and secondary fatigue in patients with Parkinson's disease. J Neurol 260:1554–1561

190. Lindqvist D, Hall S, Surova Y et al (2014) Cerebrospinal fluid inflammatory markers in Parkinson's disease – associations with depression, fatigue, and cognitive impairment. Brain Behav Immun 33:183–189

191. Pereira JR, dos Santos LV, Renata M et al (2016) IL-6 serum levels are elevated in Parkinson's disease patients with fatigue compared to patients without fatigue. J Neurol Sci 370:153–156

192. Kenangil G, Oreken DN, Ur E et al (2009) The relation of testosterone levels with fatigue and apathy in Parkinson's disease. Clin Neurol Neurosurg 111:412–414

193. Ondo W, Shinawi L, Davidson A et al (2011) Memantine for non-motor features of Parkinson's disease: a double-blind placebo controlled exploratory pilot trial. Parkinsonism Relat Disord 17:156–159

194. Rascol O, Fitzer-Attas C, Hauser R et al (2011) A double-blind, delayed-start trial of rasagiline in Parkinson's disease (the ADAGIO study): prespecified and post-hoc analysis of the need for additional therapies, change in UPDRS scores, and non-motor outcomes. Lancet Neurol 10:415–423

195. Lim TT, Kluger BM, Rodriguez RL et al (2015) Rasagiline for the symptomatic treatment of fatigue in Parkinson's disease. Mov Disord 30:1825–1830

196. Postuma R, Lang AE, Munhoz RP et al (2012) Caffeine for treatment of Parkinson disease. Neurology 79:651–659

197. Mendonça D, Menezes K, Jog M (2007) Methylphenidate improves fatigue scores in Parkinson's disease: a randomized controlled trial. Mov Disord 22:2070–2076

198. Lou JS, Dimitrova DM, Park BS et al (2009) Using modafinil to treat fatigue in Parkinson disease: a double-blind, placebo-controlled pilot study. Clin Neuropharmacol 32:305–310

199. Ondo WG, Fayle R, Atassi F et al (2005) Modafinil for daytime somnolence in Parkinson's disease: double blind controlled parallel trial. J Neurol Neurosurg Psychiatry 76:1636–1639

200. Tyne H, Taylor J, Baker GA et al (2010) Modafinil for Parkinson's disease fatigue. J Neurol 257:452–456

201. Canning C, Allen NE, Dean CM et al (2012) Home-based treadmill training for individuals with Parkinson's disease: a randomized controlled pilot trial. Clin Rehabil 26:817–826

202. Winward C, Sackley C, Meek C et al (2012) Weekly exercise does not improve fatigue levels in Parkinson's disease. Mov Disord 27:143–146

203. Ribas CG, da Silva LA, Correa MR et al (2017) Effectiveness of exergaming in improving functional balance, fatigue and quality of life in Parkinson's disease: a pilot randomized controlled trial. Parkinsonism Relat Disord 38:13–18

204. Forogh B, Rafiei M, Arbabi A et al (2017) Repeated sessions of transcranial direct current stimulation evaluation on fatigue and daytime sleepiness in Parkinson's disease. Neurol Sci 38:249–254

205. Romenets SR, Anang J, Fereshtehnejad SM et al (2015) Tango for the treatment of motor and non-motor manifestations in Parkinson's disease: a randomized control study. Complement Ther Med 23:175–184

206. Kluger BM, Rakowski D, Christian M et al (2016) Randomized, controlled trial of acupuncture for fatigue in Parkinson's disease. Mov Disord 31:1027–1032

207. Hidding U, Gulberti A, Horn A et al (2017) Impact of combined subthalamic nucleus and substantia Nigra stimulation on neuropsychiatric symptoms in Parkinson's disease patients. Parkinson Dis 2017:7306192

208. Chou KL, Taylor JL, Patil PG (2013) The MDS-UPDRS tracks motor and non-motor improvement due to subthalamic nucleus deep brain stimulation in Parkinson disease. Parkinsonism Relat Disord 19:966–969

209. Storch A, Schneider CB, Wolz M et al (2013) Nonmotor fluctuations in Parkinson disease severity and correlation with motor complications. Neurology 80:800–809

210. Kluger BM, Parra V, Jacobson C et al (2012) The prevalence of fatigue following deep brain stimulation surgery in Parkinson's disease and association with quality of life. Parkinson's Dis 2012:769506

211. Cugusi L, Solla P, Serpe R et al (2015) Effects of a Nordic walking program on motor and non-motor symptoms, functional performance and body composition in patients with Parkinson's disease. NeuroRehabilitation 37:245–254

212. Havlikova E, Rosenberger J, Nagyova I et al (2008) Clinical and sychosocial factors associated with fatigue in patients with Parkinson's disease. Parkinsonism Relat Disord 14:187–192

213. Karlsen K, Larsen JP, Tandberg E, Jorgensen K (1999) Fatigue in patients with Parkinson's disease. Mov Disord 14:237–241

Chapter 11

Clinical Trials for Sleep Disorders and Daytime Somnolence in Parkinson's Disease

Marissa N. Dean and Amy W. Amara

Abstract

Sleep disorders in Parkinson's disease (PD) are frequently encountered and can have a significant negative impact on quality of life. In this chapter, we will specifically address ongoing and completed trials for treatment of rapid eye movement (REM) sleep behavior disorder (RBD), insomnia, restless legs syndrome (RLS) and periodic limb movement disorder (PLMD), circadian rhythm disorders (CRD), nocturia, sleep disordered breathing (SDB), and excessive daytime sleepiness (EDS). Multisite, randomized, controlled clinical trials that specifically address treatment of disorders of sleep and wakefulness in the PD population have been completed in some areas (EDS) but not others (RLS). There is a need to standardize trial design as well as tools for diagnosis and assessment across studies, among other shortcomings. More large-scale, randomized, controlled clinical trials are needed to better elucidate the role of disease-specific treatments in this patient population.

Key words Sleep disorders, Parkinson's disease, Treatment, Clinical trials, Somnolence, Insomnia, REM sleep behavior disorders, Restless legs syndrome, Nocturia, Sleep disordered breathing

1 Introduction

Sleep disorders in Parkinson's disease (PD) are frequently encountered and can have a significant negative impact on quality of life [1, 2]. Insufficient sleep not only interferes with everyday routine but can also have a negative influence on motor symptoms in PD. Many clinical trials have evaluated potential treatments for nonmotor symptoms in PD, and sleep disorders have drawn increasing attention. In this chapter, we will review how clinical trials are being used to address the most common disorders of sleep and wakefulness in PD. We will specifically address ongoing and completed trials for treatment of rapid eye movement (REM) sleep behavior disorder (RBD), insomnia, restless legs syndrome (RLS) and periodic limb movement disorder (PLMD), circadian rhythm disorders (CRD), nocturia, sleep disordered breathing (SDB), and excessive daytime sleepiness (EDS).

Santiago Perez-Lloret (ed.), *Clinical Trials In Parkinson's Disease*, Neuromethods, vol. 160,
https://doi.org/10.1007/978-1-0716-0912-5_11, © Springer Science+Business Media, LLC, part of Springer Nature 2021

2 Rapid Eye Movement Sleep Behavior Disorder

RBD is an REM parasomnia in which individuals move during dream sleep (i.e., act out their dreams). Normal REM sleep is characterized by somatic muscle atonia, but in RBD there is a loss of REM atonia. RBD is common in patients with PD and, at tertiary care centers, the prevalence of RBD with clinically established PD is 39–46% [3, 4]. Although a male predominance has been reported in what has been called "idiopathic RBD," there does not appear to be a difference in prevalence based on sex among PD patients [4, 5]. RBD may precede motor symptoms in PD, or it may occur after the clinical diagnosis of PD has been made. Recognition of RBD prior to the development of observable motor symptoms has been of increasing interest and is useful in identifying prodromal PD patients [6].

2.1 Establishing a Diagnosis

The diagnosis of RBD requires a history or visualization of dream enactment along with polysomnographic evidence of REM sleep without atonia [7]. If dream enactment is not visualized with polysomnography, this history can be obtained through screening questionnaires. Many such questionnaires have been proposed, but all demonstrate low specificities. The REM behavior disorder screening questionnaire (RBDSQ) has a specificity of 63–82.8% with a sensitivity of 68–90% [8–10], while the REM sleep behavior disorder questionnaire Hong Kong (RBDQ-HK) has a specificity of 86.9% and a sensitivity of 82.2% [11]. The RBDSQ is a 10-item questionnaire, while the RBDQ-HK has 13 questions. There are two other questionnaires that assess RBD symptoms with a single question: the Mayo sleep questionnaire (MSQ) and the RBD1Q [12]. The MSQ has a specificity of 87.9% and sensitivity of 90% in PD. Although the RBD1Q has not been evaluated in PD patients [9], it has a sensitivity of 93.8% and specificity of 87.2% in an "idiopathic RBD" population [13]. The Innsbruck REM sleep behavior inventory has also been used, with similar sensitivities and specificities [13]. Although the screening questionnaires may be helpful in identifying possible RBD, the detection through identification of loss of REM atonia through polysomnography is required for definitive diagnosis.

Assessing and scoring REM without atonia (REMWA) may be accomplished by different methods. Lapierre and Montplaisir developed the first widely accepted visual scoring system in 1992, and this was validated in 80 patients in 2010 [14, 15]. The authors proposed a scoring system that requires behavioral motor manifestations as well as an increase in both tonic (>50% of 20-s epoch) and phasic chin EMG activity during REM for diagnosis of RBD. This scoring system correctly identified REMWA in 82% of patients [15]. The SINBAR group developed a more specific visual scoring system that assessed muscle groups in various combinations, and found that the combination of chin (tonic or phasic) EMG activity and bilateral flexor digitorum superficialis (FDS) phasic EMG activity provided the best diagnostic accuracy [16]. The International Classification of Sleep Disorders, third edition (ICSD-3) recommends REMWA be diagnosed when >27% of 30-s epochs during REM sleep contain chin (tonic or phasic) EMG activity and bilateral FDS phasic EMG activity, based on current evidence-based data [7]. However, visual scoring systems can be time consuming, and automated detection systems have the potential to enhance recognition and diagnosis of RBD. One such system has been developed, which utilizes chin EMG to automatically compute the REM sleep Atonia Index (RAI) [17]. The three different scoring systems were recently compared, and the diagnostic accuracy was comparable between the visual scoring systems and automated scoring system [18].

2.2 Previous Clinical Trials

There are no completed double-blind, placebo-controlled, large-scale clinical trials that have evaluated treatment efficacy for RBD in PD. Instead, patients with RBD are treated based on guidelines established for idiopathic RBD [19]. Melatonin and clonazepam are suggested for the treatment of RBD and clonazepam may also decrease the occurrence of sleep-related injury caused by RBD [19]. Clonazepam's efficacy in treatment of RBD has been reported in 86–87% of cases through several large case series [20–22]. A recent meta-analysis evaluated the efficacy of melatonin use for sleep disorders in neurodegenerative diseases and determined that melatonin improved the clinical and neurophysiological findings of RBD [23]. However, this recommendation was based on one randomized controlled trial with melatonin that included one patient with PD and RBD (one out of eight patients) [24]. Therefore, more randomized controlled trials are needed to determine the efficacy of medication treatment for RBD in PD.

In addition to clonazepam and melatonin, several other agents have been investigated for utility in RBD in small studies. For example, a Japanese group evaluated the use of ramelteon in treatment for idiopathic RBD in an open-label trial, and found that there was no clear effect on REMWA or an RBD severity scale [25]. Rivastigmine was also evaluated as a treatment for RBD in PD in a single site, double-blind, placebo-controlled, crossover pilot trial. This study suggested a reduced mean frequency of RBD episodes with rivastigmine compared to placebo in 12 participants with PD [26].

2.3 Recently Completed and Ongoing Clinical Trials

Specific treatment for RBD in PD is an increasing area of interest, as evidenced by the several recently completed and ongoing clinical trials that address this issue. In Seoul, Korea, a phase 2, double-blind, randomized, controlled trial evaluating clonazepam versus placebo in patients with RBD and Parkinsonism was completed in 2016 (ClinicalTrials.gov NCT02312908). Findings have not yet been published. There is another ongoing phase 2 clinical trial evaluating clonazepam together with prolonged-release melatonin in the treatment of RBD in PD (ClinicalTrials.gov NCT02789592). In the USA, nelotanserin, a 5-HT$_{2A}$ receptor inverse agonist, is under investigation as a treatment for RBD in the setting of dementia. This trial is currently in phase 2 assessing patients with dementia (Parkinson's Disease Dementia or Dementia with Lewy Bodies) and RBD to assess nelotanserin's efficacy in this subset of patients (ClinicalTrials.gov NCT02708186).

2.4 Future Direction on Study Design

To date, no completed large multicenter, placebo-controlled clinical trials have addressed treatment options for RBD in patients with PD, so this is an area with great potential. RBD is common in PD, can lead to injury for the patient and/or bed partner, and may precede motor-manifest PD by many years. One important consideration in clinical trial design is appropriate diagnosis of RBD for inclusion. Although many RBD questionnaires are available, none are very specific, highlighting the importance of polysomnographic diagnosis for inclusion in future clinical trials. The best way to quantify frequency and severity of RBD, as well as response to treatment may also require polysomnography, although patient diaries, wearable monitors, or ancillary (i.e., sleep quality and daytime sleepiness) scales may also have some benefit.

In addition to the need for randomized, controlled trials for therapy for RBD in PD, early recognition of patients with RBD prior to development of motor-manifest PD (i.e., prodromal PD) is also important. This patient population offers a unique opportunity to intervene to test potentially neuroprotective or disease-modifying therapies in clinical trials.

3   Insomnia

Insomnia is a subjective report of inadequate sleep quality, and should be thought of as a symptom with many possible etiologies. In PD, sleep fragmentation is the most commonly reported sleep-related complaint [27]. The prevalence of reported poor sleep in PD has a large range from 20% to 80%, however, the actual prevalence may be on the higher side [28–30]. In a longitudinal study of PD patients over 8 years, 83% experienced insomnia at one or more study visits, indicating that insomnia is a frequent nonmotor symptom associated with Parkinson's Disease, and should be adequately addressed and treated [28].

3.1 Establishing a Diagnosis

Establishing the diagnosis of insomnia is achieved through questionnaires, as insomnia is a subjective symptom. Although many questionnaires are available, only three have been recommended by the Movement Disorders Society task force for use in PD [31]. These include the Parkinson's Disease Sleep Scale (PDSS), Scales for Outcomes in Parkinson's Disease—Sleep (SCOPA-Sleep), and the Pittsburgh Sleep Quality Index (PSQI) [16]. The PDSS and SCOPA-Sleep scales were designed specifically for PD patients, with the SCOPA-Sleep being a shorter five-item questionnaire [32, 33]. The PSQI has also been used to screen for and assess the severity of insomnia in PD patients [34, 35]. Use of polysomnography is typically not indicated in insomnia unless another primary sleep disorder, such as sleep apnea, is suspected. Further, objective correlation with subjective complaints of insomnia has not been consistent [36].

3.2 Previous Clinical Trials

Designing clinical trials to assess insomnia in PD can be difficult, as the etiology for insomnia in PD may be multifactorial. Many factors can contribute to insomnia in PD, including primary sleep disorders such as restless legs syndrome and obstructive sleep apnea; PD-related motor symptoms such as tremor, dystonia, and rigidity, which can impair the ability to roll over in bed; PD medication effects; psychiatric symptoms such as anxiety, depression and hallucinations; and nocturia [37]. The Movement Disorder Society Task Force on Evidence-Based Medicine review of treatments for PD determined that there was insufficient evidence to determine whether controlled-release formulation carbidopa-levodopa, pergolide, eszopiclone, or melatonin had efficacy in treating insomnia in PD [38]. None of the studies that evaluated these medications lasted for greater than 10 weeks, so only short-term efficacy was assessed. Eszopiclone was compared to placebo in a randomized group of 30 patients with PD and insomnia, and eszopiclone trended toward superiority in improving quality of sleep and some measures of sleep maintenance, but without a significant change in total sleep time [39]. The evaluation of rotigotine on sleep disruption in PD was evaluated as an open-label, multicenter trial (SLEEP-FRAM study) and utilized item 3 of the PDSS-2 as the primary endpoint to assess sleep fragmentation [40]. After 3 months, there was significant improvement of sleep fragmentation as well as improvement in Parkinsonian nocturnal motor symptoms [40]. The ACCORD-PD study was published recently and suggests that computerized cognitive behavioral therapy for insomnia (CCBT-I) may be an effective treatment option for PD patients with insomnia [41].

3.3 Recently Completed and Ongoing Clinical Trials

Exercise has been evaluated as treatment for chronic insomnia, and is effective to improve subjective sleep complaints [42]. Currently, there is an ongoing study assessing the effects of high intensity exercise on sleep efficiency in PD patients (ClinicalTrials.gov NCT02593955). Suvorexant is also being evaluated for efficacy in treating insomnia in PD, and is currently in a phase 4 clinical trial (ClinicalTrials.gov NCT02729714).

3.4 Future Direction on Study Design

While there is a large amount of literature regarding chronic insomnia, the number of RCT for treating insomnia in PD is lacking. This may be secondary to the many hurdles that exist when evaluating and treating insomnia in PD. First, there are multiple etiologies for the sleep fragmentation experienced in PD, and the lack of a single etiology makes data interpretation from clinical trials difficult. Each PD patient may respond differently to a medication, depending on whether the underlying cause for the insomnia is the same. Secondly, differences in study design and primary endpoints (i.e., different questionnaires and objective polysomnography outcomes) make it difficult to compare studies. Finally, many studies that have been completed thus far for insomnia in PD have a small sample size. Enrollment of larger samples are needed in order to accurately assess the impact of these medications.

4 Restless Legs Syndrome and Periodic Limb Movement Disorder

Restless legs syndrome (RLS) is a sleep-related movement disorder that may be seen in patients with PD. The prevalence of RLS in PD is higher than in the general population, ranging from 3% to 21.3%, but questions remain regarding the temporal relationship between development of RLS and PD [43–46]. Some studies suggest that people with RLS may be predisposed to develop PD; however, some evidence indicates that PD is a risk factor for developing RLS, especially if RLS symptoms are severe or dopaminergic therapy is initiated [47, 48]. Periodic limb movements of sleep (PLMS) may be associated with RLS in PD patients as in the general population, but data related to this are conflicting [44, 49].

4.1 Establishing a Diagnosis

The diagnosis of RLS requires an urge to move the legs, which may be associated with an uncomfortable sensation [7]. This sensation primarily occurs with rest/inactivity, is at least partially relieved by movement, and mainly occurs at night or in the evening. For diagnosis of RLS, the ICSD III criteria require that the symptoms lead to significant distress or result in sleep impairment [7]. RLS mimics need to be ruled out before diagnosing RLS in PD. For example, in addition to the mimics seen in the general population, nocturnal restlessness is a common manifestation in the PD population, and may be easily mistaken for RLS [50, 51]. This idea is supported by a study showing that RLS symptoms are increased with wearing off in PD [52].

An objective diagnostic tool, the suggested immobilization test (SIT) has been developed for RLS [53]. Patients are instructed to remain still in a laying position for 1 h at night. Every 10 min, patients report the severity of leg discomfort on a scale of 0 (no discomfort) to 100 (extreme discomfort). Using a cutoff score of 11/100, the SIT has a sensitivity of 91% and a specificity of 72% for diagnosing RLS in PD [53]. The SIT also has a motor component in which the periodic leg movements index is recorded during the hour of testing; however, this did not differ between the RLS and non-RLS groups. Although the SIT may be a useful diagnostic tool, its utility for ruling out confounders of RLS in PD has not been evaluated. Therefore, the SIT may be a helpful adjunctive diagnostic tool in patients with RLS and PD.

PLMD is defined by the presence of periodic limb movements during sleep (PLMS), which lead to significant impairment of sleep and/or functioning [7]. Adults must have >15 PLMS per hour to fulfill the diagnosis of PLMD. PLMS are frequently seen in patients with RLS; however, this finding is not specific for RLS and PLMS can frequently be seen in isolation in the elderly population [54, 55].

4.2 Previous and Ongoing Clinical Trials

Designing a clinical trial for RLS in PD is difficult, as there are many possible confounding etiologies for RLS in the PD population. Leg motor restlessness (or akathisia) and wearing off phenomenon may mimic RLS in PD and need to be appropriately assessed when designing a clinical trial. Questionnaires are a useful screening tool to evaluate for RLS; however, there are a limited number of questionnaires that have been validated in the general population and none specifically in PD patients. The Cambridge-Hopkins questionnaire (CH-RLSq) is a short questionnaire that was found to be specific (87.2%) and sensitive (94.4%) for diagnosing RLS in the general population [56]. In addition, the Restless Legs Syndrome Diagnostic Index (RLS-DI) is a validated ten-item questionnaire that was found to be 93% sensitive and 98.9% specific for the diagnosis of RLS in the general population [57]. The Johns Hopkins telephone diagnostic interview (HTDI) was also found to be sensitive (97%) and specific (92%) for diagnosing RLS by experienced interviewers [58]. This study was further validated in a nonpatient population and found to have at least a 90% sensitivity and 91% specificity [59]. In a review of the available RLS diagnostic questionnaires, the recommended instruments were the CH-RLSq, RSL-DI, and HTDI [60]. These questionnaires, however, do not address the impact of RLS symptoms on quality of sleep or daily activities. The Restless Legs Syndrome Quality of Life questionnaire (RLSQoL) has been used to reliably assess the quality of life in patients with RLS [61, 62].

Questionnaires are also available to evaluate RLS severity. In a recent review, two questionnaires were recommended by the International Parkinson and Movement Disorder Society (MDS) task force: the International Restless Legs Scale (IRLS) and the Augmentation Severity Rating Scale (ASRS) [63]. The IRLS is a

10-item questionnaire that ranks severity on a scale of 0–4 and was found to have a high level of internal consistency, inter-examiner reliability and test-retest reliability [64]. The ASRS was created by the European RLS Study Group to assess the severity of augmentation in RLS patients, and it was shown to be both a reliable and valid instrument in the general population [65]. To summarize, questionnaires are useful in identifying patients with RLS and assessing the severity of RLS and its impact on quality of life; however, further validation studies of these questionnaires within the PD population are still needed.

To date, there have been no randomized clinical trials and there are no ongoing clinical trials that have evaluated treatment of RLS in PD patients. Treatments used for the motor symptoms of PD (levodopa, ropinirole, pramipexole, and rotigotine) have been evaluated in randomized trials and found to be effective for treating RLS, but the effects of these medications on RLS in PD have not been studied. There is potential for these medications to help both the motor PD symptoms and RLS symptoms in these patients, although this has not yet been well described in the literature.

4.3 Future Direction on Study Design

Treatment of RLS in PD patients is an area that needs further exploration. In addition to the difficulty in identifying RLS mimics in PD patients, another confounder is the potential masking of RLS symptoms with dopaminergic therapy used for treatment of PD symptoms. Further, the use of questionnaires and SIT should be further studied and validated for utility in PD.

5 Circadian Rhythm Disorders

The circadian rhythm is an approximately 24-hour biological rhythm, and circadian rhythm sleep disorders (CRSD) are disturbances of this rhythm that affect the sleep-wake cycle. CRSD in PD are an emerging area of investigation that has recently sparked more interest. There are several symptoms of PD that appear to fluctuate or worsen in correlation to different times of the day, which have been postulated to be related to circadian rhythm disorders. For example, diurnal variation of levodopa response and hallucinations are frequently encountered in PD patients. The extent of impact of CRSD in PD remains an intriguing area of investigation, which needs more research.

5.1 Establishing a Diagnosis

Circadian rhythm sleep disorders present diagnostic challenges in PD and are often overlooked. Methods for analysis of endogenous circadian markers are well correlated with the biological circadian rhythm, but are not readily available for use in the clinic. Despite this, some studies have evaluated these levels in PD patients. For example, levels of melatonin were assessed in PD patients, and those patients who were on dopaminergic therapy had delayed sleep initiation and prolongation of dim light melatonin onset (DLMO) (melatonin secretion under dim light conditions) [66]. This may imply that dopaminergic medications impact the secretion of melatonin, which in turn affects the circadian rhythm in PD patients. Another study evaluated circadian rhythm markers in PD and found elevated serum cortisol, reduced serum melatonin, and altered expression of the major core clock gene Bmal1 [67]. Further, a small case–control study investigated diurnal changes in thermoregulation in PD, finding a decrease in core-body temperature MESOR and a decrease in the nocturnal fall of core body temperature in PD patients compared to healthy controls [68]. These alterations in circadian rhythm biomarkers suggest there may be an underlying dysfunction of thermoregulation and circadian rhythm regulation in patients with PD.

5.2 Previous Clinical Trials

There are a limited number of completed clinical trials that have evaluated the efficacy of treatment for CRSD in PD patients. The few studies that evaluated the role of bright light therapy (BLT) in PD patients did not measure levels of circadian rhythm markers (such as melatonin or cortisol). These small case studies showed some improvement of depressive symptoms, sleep onset, and sleep continuity in PD patients when compared to controls [69–71].

5.3 Recently Completed and Ongoing Clinical Trials

Clinical trials in PD patients with CRSD have been focused on light therapy and exposure. The safety and efficacy of light exposure in 31 PD patients was recently evaluated in a randomized, placebo-controlled clinical trial and found to be well tolerated [72]. There was significant improvement in daytime sleepiness (as measured by the Epworth Sleepiness Scale) in the BLT group and improvement in sleep quality in both the BLT group and dim-red light therapy (control) group. Although circadian rhythm markers were not measured, it is postulated that the effects of bright light therapy act through the suprachiasmatic nucleus in the hypothalamus to improve the dysregulated sleep-wake cycle in PD.

In the USA, there are no currently enrolling clinical trials that are evaluating treatment for CRSD in PD. In France, there is one ongoing randomized clinical trial that is evaluating the effects of active light therapy (10,000 lux) in PD patients (ClinicalTrials.gov NCT02072642).

5.4 Future Direction on Study Design

Given the favorable results in a recent clinical trial evaluating the safety and efficacy of BLT in PD for treating excessive daytime sleepiness and sleep impairment, further investigation in larger samples is needed to delineate the utility of BLT on circadian rhythm disorders in PD. The addition of measurement of melatonin levels and other circadian rhythm markers could strengthen the impact of such studies. In addition, a more unified way of assessing, diagnosing and measuring the extent of CRSD in PD patients is warranted.

6   Nocturia

Nocturia is common and is reported in up to 60–80% of PD patients [73, 74]. This makes nocturia the most common urinary symptom in PD and is also one of the most commonly reported nonmotor symptoms in PD [75, 76]. Studies have suggested an increased prevalence of nocturia in PD patients with male sex, older age, older age of PD onset, and increased PD duration and severity [75–78]. The causes for nocturia in PD patients may be due to multiple factors. Detruser hyperactivity and polyuria have been reported in some PD patients with lower urinary tract symptoms (including nocturia), but this is not present in all PD patients [79]. Other factors, such as dopaminergic medications and sleep apnea may also contribute to nocturia [80].

6.1   Establishing a Diagnosis

There is no established guideline in regard to what makes nocturia "bothersome" or clinically significant. Two or more episodes of nocturia in PD patients was found to be associated with higher subjective "bother," lower PSG-defined sleep efficiency and decreased whole-night total sleep time, when compared to PD patients with 0–1 episodes of nocturia [81]. Therefore, it may be useful to consider clinically significant nocturia in PD as more than one episode of nighttime urination.

Evaluation of nocturia in PD patients has been obtained through questionnaires. The Movement Disorders Society-Unified Parkinson's Disease Rating Scale (MDS-UPDRS), the Parkinson's disease sleep scale version-2 and Scales for Outcomes in Parkinson's disease (SCOPA)—sleep all include a question related to nocturia and have been utilized in the PD population [33, 82]. Collection of information pertaining to nocturia may also be obtained through diaries or other questionnaires, such as the nocturia, nocturnal enuresis and sleep-interruptions questionnaire (NNES-Q) or the overactive bladder symptom score [83–85]. The NNES-Q and overactive bladder symptom score have not been validated in the PD population, so this is an area that needs further investigation.

6.2   Previous Clinical Trials

Few clinical trials have investigated therapies for nocturia in PD. In Japan, one group studied pergolide (a dopamine agonist) as a treatment for nocturia in PD patients, and found that when patients were switched from bromocriptine to pergolide, there was an improvement in nocturia [86]. However, pergolide is no longer available for therapeutic use for PD in the USA, as it was withdrawn from the market in 2007.

6.3   Recently Completed and Ongoing Clinical Trials

There are only a small number of clinical trials evaluating therapeutic interventions for nocturia in PD. Earlier in 2017, recruitment closed for the behavioral therapy to treat urinary symptoms in PD (BETTUR PD) study (ClinicalTrials.gov NCT01520948). The study investigators conducted a two-site randomized controlled trial that evaluated the efficacy of pelvic floor muscle exercise (PFME) therapy in PD patients with urinary symptoms. Published results are still pending. In Canada, another clinical trial is evaluating the use of fesoterodine in treatment of overactive urinary bladder (OAB) in PD patients, and nocturnal micturitions are a secondary outcome measure (ClinicalTrials.gov NCT02385500). An additional clinical trial is planned to start soon through the Veterans Affairs Office of Research and Development. This study will compare treatment with behavioral therapy (PFME) versus solifenacin in treating urinary symptoms in PD (ClinialTrials.gov NCT03149809). This will be a three-site randomized controlled phase 3 clinical trial. The primary outcome measure will be response to a questionnaire rating the degree and frequency of bladder symptoms (urgency, frequency, nocturia, and urinary incontinence).

6.4   Future Direction on Study Design

There are a limited number of clinical trials that have evaluated therapeutic treatment efficacy for PD patients with nocturia. Recent studies have begun to evaluate the use of behavioral therapy for urinary symptoms in PD, but larger trials that incorporate multiple sites are warranted. In addition, the need for a better way to quantify efficacy of treatment (apart from questionnaires) is an area that should be further explored.

7   Sleep Disordered Breathing

The prevalence of sleep disordered breathing (SDB) in PD patients ranges from 15% to 76% [37]. The patterns of SDB that have been described in PD include obstructive and restrictive pulmonary dysfunction and an abnormal response to hypercapnia [87–89]. Initially, it was postulated that PD increases the risk of development of obstructive sleep apnea (OSA); however, this was not replicated in later studies [90]. In other controlled studies, the prevalence of OSA was actually less frequent in PD patients when compared to controls [37]. PD patients may be at increased risk of developing SDB if pulmonary dysfunction is already present [87, 88]. To date, no correlation between PD motor symptoms and OSA have been validated in clinical studies, and it is suspected that the etiology for sleep apnea is similar for PD patients as it is for the general population.

7.1   Establishing a Diagnosis

The diagnosis of SDB is made with polysomnography (PSG). Questionnaires have been developed to screen for OSA in the general population; however, these have not been validated for use in PD patients. It is interesting to note that in PD patients, snoring and sleepiness were not strong predictors for sleep apnea [91]. Available questionnaires for screening for SDB include the Berlin questionnaire and STOP questionnaire [92, 93]. These questionnaires place strong emphasis on snoring and subjective sleepiness symptoms, so it is reasonable to question the utility in PD patients that may not have these associated symptoms. The STOP-BANG questionnaire additionally includes age >65 years and gender as risk factors for SDB. Because increasing age is associated with increased risk of PD and PD is more likely to affect men, this questionnaire has the potential to lead to false positives in the PD population.

7.2 Recently Completed and Ongoing Clinical Trials

In Italy, a group attempted to evaluate changes in cognition and sleepiness with treatment of OSA with CPAP in PD patients; however, there was a significantly high dropout rate (75%) that made interpretation of the results difficult [94]. In addition, there was no placebo group in this study. An additional group in the USA conducted a trial that evaluated PD patients for sleep apnea. Those with OSA had worse scores on cognitive testing. Those with OSA were then randomized to receive CPAP versus placebo CPAP [95]. The study results did not show a significant change in cognition in the CPAP-treated group, which further suggests that the cognitive changes seen in PD may not be affected by treatment of concurrent OSA, at least in the short term. A group in Canada is currently actively recruiting participants to investigate cognitive outcomes of PD patients with OSA who are treated with positive airway pressure versus nasal dilator strips (ClinicalTrials.gov NCT02209363). This randomized clinical trial has a primary outcome measure of change in global cognitive function.

A previous study completed in Canada assessed the occurrence of SDB in PD patients taking long-acting levodopa and those who were not [96]. There were only a small number of patients taking a long-acting levodopa at nighttime in this study (eight patients total), but results suggested that these patients had fewer OSA-related findings on PSG when compared to PD patients not taking a long-acting levodopa at bedtime. Currently, the effect of long-acting levodopa on OSA in PD patients is being assessed in a randomized crossover study in Canada (ClinicalTrials.gov NCT03111485). The primary outcome measure is the apnea–hypopnea index (AHI) and secondary measures include oxygenation from PSG, daytime sleepiness and cognitive function.

7.3 Future Direction on Study Design

Identifying which PD patients are at risk for sleep apnea can be difficult, as there is some evidence that the symptom profile may be different in the PD population. Two common symptoms that are seen in the general population with OSA (snoring and daytime sleepiness) may not be seen in PD patients with OSA [91]. Validation of a screening questionnaire that reliably identifies PD patients at risk for OSA is an area that needs further investigation.

8 Excessive Daytime Sleepiness

Excessive daytime sleepiness (EDS) is a common sleep disturbance witnessed in PD patients, with a prevalence of 20–60% [37]. Overall, studies have demonstrated that subjective sleepiness is more common in PD patients when compared to healthy controls [37]. There are conflicting studies suggesting when EDS is most commonly seen throughout the course of disease. Some studies suggest EDS may be more common in early nonmedicated PD patients, while more recent and larger studies found no difference between these patients and healthy controls [97–100]. Several studies have demonstrated that EDS is more common in advanced PD. For example, one study found that patients with advanced PD had shorter mean sleep latencies when compared to early PD patients, but subjective reports of sleepiness did not correlate with objective measurements [99]. Duration of PD, male sex, and anxiety are all factors that increase the risk of EDS [101, 102]. Dopaminergic medications may also increase subjective sleepiness, but no objective correlation has been found on PSG [37].

It is not surprising that EDS in PD significantly impacts daily activities. EDS has been shown to have a negative impact on quality of life, activities of daily living and caregiver burden in PD.[1, 103, 104] EDS can also affect driving performance, processing speed, and reaction time in PD patients [105, 106]. All of these reasons support the need for accurate assessment and treatment for EDS in PD patients.

8.1 Establishing a Diagnosis

Defining EDS can be difficult since subjective reports of sleepiness do not always correlate with objective findings. The Epworth sleepiness scale (ESS) has been validated in the PD population and is commonly used as a measurement for EDS [35, 107]. Other questionnaires have been used to assess EDS in PD patients, including SCOPA-SLEEP—daytime sleepiness (-DS), Parkinson's disease sleep scale (PDSS), and the MDS-UPDRS item 1.8 on daytime sleepiness, which has been shown to correlate with the ESS [32, 33, 108, 109].

Objective measures are available to quantify EDS in PD. The multiple sleep latency test (MSLT) and maintenance of wakefulness test (MWT) measure the tendency to fall asleep and the ability to stay awake, respectfully [7, 110]. Actigraphy has also been utilized, but subjective sleepiness has not consistently correlated with objective measures of napping [111, 112].

8.2 Previous Clinical Trials

Because of the significant negative impact on quality of life, it is not surprising that several randomized therapeutic trials for EDS in PD patients have already been completed. Memantine [113], modafinil [114], and bright light therapy [69] did not significantly impact

EDS when compared to placebo in PD patients. However, modafinil did show efficacy for improving subjective sleepiness in two studies and a meta-analysis of this therapy also suggests this medication can improve EDS [115–117]. Melatonin use improved scores on the general sleep disturbance scale, but not on the ESS [118]. Daytime sleepiness was assessed as a secondary outcome measure in the randomized controlled trial with atomoxetine in PD patients, and there was significant improvement of daytime sleepiness in this study [119]. Sodium oxybate treatment in a multicenter randomized controlled trial for PD patients also showed statistically significant improvement of daytime sleepiness on the ESS [120]. A more recent single site, randomized, controlled, crossover study showed that sodium oxybate improves both subjective (ESS) and objective (MSLT) measures in PD patients with EDS [121]. Caffeine has also been an area of interest in Parkinson's Disease, and when administered in a randomized controlled trial, there was no statistically significant change in the ESS, but there was a trend toward improvement [122].

8.3 Recently Completed and Ongoing Clinical Trials

Several ongoing therapeutic clinical trials address EDS in PD. For example, bavisant is being evaluated for efficacy and safety for treatment of EDS in PD patients in a multicenter, randomized, double-blind, parallel-group, and placebo-controlled phase 2 clinical trial (ClinicalTrials.gov NCT03194217). The primary outcome measure is the ESS. JZP-110 is also being evaluated in a similar study design (multicenter, randomized, double-blind, placebo-controlled phase 2 clinical trial) for efficacy using the ESS and MWT as outcomes (ClinicalTrials.gov NCT03037203). Droxidopa is being studied for its efficacy in treating daytime fatigue in PD patients using the Parkinson's Disease Fatigue Scale as the primary outcome and PDQ-39 and ESS as secondary outcomes (ClinicalTrials.gov NCT03034564).

8.4 Future Direction on Study Design

Although several studies have addressed therapy of EDS in PD, there is still much to be learned in this area. One important consideration is the utilization of similar measuring tools to make interpretation and comparison across studies possible. Additional, larger randomized controlled clinical trials of both pharmacologic and nonpharmacologic interventions are needed to assess the efficacy of proposed treatments for EDS in PD.

9   Conclusion

As interest in PD nonmotor symptoms continues to grow, treatments for sleep disturbances and daytime sleepiness have the potential to significantly improve quality of life for PD patients. Multisite, randomized, controlled clinical trials that specifically address treatment of disorders of sleep and wakefulness in the PD population have been completed in some areas (EDS) but not others (RLS). Because PD is such a heterogeneous disorder, selection by specific sleep disorder may contribute to optimization of future clinical trials in this area. There is also a need to standardize trial design as well as tools for diagnosis and assessment across studies to allow more accurate comparison between the trials. Design of clinical studies can also be informed by our increasing understanding of the importance of sleep disorders to PD disease course (i.e., PD patients with RBD are more likely to have balance and cognitive problems and psychosis over time). Across all areas of sleep and wakefulness in PD, more large-scale, randomized, controlled clinical trials are needed to better elucidate the role of disease-specific treatments in this patient population.

References

1. Gallagher DA, Lees AJ, Schrag A (2010) What are the most important nonmotor symptoms in patients with Parkinson's disease and are we missing them? Mov Disord 25(15):2493–2500. https://doi.org/10.1002/mds.23394

2. Avidan A, Hays RD, Diaz N, Bordelon Y, Thompson AW, Vassar SD, Vickrey BG (2013) Associations of sleep disturbance symptoms with health-related quality of life in Parkinson's disease. J Neuropsychiatry Clin Neurosci 25(4):319–326. https://doi.org/10.1176/appi.neuropsych.12070175

3. Neikrug AB, Maglione JE, Liu L, Natarajan L, Avanzino JA, Corey-Bloom J, Palmer BW, Loredo JS, Ancoli-Israel S (2013) Effects of sleep disorders on the non-motor symptoms of Parkinson disease. J Clin Sleep Med 9(11):1119–1129. https://doi.org/10.5664/jcsm.3148

4. Sixel-Doring F, Trautmann E, Mollenhauer B, Trenkwalder C (2011) Associated factors for REM sleep behavior disorder in Parkinson disease. Neurology 77(11):1048–1054. https://doi.org/10.1212/WNL.0b013e31822e560e

5. Plomhause L, Dujardin K, Duhamel A, Delliaux M, Derambure P, Defebvre L, Monaca Charley C (2013) Rapid eye movement sleep behavior disorder in treatment-naive Parkinson disease patients. Sleep Med 14(10):1035–1037. https://doi.org/10.1016/j.sleep.2013.04.018

6. Berg D, Postuma RB, Adler CH, Bloem BR, Chan P, Dubois B, Gasser T, Goetz CG, Halliday G, Joseph L, Lang AE, Liepelt-Scarfone I, Litvan I, Marek K, Obeso J, Oertel W, Olanow CW, Poewe W, Stern M, Deuschl G (2015) MDS research criteria for prodromal Parkinson's disease. Mov Disord 30(12):1600–1611. https://doi.org/10.1002/mds.26431

7. Medicine AAoS (2014) International classification of sleep disorders. American Academy of Sleep Medicine, Darien, IL

8. Stiasny-Kolster K, Mayer G, Schafer S, Moller JC, Heinzel-Gutenbrunner M, Oertel WH (2007) The REM sleep behavior disorder screening questionnaire—a new diagnostic instrument. Mov Disord 22(16):2386–2393. https://doi.org/10.1002/mds.21740

9. Chahine LM, Daley J, Horn S, Colcher A, Hurtig H, Cantor C, Dahodwala N (2013) Questionnaire-based diagnosis of REM sleep behavior disorder in Parkinson's disease. Mov Disord 28(8):1146–1149. https://doi.org/10.1002/mds.25438

10. Stiasny-Kolster K, Sixel-Doring F, Trenkwalder C, Heinzel-Gutenbrunner M, Seppi K, Poewe W, Hogl B, Frauscher B (2015) Diagnostic value of the REM sleep behavior disorder screening questionnaire in Parkinson's disease. Sleep Med 16(1):186–189. https://doi.org/10.1016/j.sleep.2014.08.014

11. Li SX, Wing YK, Lam SP, Zhang J, Yu MW, Ho CK, Tsoh J, Mok V (2010) Validation of a new REM sleep behavior disorder questionnaire (RBDQ-HK). Sleep Med 11(1):43–48. https://doi.org/10.1016/j.sleep.2009.06.008

12. Boeve BF, Molano JR, Ferman TJ, Smith GE, Lin SC, Bieniek K, Haidar W, Tippmann-Peikert M, Knopman DS, Graff-Radford NR, Lucas JA, Petersen RC, Silber MH (2011) Validation of the Mayo sleep questionnaire to screen for REM sleep behavior disorder in an aging and dementia cohort. Sleep Med

12(5):445–453. https://doi.org/10.1016/j.sleep.2010.12.009

13. Postuma RB, Arnulf I, Hogl B, Iranzo A, Miyamoto T, Dauvilliers Y, Oertel W, Ju YE, Puligheddu M, Jennum P, Pelletier A, Wolfson C, Leu-Semenescu S, Frauscher B, Miyamoto M, Cochen De Cock V, Unger MM, Stiasny-Kolster K, Fantini ML, Montplaisir JY (2012) A single-question screen for rapid eye movement sleep behavior disorder: a multicenter validation study. Mov Disord 27(7):913–916. https://doi.org/10.1002/mds.25037

14. Lapierre O, Montplaisir J (1992) Polysomnographic features of REM sleep behavior disorder: development of a scoring method. Neurology 42(7):1371–1374

15. Montplaisir J, Gagnon JF, Fantini ML, Postuma RB, Dauvilliers Y, Desautels A, Rompre S, Paquet J (2010) Polysomnographic diagnosis of idiopathic REM sleep behavior disorder. Mov Disord 25(13):2044–2051. https://doi.org/10.1002/mds.23257

16. Frauscher B, Iranzo A, Gaig C, Gschliesser V, Guaita M, Raffelseder V, Ehrmann L, Sola N, Salamero M, Tolosa E, Poewe W, Santamaria J, Hogl B, Group S (2012) Normative EMG values during REM sleep for the diagnosis of REM sleep behavior disorder. Sleep 35(6):835–847. https://doi.org/10.5665/sleep.1886

17. Ferri R, Rundo F, Manconi M, Plazzi G, Bruni O, Oldani A, Ferini-Strambi L, Zucconi M (2010) Improved computation of the atonia index in normal controls and patients with REM sleep behavior disorder. Sleep Med 11(9):947–949. https://doi.org/10.1016/j.sleep.2010.06.003

18. Figorilli M, Ferri R, Zibetti M, Beudin P, Puligheddu M, Lopiano L, Cicolin A, Durif F, Marques A, Fantini ML (2017) Comparison between automatic and visual scorings of REM sleep without atonia for the diagnosis of REM sleep behavior disorder in Parkinson disease. Sleep 40(2). https://doi.org/10.1093/sleep/zsw060

19. Aurora RN, Zak RS, Maganti RK, Auerbach SH, Casey KR, Chowdhuri S, Karippot A, Ramar K, Kristo DA, Morgenthaler TI, Standards of Practice C, American Academy of Sleep M (2010) Best practice guide for the treatment of REM sleep behavior disorder (RBD). J Clin Sleep Med 6(1):85–95

20. Olson EJ, Boeve BF, Silber MH (2000) Rapid eye movement sleep behaviour disorder: demographic, clinical and laboratory findings in 93 cases. Brain 123(Pt 2):331–339

21. Schenck CH, Hurwitz TD, Mahowald MW (1993) Symposium: normal and abnormal REM sleep regulation: REM sleep behaviour disorder: an update on a series of 96 patients and a review of the world literature. J Sleep Res 2(4):224–231

22. Wing YK, Lam SP, Li SX, Yu MW, Fong SY, Tsoh JM, Ho CK, Lam VK (2008) REM sleep behaviour disorder in Hong Kong Chinese: clinical outcome and gender comparison. J Neurol Neurosurg Psychiatry 79(12):1415–1416. https://doi.org/10.1136/jnnp.2008.155374

23. Zhang W, Chen XY, Su SW, Jia QZ, Ding T, Zhu ZN, Zhang T (2016) Exogenous melatonin for sleep disorders in neurodegenerative diseases: a meta-analysis of randomized clinical trials. Neurol Sci 37(1):57–65. https://doi.org/10.1007/s10072-015-2357-0

24. Kunz D, Mahlberg R (2010) A two-part, double-blind, placebo-controlled trial of exogenous melatonin in REM sleep behaviour disorder. J Sleep Res 19(4):591–596. https://doi.org/10.1111/j.1365-2869.2010.00848.x

25. Esaki Y, Kitajima T, Koike S, Fujishiro H, Iwata Y, Tsuchiya A, Hirose M, Iwata N (2016) An open-labeled trial of Ramelteon in idiopathic rapid eye movement sleep behavior disorder. J Clin Sleep Med 12(5):689–693. https://doi.org/10.5664/jcsm.5796

26. Di Giacopo R, Fasano A, Quaranta D, Della Marca G, Bove F, Bentivoglio AR (2012) Rivastigmine as alternative treatment for refractory REM behavior disorder in Parkinson's disease. Mov Disord 27(4):559–561. https://doi.org/10.1002/mds.24909

27. Gomez-Esteban JC, Zarranz JJ, Lezcano E, Velasco F, Ciordia R, Rouco I, Losada J, Bilbao I (2006) Sleep complaints and their relation with drug treatment in patients suffering from Parkinson's disease. Mov Disord 21(7):983–988. https://doi.org/10.1002/mds.20874

28. Gjerstad MD, Wentzel-Larsen T, Aarsland D, Larsen JP (2007) Insomnia in Parkinson's disease: frequency and progression over time. J Neurol Neurosurg Psychiatry 78(5):476–479. https://doi.org/10.1136/jnnp.2006.100370

29. Porter B, Macfarlane R, Walker R (2008) The frequency and nature of sleep disorders in a community-based population of patients with Parkinson's disease. Eur J Neurol 15(1):50–54. https://doi.org/10.1111/j.1468-1331.2007.01998.x

30. Tse W, Liu Y, Barthlen GM, Halbig TD, Tolgyesi SV, Gracies JM, Olanow CW, Koller WC (2005) Clinical usefulness of the Parkinson's disease sleep scale. Parkinsonism Relat Disord 11(5):317–321. https://doi.org/10.1016/j.parkreldis.2005.02.006

31. Hogl B, Arnulf I, Comella C, Ferreira J, Iranzo A, Tilley B, Trenkwalder C, Poewe W, Rascol O, Sampaio C, Stebbins GT, Schrag A, Goetz

CG (2010) Scales to assess sleep impairment in Parkinson's disease: critique and recommendations. Mov Disord 25(16):2704–2716. https://doi.org/10.1002/mds.23190

32. Chaudhuri KR, Pal S, DiMarco A, Whately-Smith C, Bridgman K, Mathew R, Pezzela FR, Forbes A, Hogl B, Trenkwalder C (2002) The Parkinson's disease sleep scale: a new instrument for assessing sleep and nocturnal disability in Parkinson's disease. J Neurol Neurosurg Psychiatry 73(6):629–635

33. Marinus J, Visser M, van Hilten JJ, Lammers GJ, Stiggelbout AM (2003) Assessment of sleep and sleepiness in Parkinson disease. Sleep 26(8):1049–1054

34. Buysse DJ, Reynolds CF 3rd, Monk TH, Berman SR, Kupfer DJ (1989) The Pittsburgh sleep quality index: a new instrument for psychiatric practice and research. Psychiatry Res 28(2):193–213

35. Johns MW (1991) A new method for measuring daytime sleepiness: the Epworth sleepiness scale. Sleep 14(6):540–545

36. Norlinah MI, Afidah KN, Noradina AT, Shamsul AS, Hamidon BB, Sahathevan R, Raymond AA (2009) Sleep disturbances in Malaysian patients with Parkinson's disease using polysomnography and PDSS. Parkinsonism Relat Disord 15(9):670–674. https://doi.org/10.1016/j.parkreldis.2009.02.012

37. Chahine LM, Amara AW, Videnovic A (2017) A systematic review of the literature on disorders of sleep and wakefulness in Parkinson's disease from 2005 to 2015. Sleep Med Rev 35:33–50. https://doi.org/10.1016/j.smrv.2016.08.001

38. Seppi K, Weintraub D, Coelho M, Perez-Lloret S, Fox SH, Katzenschlager R, Hametner EM, Poewe W, Rascol O, Goetz CG, Sampaio C (2011) The Movement Disorder Society evidence-based medicine review update: treatments for the non-motor symptoms of Parkinson's disease. Mov Disord 26(Suppl 3):S42–S80. https://doi.org/10.1002/mds.23884

39. Menza M, Dobkin RD, Marin H, Gara M, Bienfait K, Dicke A, Comella CL, Cantor C, Hyer L (2010) Treatment of insomnia in Parkinson's disease: a controlled trial of eszopiclone and placebo. Mov Disord 25(11):1708–1714. https://doi.org/10.1002/mds.23168

40. Pagonabarraga J, Pinol G, Cardozo A, Sanz P, Puente V, Otermin P, Legarda I, Delgado T, Serrano C, Balaguer E, Aguirregomozcorta M, Alvarez R, Kulisevsky JJ (2015) Transdermal Rotigotine improves sleep fragmentation in Parkinson's disease: results of the multicenter, prospective SLEEP-FRAM

study. Parkinsons Dis 2015:131508. https://doi.org/10.1155/2015/131508

41. Patel S, Ojo O, Genc G, Oravivattanakul S, Huo Y, Rasameesoraj T, Wang L, Bena J, Drerup M, Foldvary-Schaefer N, Ahmed A, Fernandez HH (2017) A computerized cognitive behavioral therapy randomized, controlled, pilot trial for insomnia in Parkinson disease (ACCORD-PD). J Clin Mov Disord 4:16. https://doi.org/10.1186/s40734-017-0062-2

42. Passos GS, Poyares DL, Santana MG, Tufik S, Mello MT (2012) Is exercise an alternative treatment for chronic insomnia? Clinics (Sao Paulo) 67(6):653–660

43. Azmin S, Khairul Anuar AM, Nafisah WY, Tan HJ, Raymond AA, Hanita O, Shah SA, Norlinah MI (2013) Restless legs syndrome and its associated risk factors in Parkinson's disease. Parkinsons Dis 2013:535613. https://doi.org/10.1155/2013/535613

44. Loo HV, Tan EK (2008) Case-control study of restless legs syndrome and quality of sleep in Parkinson's disease. J Neurol Sci 266(1–2):145–149. https://doi.org/10.1016/j.jns.2007.09.033

45. Rana AQ, Siddiqui I, Mosabbir A, Athar A, Syed O, Jesudasan M, Hafez K (2013) Association of pain, Parkinson's disease, and restless legs syndrome. J Neurol Sci 327(1–2):32–34. https://doi.org/10.1016/j.jns.2013.01.039

46. Verbaan D, van Rooden SM, van Hilten JJ, Rijsman RM (2010) Prevalence and clinical profile of restless legs syndrome in Parkinson's disease. Mov Disord 25(13):2142–2147. https://doi.org/10.1002/mds.23241

47. Calzetti S, Angelini M, Negrotti A, Marchesi E, Goldoni M (2014) A long-term prospective follow-up study of incident RLS in the course of chronic DAergic therapy in newly diagnosed untreated patients with Parkinson's disease. J Neural Transm (Vienna) 121(5):499–506. https://doi.org/10.1007/s00702-013-1132-8

48. Wong JC, Li Y, Schwarzschild MA, Ascherio A, Gao X (2014) Restless legs syndrome: an early clinical feature of Parkinson disease in men. Sleep 37(2):369–372. https://doi.org/10.5665/sleep.3416

49. Prudon B, Duncan GW, Khoo TK, Yarnall AJ, Anderson KN (2014) Primary sleep disorder prevalence in patients with newly diagnosed Parkinson's disease. Mov Disord 29(2):259–262. https://doi.org/10.1002/mds.25730

50. Gjerstad MD, Tysnes OB, Larsen JP (2011) Increased risk of leg motor restlessness but not RLS in early Parkinson disease. Neurology 77(22):1941–1946. https://doi.org/10.1212/WNL.0b013e31823a0cc8

51. Suzuki K, Miyamoto M, Miyamoto T, Tatsumoto M, Watanabe Y, Suzuki S, Iwanami M, Sada T, Kadowaki T, Numao A, Trenkwalder C, Hirata K (2012) Nocturnal disturbances and restlessness in Parkinson's disease: using the Japanese version of the Parkinson's disease sleep scale-2. J Neurol Sci 318(1–2):76–81. https://doi.org/10.1016/j.jns.2012.03.022

52. Peralta CM, Wolf E, Seppi K, Wenning G, Hogl B, Poewe W (2005) Restless legs in idiopathic Parkinson's disease. Mov Disord 20:364

53. De Cock VC, Bayard S, Yu H, Grini M, Carlander B, Postuma R, Charif M, Dauvilliers Y (2012) Suggested immobilization test for diagnosis of restless legs syndrome in Parkinson's disease. Mov Disord 27(6):743–749. https://doi.org/10.1002/mds.24969

54. Ancoli-Israel S, Kripke DF, Klauber MR, Mason WJ, Fell R, Kaplan O (1991) Periodic limb movements in sleep in community-dwelling elderly. Sleep 14(6):496–500

55. Montplaisir J, Boucher S, Poirier G, Lavigne G, Lapierre O, Lesperance P (1997) Clinical, polysomnographic, and genetic characteristics of restless legs syndrome: a study of 133 patients diagnosed with new standard criteria. Mov Disord 12(1):61–65. https://doi.org/10.1002/mds.870120111

56. Allen RP, Burchell BJ, MacDonald B, Hening WA, Earley CJ (2009) Validation of the self-completed Cambridge-Hopkins questionnaire (CH-RLSq) for ascertainment of restless legs syndrome (RLS) in a population survey. Sleep Med 10(10):1097–1100. https://doi.org/10.1016/j.sleep.2008.10.007

57. Benes H, Kohnen R (2009) Validation of an algorithm for the diagnosis of restless legs syndrome: the restless legs syndrome-diagnostic index (RLS-DI). Sleep Med 10(5):515–523. https://doi.org/10.1016/j.sleep.2008.06.006

58. Hening WA, Allen RP, Thanner S, Washburn T, Heckler D, Walters AS, Earley CJ (2003) The Johns Hopkins telephone diagnostic interview for the restless legs syndrome: preliminary investigation for validation in a multi-center patient and control population. Sleep Med 4(2):137–141

59. Hening WA, Allen RP, Washburn M, Lesage S, Earley CJ (2008) Validation of the Hopkins telephone diagnostic interview for restless legs syndrome. Sleep Med 9(3):283–289. https://doi.org/10.1016/j.sleep.2007.04.021

60. Walters AS, Frauscher B, Allen R, Benes H, Chaudhuri KR, Garcia-Borreguero D, Lee HB, Picchietti DL, Trenkwalder C, Martinez-Martin P, Stebbins GT, Schrag A, Scales MDSCoR (2014) Review of diagnostic instruments for the restless legs syndrome/Willis-Ekbom disease (RLS/WED): critique and recommendations. J Clin Sleep Med 10(12):1343–1349. https://doi.org/10.5664/jcsm.4298

61. Abetz L, Arbuckle R, Allen RP, Mavraki E, Kirsch J (2005) The reliability, validity and responsiveness of the restless legs syndrome quality of life questionnaire (RLSQoL) in a trial population. Health Qual Life Outcomes 3:79. https://doi.org/10.1186/1477-7525-3-79

62. Abetz L, Vallow SM, Kirsch J, Allen RP, Washburn T, Earley CJ (2005) Validation of the restless legs syndrome quality of life questionnaire. Value Health 8(2):157–167. https://doi.org/10.1111/j.1524-4733.2005.03010.x

63. Walters SA, Frauscher B, Allen R, Benes H, Ray Chaudhuri K, Garcia-Borreguero D, Lee H, Picchietti LD, Trenkwalder C, Martinez-Martin P, Schrag A, Stebbins G (2014) Review of severity rating scales for restless legs syndrome: critique and recommendations. Mov Disord 1(4):317–324. https://doi.org/10.1002/mdc3.12088

64. Walters AS, LeBrocq C, Dhar A, Hening W, Rosen R, Allen RP, Trenkwalder C, International Restless Legs Syndrome Study G (2003) Validation of the international restless legs syndrome study group rating scale for restless legs syndrome. Sleep Med 4(2):121–132

65. Garcia-Borreguero D, Kohnen R, Hogl B, Ferini-Strambi L, Hadjigeorgiou GM, Hornyak M, de Weerd AW, Happe S, Stiasny-Kolster K, Gschliesser V, Egatz R, Cabrero B, Frauscher B, Trenkwalder C, Hening WA, Allen RP (2007) Validation of the augmentation severity rating scale (ASRS): a multicentric, prospective study with levodopa on restless legs syndrome. Sleep Med 8(5):455–463. https://doi.org/10.1016/j.sleep.2007.03.023

66. Bolitho SJ, Naismith SL, Rajaratnam SM, Grunstein RR, Hodges JR, Terpening Z, Rogers N, Lewis SJ (2014) Disturbances in melatonin secretion and circadian sleep-wake regulation in Parkinson disease. Sleep Med 15(3):342–347. https://doi.org/10.1016/j.sleep.2013.10.016

67. Breen DP, Vuono R, Nawarathna U, Fisher K, Shneerson JM, Reddy AB, Barker RA (2014) Sleep and circadian rhythm regulation in early Parkinson disease. JAMA Neurol 71(5):589–595. https://doi.org/10.1001/jamaneurol.2014.65

68. Zhong G, Bolitho S, Grunstein R, Naismith SL, Lewis SJ (2013) The relationship between thermoregulation and REM sleep behaviour disorder in Parkinson's disease. PLoS One

8(8):e72661. https://doi.org/10.1371/journal.pone.0072661

69. Paus S, Schmitz-Hubsch T, Wullner U, Vogel A, Klockgether T, Abele M (2007) Bright light therapy in Parkinson's disease: a pilot study. Mov Disord 22(10):1495–1498. https://doi.org/10.1002/mds.21542

70. Willis GL, Moore C, Armstrong SM (2012) A historical justification for and retrospective analysis of the systematic application of light therapy in Parkinson's disease. Rev Neurosci 23(2):199–226. https://doi.org/10.1515/revneuro-2011-0072

71. Willis GL, Turner EJ (2007) Primary and secondary features of Parkinson's disease improve with strategic exposure to bright light: a case series study. Chronobiol Int 24(3):521–537. https://doi.org/10.1080/07420520701420717

72. Videnovic A, Klerman EB, Wang W, Marconi A, Kuhta T, Zee PC (2017) Timed light therapy for sleep and daytime sleepiness associated with Parkinson disease: a randomized clinical trial. JAMA Neurol 74(4):411–418. https://doi.org/10.1001/jamaneurol.2016.5192

73. Martinez-Martin P, Schapira AH, Stocchi F, Sethi K, Odin P, MacPhee G, Brown RG, Naidu Y, Clayton L, Abe K, Tsuboi Y, MacMahon D, Barone P, Rabey M, Bonuccelli U, Forbes A, Breen K, Tluk S, Olanow CW, Thomas S, Rye D, Hand A, Williams AJ, Ondo W, Chaudhuri KR (2007) Prevalence of nonmotor symptoms in Parkinson's disease in an international setting; study using nonmotor symptoms questionnaire in 545 patients. Mov Disord 22(11):1623–1629. https://doi.org/10.1002/mds.21586

74. Winge K, Skau AM, Stimpel H, Nielsen KK, Werdelin L (2006) Prevalence of bladder dysfunction in Parkinsons disease. Neurourol Urodyn 25(2):116–122. https://doi.org/10.1002/nau.20193

75. Breen KC, Drutyte G (2013) Non-motor symptoms of Parkinson's disease: the patient's perspective. J Neural Transm (Vienna) 120(4):531–535. https://doi.org/10.1007/s00702-012-0928-2

76. Cheon SM, Ha MS, Park MJ, Kim JW (2008) Nonmotor symptoms of Parkinson's disease: prevalence and awareness of patients and families. Parkinsonism Relat Disord 14(4):286–290. https://doi.org/10.1016/j.parkreldis.2007.09.002

77. Rana AQ, Vaid H, Akhter MR, Awan NY, Fattah A, Cader MH, Hafez K, Rana MA, Yousuf MS (2014) Prevalence of nocturia in Parkinson's disease patients from various ethnicities. Neurol Res 36(3):234–238. https://doi.org/10.1179/1743132813Y.0000000264

78. Spica V, Pekmezovic T, Svetel M, Kostic VS (2013) Prevalence of non-motor symptoms in young-onset versus late-onset Parkinson's disease. J Neurol 260(1):131–137. https://doi.org/10.1007/s00415-012-6600-9

79. Xue P, Wang T, Zong H, Zhang Y (2014) Urodynamic analysis and treatment of male Parkinson's disease patients with voiding dysfunction. Chin Med J 127(5):878–881

80. Picillo M, Erro R, Amboni M, Longo K, Vitale C, Moccia M, Pierro A, Scannapieco S, Santangelo G, Spina E, Orefice G, Barone P, Pellecchia MT (2014) Gender differences in non-motor symptoms in early Parkinson's disease: a 2-years follow-up study on previously untreated patients. Parkinsonism Relat Disord 20(8):850–854. https://doi.org/10.1016/j.parkreldis.2014.04.023

81. Vaughan CP, Juncos JL, Trotti LM, Johnson TM 2nd, Bliwise DL (2013) Nocturia and overnight polysomnography in Parkinson disease. Neurourol Urodyn 32(8):1080–1085. https://doi.org/10.1002/nau.22365

82. Trenkwalder C, Kohnen R, Hogl B, Metta V, Sixel-Doring F, Frauscher B, Hulsmann J, Martinez-Martin P, Chaudhuri KR (2011) Parkinson's Disease Sleep Scale—validation of the revised version PDSS-2. Mov Disord 26(4):644–652. https://doi.org/10.1002/mds.23476

83. Bing MH, Moller LA, Jennum P, Mortensen S, Lose G (2006) Validity and reliability of a questionnaire for evaluating nocturia, nocturnal enuresis and sleep-interruptions in an elderly population. Eur Urol 49(4):710–719. https://doi.org/10.1016/j.eururo.2005.11.034

84. Homma Y, Yoshida M, Seki N, Yokoyama O, Kakizaki H, Gotoh M, Yamanishi T, Yamaguchi O, Takeda M, Nishizawa O (2006) Symptom assessment tool for overactive bladder syndrome—overactive bladder symptom score. Urology 68(2):318–323. https://doi.org/10.1016/j.urology.2006.02.042

85. Romenets SR, Wolfson C, Galatas C, Pelletier A, Altman R, Wadup L, Postuma RB (2012) Validation of the non-motor symptoms questionnaire (NMS-quest). Parkinsonism Relat Disord 18(1):54–58. https://doi.org/10.1016/j.parkreldis.2011.08.013

86. Kuno S, Mizuta E, Yamasaki S, Araki I (2004) Effects of pergolide on nocturia in Parkinson's disease: three female cases selected from over 400 patients. Parkinsonism Relat Disord 10(3):181–187. https://doi.org/10.1016/j.parkreldis.2003.08.001

87. Monteiro L, Souza-Machado A, Valderramas S, Melo A (2012) The effect of levodopa on pulmonary function in Parkinson's disease: a systematic review and meta-analysis.

Clin Ther 34(5):1049–1055. https://doi.org/10.1016/j.clinthera.2012.03.001

88. Sabate M, Rodriguez M, Mendez E, Enriquez E, Gonzalez I (1996) Obstructive and restrictive pulmonary dysfunction increases disability in Parkinson disease. Arch Phys Med Rehabil 77(1):29–34

89. Seccombe LM, Giddings HL, Rogers PG, Corbett AJ, Hayes MW, Peters MJ, Veitch EM (2011) Abnormal ventilatory control in Parkinson's disease—further evidence for non-motor dysfunction. Respir Physiol Neurobiol 179(2–3):300–304. https://doi.org/10.1016/j.resp.2011.09.012

90. Sabate M, Gonzalez I, Ruperez F, Rodriguez M (1996) Obstructive and restrictive pulmonary dysfunctions in Parkinson's disease. J Neurol Sci 138(1–2):114–119

91. Trotti LM, Bliwise DL (2010) No increased risk of obstructive sleep apnea in Parkinson's disease. Mov Disord 25(13):2246–2249. https://doi.org/10.1002/mds.23231

92. Chung F, Yegneswaran B, Liao P, Chung SA, Vairavanathan S, Islam S, Khajehdehi A, Shapiro CM (2008) STOP questionnaire: a tool to screen patients for obstructive sleep apnea. Anesthesiology 108(5):812–821. https://doi.org/10.1097/ALN.0b013e31816d83e4

93. Netzer NC, Stoohs RA, Netzer CM, Clark K, Strohl KP (1999) Using the Berlin questionnaire to identify patients at risk for the sleep apnea syndrome. Ann Intern Med 131(7):485–491

94. Terzaghi M, Spelta L, Minafra B, Rustioni V, Zangaglia R, Pacchetti C, Manni R (2017) Treating sleep apnea in Parkinson's disease with C-PAP: feasibility concerns and effects on cognition and alertness. Sleep Med 33:114–118. https://doi.org/10.1016/j.sleep.2017.01.009

95. Harmell AL, Neikrug AB, Palmer BW, Avanzino JA, Liu L, Maglione JE, Natarajan L, Corey-Bloom J, Loredo JS, Ancoli-Israel S (2016) Obstructive sleep apnea and cognition in Parkinson's disease. Sleep Med 21:28–34. https://doi.org/10.1016/j.sleep.2016.01.001

96. Gros P, Mery VP, Lafontaine AL, Robinson A, Benedetti A, Kimoff RJ, Kaminska M (2016) Obstructive sleep apnea in Parkinson's disease patients: effect of Sinemet CR taken at bedtime. Sleep Breath 20(1):205–212. https://doi.org/10.1007/s11325-015-1208-9

97. Buskova J, Klempir J, Majerova V, Picmausova J, Sonka K, Jech R, Roth J, Ruzicka E (2011) Sleep disturbances in untreated Parkinson's disease. J Neurol 258(12):2254–2259. https://doi.org/10.1007/s00415-011-6109-7

98. Dhawan V, Dhoat S, Williams AJ, Dimarco A, Pal S, Forbes A, Tobias A, Martinez-Martin P, Chaudhuri KR (2006) The range and nature of sleep dysfunction in untreated Parkinson's disease (PD). A comparative controlled clinical study using the Parkinson's disease sleep scale and selective polysomnography. J Neurol Sci 248(1–2):158–162. https://doi.org/10.1016/j.jns.2006.05.004

99. Wienecke M, Werth E, Poryazova R, Baumann-Vogel H, Bassetti CL, Weller M, Waldvogel D, Storch A, Baumann CR (2012) Progressive dopamine and hypocretin deficiencies in Parkinson's disease: is there an impact on sleep and wakefulness? J Sleep Res 21(6):710–717. https://doi.org/10.1111/j.1365-2869.2012.01027.x

100. Simuni T, Caspell-Garcia C, Coffey C, Chahine LM, Lasch S, Oertel WH, Mayer G, Hogl B, Postuma R, Videnovic A, Amara AW, Marek K, Investigators PSWgobotP (2015) Correlates of excessive daytime sleepiness in de novo Parkinson's disease: a case control study. Mov Disord 30(10):1371–1381. https://doi.org/10.1002/mds.26248

101. Borek LL, Kohn R, Friedman JH (2006) Mood and sleep in Parkinson's disease. J Clin Psychiatry 67(6):958–963

102. Stavitsky K, Neargarder S, Bogdanova Y, McNamara P, Cronin-Golomb A (2012) The impact of sleep quality on cognitive functioning in Parkinson's disease. J Int Neuropsychol Soc 18(1):108–117. https://doi.org/10.1017/S1355617711001482

103. Havlikova E, van Dijk JP, Nagyova I, Rosenberger J, Middel B, Dubayova T, Gdovinova Z, Groothoff JW (2011) The impact of sleep and mood disorders on quality of life in Parkinson's disease patients. J Neurol 258(12):2222–2229. https://doi.org/10.1007/s00415-011-6098-6

104. Ozdilek B, Gunal DI (2012) Motor and non-motor symptoms in Turkish patients with Parkinson's disease affecting family caregiver burden and quality of life. J Neuropsychiatry Clin Neurosci 24(4):478–483. https://doi.org/10.1176/appi.neuropsych.11100315

105. Naismith SL, Terpening Z, Shine JM, Lewis SJ (2011) Neuropsychological functioning in Parkinson's disease: differential relationships with self-reported sleep-wake disturbances. Mov Disord 26(8):1537–1541. https://doi.org/10.1002/mds.23640

106. Uc EY, Rizzo M, Anderson SW, Sparks JD, Rodnitzky RL, Dawson JD (2006) Driving with distraction in Parkinson disease. Neurology 67(10):1774–1780. https://doi.org/10.1212/01.wnl.0000245086.32787.61

107. Hagell P, Broman JE (2007) Measurement properties and hierarchical item structure of the Epworth sleepiness scale in Parkinson's disease. J Sleep Res 16(1):102–109. https://doi.org/10.1111/j.1365-2869.2007.00570.x

108. Goetz CG, Tilley BC, Shaftman SR, Stebbins GT, Fahn S, Martinez-Martin P, Poewe W, Sampaio C, Stern MB, Dodel R, Dubois B, Holloway R, Jankovic J, Kulisevsky J, Lang AE, Lees A, Leurgans S, LeWitt PA, Nyenhuis D, Olanow CW, Rascol O, Schrag A, Teresi JA, van Hilten JJ, LaPelle N, Movement Disorder Society URTF (2008) Movement Disorder Society-sponsored revision of the unified Parkinson's disease rating scale (MDS-UPDRS): scale presentation and clinimetric testing results. Mov Disord 23(15):2129–2170. https://doi.org/10.1002/mds.22340

109. Horvath K, Aschermann Z, Acs P, Bosnyak E, Deli G, Pal E, Janszky J, Faludi B, Kesmarki I, Komoly S, Bokor M, Rigo E, Lajtos J, Klivenyi P, Dibo G, Vecsei L, Takats A, Toth A, Imre P, Nagy F, Herceg M, Kamondi A, Hidasi E, Kovacs N (2014) Is the MDS-UPDRS a good screening tool for detecting sleep problems and daytime sleepiness in Parkinson's disease? Parkinsons Dis 2014:806169. https://doi.org/10.1155/2014/806169

110. Littner MR, Kushida C, Wise M, Davila DG, Morgenthaler T, Lee-Chiong T, Hirshkowitz M, Daniel LL, Bailey D, Berry RB, Kapen S, Kramer M, Standards of Practice Committee of the American Academy of Sleep M (2005) Practice parameters for clinical use of the multiple sleep latency test and the maintenance of wakefulness test. Sleep 28(1):113–121

111. Kotschet K, Johnson W, McGregor S, Kettlewell J, Kyoong A, O'Driscoll DM, Turton AR, Griffiths RI, Horne MK (2014) Daytime sleep in Parkinson's disease measured by episodes of immobility. Parkinsonism Relat Disord 20(6):578–583. https://doi.org/10.1016/j.parkreldis.2014.02.011

112. Bolitho SJ, Naismith SL, Salahuddin P, Terpening Z, Grunstein RR, Lewis SJ (2013) Objective measurement of daytime napping, cognitive dysfunction and subjective sleepiness in Parkinson's disease. PLoS One 8(11):e81233. https://doi.org/10.1371/journal.pone.0081233

113. Ondo WG, Shinawi L, Davidson A, Lai D (2011) Memantine for non-motor features of Parkinson's disease: a double-blind placebo controlled exploratory pilot trial. Parkinsonism Relat Disord 17(3):156–159. https://doi.org/10.1016/j.parkreldis.2010.12.003

114. Ondo WG, Fayle R, Atassi F, Jankovic J (2005) Modafinil for daytime somnolence in Parkinson's disease: double blind, placebo controlled parallel trial. J Neurol Neurosurg Psychiatry 76(12):1636–1639. https://doi.org/10.1136/jnnp.2005.065870

115. Rodrigues TM, Castro Caldas A, Ferreira JJ (2016) Pharmacological interventions for daytime sleepiness and sleep disorders in Parkinson's disease: systematic review and meta-analysis. Parkinsonism Relat Disord 27:25–34. https://doi.org/10.1016/j.parkreldis.2016.03.002

116. Adler CH, Caviness JN, Hentz JG, Lind M, Tiede J (2003) Randomized trial of modafinil for treating subjective daytime sleepiness in patients with Parkinson's disease. Mov Disord 18(3):287–293. https://doi.org/10.1002/mds.10390

117. Hogl B, Saletu M, Brandauer E, Glatzl S, Frauscher B, Seppi K, Ulmer H, Wenning G, Poewe W (2002) Modafinil for the treatment of daytime sleepiness in Parkinson's disease: a double-blind, randomized, crossover, placebo-controlled polygraphic trial. Sleep 25(8):905–909

118. Dowling GA, Mastick J, Colling E, Carter JH, Singer CM, Aminoff MJ (2005) Melatonin for sleep disturbances in Parkinson's disease. Sleep Med 6(5):459–466. https://doi.org/10.1016/j.sleep.2005.04.004

119. Weintraub D, Mavandadi S, Mamikonyan E, Siderowf AD, Duda JE, Hurtig HI, Colcher A, Horn SS, Nazem S, Ten Have TR, Stern MB (2010) Atomoxetine for depression and other neuropsychiatric symptoms in Parkinson disease. Neurology 75(5):448–455. https://doi.org/10.1212/WNL.0b013e3181ebdd79

120. Ondo WG, Perkins T, Swick T, Hull KL Jr, Jimenez JE, Garris TS, Pardi D (2008) Sodium oxybate for excessive daytime sleepiness in Parkinson disease: an open-label polysomnographic study. Arch Neurol 65(10):1337–1340. https://doi.org/10.1001/archneur.65.10.1337

121. Buchele F, Hackius M, Schreglmann SR, Omlor W, Werth E, Maric A, Imbach LL, Hagele-Link S, Waldvogel D, Baumann CR (2017) Sodium oxybate for excessive daytime sleepiness and sleep disturbance in Parkinson disease: a randomized clinical trial. JAMA Neurol 75:114. https://doi.org/10.1001/jamaneurol.2017.3171

122. Postuma RB, Lang AE, Munhoz RP, Charland K, Pelletier A, Moscovich M, Filla L, Zanatta D, Rios Romenets S, Altman R, Chuang R, Shah B (2012) Caffeine for treatment of Parkinson disease: a randomized controlled trial. Neurology 79(7):651–658. https://doi.org/10.1212/WNL.0b013e318263570d

Chapter 12

Clinical Trials on Management of Pain in Parkinson's Disease

Azman Aris, Katarina Rukavina, Raquel Taddei, Alexandra Rizos, Anna Sauerbier, and K. Ray Chaudhuri

Abstract

Pain is highly prevalent in Parkinson's disease (PD). It is one of the most common nonmotor symptoms of PD, and it may precede motor symptoms by several years. There are various types of pain described in PD, reflecting its heterogeneity and multifactorial origin. Pain in PD has been reported to differ between gender, early and advanced stages of PD, motor and nonmotor subtypes, and to fluctuate during *on* and *off* states. For the management of pain, it is crucial to separate PD-related pain from pain of other origins. Pain has a negative impact on the overall quality of life in PD patients. Early recognition and holistic assessment of pain are therefore important for personalized medicine delivery.

When conducting a clinical trial in pain, study methodology and design need to be tailored to the multifaceted features of PD pain. The used assessment tools ideally have to be specific and sensitive to these features.

In this book chapter, we discuss and outline the evaluation of PD-related pain, review PD general and specific pain assessment tools, and discuss relevant clinical trials addressing treatment of pain in PD.

Key words Pain, Parkinson's disease, Clinical trials, Treatment, Nociception

1 Introduction

Parkinson's disease (PD) is a complex neurodegenerative disorder, traditionally recognized by its cardinal motor symptoms of bradykinesia, resting tremor, rigidity, and postural instability. Equally burdensome are the nonmotor symptoms (NMS), ranging from cognitive, neuropsychiatric, somatosensory, and autonomic domains, which have now been increasingly identified and have become of focus in PD care.

In the early description of PD, James Parkinson reported "rheumatic pain extending from the arms to the fingers" in his patients [1]. It is now recognized that pain is among the first NMS

Santiago Perez-Lloret (ed.), *Clinical Trials In Parkinson's Disease*, Neuromethods, vol. 160,
https://doi.org/10.1007/978-1-0716-0912-5_12, © Springer Science+Business Media, LLC, part of Springer Nature 2021

together with hyposmia and sleep disturbances [2]. Along with depression and anxiety, pain has been shown to adversely affect quality of life (QoL) [3–5] and has been regarded as the one of the most bothersome NMS of PD [6]. It is one of the main causes of hospitalization and institutionalization among PD patients and may increase financial burden by quadruple [7]. Despite its major impact on patients' QoL and cost of healthcare, PD pain is still not routinely addressed in daily clinical practice. A previous survey showed that more than 40% of patients did not voluntarily declare this to their clinicians [8]. This could be due to patients' unawareness that painful symptoms could be related to PD [9], as well as possible lack of understanding of the treating clinicians to the nature of PD-related pain itself.

2 Pain in Parkinson's Disease

2.1 Prevalence

Epidemiological studies utilizing different methodologies reported the prevalence of pain in PD to range between 30% and 85% [5, 10–12]. The DoPaMiP (*Douleur et maladie de Parkinson en Midi-Pyrenees*) survey in southwest France, examining the prevalence of chronic pain in 450 patients with PD, found that 62% of PD patients had at least one form of chronic pain, twice as high as the age and sex-matched patients who had other chronic conditions [13]. The high prevalence of PD pain is postulated due to heighten sensitivity to pain among PD patients as compared to healthy individuals [14]. The key characteristics of a few controlled and uncontrolled observational studies on the prevalence of pain in PD are shown in Table 1. A recent meta-analysis involving 26 studies on pain perception in PD, has revealed that PD patients had lower pain threshold during unmedicated *off* states which was attenuated during dopamine-medicated *on* states, irrespective of age, PD duration, or PD severity [37]. This finding at least partly might explain the presence of hyperalgesia in PD and implicates dopamine deficiency as a potential underlying mechanism. There are numerous factors that contribute to this noticeable discrepancy in the prevalence rate of pain in PD reported among the studies. These include variation in description of PD pain, variation in studied patient population, lack of standardized or validated pain measurement tools, and differences in methods of data collection and study design [11, 12]. While majority of studies assessed all types of PD pain, some opted to explore only specific types of pain (e.g., shoulder pain, back pain, or dystonia) [15, 18]. Furthermore, the different studies utilized different time frame when assessing pain ranging from 1 week to 1 month [21]. In addition, some reported pain prevalence as part of a pilot study for a screening questionnaire on NMS of which pain was not the primary outcome measured [19]. A standardized definition of PD-related pain, spec-

Table 1
Key characteristics of controlled and uncontrolled observational studies on the prevalence of pain in PD

Study type	Year	Study	Study design	Pain classification	Sample size	Prevalence (%)
Controlled studies	2010	Madden and Hall [15]	Cross-sectional retrospective study	Shoulder pain	25	80.0 PD vs. 40.0 controls
	2009	Ehrt et al. [16]	Cross-sectional cohort study	All types of pain	227	67.0 PD vs. 39.0 controls
	2008	Defazio et al. [17]	Case–control study	All types of pain	402	69.9 PD vs. 62.8 controls
	2008	Negre-Pages et al. [13]	Cross-sectional survey	All types of pain; PD-related, PD-unrelated pain	450	61.8 PD chronic pain
	2007	Broetz et al. [18]	Case–control study	Back pain	101	74.0 PD vs. 27.0 controls
	2006	Chaudhuri et al. [19]	Cross-sectional study	All types of pain	123	27.6 PD vs. 30.2 controls
	2006	Etchepare et al. [20]	Cross-sectional study	Back pain	104	59.6 PD vs. 23.0 controls
	2004	Quittenbaum and Grahn [21]	Cross-sectional study	All types of pain	57	68.4 PD vs. 52.6 controls
	1976	Snider et al. [22]	Cross-sectional study	Sensory symptoms	101	43.0 PD vs. 8.0 controls

(continued)

Table 1
(continued)

Study type	Year	Study	Study design	Pain classification	Sample size	Prevalence (%)
Open studies	2016	Ozturk et al. [23]	Observational cohort study	Chronic pain	113	64.6
	2013	Rana et al. [24]	Cross-sectional study	All types of pain	121	66.0
	2012	Wen et al. [25]	Cross-sectional study	Unexplained pain	901	29.9
	2011	Hanagasi et al. [26]	Prospective study	All types of pain	96	64.9
	2011	Santos-Garcia et al. [27]	Cross-sectional study	All types of pain	159	72.3
	2010	Chaudhuri et al. [8]	Cross-sectional study	Any pain (unrelated to other causes)	242	45.9
	2009	Beiske et al. [10]	Cross-sectional study	All types of pain	176	83.0
	2008	Silva EG, Viana MA, Quagliato EM [28]	Prospective study	All types of pain	50	56.0
	2008	Stamey et al. [29]	Retrospective analysis	Shoulder pain	309	35.0
	2007	Martinez-Martin et al. [30]	Cross-sectional study	Any pain (unrelated to other causes)	545	28.8
	2007	Sullivan et al. [31]	Cross-sectional survey	All types of pain	100	35.0
	2006	Lee et al. [32]	Cross-sectional study	PD-related, PD-unrelated pain	123	68.4
	2006	Tinazzi et al. [33]	Case–control study	Any pain	117	40.0
	1998	Clifford et al. [34]	Cross-sectional study	Burning mouth	115	25.0
	1986	Goetz et al. [35]	Cross-sectional study	All types of pain	95	46.0
Reviews	2013	Rana et al. [36]	Literature review	All types of pain	15,636	59.8
	2011	Broen et al. [11]	Systemic review	All types of pain	n.a.	67.6

n.a. not available

ified time frame of measurement, utilizing internationally standardized validated tools within a defined PD study population, and ideally with multicenter involvement are essential requirements to capture the true incidence and prevalence rates of pain in PD.

2.2 Mechanisms of Pain

Pain is defined by IASP (*International Association for the Study of Pain*) as unpleasant sensory sensation or emotional experience that typically resulted from underlying actual or potential tissue injuries. Pain can present when no actual harm is being done to the body. It can either be acute or chronic in nature.

Dopamine is known to influence pain modulation within the nervous system, however the exact pathomechanisms remain elusive as the relationship of dopaminergic status and pain is complex, encompassing numerous central, spinal, and peripheral pain pathways as well as various other neurotransmitters and peptides [38, 39]. Several neurophysiological studies have suggested that there are lower pain thresholds in PD patients [40–42], rendering them more susceptible to pain. However, these data are contradicting as some studies have reported that PD patients have a higher pain threshold instead [43, 44]. Other mechanisms have also been proposed including abnormal nociceptive flexion reflex and abnormal laser evoked potential which contribute to increased pain perception in PD patients [45–48]. Furthermore, there are anatomical evidence suggesting epidermal fibre loss on skin biopsies of PD patients that cause peripheral deafferentiation [44], as well as deposition of alpha-synuclein inclusions in the spinal cord lamina I neurons that contribute to abnormal peripheral pain processing and increased pain susceptibility in PD [49].

2.3 Characteristics

As aforementioned, separating PD pain and pain of other origin is of great challenge as the two often overlap and synergistically influence each other. A recent study examining the association between pain and motor and NMS in PD has reported that pain in PD is more frequent in subjects with medical conditions predisposing to painful symptoms [50]. Allen et al. reported that pain is experienced at least once in a month in most patients with PD and pain interferes with daily activities [51]. Moreover, a higher pain frequency and pain that interferes with work particularly rigidity, are found to associate with functional impairments in PD. As pain can be influenced by various factors such as PD motor symptoms, its treatment and medication side effects, plus considering the high prevalence of chronic pain in the general population itself, it is important to determine whether the pain is related to PD or not. The timing of occurrence and the patient's motor and nonmotor status may provide evidence for an association with PD. A substantial proportion of PD pain has been shown to be related to motor fluctuations and dyskinesia secondary to dopaminergic therapy

[52–54]. In general, pain that responds to dopaminergic therapy or worsens with motor fluctuation is likely to be PD in origin.

Pain can potentially present early in PD, from several years prior to the clinical onset of motor symptom to the palliative stage of the condition [55–57]. Based on a clinicopathological study, premotor pain has been reported to be the first presenting complain of PD in up to 15% of patients [58]. Previous studies have reported that patients with PD pain were younger than those without pain [13, 35]. However, more recent studies did not find age at disease onset a predictor of pain in PD [32, 50]. In comparison to men, women tend to experience more PD pain [5, 10, 50, 59], which is possibly due to lower pain threshold among women and the difference in pain coping strategy [60, 61]. However, some studies failed to report similar finding [13, 24].While risk factors associated with pain have shown controversial results with uncertainty on the negative influence of female gender and lower mean age on the development of pain in PD, the strongest associated main risk factor to date is the presence of motor fluctuation especially the wearing-*off* phenomenon, and the severity of the motor symptoms [62], suggesting a strong link between motor and nonmotor phenomenon with this respect. PD pain also has been shown to worsen in advanced stages of PD in parallel to depleted dopamine reserve. In addition, at this later stage, PD is more often than not complicated with other debilitating issues such as more prominent and severe motor symptoms, worsening cognitive impairment, skeletal deformity and other medical comorbidities [50]. PD-related pain has also been shown to cause increased anxiety and worsened sleep quality [63]. An association between the commonly arising depression in PD patients and pain, however, has so far failed to be proven [62]. The key characteristics of PD pain are listed in Table 2.

2.4 Classification of PD-Related Pain

Despite of many attempts, to date there is no absolute consensus on the classification of PD-related pain. Firstly, the Ford classification is widely used for its easiness and simplicity which categorizes PD pain into five main subtypes: musculoskeletal, dystonic, neuro-

Table 2
Key features of PD-related pain

- The majority show improvement with dopaminergic therapy
- More prevalent among females and also more common in late stage of PD
- Often worse on the side initially or more severely affected by PD motor symptoms
- Might fluctuates during *on* and *off* states
- Treatable and treatment of pain improves QoL

PD Parkinson's disease, *QoL* quality of life

pathic/radicular, central/primary, and akathisia-related pain [64] as listed in Table 3.

Secondly, a significant proportion of PD-related pain is caused by underlying motor fluctuations and dyskinesia secondary to dopaminergic therapies, and there have been several attempts to characterize PD pain centered on this concept [7, 53, 65]. The Nonmotor Fluctuations in PD (NoMoFlu-PD) study explored the burden and correlation of NMS with motor complications in PD

Table 3
Ford classification of PD-related pain [64]

Category	Characteristics
Musculoskeletal	• Muscle and/or joint pain, inflammation, bone deformity, reduced joint mobility, and abnormal posture • Associated with muscle rigidity and can improve with levodopa therapy
Dystonic	• Associated with abnormal postures and can improve with levodopa therapy
Neuropathic/radicular	• Peripheral neuropathic pain: restricted to the territory of the affected nerve or nerve root
Central or primary	• Neuropathic pain that is not restricted to the affected nerve or nerve root • Varies with the medication cycle as a nonmotor fluctuation • Pain may have an autonomic character, with visceral sensations • Not associated with rigidity, dystonia, or musculoskeletal or structural lesions
Akathisia	• Subjective sensation of restlessness and an inability to remain still • Can vary with medication and improve with levodopa therapy

PD Parkinson's disease

and reported that almost all NMS including pain, fluctuated with motor oscillations and were more frequent and severe in *off* compared to *on* state [53]. In this context, PD pain was indirectly categorized as *on*-pain, *on-off* related pain and *off*-related pain. Pain was noted to confine to *off* states or increased during *off* in most patient with painful conditions, and these pain fluctuations were also found to be associated with poor health-related QoL.

Thirdly, Sauerbier et al. have recently proposed that there are seven distinct nonmotor dominant phenotypes within PD, one being the Park-pain subtype [66]. This subtype expresses various pain syndromes that dominate the clinical picture, such as the previously described by Quinn et al. as "painful Parkinson's disease" phenotype [67] and by Wallace and Chaudhuri who described an unexplained lower limb syndrome in PD [68]. Patients with the Park Pain subtype are at higher risk to develop disproportionate pain during the progression of PD compared to the motor severity of disease.

Furthermore, pain in PD can be classified into nociceptive and neuropathic pain. The nociceptive PD pain is related to motor state and typically worsen during *off* time. These include painful *off* dystonia, diphasic and peak-dose dyskinesia, *off*-related rigidity, day or night cramps, and postural-related pain such as *Pisa syndrome* and camptocormia. On another spectrum, neuropathic PD pain refers to unpleasant sensory experience such as aching, numbness, burning, stabbing, and paresthesia that is often localized to deep abdominal viscera, upper thigh, oral, and genital regions [69]. Based on this classification, in addition to acute and chronic concept of pain, as well as to its distribution (*focal, regional, and radicular*) and response to dopaminergic therapy (*dopaminergic or nondopaminergic related*), Chaudhuri and Schapira proposed another clinically relevant overall classification of pain in PD that includes musculoskeletal pain, PD-related chronic pain, fluctuation-related pain, nocturnal pain, coat-hanger pain, orofacial pain, and peripheral limb or abdominal pain [7].

In addition, Wasner and Deuschl proposed a four-tier taxonomy characterization to address the complexity of PD pain [65]. Pain is classified firstly by recognizing it as either or not directly related to PD. Pain that is related to PD is then further subcategorized into nociceptive (*musculoskeletal, visceral, or cutaneous in origin*), neuropathic (*peripheral or central*) and miscellaneous (e.g., *preceding PD, linked to restless leg syndrome, akathisia, depression-linked pain*). All other kinds of pain are categorized as unrelated to PD, for example pain caused by other pain syndrome such as osteoarthritis.

2.5 Assessment of PD-Related Pain

In order to holistically assess pain in PD and ensure that important details are not missed, it is recommended to apply structured and focused validated tools. This is not only important in clinical practice but also in the context of a clinical trials in which applying a standardized and validated assessment tool is one of the keys in achieving high quality data and reliable end results. The International Parkinson and Movement Disorders Society (IPMDS) took the initiative to commission a writing committee, consisting of experts in the field of movement disorder, pain, and clinimetric assessment from Europe and America to critically appraise rating scales for pain in PD [70]. The committee systematically reviewed tools that have had been used in PD, focusing on those that evaluate pain intensity or allow for pain syndromic classification. Eleven tools met the committee's preset criteria, and the dimensions of the tools and the recommendation of their usage in PD are shown in Table 4.

In the following section, we describe a few of these tools that are commonly used in PD including the Brief Pain Inventory (BPI), the McGill Pain Questionnaire (MPQ), the Visual Analog Scale (VAS), the Numeric Rating Scale (NRS), and the King's Parkinson Pain Scale (KPPS).

The BPI is a self-administered questionnaire that allows patients to rate the severity of their pain and the degree of its interference with their daily activities [72]. It has two forms, a short form and a long form. The BPI is composed of series of items on a 11-point scale (0–10) with four questions focus on pain intensity (worst, least, average, and current), seven questions address the interference of pain with general life activities and the remaining seven questions deal with the interference of pain with specific activities. While the short form address the past 24 h the long form focus on the past week. The time to complete the scale is estimated between 5 and 10 min, respectively. In contrast to the short-form BPI, the long-form BPI includes questions on demographics (date of birth, marital status, education, employment), pain history, aggravating and relieving factors, treatment and medication, pain quality, and response to treatment. The BPI considers sensory, emotional, and functional aspects of the pain experience, making it sensitive to changes in pain in relation to pharmacological, physical, and psychological interventions. While the long-form BPI is suitable to be used as baseline measure as it covers broader aspects of pain, the simplicity of the short-form BPI and shorter timing to complete makes it more favorable in situations where pain is assessed on a daily basis such as in a clinical trial. The scale was considered adequately reliable based on good internal consistency and test–retest reliability (Cronbach alpha reliability ranges from 0.77 to 0.91) [82, 83]. Initially designed to assess pain in cancer patients, the BPI has been validated in many languages and has now been widely employed in a wide range of medical conditions including osteoarthritis, postoperative pain, low-back pain, and neuropathy [83–86]. This scale has been utilized, despite not being validated, in PD studies [10, 87–89].

Table 4

Pain rating scales and questionnaires that are commonly used to assess pain in Parkinson's disease [70]

Scale/questionnaire		Method of administration	Estimated time for administration (min)	Time period assessed	Intensity	Frequency	Type of pain	Validated for PD	IPMDS recommendation [70]	Comments
Specific scale for PD pain	KPPS [71]	HCP	20	Past 1 month	Y	Y	All types	Y	Recommended for pain intensity rating, suggested for syndromic aspects	Allow for quantification of intensity and frequency of pain
General pain scales used in PD	BPI short form [72]	Self	5–10	Past 24 h	Y	N	All types	N	Recommended with caution	Allow for quantification of pain intensity and pain interference in daily activities
	MPQ long form [73]	Self	15–30	n/s	Y	N	All types	N	Recommended with caution	Allow for quantification of sensory-discriminative, affective, and evaluative aspects of pain
	MPQ short form [74]	Self	5–10	n/s	Y	N	All types	N	Recommended with caution	Allow for quantification of sensory-discriminative, affective, and evaluative aspects of pain
	VAS [75, 76]	Self	0–5	n/s	Y	N	All types	N	Recommended with caution	Inappropriate to assess intermittent pain
	NRS [75]	Self	0–5	n/s	Y	N	All types	N	Recommended with caution	Inappropriate to assess intermittent pain
	LANSS [77]	Self and HCP	5–10	Past 1 week	N	N	Neuropathic	N	Suggested for syndromic aspects	
	NPSI [78]	Self	5–10	Past 24 h	Y	Y	Neuropathic	N	Recommended with caution	Enables the characterization of clusters of symptoms
	Pain DETECT [79]	Self/HCP	0–5	Present time	Y	N	Neuropathic / nociceptive	N	Suggested for syndromic aspects	
	Pain-o-meter [80]	Self	5	n/s	Y	N	Sensory/affective components	N	Recommended with caution	
	DN4 [81]	HCP	5–10	n/s	N	N	Neuropathic	N	Recommended with caution	

PD Parkinson's disease, HCP health care professional, n/s not specified, N no, IPMDS International Parkinson and Movement Disorder Society, KPPS King's Parkinson's Pain Scale, BPI Brief Pain Inventory, MPQ McGill Pain Questionnaire, VAS Visual Analog Scale, LANSS Leeds Assessment of Neuropathic Symptoms and Signs, NPSI

The MPQ is often referred to as the gold standard for pain measurement. This self-administered questionnaire can be used to measure different qualities of the subjective pain experience [73]. It consists of four main sections, with the first section showing a body map used for localizing pain. The second section comprises classes of words (total of 78), divided into 20 subitems that describe the sensory (item 1–10), affective (item 11–15), evaluative (item 16), and miscellaneous (item 17–20) aspects of pain. The third section evaluates changes over time and the fourth section characterizes pain intensity using a five-point scale. This questionnaire requires the respondent to choose one word from each subitem that fits their present pain. The interviewer is obliged to define any words that the respondent does not understand. The estimated time to complete the MPQ varies between 5 and 20 min, depending on the short (contains only 15 words) or long version [74, 90]. There is no specified time frame for assessment. Although it is widely available in many languages, they may not exactly match the words on the original English version; therefore, caution is advised in comparing populations of different languages and cultural groups [70, 90]. In addition, the MPQ requires a good understanding of some vocabulary, which may not be appropriate for low literacy respondents.

The VAS is one of most frequently used scales to measure pain severity in various medical conditions including PD [70]. The VAS consists of a linear line of 100 mm in length that is usually anchored with descriptors of "no pain" and "worst pain" at each end, respectively [75]. The respondent is asked to mark on the line that best indicates his or her pain. The Numeric Rating Scale (NRS) is another version of VAS that has 11-points (0 = "no pain" to 10 = "worst pain possible"), where respondent can choose which number best describes his or her pain intensity [75]. Several VAS or NRS can be combined to form a Likert type pain scale. The time frame is flexible and can be set accordingly, ranging from the present time to the previous week or month. These scales can be completed in less than a minute, provided the respondents understand the concept of marking crosses along a linear line which can be a challenge in a PD patient with severe motor problem. However, the VAS and NRS, may be inappropriate to assess intermittent pain, such as pain that worsens during *off* period.

Although the above described questionnaires and scales can be applied when measuring PD-related pain in general, they cannot distinguish between different pain syndromes encountered in PD and do not elucidate PD-related pain classification and treatment. Their usage so far has not been validated in PD, rendering the IPMDS writing committee on Rating Scales to recommend their usage in PD with caution [70].

Specifically designed and developed for assessing burden of PD-related pain, the KPPS consists of 14 items grouped into seven different phenotypes of pain or domains (musculoskeletal, chronic, fluctuation-related, nocturnal, orofacial, discoloration/swelling/edema, and radicular pain) [71]. Each item is scored by severity (0–3) multiplied by frequency (0–4), resulting in a sub-score of 0–12, with a total possible score of 0–168. It evaluates the frequency and severity of various pain phenotypes that are commonly encountered in PD patients. This scale addresses localization, intensity, and frequency as well as its relationship with motor fluctuations or musculoskeletal pain. The KPPS is an interviewer-rated scale subjected to a patient, and if needed, aided by the carer. The time frame is the previous month and the completion time is approximately between 10 and 15 min [70]. The clinimetric quality of the scale is considered adequate based on good quality of data (no missing data reported), acceptable internal consistency (Cronbach alpha reliability 0.78) and excellent interrater and test–retest reliabilities [71]. With these properties, the authors advocated that the scale allows for adequate responsiveness for long-term longitudinal observations and clinical trials. The scale's validity was achieved from statistically significant correlation with other pain measures such as with VAS total score (Spearman coefficient, rs = 0.55), item 4, 9, and 12 of the PDSS-2 (related to restless leg syndrome; rs = 0.54, related to difficulty turning in bed; rs = 0.58 and painful posturing in early morning: rs = 0.58 respectively), and with pain items from both the EuroQoL-5 dimensions (EQ-5D) and NMSS [71]. Despite offering a simple and convenient way of assessing the frequency and intensity of pain syndromes in PD, KPPS raters however need to be trained to recognize categorical pain syndromes as outlined in the scale [70]. KPPS is currently the only pain assessment tool that has been validated and recommended for pain intensity and suggested for syndromic classification by the IPMDS review committee to be used in PD [70]. To date, KPPS has been utilized in several international multicenter randomized controlled trials [91, 92] as well as in smaller studies looking at the various aspects of PD-related pain including health-related QoL [93], neuroanatomical function [94], genetic [95], and phenotypic variants [68].

The **NMS Questionnaire (NMS-Quest)** and **NMS Scale (NMSS)** are validated patient completed and health professional completed (respectively) tools which include a question on pain in PD [19, 96]. While the NMS-Quest only screens for NMS in a simple yes/no question, the NMSS grades each question, including pain, by severity and frequency (burden).

3   Clinical Trials on Management of Pain in PD

In the following section, we will focus on published clinical trials addressing pain in PD and analyze their settings and designs. We revise the current approach to the treatment of pain with different

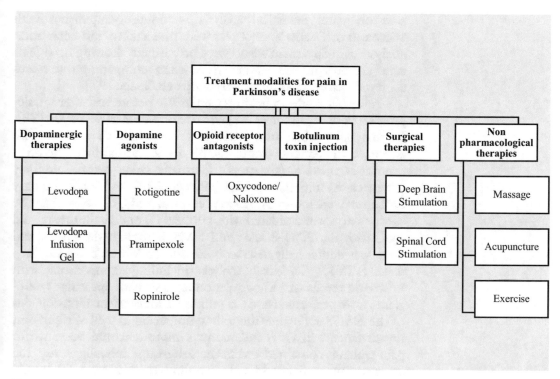

Fig. 1 Treatment modalities of pain in Parkinson's disease

treatment options and address pain as a unique syndrome with no further subdivision. The treatment modalities of pain in PD are summarized in Fig. 1.

3.1 Studies on Dopaminergic Therapies in PD-Related Pain

Dopaminergic therapies have been shown in several studies to relieve pain in PD. Some of the major trials are discussed as follow:

3.1.1 Levodopa/ Intrajejunal Levodopa Infusion

A cross-sectional study in late stages of PD including a total of 42 patients (20 patients with late stage PD, 22 advanced PD patients treated with deep brain stimulation, with a mean age of 78.8 ± 5.3 years) assessed as one of the primary objectives the response of pain following an acute levodopa challenge (median dose of 315 mg) [97]. The study took place in the outpatient clinic of a tertiary hospital in Lisbon, Portugal and included 20 late stage PD patients with either a Schwab and England score (S&E) < 50 (in the *on* state) or a Hoehn and Yahr stage (H&Y) > 3 (in the *on* state) and 20 advanced PD patients treated with deep brain stimulation. The assessment of pain alongside the other evaluated NMS was performed using the MDS-UPDRS part I, the NMSS, the Neuropsychiatric Inventory test 12-items and the Geriatric Depression Scale (GDS). The patients were first assessed in the *off* state and then 60–90 min after levodopa administration in their best *on* state. Statistical analysis including comparisons with Wilcoxon or Fischer's exact test were undertaken and correlation analysis with Spearman's Rho were performed, showing no significant changes in the feature pain when comparing the scores achieved before and after the levodopa challenge [97].

A further study addressing pain in PD before and after intrajejunal levodopa infusion (IJLI) was performed in 2007/2008 [98]. In this open label prospective and observational study on 22 advanced PD patients (mean age 58.6 ± 9.1 years) from the three centers in the UK (six cases), Germany (nine cases), and Italy (seven cases) initiated on IJLI between 2007 and 2008. Primary outcomes were motor as well as nonmotor and QoL score changes. Motor status was evaluated with UPDRS III and IV, health-related QoL with the PDQ-8 scale and NMS state with the NMSS and sleep was additionally assessed with the Parkinson's disease sleep scale (PDSS). Statistical analysis including comparisons with Wilcoxon test as well as correlation analysis with Spearman coefficients were performed and a significant benefit of the nine domains of the NMSS scale after the follow-up period as well as significant improvements in QoL and motor symptoms could be seen (for pain which is contained within the subgroup "miscellaneous," the score dropped from 14.1 at baseline to 6.4 after follow-up, $p = 0.0004$) [98].

Further studies on the effect of levodopa on NMS in PD, when administered as an intrajejunal infusion are summarized in Table 5.

A further open-label, prospective, observational study, called EuroInf was conducted comparing the effect of IJLI with apomorphine infusion in advanced PD stages [103]. This 6-month study compared 43 patients on apomorphine (48.8% males, age 62.3 ± 10.6 years) and 44 on IJLI (56.8% males, age 62.7 ± 9.1 years). Primary outcomes were the changes in total motor, nonmotor, and quality-of-life scores. Statistical analysis using Student's t test for parametric and Wilcoxon's/Mann–Whitney's for nonparametric data was performed and a significantly higher improvement of pain, regarded as a subtopic of the NMSS part 8 ("miscellaneous") was found only in the IJLI treated group (baseline score of 14.66, follow-up score of 9.68 $p = 0.0008$) [103].

3.1.2 Dopamine Agonists

The RECOVER study was a double-blind, placebo-controlled trial investigating the effect of rotigotine (2–16 mg/24 h) compared to placebo in a total 287 PD patients (mean age 64.4 ± 10.6) within 49 centers in 12 countries between 2007 and 2008 [108]. The subjects were randomized to the intervention drug or placebo (2:1) and primary outcomes defined as motor and sleep changes, assessed by means of the Unified Parkinson's Disease Rating Scale (UPDRS) Part III and the modified version of the original 15-item Parkinson's Disease Sleep Scale (PDSS-2). Within the secondary

Table 5
Overview of current studies on intrajejunal levodopa infusion (IJLI) for NMS including pain in PD [99]

Study	Subjects (*n*)	Study design	Utilized scales
Honig et al., 2009 [98]	22	Open, multicenter (UK, Germany, Italy, five centers)	UPDRS III, UPDRS IV, NMSS, PDQ39/PDQ8
Fasano et al., 2012 [100]	14	Open, two-center, retrospective (Italy, two centers)	UPDRS II, UPDRS III, UPDRS IV, NMSS, PDQ39/PDQ8
Reddy et al., 2014 [101]	22	Open, single center (UK)	UPDRS III, UPDRS IV, NMSS, PDQ39/PDQ8
Caceres-Redondo et al., 2014 [102]	29	Open, single center (Spain)	UPDRS II, UPDRS III, UPDRS IV, NMSS, PDQ39/PDQ8
Martinez-Martin et al., 2015 [103]	43	Open, multicenter, prospective (Spain, UK, Austria, Italy, Sweden, Denmark, Germany, Slovenia, 13 centers)	UPDRS III, UPDRS IV, NMSS, PDQ39/PDQ8
Bohlega et al., 2015 [104]	20	Open, single center, prospective (Saudi Arabia)	UPDRS III, NMSS, PDQ39/PDQ8
Antonini et al., 2013 [105]	73	Open, multicenter, routine care (Australia, Austria, Belgium, Bulgaria, Czech Republic, Denmark, France, Germany, Greece, Ireland, Italy, Netherlands, Norway, Romania, Slovenia, Spain, Switzerland, the UK, 75 centers)	UPDRS II, UPDRS III, NMSS, PDQ39/PDQ8
Sensi et al., 2014 [106]	28	Open, single center (Italy)	UPDRS III, UPDRS IV, NMSS, PDQ39/PDQ8
Zibetti et al., 2014 [107]	59	Open, single center, retrospective (Italy)	UPDRS II, UPDRS III, UPDRS IV, PDQ39/PDQ8

NMSS Nonmotor Symptom Scale, *PD* Parkinson's disease, *PDQ* Parkinson's Disease Questionnaire, *UPDRS* Unified Parkinson's Disease Rating Scale

outcomes pain was assessed among others, including assessments by means of Nocturnal Akinesia, Dystonia, and Cramps Score (NADCS), Parkinson's Disease Nonmotor Symptom scale (NMSS), the Beck Depression Inventory (BDI-II), an 11-point Likert pain scale, the short-form Parkinson's Disease Questionnaire

(PDQ-8) and the UPDRS Part II and UPDRS Part IV domains. The results found an improvement of pain assessed by the Likert pain scale in PD patients treated with rotigotine when compared to placebo (baseline score of 2.8, follow-up score of −0.9 (2.2), $p < 0.01$) [108]. A post-hoc analysis published in 2014 analyzed the topic pain more thoroughly. The Likert pain scale score change from baseline to end of treatment was assessed in patients with "any pain" at baseline, which was defined as those who recorded a baseline Likert pain scale score ≥ 1 ($N = 267$) and two further subgroups were defined from this group: patients reporting "mild pain" (Likert pain scale score 1–3), and those reporting "moderate-to-severe pain" (Likert pain scale score ≥ 4) at baseline. Statistical analysis using analysis of covariance (ANCOVA) were performed and significant changes in Likert pain scores from baseline to end of treatment under rotigotine treatment compared with placebo were found (mean treatment difference of −0.88 (−1.56, −0.19), $p = 0.013$). This was also found in the subgroup of patients with "moderate-to-severe" pain (−1.38 (−2.44, −0.31), $p = 0.012$), but not encountered in patients with "mild" pain.

The DOLORES study was a double-blind placebo-controlled study analyzing the effect of rotigotine vs. placebo over a period of 12 weeks on a total of 68 PD patients (mean age 65.3 ± 13.8 years in the placebo and 66.5 ± 11.9 years in the rotigotine group) randomized 1:1 to receive either of the interventions [91]. The primary outcome was the change from baseline to end of maintenance in pain severity, assessed by an 11-point Likert pain scale. Statistical analysis using analysis of covariance (ANCOVA) were performed and a numerical improvement in the average pain severity on the Likert pain scale was observed in favor of rotigotine treatment (treatment difference, −0.76 (−1.87 to 0.34) $p = 0.172$) [91].

With regard to the effect of other dopamine agonists on PD associated pain, the current literature is sparse. Pramipexole has been postulated as inducing complex-regional pain syndrome in a recently published case-report of a levodopa treated patient, who developed CRPS and dystonia after initiating pramipexole [109]. Other studies however, have reported benefits on pain in PD patients treated with pramipexole when compared to placebo over a 3-month period [110] and single-case reports have shown benefits of this dopamine agonist for the uncommonly arising burning mouth syndrome [111]. With regard to ropinirole, a long-term open label study on 419 patients, showed a potential enhancement of back pain by 14% and an increase in abdominal pain in the treated population in a safety and tolerability study [112, 113]. With regard to specific subtypes of pain, dopaminergic therapy has shown to often be effective in the treatment of dystonia-related musculoskeletal pain [6, 12]. Safinamide, a novel and recently approved agent to treat motor complications in PD with a double mechanism of action targeting not only the dopaminergic but also the glutamatergic neurotransmission, has shown to exert an action on pain as well, reducing the number of concomitant pain treatments as well as improving the QoL measured by means of the PDQ-39 scale when compared with placebo [69].

3.2 Studies on Opioid Receptor Antagonists in PD-Related Pain

The PANDA trial, a double-blind, randomized, placebo-controlled study compared the effect of oxycodone (5 mg) and naloxone (2.5 mg) versus placebo over a period of 16 weeks in 202 PD patients (mean age 66.7 ± 8.9) and 88 patients received the intervention drug while 106 were randomized to receive placebo [92]. The primary outcome was to assess whether there was a superiority of the treatment arm versus placebo for average 24-h pain score in the 7 days preceding the week 16 clinic visit, 24-h pain was assessed on a Likert-pain-11-point numerical rating scale. Statistical analysis using Wilcoxon test and ANCOVA analysis were performed. The results of this trial however failed to show a statistically significant of the intervention arm vs placebo on the primary endpoint of improved average 24-h pain scores.

Another smaller-scale, single-center study analyzing the efficacy and safety of prolonged release oxycodone and naloxone in 16 chronic pain PD patients (mean age 71.6 ± 8.9 years) over 2 months, had as primary endpoint the rate of favorable response to the intervention, defined as the proportion of patients who achieved a ≥30% reduction in pain intensity score from baseline to end-of trial visit. Assessments were performed by means of NRS and BPI and were complemented by the clinical global impression of change (CGI-C) and three visits including one after 1 month were performed. The results of this trail found significant changes of NRS scores at V1 and V2 compared to the baseline (ANOVA, $p < 0.05$), with no significant changes found in BPI scores across visits [88].

3.3 Studies on Botulinum Toxin in PD-Related Pain

The use of Botulinum toxin (BTX) injections for the treatment of pain in PD has been assessed in a recent retrospective chart review, including 160 patients with Parkinsonism (probable idiopathic PD was the diagnosis in 117 patients). Primary outcome was the response to BTX assessed using a 5-point subjective Clinical Global Impression (CGI). Considering pain as the indication, 81% of all patients with PD reported subjective benefits (score of +1 or greater) after the first BTX injections, with 53.4% reporting they had very much improved [62].

An uncontrolled study on 30 PD patients (median age 58 years) suffering from *off* dystonia-related pain of the foot assessed the use of BTX injections (median dose of 40 UI) into the calf muscles. The clinical features were assessed using the UPDRS scales and the location, duration, and severity of the pain by means of the MQP. The result showed an overall improvement of pain ($r = -0.7$, $p = 0.001$) [114].

A therapy approach of BTX for the treatment of pain in PD can be considered in focal dystonia cases, with no placebo-controlled trials assessing its effect to date but literature evidence showing benefit of its use in PD patients.

3.4 Studies on Spinal Cord Stimulation in PD-Related Pain

The effect of this invasive approach for the treatment of pain in PD has been studied in a case series of three PD patients with intractable pain, assessed by the pain VAS and the widespread pain index (WPI). A multielectrode plate was inserted into the epidural space at T8-L1 level by partial laminectomy and regular follow-ups followed up to a year after the intervention. The VAS score significantly declined, but the WPI did not show a significant improvement. After the 1-year follow-up, also motor symptoms showed a significant improvement among the three cases [115].

Spinal cord stimulation is supposed to be most effective for the treatment of chronic pain, especially sympathetic-mediated neuropathic pain [116, 117]. The mechanisms of action underlying the effect of this intervention on pain transmission are complex, with up to ten different mechanisms thought to take place in the improvement of pain [118].

3.5 Studies on Deep Brain Stimulation in PD-Related Pain

While deep brain stimulation (DBS) is performed in advanced stages of PD mainly to treat motor symptoms of the condition, several studies have also shown its benefit for the treatment of pain in PD. The most frequent targets for the treatment of PD-related motor symptoms are the nucleus subthalamicus and the globus pallidus internus [117], with studies showing a reduction in pain, especially in *off* state, after STN-DBS in 87% of the PD patients [119]. The current studies and outcomes of DBS on pain in PD are listed in Table 6 [117].

The response rates in the distinct pain subtypes after DBS have been estimated as 100% for dystonic, 92% for central, 63% for neuritic/radicular, and 61% for musculoskeletal pain. This beneficial effect on pain has also shown to be maintained over a long-term period, as assessed by several studies, where reductions of the VAS pain score was maintained after 12 month follow-up [132] or were even reported after 24 months in a prospective study on 18 PD patients [127] and after 8 years in a prospective study on 16 PD patients [134]. With regard to the effect of pallidotomy on pain [117], one study that reported on the effect of posteroventral pallidotomy on 38 PD patients with bradykinesia, tremor, and muscle pain showed a significant reduction in dystonia and/or pain from 63% to 32% after this intervention [136]. A long-term maintenance of this effect was showed in a study on unilateral pallidotomies performed among 21 PD patients suffering from PD-related pain and assessing them after 6 weeks and after 1 year, by a pain ordinal scale [135].

Table 6
Summary of evidence of DBS for the treatment of pain in PD. [117]

Author	Subjects (n)	Assessment tools for pain	Type of pain	Body distribution	Treatment	Main findings
Loher et al., 2002 [120]	16	Ordinal severity scale	n.a.	Neck, trunk, upper and lower extremities	GPi-DBS (unilateral, bilateral)	Pain and dysesthesia improved after 3–5 days after surgery and sustained at 1 year of FU
Witjas et al., 2007 [121]	40	NMF Questionnaire	n.a.	n.a.	STN-DBS (bilateral)	Improvement of pain symptoms after 12 months
Kim et al., 2008 [119]	29	Ordinal severity scale	Dystonic, central, neuritic/radicular, musculoskeletal	Head, neck, upper and lower extremities, trunk	STN-DBS (unilateral, bilateral)	Improvement of pain symptoms after 3 months
Gierthmühlen et al., 2010 [122]	17	Pain-DETECT Questionnaire	Neuropathic, nociceptive	Hands, back, upper and lower extremities, neck	SNT-DBS (bilateral) and levodopa	Improvement of pain symptoms and intensity, and changes in CDT, WDT, and TSL with ON-stim. No influences of SNT-DBS on pain threshold
Maruo et al., 2011 [123]	17	QST	–	–	STN-DBS (bilateral)	CDT and WDT were lower with On-Stim. No differences in CPT or HPT.
Spielberger et al., 2011 [124]	15	n.s.	n.a.	n.a.	STN-DBS (bilateral) and levodopa	No significant changes in CDT, WDT, CPT, or HPT with ON-stim
Ciampi de Andrade et al., 2012 [125]	25	B-PV MPQ-*short form* B-PV BPI B-PV DN4 B-PV NPSI	Chronic dystonic, musculoskeletal	n.a.	STN-DBS (bilateral)	No changes in MDT, VDT with ON-stim; increased MPT, HPT but CPT decreased with ON-stim; WDT decreased and CDT increased. VAS score in SuH and InC was reduced with ON-stim.

(continued)

Table 6
(continued)

Author	Subjects (*n*)	Assessment tools for pain	Type of pain	Body distribution	Treatment	Main findings
Oshima et al., 2012 [126]	163	VAS	Musculoskeletal, dystonic, somatic PD related, central, neuritic/radicular	Face, neck, upper and lower limb, abdomen, back	STN-DBS (bilateral)	Improvement in pain intensity after 2 weeks postsurgery (VAS scale decreased by 75%), 6 months (by 69%), and 12 months (by 80%) with ON-stim
Kim et al., 2012 [127]	21	Ordinal severity scale	Dystonic, musculoskeletal, neuritic/radicular, and central	Head, neck, upper and lower extremities	STN-DBS (unilateral, bilateral)	Improvement in pain symptoms after 3 and 24 months
Wolz et al., 2012 [128]	34	VAS	n.a.	n.a.	STN-DBS (bilateral)	No changes in pain with ON-stim
Sürücü et al., 2013 [129]	14	Ordinal severity scale	Dystonic, musculoskeletal, neuritic/radicular, and central	Neck, abdomen/viscera, arm, leg, lumber spine, multifocal	STN-DBS vs. levodopa	Eight patients with ON-levodopa showed improvement in pain. Greater improvements were observed with ON-stim, with long-lasting effects (41 months).
Dellapina et al., 2012 [130]	16	DN4 VAS NPSI	Neuropathic, nociceptive	Upper and lower limb, trunk	STN-DBS (bilateral)	Significantly increased HPT, reduced pain and pain-induced cerebral activity in the somatosensory cortex and cerebellum in patients with pain with ON-stim. Stim had no effect in pain-free patients.
Marques et al., 2013 [131]	19	Thermal and mechanical stimuli	Central	Upper limb, hand	STN-DBS (bilateral)	MPT and MPTo increased with ON-stim and ON-levodopa, compared with OFF state, condition

Author	Subjects (n)	Assessment tools for pain	Type of pain	Body distribution	Treatment	Main findings
Pellaprat et al., 2014 [132]	58	MPQ	n.a.	Head, neck, trunk, upper and lower limb	STN-DBS (bilateral)	ON-stim decreased pain symptoms after 12 months (19 patients were pain-free)
Cury et al., 2014 [133]	41	MPQ-*Short form* BPI DN4 NPSI	Musculoskeletal, dystonic, radicular/neuropathic, central	Head, neck, back, upper and lower limb	STN-DBS	Decrease in pain intensity with ON-stim. The highest response was in dystonic pain, followed by musculoskeletal. Central pain and neuropathic pain were not influenced by treatment
Jung et al., 2015 [134]	24	Ordinal severity scale	Dystonic, musculoskeletal, neuritic/radicular, central	Head, neck, trunk, upper and lower extremities	STN-DBS (unilateral, bilateral)	Improvement in pain symptoms in 83% of patients, with long-term effects at 8 years
Honey et al., 1999 [135]	21	Ordinal severity scale	Somatic exacerbated by the PD, musculoskeletal, dystonic, dysesthetic	Upper and lower limbs	Unilateral pallidotomy	Significant reduction in overall pain scores at 6 weeks and 1 year following pallidotomy

B-PV Brazilian-Portuguese Version, *MPQ* McGill Pain Questionnaire, *BPI* Brief Pain Inventory, *NPSI* Neuropathic Pain Symptoms Inventory, *DN4* Douleur Neuropathique 4, *NMF* nonmotor fluctuation, *CP* chronic pain, *QST* quantitative sensory testing, *CDT* cold detection threshold, *CPT* cold pain threshold, *DBS* deep brain stimulation, *FU* follow-up, *GPi* globus pallidus internus, *InC* infrathreshold cold stimulation, *HPT* heat pain threshold, *MPT* mechanical pain threshold, *MPTo* mechanical pain tolerance, *n.s.* not specified, *n.a.* not available, *OFF-levodopa* PD patient in "off" period, *OFF-stim* DBS turned off, *ON-levodopa* PD patient in "on" period, *ON-stim* DBS turned on, *stim* DBS stimulation, *STN* subthalamic nucleus, *SuH* suprathreshold heat stimulation, *TSL* thermal sensory limen, *VDT* vibration detection threshold, *WDT* warm detection threshold

3.6 Studies on Other Nonpharmacological Therapies

Studies on physical exercise and its effect on pain have so far not been performed with pain as primary outcome measure but showed some promising results in the subanalysis of the effect on pain. One study assessed QoL as a primary outcome in 20 PD patients (mean age 61.5 ± 9.8 years) after a program of physical activity of 36 group sessions of aerobic conditioning and muscular strengthening [137]. The patients were assessed with the Nottingham Health Profile (NHP) which also includes the subitem pain. In relation to pain, they found a strong tendency for improvement ($p = 0.06$).

A further randomized and controlled trial allocated 90 PD patients to three different treatment groups (flexibility and relaxation, walking and Nordic walking). The included patients participated during 6-months with three exercise sessions per week, each lasting 70 min. The primary outcome measures included characteristics of walking and gait, PD disability (Parkinson-specific disability (UPDRS)) and quality of life (PDQ 39). Pain in neck, arms, hands, back, iliosacral joint, hip, knees, feet, and toes was assessed with a VAS.

With regard to pain, they found a significant improvement in pain of the neck, back, hand, iliosacral joint, hip and legs ($p < 0.001$) when comparing to baseline. The pain of the back, hands and legs were more eased by walking and Nordic walking than by the flexibility and relaxation programmes [138].

Other alternative medicine procedures including traditional Chinese medicine with the application of herbs or vitamins, as well as massage therapy and acupuncture have so far not been widely studied, with only one group reporting in a setting of a case series study, the beneficial effect of a 30-min session of traditional Japanese massage on ten PD cases (mean age, 69.6 ± 7.7 years) assessed by means of a VAS [139].

A further nonblinded pilot study on the safety, tolerability and efficacy of acupuncture in PD was performed. In total, 20 PD patients (mean age 68 years) were included and received between 10 and 16 acupuncture sessions. The protocol included the Sickness Impact Profile (SIP), UPDRS, H&Y, S&E, Beck Anxiety Inventory (BAI), Beck Depression Inventory (BDI), quantitative motor tests, and a subjective patient questionnaire. The results showed a subjective improvement of symptoms including pain [140]. Some of the current studies looking at nonpharmacological approach to pain in PD are listed in Table 7.

Table 7
Summary of evidence of other nonpharmacological treatments for pain in PD

Author	Subjects (n)	Assessment tools for pain	Type of pain	Body distribution	Treatment	Main findings
Shulman et al., 2002 [140]	20	Patient Questionnaire	n.a.	n.a.	Acupuncture	Improvement in pain symptoms after completing the acupuncture protocol
Rodrigues de Paula et al., 2006 [137]	20	NHP	n.a.	n.a.	Exercise programs	Improvement in pain symptoms after 12-week training program (tendency toward significant values)
Donoyama and Ohkoshi, 2012 [139]	10	VAS	Musculoskeletal	Whole body	Traditional Japanese massage	Improvement in pain symptoms after a single massage session (30 min)
Reuter et al., 2011 [138]	90	VAS	n.a.	Neck, arms, hands, back, iliosacral joint, hip, knees, feet, toes	Nordic walking, walking, flexibility exercises	Improvement in pain symptoms after a 6-month training program

n.a. not available, *NHP* Nottingham Health Profile, *VAS* Visual Analog Scale

4   Conclusions

Pain is an important NMS in PD. In order to improve PD pain management and treatment options, a profound understanding on how pain affects patients with PD and vice versa is important.

In this chapter, we explored pain in the context of PD, its prevalence, clinical characteristics, available classifications, and assessment tools. Furthermore, we reviewed relevant clinical trials on pain in PD and summarized the key information (study design, outcome measures, etc.).

To date, studies on pain in PD are difficult to compare as they vary in methodologies, study designs and assessment tools. In many clinical trials pain in PD is still assessed as a general entity utilizing general pain assessment tools. In future, validated PD pain specific assessment tools should be applied to capture the unique characteristics of pain in PD. This chapter shall help with conducting future clinical trials on pain in PD.

References

1. Garcia-Ruiz PJ, Chaudhuri KR, Martinez-Martin P (2014) Non-motor symptoms of Parkinson's disease. A review...from the past. J Neurol Sci 338:30–33
2. Baig F, Lawton M, Rolinski M et al (2015) Delineating nonmotor symptoms in early Parkinson's disease and first-degree relatives. Mov Disord 30:1759–1766
3. Classen J, Koschel J, Oehlwein C et al (2017) Nonmotor fluctuations: phenotypes, pathophysiology, management, and open issues. J Neural Transm (Vienna) 124:1029–1036
4. Schrag A (2006) Quality of life and depression in Parkinson's disease. J Neurol Sci 248:151–157
5. Valkovic P, Minar M, Singliarova H et al (2015) Pain in Parkinson's disease: a cross-sectional study of its prevalence, types, and relationship to depression and quality of life. PLoS One 10:e0136541
6. Skogar O, Lokk J (2016) Pain management in patients with Parkinson's disease: challenges and solutions. J Multidiscip Healthc 9:469–479
7. Chaudhuri KR, Schapira AH (2009) Non-motor symptoms of Parkinson's disease: dopaminergic pathophysiology and treatment. Lancet Neurol 8:464–474
8. Chaudhuri KR, Prieto-Jurcynska C, Naidu Y et al (2010) The nondeclaration of nonmotor symptoms of Parkinson's disease to health care professionals: an international study using the nonmotor symptoms questionnaire. Mov Disord 25:704–709
9. Mitra T, Naidu Y, Martinez-Martin P (2008) The non declaration of non motor symptoms of Parkinson's disease to healthcare professionals. An international survey using the NMSQuest. In: 6th international congress on mental dysfunctions and other non-motor features in Parkinson's disease and related disorders. Dresden, October
10. Beiske AG, Loge JH, Ronningen A et al (2009) Pain in Parkinson's disease: prevalence and characteristics. Pain 141:173–177
11. Broen MP, Braaksma MM, Patijn J et al (2012) Prevalence of pain in Parkinson's disease: a systematic review using the modified QUADAS tool. Mov Disord 27:480–484
12. Ha AD, Jankovic J (2012) Pain in Parkinson's disease. Mov Disord 27:485–491
13. Negre-Pages L, Regragui W, Bouhassira D et al (2008) Chronic pain in Parkinson's disease: the cross-sectional French DoPaMiP survey. Mov Disord 23:1361–1369
14. Cury RG, Galhardoni R, Fonoff ET et al (2016) Sensory abnormalities and pain in Parkinson disease and its modulation by treatment of motor symptoms. Eur J Pain 20:151–165
15. Madden MB, Hall DA (2010) Shoulder pain in Parkinson's disease: a case-control study. Mov Disord 25:1105–1106
16. Ehrt U, Larsen JP, Aarsland D (2009) Pain and its relationship to depression in Parkinson disease. Am J Geriatr Psychiatry 17:269–275
17. Defazio G, Berardelli A, Fabbrini G et al (2008) Pain as a nonmotor symptom of

Parkinson disease: evidence from a case-control study. Arch Neurol 65:1191–1194

18. Broetz D, Eichner M, Gasser T et al (2007) Radicular and nonradicular back pain in Parkinson's disease: a controlled study. Mov Disord 22:853–856

19. Chaudhuri KR, Martinez-Martin P, Schapira AH et al (2006) International multicenter pilot study of the first comprehensive self-completed nonmotor symptoms questionnaire for Parkinson's disease: the NMSQuest study. Mov Disord 21:916–923

20. Etchepare F, Rozenberg S, Mirault T et al (2006) Back problems in Parkinson's disease: an underestimated problem. Joint Bone Spine 73:298–302

21. Quittenbaum BH, Grahn B (2004) Quality of life and pain in Parkinson's disease: a controlled cross-sectional study. Parkinsonism Relat Disord 10:129–136

22. Snider SR, Fahn S, Isgreen WP et al (1976) Primary sensory symptoms in parkinsonism. Neurology 26:423–429

23. Ozturk EA, Gundogdu I, Kocer B et al (2016) Chronic pain in Parkinson's disease: frequency, characteristics, independent factors, and relationship with health-related quality of life. J Back Musculoskelet Rehabil. https://doi.org/10.3233/BMR-160720

24. Rana A, Saeed U, Masroor MS et al (2013) A cross-sectional study investigating clinical predictors and physical experiences of pain in Parkinson's disease. Funct Neurol 28:297–304

25. Wen HB, Zhang ZX, Wang H et al (2012) Epidemiology and clinical phenomenology for Parkinson's disease with pain and fatigue. Parkinsonism Relat Disord 18(Suppl 1):S222–S225

26. Hanagasi HA, Akat S, Gurvit H et al (2011) Pain is common in Parkinson's disease. Clin Neurol Neurosurg 113:11–13

27. Santos-Garcia D, Abella-Corral J, Aneiros-Diaz A et al (2011) Pain in Parkinson's disease: prevalence, characteristics, associated factors, and relation with other non motor symptoms, quality of life, autonomy, and caregiver burden. Rev Neurol 52:385–393

28. Silva EG, Viana MA, Quagliato EM (2008) Pain in Parkinson's disease: analysis of 50 cases in a clinic of movement disorders. Arq Neuropsiquiatr 66:26–29

29. Stamey W, Davidson A, Jankovic J (2008) Shoulder pain: a presenting symptom of Parkinson disease. J Clin Rheumatol 14:253–254

30. Martinez-Martin P, Schapira AH, Stocchi F et al (2007) Prevalence of nonmotor symptoms in Parkinson's disease in an international setting; study using nonmotor symptoms questionnaire in 545 patients. Mov Disord 22:1623–1629

31. Sullivan KL, Ward CL, Hauser RA et al (2007) Prevalence and treatment of nonmotor symptoms in Parkinson's disease. Parkinsonism Relat Disord 13:545

32. Lee MA, Walker RW, Hildreth TJ et al (2006) A survey of pain in idiopathic Parkinson's disease. J Pain Symptom Manag 32:462–469

33. Tinazzi M, Del Vesco C, Fincati E et al (2006) Pain and motor complications in Parkinson's disease. J Neurol Neurosurg Psychiatry 77:822–825

34. Clifford TJ, Warsi MJ, Burnett CA et al (1998) Burning mouth in Parkinson's disease sufferers. Gerodontology 15:73–78

35. Goetz CG, Tanner CM, Levy M et al (1986) Pain in Parkinson's disease. Mov Disord 1:45–49

36. Rana AQ, Kabir A, Jesudasan M et al (2013) Pain in Parkinson's disease: analysis and literature review. Clin Neurol Neurosurg 115:2313–2317

37. Thompson T, Gallop K, Correll CU et al (2017) Pain perception in Parkinson's disease: a systematic review and meta-analysis of experimental studies. Ageing Res Rev 35:74–86

38. Jarcho JM, Mayer EA, Jiang ZK et al (2012) Pain, affective symptoms, and cognitive deficits in patients with cerebral dopamine dysfunction. Pain 153:744–754

39. Wood PB (2008) Role of central dopamine in pain and analgesia. Expert Rev Neurother 8:781–797

40. Avenali M, Tassorelli C, De Icco R et al (2017) Pain processing in atypical Parkinsonisms and Parkinson disease: a comparative neurophysiological study. Clin Neurophysiol 128:1978–1984

41. Djaldetti R, Shifrin A, Rogowski Z et al (2004) Quantitative measurement of pain sensation in patients with Parkinson disease. Neurology 62:2171–2175

42. Zambito Marsala S, Tinazzi M, Vitaliani R et al (2011) Spontaneous pain, pain threshold, and pain tolerance in Parkinson's disease. J Neurol 258:627–633

43. Gierthmuhlen J, Schumacher S, Deuschl G et al (2010) Somatosensory function in asymptomatic Parkin-mutation carriers. Eur J Neurol 17:513–517

44. Nolano M, Provitera V, Estraneo A et al (2008) Sensory deficit in Parkinson's disease: evidence of a cutaneous denervation. Brain 131:1903–1911

45. Mylius V, Engau I, Teepker M et al (2009) Pain sensitivity and descending inhibition of pain in Parkinson's disease. J Neurol Neurosurg Psychiatry 80:24–28

46. Perrotta A, Sandrini G, Serrao M et al (2011) Facilitated temporal summation of pain at spinal level in Parkinson's disease. Mov Disord 26:442–448

47. Schestatsky P, Kumru H, Valls-Sole J et al (2007) Neurophysiologic study of central pain in patients with Parkinson disease. Neurology 69:2162–2169

48. Tinazzi M, Del Vesco C, Defazio G et al (2008) Abnormal processing of the nociceptive input in Parkinson's disease: a study with CO_2 laser evoked potentials. Pain 136:117–124

49. Braak H, Sastre M, Bohl JR et al (2007) Parkinson's disease: lesions in dorsal horn layer I, involvement of parasympathetic and sympathetic pre- and postganglionic neurons. Acta Neuropathol 113:421–429

50. Defazio G, Antonini A, Tinazzi M et al (2017) Relationship between pain and motor and non-motor symptoms in Parkinson's disease. Eur J Neurol 24:974–980

51. Allen NE, Wong CM, Canning CG et al (2016) The association between Parkinson's disease motor impairments and pain. Pain Med 17:456–462

52. Franke C, Storch A (2017) Nonmotor fluctuations in Parkinson's disease. Int Rev Neurobiol 134:947–971

53. Storch A, Schneider CB, Wolz M et al (2013) Nonmotor fluctuations in Parkinson disease: severity and correlation with motor complications. Neurology 80:800–809

54. Yakovleva LA, Zalyalova ZA, Altunbaev RA (2015) Pain fluctuations in Parkinson's disease. Zh Nevrol Psikhiatr Im S S Korsakova 115:41–44

55. Lin CH, Wu RM, Chang HY et al (2013) Preceding pain symptoms and Parkinson's disease: a nationwide population-based cohort study. Eur J Neurol 20:1398–1404

56. Schrag A, Horsfall L, Walters K et al (2015) Prediagnostic presentations of Parkinson's disease in primary care: a case-control study. Lancet Neurol 14:57–64

57. Tolosa E, Compta Y, Gaig C (2007) The premotor phase of Parkinson's disease. Parkinsonism Relat Disord 13(Suppl):S2–S7

58. O'sullivan SS, Williams DR, Gallagher DA et al (2008) Nonmotor symptoms as presenting complaints in Parkinson's disease: a clinicopathological study. Mov Disord 23:101–106

59. Georgiev D, Hamberg K, Hariz M et al (2017) Gender differences in Parkinson's disease: a clinical perspective. Acta Neurol Scand 136:570–584

60. Greenspan JD, Craft RM, Leresche L et al (2007) Studying sex and gender differences in pain and analgesia: a consensus report. Pain 132(Suppl 1):S26–S45

61. Keogh E, Denford S (2009) Sex differences in perceptions of pain coping strategy usage. Eur J Pain 13:629–634

62. Bruno VA, Fox SH, Mancini D et al (2016) Botulinum toxin use in refractory pain and other symptoms in Parkinsonism. Can J Neurol Sci 43:697–702

63. Rana AQ, Qureshi ARM, Kachhvi HB et al (2016) Increased likelihood of anxiety and poor sleep quality in Parkinson's disease patients with pain. J Neurol Sci 369:212–215

64. Ford B (1998) Pain in Parkinson's disease. Clin Neurosci 5:63–72

65. Wasner G, Deuschl G (2012) Pains in Parkinson disease—many syndromes under one umbrella. Nat Rev Neurol 8:284–294

66. Sauerbier A, Rosa-Grilo M, Qamar MA et al (2017) Nonmotor subtyping in Parkinson's disease. Int Rev Neurobiol 133:447–478

67. Quinn NP, Koller WC, Lang AE et al (1986) Painful Parkinson's disease. Lancet 1:1366–1369

68. Wallace VC, Chaudhuri KR (2014) Unexplained lower limb pain in Parkinson's disease: a phenotypic variant of "painful Parkinson's disease". Parkinsonism Relat Disord 20:122–124

69. Bonuccelli U (2015) Effects of safinamide on motor complications and pain in advancing Parkinson's disease – post hoc analyses of pivotal trials. Eur Neurol Rev 10(2):176–181

70. Perez-Lloret S, Ciampi De Andrade D, Lyons KE et al (2016) Rating scales for pain in Parkinson's disease: critique and recommendations. Mov Disord Clin Pract 3:527–537

71. Chaudhuri KR, Rizos A, Trenkwalder C et al (2015) King's Parkinson's disease pain scale, the first scale for pain in PD: an international validation. Mov Disord 30:1623–1631

72. Cleeland CS, Ryan KM (1994) Pain assessment: global use of the Brief Pain Inventory. Ann Acad Med Singap 23:129–138

73. Melzack R (1975) The McGill Pain Questionnaire: major properties and scoring methods. Pain 1:277–299

74. Melzack R (1987) The short-form McGill Pain Questionnaire. Pain 30:191–197

75. Breivik H, Borchgrevink PC, Allen SM et al (2008) Assessment of pain. Br J Anaesth 101:17–24

76. Huskisson E (ed) (1983) Visual analogue scales, pain measurement and assessment. Raven Press, New York, NY

77. Bennett M (2001) The LANSS Pain Scale: the Leeds assessment of neuropathic symptoms and signs. Pain 92:147–157

78. Bouhassira D, Attal N, Fermanian J et al (2004) Development and validation of the

Neuropathic Pain Symptom Inventory. Pain 108:248–257

79. Freynhagen R, Baron R, Gockel U et al (2006) painDETECT: a new screening questionnaire to identify neuropathic components in patients with back pain. Curr Med Res Opin 22:1911–1920

80. Gaston-Johansson F (1996) Measurement of pain: the psychometric properties of the Pain-O-Meter, a simple, inexpensive pain assessment tool that could change health care practices. J Pain Symptom Manag 12:172–181

81. Bouhassira D, Attal N, Alchaar H et al (2005) Comparison of pain syndromes associated with nervous or somatic lesions and development of a new neuropathic pain diagnostic questionnaire (DN4). Pain 114:29–36

82. Atkinson TM, Mendoza TR, Sit L et al (2010) The Brief Pain Inventory and its "pain at its worst in the last 24 hours" item: clinical trial endpoint considerations. Pain Med 11:337–346

83. Mendoza T, Mayne T, Rublee D et al (2006) Reliability and validity of a modified Brief Pain Inventory short form in patients with osteoarthritis. Eur J Pain 10:353–361

84. Gjeilo KH, Stenseth R, Wahba A et al (2007) Validation of the brief pain inventory in patients six months after cardiac surgery. J Pain Symptom Manag 34:648–656

85. Konno S, Oda N, Ochiai T et al (2016) Randomized, double-blind, placebo-controlled phase III trial of duloxetine monotherapy in japanese patients with chronic low back pain. Spine (Phila Pa 1976) 41:1709–1717

86. Lin XJ, Yu N, Lin XG et al (2016) A clinical survey of pain in Parkinson's disease in Eastern China. Int Psychogeriatr 28:283–289

87. Djaldetti R, Yust-Katz S, Kolianov V et al (2007) The effect of duloxetine on primary pain symptoms in Parkinson disease. Clin Neuropharmacol 30:201–205

88. Madeo G, Schirinzi T, Natoli S et al (2015) Efficacy and safety profile of prolonged release oxycodone in combination with naloxone (OXN PR) in Parkinson's disease patients with chronic pain. J Neurol 262:2164–2170

89. Rana AQ, Siddiqui I, Mosabbir A et al (2013) Association of pain, Parkinson's disease, and restless legs syndrome. J Neurol Sci 327:32–34

90. Burckhardt CS, Jones KD (2003) Adult measures of pain: the McGill Pain Questionnaire (MPQ), Rheumatoid Arthritis Pain Scale (RAPS), Short-Form McGill Pain Questionnaire (SF-MPQ), Verbal Descriptive Scale (VDS), Visual Analog Scale (VAS), and West Haven-Yale Multidisciplinary Pain Inventory (WHYMPI). Arthritis Care Res 49:S96–S104

91. Rascol O, Zesiewicz T, Chaudhuri KR et al (2016) A randomized controlled exploratory pilot study to evaluate the effect of rotigotine transdermal patch on Parkinson's disease-associated chronic pain. J Clin Pharmacol 56:852–861

92. Trenkwalder C, Chaudhuri KR, Martinez-Martin P et al (2015) Prolonged-release oxycodone-naloxone for treatment of severe pain in patients with Parkinson's disease (PANDA): a double-blind, randomised, placebo-controlled trial. Lancet Neurol 14:1161–1170

93. Martinez-Martin P, Manuel Rojo-Abuin J, Rizos A et al (2017) Distribution and impact on quality of life of the pain modalities assessed by the King's Parkinson's disease pain scale. NPJ Parkinsons Dis 3:8

94. Polli A, Weis L, Biundo R et al (2016) Anatomical and functional correlates of persistent pain in Parkinson's disease. Mov Disord 31:1854–1864

95. Lin CH, Chaudhuri KR, Fan JY et al (2017) Depression and catechol-O-methyltransferase (COMT) genetic variants are associated with pain in Parkinson's disease. Sci Rep 7:6306

96. Chaudhuri KR, Martinez-Martin P, Brown RG et al (2007) The metric properties of a novel non-motor symptoms scale for Parkinson's disease: results from an international pilot study. Mov Disord 22:1901–1911

97. Fabbri M, Coelho M, Guedes LC et al (2017) Response of non-motor symptoms to levodopa in late-stage Parkinson's disease: results of a levodopa challenge test. Parkinsonism Relat Disord 39:37–43

98. Honig H, Antonini A, Martinez-Martin P et al (2009) Intrajejunal levodopa infusion in Parkinson's disease: a pilot multicenter study of effects on nonmotor symptoms and quality of life. Mov Disord 24:1468–1474

99. Wirdefeldt K, Odin P, Nyholm D (2016) Levodopa-carbidopa intestinal gel in patients with Parkinson's disease: a systematic review. CNS Drugs 30:381–404

100. Fasano A, Ricciardi L, Lena F et al (2012) Intrajejunal levodopa infusion in advanced Parkinson's disease: long-term effects on motor and non-motor symptoms and impact on patient's and caregiver's quality of life. Eur Rev Med Pharmacol Sci 16:79–89

101. Reddy P, Martinez-Martin P, Brown RG et al (2014) Perceptions of symptoms and expectations of advanced therapy for Parkinson's disease: preliminary report of a Patient-Reported Outcome tool for Advanced Parkinson's disease (PRO-APD). Health Qual Life Outcomes 12:11

102. Caceres-Redondo MT, Carrillo F, Lama MJ et al (2014) Long-term levodopa/carbidopa intestinal gel in advanced Parkinson's disease. J Neurol 261:561–569

103. Martinez-Martin P, Reddy P, Katzenschlager R et al (2015) EuroInf: a multicenter comparative observational study of apomorphine and levodopa infusion in Parkinson's disease. Mov Disord 30:510–516

104. Bohlega S, Abou Al-Shaar H, Alkhairallah T et al (2015) Levodopa-carbidopa intestinal gel infusion therapy in advanced Parkinson's disease: single Middle Eastern Center experience. Eur Neurol 74:227–236

105. Antonini A, Odin P, Opiano L et al (2013) Effect and safety of duodenal levodopa infusion in advanced Parkinson's disease: a retrospective multicenter outcome assessment in patient routine care. J Neural Transm 120:1553–1558

106. Sensi M, Preda F, Trevisani L et al (2014) Emerging issues on selection criteria of levodopa carbidopa infusion therapy: considerations on outcome of 28 consecutive patients. J Neural Transm (Vienna) 121:633–642

107. Zibetti M, Merola A, Artusi CA et al (2014) Levodopa/carbidopa intestinal gel infusion in advanced Parkinson's disease: a 7-year experience. Eur J Neurol 21:312–318

108. Trenkwalder C, Kies B, Rudzinska M et al (2011) Rotigotine effects on early morning motor function and sleep in Parkinson's disease: a double-blind, randomized, placebo-controlled study (RECOVER). Mov Disord 26:90–99

109. Park D (2017) Pramipexole-induced limb dystonia and its associated complex regional pain syndrome in idiopathic Parkinson's disease: a case report. Medicine 96:e7530

110. Letvinenko IV, Odinak MM, Mogil'naia VI (2008) Pain and depression in Parkinson's disease: new therapeutic possibilities of pramipexole. Zh Nevrol Psikhiatr Im S S Korsakova 108:36–38

111. Stuginski-Barbosa J, Rodrigues GG, Bigal ME et al (2008) Burning mouth syndrome responsive to pramipexol. J Headache Pain 9:43–45

112. Makumi CW, Asgharian A, Ellis J et al (2016) Long-term, open-label, safety study of once-daily ropinirole extended/prolonged release in early and advanced Parkinson's disease. Int J Neurosci 126:30–38

113. Nashatizadeh MM, Lyons KE, Pahwa R (2009) A review of ropinirole prolonged release in Parkinson's disease. Clin Interv Aging 4:179–186

114. Pacchetti C, Albani G, Martignoni E et al (1995) "Off" painful dystonia in Parkinson's disease treated with botulinum toxin. Mov Disord 10:333–336

115. Nishioka K, Nakajima M (2015) Beneficial therapeutic effects of spinal cord stimulation in advanced cases of Parkinson's disease with intractable chronic pain: a case series. Neuromodulation 18:751–753

116. Fenelon G, Goujon C, Gurruchaga JM et al (2012) Spinal cord stimulation for chronic pain improved motor function in a patient with Parkinson's disease. Parkinsonism Relat Disord 18:213–214

117. Geroin C, Gandolfi M, Bruno V et al (2016) Integrated approach for pain management in Parkinson disease. Curr Neurol Neurosci Rep 16:28

118. Oakley JC, Prager JP (2002) Spinal cord stimulation: mechanisms of action. Spine (Phila Pa 1976) 27:2574–2583

119. Kim HJ, Paek SH, Kim JY et al (2008) Chronic subthalamic deep brain stimulation improves pain in Parkinson disease. J Neurol 255:1889–1894

120. Loher TJ, Burgunder JM, Weber S et al (2002) Effect of chronic pallidal deep brain stimulation on off period dystonia and sensory symptoms in advanced Parkinson's disease. J Neurol Neurosurg Psychiatry 73:395–399

121. Witjas T, Kaphan E, Regis J et al (2007) Effects of chronic subthalamic stimulation on nonmotor fluctuations in Parkinson's disease. Mov Disord 22:1729–1734

122. Gierthmuhlen J, Arning P, Binder A et al (2010) Influence of deep brain stimulation and levodopa on sensory signs in Parkinson's disease. Mov Disord 25:1195–1202

123. Maruo T, Saitoh Y, Hosomi K et al (2011) Deep brain stimulation of the subthalamic nucleus improves temperature sensation in patients with Parkinson's disease. Pain 152:860–865

124. Spielberger S, Wolf E, Kress M et al (2011) The influence of deep brain stimulation on pain perception in Parkinson's disease. Mov Disord 26:1367–1368. author reply 1368–1369

125. Ciampi De Andrade DLJ, Galhardoni R et al (2012) Subthalamic deep brain stimulation modulates small fiber-dependent sensory thresholds in Parkinson's disease. Pain 153:1107–1113

126. Oshima H, Katayama Y, Morishita T et al (2012) Subthalamic nucleus stimulation for attenuation of pain related to Parkinson disease. J Neurosurg 116:99–106

127. Kim HJ, Jeon BS, Lee JY et al (2012) The benefit of subthalamic deep brain stimulation for pain in Parkinson disease: a 2-year follow-up study. Neurosurgery 70:18–23. discussion 14–23

128. Wolz M, Hauschild J, Koy J et al (2012) Immediate effects of deep brain stimulation of the subthalamic nucleus on nonmotor symptoms in Parkinson's disease. Parkinsonism Relat Disord 18:994–997

129. Surucu O, Baumann-Vogel H, Uhl M et al (2013) Subthalamic deep brain stimulation versus best medical therapy for L-dopa responsive pain in Parkinson's disease. Pain 154:1477–1479

130. Dellapina E, Ory-Magne F, Regragui W et al (2012) Effect of subthalamic deep brain stimulation on pain in Parkinson's disease. Pain 153:2267–2273

131. Marques A, Chassin O, Morand D et al (2013) Central pain modulation after subthalamic nucleus stimulation: a crossover randomized trial. Neurology 81:633–640

132. Pellaprat J, Ory-Magne F, Canivet C et al (2014) Deep brain stimulation of the subthalamic nucleus improves pain in Parkinson's disease. Parkinsonism Relat Disord 20:662–664

133. Cury RG, Galhardoni R, Fonoff ET et al (2014) Effects of deep brain stimulation on pain and other nonmotor symptoms in Parkinson disease. Neurology 83:1403–1409

134. Jung YJ, Kim HJ, Jeon BS et al (2015) An 8-year follow-up on the effect of subthalamic nucleus deep brain stimulation on pain in Parkinson disease. JAMA Neurol 72:504–510

135. Honey CR, Stoessl AJ, Tsui JK et al (1999) Unilateral pallidotomy for reduction of parkinsonian pain. J Neurosurg 91:198–201

136. Laitinen LV, Bergenheim AT, Hariz MI (1992) Leksell's posteroventral pallidotomy in the treatment of Parkinson's disease. J Neurosurg 76:53–61

137. Rodrigues De Paula F, Teixeira-Salmela LF, Coelho De Morais Faria CD et al (2006) Impact of an exercise program on physical, emotional, and social aspects of quality of life of individuals with Parkinson's disease. Mov Disord 21:1073–1077

138. Reuter IMS, Leone P, Kaps M, Oechsner M, Engelhardt M (2011) Effects of a flexibility and relaxation programme, walking, and nordic walking on Parkinson's disease. J Aging Res 2011:Article ID 232473

139. Donoyama N, Ohkoshi N (2012) Effects of traditional Japanese massage therapy on various symptoms in patients with Parkinson's disease: a case-series study. J Alternat Complement Med 18:294–299

140. Shulman LM, Wen X, Weiner WJ et al (2002) Acupuncture therapy for the symptoms of Parkinson's disease. Mov Disord 17:799–802

Chapter 13

Clinical Trials for Orthostatic Hypotension in Parkinson's Disease and Other Synucleinopathies

Jose-Alberto Palma and Horacio Kaufmann

Abstract

Neurogenic orthostatic hypotension (nOH) is one of the most debilitating nonmotor symptoms in patients with Parkinson disease and other synucleinopathies. Patients with Parkinson disease and nOH suffer from more hospitalizations, emergency room visits, more telephone calls and e-mails to providers, and have a significantly shorter survival compared to patients with Parkinson disease and no nOH. Overall, health-related costs in patients with Parkinson disease and OH are 2.5-fold higher compared to patients with Parkinson disease without OH. Therefore, the development of effective therapies for patients with Parkinson disease and nOH should be a research priority. In recent years, better understanding of the pathophysiology of nOH has resulted in the identification of novel therapeutic targets and the development and approval of effective drug therapies, such as midodrine and droxidopa. We here review the design and endpoint selection for clinical trials of nOH in patients with Parkinson disease and other synucleinopathies, recapitulate the results of completed and ongoing clinical trials for nOH, and discuss common challenges and their potential remedies.

Key words Parkinson's disease, Orthostatic hypotension, Synucleinopathies, Clinical trials, Autonomic nervous system

1 Introduction

In his 1817 seminal essay, James Parkinson captured with astonishing clarity numerous clinical aspects of Parkinson disease (PD), including key elements of the natural history of the disease, several of the prominent motor features, and some of its nonmotor autonomic features, such as constipation, urinary dysfunction, and dysphagia [1]. He was not able to describe orthostatic hypotension (OH), though. Indeed, Parkinson did not have the opportunity to carefully question or even examine his patients, as five of the six cases included in his essay were only casually observed on the street.

It was not until the mid-twentieth century when OH was reported in patients with PD. In some of these cases, L-dopa was identified as the main culprit of the OH [2, 3]. In others, many

Santiago Perez-Lloret (ed.), *Clinical Trials In Parkinson's Disease*, Neuromethods, vol. 160,
https://doi.org/10.1007/978-1-0716-0912-5_13, © Springer Science+Business Media, LLC, part of Springer Nature 2021

with neuropathology-proven PD, OH was present before the emergence of parkinsonian deficits and, therefore, a common neuropathological process, independent of L-dopa, was apparent [4–8].

Today, dysfunction of the autonomic nervous system—causing neurogenic OH (nOH) among other signs and symptoms—is a well-known characteristic nonmotor feature of patients with PD and other synucleinopathies, occurring at all stages of the disease. Moreover, there is strong evidence indicating that it can be the one of the earliest, prodromal, manifestations of the disease, occurring years—and sometimes decades—before any of the defining motor characteristics are evident [9, 10].

OH is among the most debilitating nonmotor symptoms, which reduces quality of life in affected patients. OH is associated with increased morbidity, particularly syncope and falls [11, 12]. OH is a common reason for, or contributor to, hospitalization in elderly patients. Patients with PD and nOH generates more hospitalizations, emergency room visits, more telephone calls and e-mails to doctors, and have a shorter survival than those with PD and no nOH [13, 14]. Overall, the health-related cost in patients with PD and OH is 2.5-fold higher compared to patients with PD without OH [14]. Hence, developing effective therapies for nOH should be a research priority.

In the last decades, improved understanding of the pathophysiology of nOH in synucleinopathies has led to the identification of therapeutic targets and the development and approval of effective drug therapies, such as midodrine and droxidopa.

Here we review the design and conductance of clinical trials in patients with PD and other synucleinopathies with nOH, summarize the results of the most recently completed and ongoing trials, and discuss challenges, bottlenecks, and potential remedies.

2   Epidemiology and Burden of nOH

Orthostatic hypotension (OH) is a sustained fall in blood pressure (BP) on standing. The current definition of OH, based on expert consensus [15], is a fall of at least 20 mmHg in systolic BP or 10 mmHg in diastolic BP within 3 min of standing or upright tilt. OH can impair perfusion to organs above the heart, most notably the brain, resulting in symptoms of tissue hypoperfusion. Symptoms can be very disabling, have a profound impact on a patient's quality of life, and increase morbidity and mortality [16, 17].

OH is a frequent problem in the general population, particularly in the frail elderly [15, 17–19]. The overall prevalence of OH in patients over age 65 is ~20% [20]. OH can be due to a variety of medical conditions, such as intravascular volume depletion, blood pooling (i.e., varicose veins [21]), severe anemia, medications (e.g., nitrates, calcium channel blockers, alpha adrenergic blockers, tricyclic antidepressants, opioids), alcohol, and physical deconditioning. OH is present in 25% of patients evaluated in the emergency department for syncope [22]. The estimate of OH-related hospitalization is 36 per 100,000 adults, and can be as high as 233 per 100,000 patients >75 years of age, with an overall in-hospital mortality rate of 0.9% [18]. In inpatient series, the prevalence of OH in elderly patients is as high as 60% [23, 24]. OH is an independent predictor of mortality [25–27].

In patients with PD and other synucleinopathies [dementia with Lewy bodies, multiple system atrophy (MSA), and pure autonomic failure (PAF)], in which there is abnormal accumulation of α-synuclein (αSyn) in the nervous system, OH is due to reduced norepinephrine release from postganglionic sympathetic nerves, resulting in defective vasoconstriction when in the upright posture [15]. This is referred to as *neurogenic* OH (nOH) [28, 29].

In contrast to regular OH, nOH is an orphan condition as it affects less than 200,000 people in the USA, although the true prevalence of nOH may be underestimated, as BP is not always measured in the upright posture. In cross-sectional studies, between 30 and 50% of patients with PD have nOH, but less than a third of those patients are symptomatic, that is, only 16% of patients with PD have symptomatic nOH [30–33]. The prevalence of nOH in PD increases with age and disease duration [30]. In DLB the prevalence of OH is higher, 50–60% [33, 34]. In patients with MSA, the diagnostic criteria for nOH are a fall of 30 mmHg in systolic or 15 mmHg in diastolic BP [35]. Even according to these more stringent criteria, 70–80% of patients with MSA have nOH [36, 37]. Complicating nOH management is arterial hypertension when supine (SH), which occurs in 50–70% of patients with PD and nOH [30, 38–40].

As nOH is a relatively rare disorder with few treatment options, limited information identifying burden of illness or cost-effectiveness of treatment is available. However, falls and fall-related injuries associated with nOH may represent a substantial burden of disease for individual patients as well as the health care system in general. Patients with nOH, or suboptimal nOH management, may limit their daily activities and experience a decreased quality of life because of the increased potential for falls and fall-related injuries. The resultant health care costs from fall events are likely to represent a significant economic burden on the US health care system.

3 Pathophysiology: Key to Developing New Treatments for nOH

Knowledge of the pathophysiology and neural pathways involved in nOH is key to developing effective treatments. Normally, unloading of the baroreceptors by standing up eventually triggers norepinephrine release from sympathetic postganglionic nerves activating alpha-adrenergic receptors in the blood vessels causing vasoconstriction, which maintains BP in the standing position. Norepinephrine release is mediated by activation of central autonomic pathways, preganglionic sympathetic cholinergic neurons, and postganglionic sympathetic noradrenergic neurons. This norepinephrine-mediated compensatory vasoconstriction is absent or attenuated in patients with synucleinopathies resulting in nOH.

Therefore, nOH is best understood as a neurotransmitter disorder. In patients with PD and other synucleinopathies, dopamine deficiency in the nigrostriatal pathway causes the motor abnormalities, while impaired release of the neurotransmitter norepinephrine from sympathetic postganglionic neurons causes nOH (Fig. 1).

The site of the "autonomic lesion" in the baroreflex pathways responsible for nOH is different in patients with PD and DLB (Lewy body disorders) versus patients with MSA. In patients with Lewy body disorders, nOH is predominantly due to degeneration of postganglionic sympathetic neurons. There is robust imaging and neuropathological data showing that postganglionic peripheral sympathetic neurons innervating the myocardium are functionally affected due to αSyn deposits and fiber loss [41, 42]. Sympathetic fibers innervating blood vessels are also affected. This results in impaired norepinephrine release and defective vasocon-

Motor dysfunction: Disorder of dopaminergic neurotransmission

Neurogenic OH: Disorder of noradrenergic neurotransmission

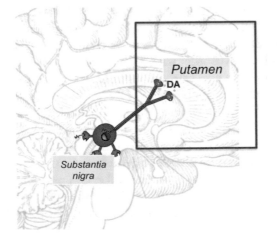

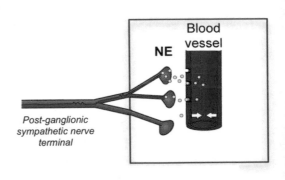

Fig. 1 Neurogenic orthostatic hypotension as a neurotransmitter disorders in Parkinson disease. Motor dysfunction in Parkinson disease and other synucleinopathies is mostly due to a dopaminergic deficit in basal ganglia neurons. Dopamine deficit can be treated with oral administration of the dopamine precursor L-dopa or with direct dopaminergic agonists. Defective vasoconstriction leading to neurogenic orthostatic hypotension is due to impaired release of norepinephrine from postganglionic sympathetic terminals. Norepinephrine deficit can be treated with oral administration of the norepinephrine precursor droxidopa or with direct adrenergic agonists like midodrine, among other agents

striction upon standing causing the BP to fall [32, 42]. Indeed, plasma norepinephrine levels are lower in patients with PD and nOH than in those without nOH [43].

In patients with MSA, nOH is caused by degeneration of CNS neurons involved in baroreflex control, while only a minority (<30%) of patients have degeneration of postganglionic sympathetic nerves [44–48]. Norepinephrine release from peripheral sympathetic terminals is also blunted but the nerves themselves are mostly spared [41]. Plasma norepinephrine levels are mostly normal in patients with MSA [49] reflecting intact sympathetic neurons. These differences between Lewy body disorders and MSA are relevant to explain the different pressor responses to norepinephrine enhancing medications, and can be used to develop novel molecules targeting autonomic dysfunction due to CNS (i.e., MSA) versus peripheral (i.e., Lewy body disorders) degeneration.

The pathophysiology of nSH also differs in patients with Lewy body disorders versus MSA. In patients with MSA, residual sympathetic neurovascular tone may contribute to nSH. In contrast, nSH in patients with Lewy body disorders has a mechanism other than increased sympathetic outflows, yet to be fully identified [50, 51]. Patients with nOH who also have nSH are less likely to develop symptomatic nOH after 3-min standing [30] as their BP in the standing position is above the lower level of cerebral autoregulatory capacity. Cerebral vasomotor reactivity is preserved in patients with PD, which may contribute to lack of symptoms of cerebral hypoperfusion during hypotension [52, 53].

4 Challenges in Clinical Trials for nOH

nOH is currently considered a rare disease, as it affects <200,000 subjects in the USA. Clinical research in rare diseases face many challenges [54]. For instance, the natural history of nOH in PD and related disorders is poorly understood and the ability to detect clinically meaningful outcomes requires understanding of its occurrence, variability, both of which contribute to difficulties in powering a study. Assembling a large cohort of patients requires a multicenter design with multiple treatment centers in several countries. An added difficulty is the "background noise" caused by wide BP variability of patients with nOH. Patients report that symptom severity and BP readings vary from day-to-day and fluctuate throughout the day. The morning hours tend to be most difficult as OH symptoms are aggravated by intravascular volume loss overnight [55]. Meals, particularly carbohydrate-rich, lead to splanchnic vasodilatation and postprandial hypotension (i.e., fall in BP within 2 h of eating). Temperature, dehydration and physical activities can also induce variations in BP.

Patient-reported outcomes have limitations in patients with nOH and reliance on symptoms alone may not always be an accurate indicator of tissue hypoperfusion. Symptoms of nOH can be nonspecific, including fatigue and difficultly concentrating and may mimic the levodopa "off" state in PD patients. Conversely, patients may have difficultly distinguishing symptoms of nOH from other causes of lightheadedness [30]. Furthermore, nOH is frequently associated with progressive neurodegenerative disorders; failure to demonstrate long-term improvement could conceivably be due to worsening of the underlying neurodegenerative disorder rather than failure of the active agent.

Finally, the response to placebo can at times be quite impressive in some patients with nOH. The reason for this is unclear, but it may be related to a Hawthorne effect, that is, patients are particularly compliant with nonpharmacologic measures to treat nOH (liberalization of salt and water, avoiding carbohydrates and alcohol, etc.) because they are participating in a clinical trial. Unfortunately, individual use of these countermeasures is not thoroughly monitored in clinical trials of drugs for nOH. Also, whether there is substantial central or peripheral norepinephrine release in response to placebo in patients with PD (as it is the case with striatal dopamine [56]) is unknown, but definitely worth studying.

5   Patient-Reported Outcomes for Clinical Trials of nOH

The US FDA strongly advocates the use of patient-reported outcomes in drug trials. Identifying clinically meaningful outcomes is a challenge. The outcome should measure how patients feel, function and survive. These patient-reported outcome measures are increasingly taken into consideration by *Voice of the Patient* reports from patient focused meetings organized and conducted by the FDA [57]. These reports identify the most important consequences of the disease and may suggest aspects for the disease that are most readily measured.

In this regard, it is important to take into consideration that not all patients with nOH are symptomatic. Typical symptoms of nOH are lightheadedness, dizziness, blurry vision, and, when the fall in BP is pronounced, loss of consciousness and postural tone (syncope). Symptoms occur *only* when standing, less frequently when sitting, and abate when lying down. Symptoms of cerebral hypoperfusion emerge when BP standing falls below the lower limit of the cerebral autoregulatory range. In patients with PD, this usually occurs when mean BP standing is below 75 mmHg, which corresponds to ~90/60 mmHg (systolic/diastolic) at the heart level [30]. Symptoms of nOH typically disappear after the patient resumes the sitting or lying position because cerebral blood flow is restored to levels above the lower limit of autoregulatory capacity.

The chronic nature of nOH allows remarkable adaptive changes in cerebral autoregulatory mechanisms [58]. Indeed, patients with nOH are frequently able to tolerate wide swings in BPs and often remain conscious at pressures that would otherwise induce syncope in healthy subjects [59]. Symptoms of nOH can be nonspecific, including fatigue and difficultly concentrating and may sometimes mimic a levodopa "off" motor state in PD patients. In these cases, the diagnosis of nOH may be missed unless BP is measured in the standing position. Conversely, it is important to realize that in patients with PD postural lightheadedness mimicking nOH may be caused by abnormal postural reflexes, vestibular deficits or orthostatic tremor [60]. Ideally, patient-reported outcomes used in clinical trials of nOH should take into consideration all the above factors.

There are several patient-reported questionnaires that include items related to cardiovascular autonomic dysfunction [e.g., the Composite Symptoms Autonomic Score (CASS) [61], the Scales for Outcomes in Parkinson's disease—Autonomic (SCOPA-AUT) [62], and the Autonomic Symptom Profile (ASP) [63]]. Other global quality of life questionnaires were developed for patients with autonomic failure [e.g., Multiple System Atrophy Quality of Life scale (MSA-QoL) [64, 65]]. However, these instruments validated in other clinical settings are not optimized for patients with nOH as none of these questionnaires specifically quantify the impact of nOH on daily activities.

To overcome this, our group developed a novel clinical rating scale specifically for patients with nOH, the Orthostatic Hypotension Questionnaire (OHQ) [66]. The OHQ is a self-reported, validated symptom assessment tool made up of two components: (a) the OH symptom assessment (OHSA) scale, with six separate items to measure the presence and severity of symptoms, and (b) the OH daily activity scale (OHDAS) with four individual items to measure the impact of orthostatic symptoms on daily activities. The OHQ uses a scale from 0 (no symptoms/no interference) to 10 (worst possible/complete interference) and asks the patient to rate their symptoms for the previous week. Each of the ten individual items within the OHQ can be used individually or as a composite summing together the total score of all items (Fig. 2).

The psychometric properties of the OHQ were evaluated using data from a phase IV multicenter, randomized, placebo-controlled, crossover trial to assess the clinical benefit of the alpha-adrenergic agonist, midodrine, in a group of patients with synucleinopathies and nOH [67]. The OHQ was shown to accurately evaluate the severity of symptoms and the functional impact of nOH as well as assess the efficacy of treatment [66]. The OHQ was later validated for routine clinical use [68]. Both item 1 of the OHSA scale (symptoms of dizziness/lightheadedness) and the composite overall

OHQ and Composite Score

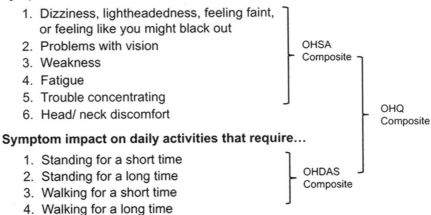

Symptoms

1. Dizziness, lightheadedness, feeling faint, or feeling like you might black out
2. Problems with vision
3. Weakness
4. Fatigue
5. Trouble concentrating
6. Head/ neck discomfort

OHSA Composite

Symptom impact on daily activities that require...

1. Standing for a short time
2. Standing for a long time
3. Walking for a short time
4. Walking for a long time

OHDAS Composite

OHQ Composite

Fig. 2 Orthostatic Hypotension Questionnaire (OHQ). The orthostatic hypotension questionnaire (OHQ) is a validated symptom assessment tool made up of two components: (a) the OH symptom assessment (OHSA) scale, with six separate items, and (b) the OH daily activity scale (OHDAS) with four individual items. The OHQ uses a scale from 0 (no symptoms/no interference) to 10 (worst possible/complete interference) and asks the patient to rate their symptoms for the previous week

OHQ score have been used as primary endpoints in phase 3 clinical trials.

The OHQ has several strengths. First, the OHQ focuses on the full range of symptoms relevant to patients with nOH. Second, the assessment of activities is specific for the most prominent activity impairments imposed by OH, standing or walking for short or long periods of time. Third, the OHQ can accurately measure the symptoms and impact of nOH in a valid and reliable way and, of particular importance for an outcome measure used to assess the impact of treatment interventions, can appropriately detect change over time. Finally, it is brief, making it quick and easy for the patient to complete with minimal patient burden, a common concern regarding patient reported outcomes in clinical trials.

A limitation of the OHQ is that it has a high rate of false positives when used alone, that is, without accompanying BP readings. Patients complaining of "lightheadedness" due to causes other than OH (e.g., vertigo, hypoglycemia, anxiety, and psychosomatic disorders) can score high on the OHQ. Of course, using the OHQ only in populations who have OH confirmed by BP measurements in the supine and standing position theoretically solves this problem. Another minor limitation is that the OHQ asks the patient to rate his symptoms of OH *in the last week*. Therefore, its properties to accurately measure short-term changes in symptoms (e.g., before and 1 h after administration of medication) are not known.

The OHQ has been translated to French, Spanish, German, and Italian. The OHQ underwent translation and linguistic validation that involved two forward translations, forward translation reconciliation, two back translations, and back translation review.

Another patient- and clinician-reported endpoints frequently used as secondary outcome measure in clinical trials of nOH is the Clinical Global Impression (CGI), a tool initially developed for mental health research [69], with two subscales. The CGI-Severity is a seven-point scale scored from 1 (no symptoms) to 7 (severe symptoms), and the CGI-Improvement is a seven-point scale scored from 1 (very much improved) to 7 (very much worse). The patient and the clinician can rate both scales.

6 Objective Outcome Measures for Clinical Trials of nOH

6.1 In-Office Blood Pressure Readings

The most obvious objective outcome measure to be used in clinical trials of nOH is blood pressure. The absolute BP after 1-min or after 3-min of active standing after drug administration was used in several clinical trials of nOH. How long after drug administration should BP be measured will depend on the duration of action of the drug. For instance, in clinical trials of midodrine, the 1 h post-dose BP after 1 min was used as objective endpoint [67]. Because drugs for nOH increase BP regardless of the position of the patient, other trials have used seated BP (with the patient seated in a chair with his/her feet on the floor) 1 h postdose, instead of standing BP, as endpoint [70]. This can be particularly advantageous in order to enroll patients with severely refractory nOH in whom standing for more than a few seconds might be challenging. Of note, the change in BP from the supine to the standing position (i.e., ΔBP) can be deceiving and should not be used as endpoint, as drugs for nOH typically increase the BP regardless of the position of the subject and, therefore, the ΔBP could be similar when comparing ΔBP before vs. after drug administration, or ΔBP in active agent vs. placebo.

Of note, when measuring BP in clinical trials of nOH it is important that subjects remain blind to the BP results, as it is not unusual that patients report feeling lightheaded after seeing in the BP monitor screen that his/her BP values are low (or if an alarm alerting of low BP sounds), but not before.

Finally, some trials of nOH used the time to symptoms upon head-up tilt as a clinical patient-reported endpoint [71]. The primary outcome was the time since the tilt table is positioned in the 60° angle until the patient experiences symptoms of syncope or near-syncope and asks to be tilted down. This endpoint has obvious advantages. Because inability to remain in the standing position is one of the main disability of patients with nOH, determining the clinical relevance of the results is straightforward. It might be difficult to establish what a 1- or 2-point reduction in the OHQ, or an increase in 5–10 mmHg in standing blood pressure represent in terms of functional impairment. In contrast, *time to symptoms after standing* is easier to understand and represents a clear-cut measure of functional change: that is, "the subject was able to remain in the standing position without symptoms for 20 s with the placebo, and for 2 min with the active agent."

6.2 Ambulatory Blood Pressure Monitoring

In patients with nOH, symptom severity and BP readings vary from day-to-day and fluctuates throughout the day and office BP readings do not fully characterize the circadian variation of BP [72]. In clinical practice, ambulatory BP monitoring (ABPM) is often used to evaluate the response to treatment with pressor agents throughout a typical day [72]. In addition to capturing BP variations while awake, ABPM can be used to assess nocturnal BP, which is often elevated in patients (i.e., supine hypertension) in patients with nOH. SH is often worsened by the very therapies used to treat daytime nOH, which creates a difficult clinical dilemma [39, 40, 51, 73]. So far, only one clinical trial examined AMBP profiles in patients with nOH receiving droxidopa and compared them to the AMBP profile of the same patients following a washout period off droxidopa [74]. The endpoints were mean 24-h SBP and mean 24-h DBP and mean nocturnal SBP and DBP.

6.3 Syncope and Presyncope

The drop in BP upon standing in patients with nOH results in cerebral hypoperfusion, which, when severe, can cause near fainting and syncope. Because this, in turn, can result in falls, both syncope and falls could be potentially used as a surrogate marker of nOH in clinical trials.

Using imminent syncope/presyncope as an endpoint is feasible in the in-office setting. Indeed, a phase-4 tilt-table placebo-controlled trial with midodrine used time to symptoms of syncope or presyncope during passive head-up tilt as the main outcome measure [42]. As discussed above, this is a patient-reported outcome that represents a straightforward functional improvement.

6.4 Falls

There two limitations to using falls as outcome measure in trials of nOH. The first one is that not all falls in patients with PD are caused by nOH. The underlying cause of falls in PD may be varied, complex, and multifactorial. Risk factors include older age, female sex, polypharmacy, fear of falling, depression, alcohol use, visual impairment, muscle weakness, use of an assistive device, OH, and cardiac arrhythmia, among others. PD-specific risk factors include prior falls, worse disease severity, medications for PD (higher daily levodopa dosage, dopamine agonist, or anticholinergic use), slow or shuffling gait, freezing of gait, postural instability (balance impairment), flexed posture, axial rigidity, dyskinesia, cognitive impairment, urinary incontinence, and deep brain stimulation. In a study that analyzed direct causes of falls, sudden falls were most common (31%), followed by freezing and festination (20%), neurologic and sensory disturbances (mostly vertigo; 12%), postural instability (11%), OH (4%), and severe dyskinesia (4%); 6.2% of falls were unclassified [75–77]. Therefore, using falls as an outcome measure in clinical trials for nOH in PD and related disorders would require a complex statistical analysis to take into consideration all the other potential causes of falls.

The second limitation is that the retrospective self-report of the number and severity of falls or syncopal episodes relies on the patient's recall, which is prone to bias. Patients may remember the most severe syncope or falls resulting in fractures or other injuries, but overlook minor events without consequences. Even when prospectively filling fall/syncope diaries, patient many not remember to write down the events. Moreover, patients with PD have been shown to significantly underreport adverse events related to medications [78]. This limitation could be potentially overcome with the use of portable devices (e.g., accelerometers or mobile apps) that can prospectively detect and record the number of falls without requiring the patient's intervention, as will be later discussed. In placebo-controlled trials, it is also necessary to perform a stratified randomization based on frequency of falls. These same limitations apply when using falls as outcome measure in other neurological disorders [79].

Analysis of falls reported as an adverse event during the early clinical development of droxidopa suggested that droxidopa might reduce falls in patients with nOH. Studies NOH301 [80] and NOH302 [81] enrolled patients with symptomatic nOH caused by synucleinopathies. After open-label droxidopa dose titration, responders were washed out for 1 week and then randomized to either droxidopa or placebo for 1 week (NOH301) [80] or continued on droxidopa for 1 week and then randomized to either droxidopa or placebo for 2 weeks (NOH302) [81]. When the combined data from the blinded phases of both trials were examined, there was 1 (0.8%) adverse event of fall in the droxidopa group compared to 9 (6.8%) adverse events of falls in the placebo group.

Study NOH306 was a 10-week, phase 3, randomized, placebo-controlled, double-blind trial of droxidopa in patients with PD and symptomatic nOH that included assessments of falls as a key secondary endpoint [82–84]. The principal analysis consisted of a comparison of the rate of patient-reported falls from randomization to end of study in droxidopa versus placebo groups. To do so, patients were instructed to record in a daily electronic diary all of their falls, defined as "unexpectedly coming to rest on the ground, floor, or a lower level from where the patient started." If patients reported a fall on a particular day, the electronic diary would ask how many times they fell that day. The device would then ask the patient a series of questions based on the worst fall of the day, including whether they experienced freezing, were lightheaded, or lost consciousness just before the fall. Patients could choose none, one, or more than one of these options. The 92 patients receiving droxidopa reported 308 falls, and the 105 patients receiving placebo reported 908 falls. In the droxidopa group, the fall rate was 0.4 falls per patient-week; in the placebo group, the rate was 1.05 falls per patient-week, yielding a relative risk reduction of 77%. Fall-related injuries occurred in 16.7% of droxidopa-treated patients and 26.9% of placebo-treated patients. Therefore, treatment with droxidopa appears to reduce falls in PD patients with symptomatic nOH [82]. The limitations of this approach, as mentioned before, include the fact that the randomization was not stratified based on frequency of falls, and that the recording of falls relied on the patients' ability to remember and describe the event.

6.5 Development of Biometric Monitoring Devices for Clinical Trials of nOH

There is increasing interest in the use of biometric monitoring devices (BMD) as a path to accurately and objectively reflect the disease process and status of patients during their daily lives to be used as endpoints in clinical trials [85, 86]. In the case of patients with nOH the most straightforward application of BMD would be that of providing measurements of BP in conjunction with their position (i.e., sitting, standing, flat) at specific times, or even constantly. This, of course, would have to be done wirelessly with smaller, lighter and more convenient devices than the currently available ABPM. Other potential applications of BMD to be used in clinical trials of nOH include measurement of falls, mobility, study drug adherence, and quality of life surrogates such as socialization. This could be potentially accomplished with biosensors measuring biological responses using wearable sensors such as watches, smartphones, clothing, implants, ingestible sensors, or remote biosensors in specific locations (beds, chairs, doors). Medication adherence is a low as 17% in older subjects with cognitive impairment [87]. Moreover, the use of in-home monitoring data may reduce sample sizes required in clinical trials [88]. However, before BMD can be used in clinical trials a rigorous path of validation must be followed [86].

7   Design of Clinical Trials of nOH

The length of the trial should be also taken into consideration when deciding the design. Because the endpoints of clinical trials for nOH can be measured in a relatively short length of time (e.g., SBP after 1 min of standing 1 h postdose), the trial can be finished a relatively short period of time, in contrast to clinical trials of symptoms that may take several weeks/months to change (e.g., depression) or clinical trials of disease modification.

The gold standard design for clinical trials of nOH remains the parallel, double-blind, randomized, placebo-controlled trial. This

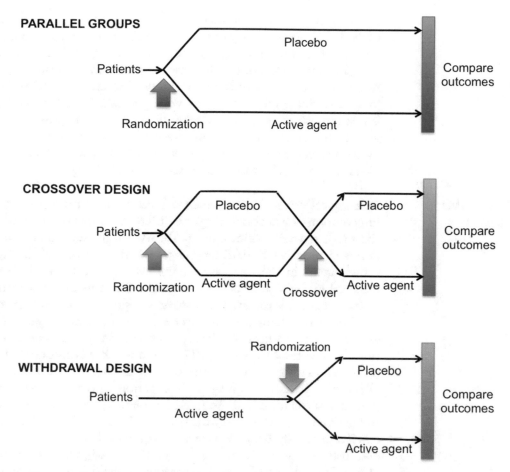

Fig. 3 Designs of clinical trials for neurogenic orthostatic hypotension. The gold standard design for clinical trials of nOH remains the double-blind, randomized, parallel-group, placebo-controlled trial. This design allows for easier interpretation of the findings and eliminates several known sources of bias. Other designs, however, are used frequently in clinical trials of nOH, including the crossover design and the withdrawal design (see text for details)

design allows for easier interpretation of the findings and eliminates several known sources of bias. An easily interpretable study that clarifies both the efficacy and safety profile of a drug for nOH allows for straightforward interpretation by patients, regulatory agencies, clinicians, and payers. Parallel designs are challenging and, frequently, relatively short-acting mechanism of action of nOH drugs allows for other designs (e.g., crossover design, withdrawal design, n-of-1 design) (Fig. 3).

7.1 Parallel Design

In a typical parallel study, each subject receives only one intervention or placebo during the trial. In the simplest parallel design, subjects are randomized to either A (e.g., active agent at a certain dose) or B (e.g., placebo or active agent at a different dose) and remain receiving that compound for the whole duration of the trial. Although this is the most commonly used design in clinical trials, it has several limitations. It typically requires large sample sizes compared to other designs (e.g., crossover, see below) [30]. Also, in clinical trials of nOH, some subjects may not want to participate in a trial if there is a probability of receiving placebo for a long time. Some subjects may be willing to take the risk if the trial is relatively short (e.g., 1–2 weeks) or they have the possibility of getting the active drug at some point during or after the trial. However, recruitment for long (e.g., 1-month) parallel studies for nOH with a placebo arm may be extremely challenging.

7.2 Crossover Design

In a typical crossover trial, each subject receives more than one intervention or placebo during the different periods of the trial. In the simplest crossover design, a AB:BA design, subjects are randomized to either the AB sequence, where they receive A (e.g., active agent) in the first period, followed by treatment B (e.g., placebo) in the second period, or to the BA sequence, where the treatment order is reversed. Crossover designs are ideal for chronic conditions that remain stable over a period of months, years and are unlikely to progress (or improve) significantly over a period of time, as it is the case with nOH. There is a key consideration in crossover designs, which is the carryover, that is, the lingering effect of a treatment given in one period into the subsequent period of the clinical trial. To mitigate the carryover effect, a washout period (i.e., a period of time between A and B when the patient receives no study drug) is typically required. This is the simplest and usually favored method to reduce or eliminate carryover effects. If the effect of drug A is reasonable rapid and the outcome is related to the physiological concentration of the drug, then using a washout period is a straightforward strategy. However, other situations might be more complicated if pharmacologic effects persists beyond the physical elimination of the drug. Placebo-controlled crossover designs were used in a phase 2 trial of droxidopa [89], phase 3 and phase 4 trials of midodrine [67, 71], phase 2 trials of atomoxetine [70, 90, 91], and acarbose for postprandial hypotension [92]. This designed was recently used in a trial of a servo-controlled inflatable abdominal binder vs. midodrine [93].

7.3 Withdrawal Design

There are several variations of a withdrawal design. In the withdrawal design used in clinical trials of nOH, participants received the active agent (Period 1) and followed for a fixed length of time. Responders are then randomized to receive active agent or placebo in a second period (Period 2). Period 1 and Period 2 do not have to be of equal length. Period 1 should be long enough to allow for the detection of a change in the selected outcome measure and Period 2 is chosen to be long enough to eliminate (or wash out) any symptomatic effect of the treatment from Period 1 in subjects receiving placebo so that they return to their baseline. This withdrawal design was used in the NOH302 clinical trial of droxidopa [81]. It has two advantages: the first is that Period 1 can identify responders that will be then randomized during Period 2. Also, the exposure of patients to placebo during Period 2 may be shorter than in a randomized placebo-controlled trial. The main problem is that there is no blinding with respect to the treatment received during Period 1. In addition to the obvious biases that can result, participant retention may become a problem in the placebo group during Period 2, particularly if it is lengthy, as participants will be aware that they are not receiving the active agent.

7.4 N-of-1 Design

N-of-1 trials are designed to determine which treatment (or dosage) is more effective for a particular individual of interest [94]. Usually, this design involves assigning two or more treatments (or dosages) to an individual in a random order. This design can be useful when deciding a clinical strategy for a particular patient [95]. However, due to its small sample size ($n = 1$) these designs have a limited application in the research/drug approval setting. A potential example of n-of-1 trial could be inferred from the design used in a clinical trial with midodrine [67]. In this trial, 25 patients were studied on 6 days. On day 1 and day 6 they received no medication. They were then randomized to receive, on successive days, placebo, midodrine 2.5 mg, midodrine 5 mg, or midodrine 10 mg. The endpoint was the 1 h postdose BP after 1 min standing. In the case of a single patient, this design can be the ultimate tool for personalized clinical medicine to determine the most appropriate medication (e.g., midodrine vs. droxidopa vs. atomoxetine) or dosage (e.g., midodrine 2.5 mg vs. 5 mg. vs. 10 mg).

7.5 Enrichment Strategies

Enrichment strategies for drug development studies might be useful to focus and accelerate the process of drug evaluation. When properly selected, these strategies identify a study population most likely to demonstrate the candidate drug's effects and provide evidence that the drug is effective. If the drug is ineffective, such strategies will not falsely suggest effectiveness. There are three enrichment strategies in clinical trials of nOH: (1) practical enrichment, that is, decreasing the heterogeneity of the population to reduce variability; (2) prognostic enrichment, that is, identifying high risk subjects more likely to experience the outcomes that are being measured; (3) predictive enrichment, that is, selecting subjects that respond to the candidate treatment.

Practical enrichment strategies include prospectively identifying likely compliant subjects who will not drop out of the study, excluding subjects who are too unstable with respect to the signs and symptoms of the disease (e.g., subjects with mild, occasional, unsustained fall in BP), identifying subjects unlikely to survive long enough to be evaluated (e.g., subjects with MSA and nOH with a disease duration longer than 3 years), and excluding subjects using medications or devices with the same effect as the candidate drug (e.g., using high pressure compression stockings in clinical trials of pharmacologic agents against nOH).

Prognostic enrichment strategies include identifying and enrolling patients at high risk of developing the study endpoint (e.g., subjects with an OHQ score of at least 3 points, and who are, therefore, susceptible to have a measurable and clinical significant improvement in the OHQ score after receiving the study drug).

Predictive enrichment strategies include identifying and enrolling patients that respond to the study drug. This strategy was used all phase 3 clinical trials with droxidopa [74, 80–84]. The trials had an initial dose optimization period with a forced upward titration from 100 to 600 mg three times daily. Dose titration lasted a maximum of 14 days with patients receiving escalating dosages in 100 mg three times daily increments until (a) becoming asymptomatic (with a score of 0 in the item 1 of the OHQ [66]), (b) developing sustained supine hypertension >180/110 mmHg, (c) reaching the maximum dose of 600 mg three/times day; or (d) patient complains of intolerable side effects.

Responders were defined as patients that had at least a 1-point improvement in item 1 of the OHQ (dizziness/lightheadedness) in conjunction with a 10 mmHg or more increase in standing systolic BP. Only responders were allowed to continue in the double-blind, placebo-controlled phase of the trials.

7.6 Telemedicine

Because clinical trials of nOH typically have patient-reported outcomes as primary endpoints (e.g., OHQ) it is conceivable that the patient may not have to come to the clinical center for a clinical trial visit. Similarly, if BP is used a secondary outcome measure, the own patient can easily measure his/her BP in the supine and standing positions at home. These circumstances make telemedicine study visits feasible in clinical trials of nOH. Telemedicine has been proven feasible in clinical trials of other disorders and may overcome geographic barriers to clinical trial participation.

Some caveats should be taken into consideration. In clinical trials where BP readings are required, the clinical trial personnel should supervise the study subject while he/she is measuring his/her BP. This can be achieved with a videoconference call. However, not every videoconferencing tools are secure and compliant with patient health data protection laws (e.g., Health Insurance Portability and Accountability Act of 1996, HIPAA in the USA). Video conferencing tools in clinical trials must protect data privacy in that all audio/video communication is securely encrypted and transmitted from point-to-point such that even the video conferencing company providing the services does not have access to any identifiable health information that may be communicated. The company must offer a law-required agreement where the company agrees to be responsible for keeping all the patient information secure and to immediately report any breath of personal health information. There is the possibility of recording and storing the videoconferences with the patients for data quality assurance and auditing. Also, BP readings should be as reliable and reproducible as possible. This can be achieved by giving all patients the same BP machine, and by ensuring that all the BP readings required throughout the trial are performed at the same time to avoid too much BP variability (e.g., 30-min before lunch to avoid postprandial BP effects) and in the same place (e.g., bedroom). Telemedicine assessments were used in a clinical trial of carbidopa for BP variability in patients with familial dysautonomia (ClinicalTrials.gov NCT02553265).

8   Recruitment Strategies of Patients with nOH

Advancing drug development for rare disease such as nOH requires cooperation and collaboration among different stakeholders. An essential area for success of drug development is timely and expeditious identification and recruitment of research participants for clinical trials. Specialized treatment centers provide the traditional sites for recruiting patients for clinical trials in rare diseases. Increasingly, patient themselves and patient-advocacy groups or foundations are becoming well-informed sources for patient recruitment. The development of patient registries is a key element to support a clinical research program. Patient registries contain contact information, diagnosis, and limited demographics facilitating identification of potential participants from various backgrounds. A recent and successful example of patient registry is the Global Multiple System Atrophy Registry (GLOMSAR), and initiative of the National Institutes of Health Rare Disease Clinical Autonomic Consortium in partnership with the Multiple System Atrophy Coalition, a patient-advocacy group. The information in GLOMSAR is used to alert registrants (e.g., via e-mail or phone calls) about research opportunities of potential interest to them, including clinical trials for new drugs against nOH. Patient-advocacy groups and foundations can also become a sophisticated resource regarding research activities regarding their disorders of interest. So far, there is no a specific patient-advocacy group or foundation focusing exclusively on nOH. In this case, organizations of patients with PD or MSA become the major supporters for patient recruitment for clinical trials of nOH. Accelerating enrolment through collaboration with patient groups may substantially shorten drug development time. Web-based technologies, including social media, offer a potentially powerful approach to recruiting as well [96].

The National Institutes of Health (NIH) has several programs that help support the development of clinical research for rare diseases. One of them, the Rare Diseases Clinical Research Network (RDCRN) provides support for clinical studies and facilitates collaboration, study enrolment and data sharing. The RDCRN includes 22 consortia consisting of multidisciplinary teams collaborating with representatives of patient advocacy groups involved in over 200 rare diseases. One of these 22 consortia is the Autonomic Disorders Consortium (ADC), comprised by the New York University Dysautonomia Center (New York, NY), Mayo Clinic (Rochester, MN), Beth Israel Deaconess Medical Center (Boston, MA), Vanderbilt University (Nashville, TN) and the Intramural Neurocardiology Section of the NIH (Bethesda, MD). The RDCRN ADC has several clinical studies including GLOMSAR, as well as the Natural History of the Synucleinopathies Study, a prospective natural history study to define the evolution, risk factors and biomarkers of patients with nOH. These are great resources for patient recruitments for clinical trials of nOH.

Many clinical trials are performed internationally to enroll an adequate number of participants. Therefore, special considerations should be made to address the complexities of conducting clinical trials across languages and cultures, recruitment, trial conduct and translation, and validation of patient-reported outcome measures.

9   Clinical Trials of Therapies for nOH

We here review clinical trials for nOH of nonpharmacologic and pharmacologic therapies, with emphasis on double-blind, placebo-controlled trials. We specify their population, outcome measures, main results, and potential biases.

9.1 Non-pharmacologic Measures

Nonpharmacologic measures are often the first step in the management of patients with nOH. That water drinking significantly and rapidly increases BP in patients with nOH was first studied in an open-label fashion in 28 patients with MSA and 19 with PAF, as well as in 19 healthy controls. After 30 min of drinking 500 ml of tap water, seated SBP/DBP increased $33 \pm 5/16 \pm 3$ mmHg in patients with MSA and $37 \pm 7/14 \pm 3$ mmHg in patients with PAF ($p < 0.0001$ vs. baseline in both disorders). The increase in SBP was sustained for >60-min [97, 98]. Symptoms of nOH were not assessed.

The first attempt to systematically study nonpharmacologic measures in patients with PD and OH was a 3-week open-label trial of 17 patients with PD and OH [99]. The trial design and endpoint were not well defined. The primary outcome measure appeared to be the ΔSBP and ΔDBP after active standing (unclear if after 1 or 3 min). The orthostatic domain of the COMPASS scale

[61] was used as patient-reported outcome. Patients were then instructed to follow 12 nonpharmacologic measures (including but not limited to drinking 1250 ml of water daily, taking 10–20 g of salt daily, elevating the head of the bed 10–15 cm, using thigh-high 30 mmHg compression stockings, having frequent small meals, and avoiding alcohol) that were explained to the patient and were printed in a sheet and given to the patient. Three weeks later, the ΔBP was measured again and the COMPASS was readministered. There were not significant changes in ΔBP or in the COMPASS after the 3-week open-label treatment period. The lack of significant differences is not surprising for a number of reasons. Patients with OH often perceive nonpharmacologic measures as inconvenient. Liberalizing water intake increases urine output and therefore urinary frequency. Compression garments must be tight to be effective and, consequently, are challenging to put on. Sleeping with the head of the bed raised 30–45° is only effectively accomplished with an electric bed or mattress, rather than just extra pillows. Therefore, even though the reported compliance of the nonpharmacologic measures was ~70%, it is likely that patients applied them half-heartedly if no one was monitoring them. Overcoming this limitation can be challenging, but remote biometric monitoring devices could be potentially helpful. Second, as discussed before, the COMPASS is not specifically validated to measure symptoms of OH, so its use is quite questionable. It must be said, however, that the OHQ had not been developed yet at the time of this trial.

The second trial studying nonpharmacologic measures enrolled 15 patients with PD and OH who were randomized to either use a compressive abdominal binder (20 mmHg of abdominal pressure) or a sham abdominal binder (3 mmHg of abdominal pressure) during a head-up tilt procedure [100]. The primary endpoint was the mean BP after 3-min of head-up tilt. Use of the compression abdominal binder elicited an increase in 7.7 mmHg compared to a decrease in −2.7 mmHg of the sham abdominal binder. This was followed by a 4-week open-label phase where all patients used the compression abdominal binder, which resulted in a reduction of orthostatic symptoms denoted by a reduction of the OHQ of 2.2 points compared to baseline ($p = 0.003$) [100].

The third and most recent trial used a much more sophisticated, servo-assisted splanchnic compression device (40 mmHg of abdominal pressure), which actives only upon standing up [93]. Twenty-three patients with nOH (three of which had PD) were randomized, in a crossover manner to receive either a single oral dose of placebo, midodrine 2.5–10 mg, placebo combined with abdominal binder (40 mmHg), or midodrine 2.5–10 mg combined with abdominal binder (40 mmHg). The primary outcome was the change from baseline in orthostatic tolerance, defined as the AUC of upright SBP calculated by the trapezoidal rule

(ΔAUC_{SBP}: upright SBP multiplied by standing time). This is a composite score that integrates both the standing time and the upright SBP [91]. On standing, inflation of the abdominal binder and midodrine produced a similar significant ΔAUC_{SBP} (i.e., improved orthostatic tolerance) compared to placebo (195 ± 35 and 197 ± 41 vs. 19 ± 38 mmHg × min on the placebo day; $p = 0.019$ and $p = 0.010$, respectively). The same number of patients in the binder and midodrine groups ($n = 14$) were able to stand for the full 10 min of the posttreatment orthostatic test, whereas only ten did in the placebo group. In clinical practice, the effectiveness of abdominal binders in patients with nOH is limited by their difficulty of use and low compliance. Waist-high compression stockings require substantial time and significant effort to put on, and once on it is impractical for patients to take them off even for periods when they are no longer needed. Elastic abdominal binders are easier to put on and off, but are difficult to apply at the required compression level for them to be effective (20–40 mmHg). Most patients may require assistance applying them, making it difficult to selectively use abdominal compression only while standing. The development of an automated abdominal binder has several advantages [93]: (a) it can apply effective abdominal compression (manually or automatically) only when needed (when the patient stands) and, (b) the servo-controlled design automatically inflating and deflating results in the maintenance of a predefined fixed compression level. This latter feature would avoid excessive, and potentially harmful, increases in intraabdominal pressure when performing physiological maneuvers such as coughing, sneezing, breathing deeply, etc.; or while bending over or lifting heavy objects. Additional trials with the automated servo-assisted splanchnic binder are underway (ClinicalTrials.gov NCT02429557).

9.2 Fludrocortisone

Even though there have been limited studies and that it is not specifically FDA-approved for this indication, fludrocortisone is frequently used for the treatment of nOH. Fludrocortisone (9α-fluorocortisol) is a synthetic mineralocorticoid that increases renal sodium and water reabsorption, thus expanding intravascular volume and increasing BP in all positions. Fludrocortisone exacerbates supine hypertension and end organ target damage (left ventricular hypertrophy and renal failure) and increases the risk of all-cause hospitalization in patients with nOH [101]. Additional, frequent side effects include hypokalemia and ankle edema [102, 103]. To reduce the risk of hypokalemia, patients taking fludrocortisone are frequently instructed to eat potassium-rich foods and to take potassium chloride 10–20 mEq/day. The dosage of fludrocortisone should not be higher than 0.2 mg/day. Higher dosages are rarely more effective but intensify side effects. Clinical effects of fludrocortisone usually require ~10 days of treatment.

There are no placebo-controlled clinical studies evaluating the efficacy of fludrocortisone in patients with nOH. There are two crossover trials that compared fludrocortisone to other active agent in patients with PD and OH.

In the first one, 13 patients with PD and OH were randomized to receive either fludrocortisone 0.1 mg/day or domperidone 30 mg/day for 3 weeks, and, after a 1-week washout period, were crossed over to receive the other medication for another 3 weeks [99]. The ΔSBP after 3-min, ΔDBP after 3-min, and the maximum reduction in BP within 5-min as well as the COMPASS and the CGI were measured as endpoints. There were no statistical differences in any of the BP measurements after taking the medications, even though there were significant improvements in the COMPASS and the CGI scores.

The second, more recent study was an oligocenter, double-blind, randomized trial where nine patients were randomized to receive either fludrocortisone 0.2 mg/day or pyridostigmine 60 mg/day for 14 days, and then crossed over to the other agent, with a 21-day washout period in between. For unclear reasons, the primary outcome measure was the ΔDBP within 3-min of active standing, and not the ΔSBP. Symptomatic burden was measured with the OHSA. When on fludrocortisone, patients had a statistically significant less severe pronounced ΔDBP within 3-min of active standing than at baseline (post hoc comparisons $p = 0.0016$), but not when compared to pyridostigmine. There were no differences in ΔSBP ($p = 0.26$) or in the OHSA score either.

The lack of statistically significant results is not surprising because, as explained before, pressor medications are expected to increase BP regardless of the position of the patient, and, therefore, the ΔBP (i.e., the difference between the BP in the sitting position and the BP after a certain time in the standing position) should remain essentially unchanged before and after active agent, even though one agent might be more successful than the other to increase BP. This explains why, in the first trial [99], there were symptomatic improvements, but no improvement in the change in BP. And also why, in the second trial [104], the minimum mean (absolute) BP within 3-min of active standing was significantly higher in patients taking fludrocortisone (83.9 ± 17.3 mmHg) vs. pyridostigmine (69.9 ± 16.4 mmHg) ($p = 0.002$ post hoc comparisons). Therefore, using absolute BP values instead of delta would have been much more appropriate endpoints.

9.3 Midodrine

Midodrine is an oral α_1-adrenoceptor agonist that induces vasoconstriction and increases BP [67, 71, 105, 106]. Midodrine is, with droxidopa, the only FDA-approved drug for the treatment of symptomatic nOH. Midodrine raises BP in the standing, sitting, and supine positions and its pressor effect is noticeable ~30–45 min after consumption, reaching a maximum after ~1 h, and persists for a total of 2–3 h. Treatment should begin with a 2.5 or 5 mg dose, which can then be increased up to 10 mg to be taken up to three times a day. Supine hypertension is common; hence, patients should not take midodrine less than 3–4 h before bedtime. Other side effects owing to activation of α1-adrenergic receptors are pilo-erection ("goosebumps"), itching of the scalp, and urinary retention. Midodrine has no effect on heart rate as it does not activate β-adrenoreceptors and, given its poor diffusion across the blood–brain barrier, has no CNS adverse effects [107].

Early, open-label studies showed that midodrine (25–40 mg/day) significantly increased the average mean BP in the standing position and improvement in orthostatic tolerance after 7 days of treatment in patients with nOH compared to baseline (i.e., no treatment), but only in a subgroup of subjects [108, 109].

The first phase 3 randomized placebo-controlled trial of midodrine for nOH enrolled 97 subjects, 22 of which had PD, 20 had PAF, and 18 had MSA [105]. The design of the trial included a 1-week single-blind placebo period. After this period, patients were randomly assigned to four treatment groups: placebo, midodrine 2.5 mg, midodrine 5 mg, or midodrine 10 mg, administered orally three times/day for 3 or 4 weeks. The primary outcomes were standing SBP (unclear after how long after standing) and symptoms of OH measured dichotomically (improvement yes/no) compared to the patient's baseline, measured 1 h postdose. Midodrine significantly ($p < 0.001$) increased standing SBP by 22 mmHg compared to only 3 mmHg in the placebo group. Midodrine also increased standing DBP by 15 mmHg ($p < 0.05$), with no changes in supine or standing heart rate. Dizziness/light-headedness, syncope, depression, and standing time also improved significantly in patients taking midodrine compared to placebo [105].

The largest phase 3 trial of midodrine for nOH enrolled 171 subjects in a randomized double-blind, placebo-controlled multi-center study [106]. The study started with a 1-week single-blind placebo period, followed by a 3-week double-blind (midodrine 10 mg three times/day or placebo) period, and a 2-week single-blind placebo period. The primary endpoints were standing BP (not defined after how long standing) and symptoms of lighthead-edness using a visual analog scale (0–10). It was unclear how long after the medication dose were the outcomes measured. Secondary endpoint was a symptom composite score of quality of life. Overall, midodrine induced significant increases in standing BP in all study visits, and a significant reduction in symptoms of lightheadedness, but only in the final visits. The most frequently reported adverse events were piloerection (goose bumps), pruritus, and urinary retention [106].

A double-blind, randomized, dose-response, placebo-controlled, crossover trial enrolled 25 patients with nOH. In this trial, subjects were studied on 6 days. On day 1 and day 6 they received no medication. They were then randomized to receive, on successive days, placebo, midodrine 2.5 mg, midodrine 5 mg, or midodrine 10 mg. The endpoint was the 1 h postdose BP after 1-min standing. Midodrine resulted in significantly increased standing SBP, peaking at 1 h postdose. The global improvement of symptoms score also improved significantly with midodrine [67].

The results of these trials resulted in accelerated approval of midodrine by the FDA for the treatment of OH in 1996. However, because the FDA considered that the increase of standing BP is a surrogate maker of effectiveness that would likely correspond to a clinical benefit and required postmarketing studies to confirm that midodrine provides a clinical benefit for patients with nOH.

To that end, a phase 4 tilt-table, double-blind, placebo-controlled, randomized, crossover, multicenter study enrolled 19 patients [71]. Patients underwent an open-label 4-week period where patients received their usual midodrine dose (2.5–15 mg/day). Midodrine was then stopped for 1 day. The next day, patients were randomized to receive either midodrine or placebo and undergo a tilt-table test. The following day, patients were crossed over to the other study drug. The primary endpoint was time to syncope or near-syncope using a 45-min head-up position with a tilt-table at 1 h postdose of midodrine. The time to symptoms in patients receiving midodrine was significantly longer than when receiving placebo (27 min vs. 18 min, $p = 0.0013$), thus confirming the symptomatic benefit of midodrine in patients with nOH (Table 1).

9.4   Droxidopa

On February 18, 2014, the US Food and Drug Administration (FDA) approved droxidopa (Northera®), an orally active synthetic precursor of norepinephrine, for the treatment of symptomatic neurogenic orthostatic hypotension (nOH). It was the first new drug approval for nOH in almost 20 years. Two years before, the FDA Cardiovascular and Renal Drugs Advisory Committee had voted 7 out of 13 in favor of approval, but the FDA requested more data. With the results of a new clinical trial, a second Committee meeting voted almost unanimously to approve droxidopa (16 in favor out of 17). Droxidopa was approved for the treatment of symptomatic nOH in patients with primary autonomic failure due to PD, PAF, MSA, dopamine beta-hydroxylase deficiency, and nondiabetic autonomic neuropathies. *Nondiabetic autonomic neuropathy* is a broad term that includes rare autoimmune, genetic, and other autonomic neuropathies.

Droxidopa is not a new compound. It was first synthesized in 1919 by German chemists [110] who thought it could be a catecholamine precursor. In the late 1940s, Blaschko and colleagues in England showed that droxidopa could be converted to norepinephrine, in vivo, and that this step required the action of the

Table 1
Placebo-controlled clinical studies with midodrine in patients with neurogenic OH

Study	Subjects (N)	Arms	Design	Primary endpoints	Results
Jankovic et al. 1993 [105]	97	Placebo three times/day Midodrine 2.5 mg three times/day Midodrine 5 mg three times/ day Midodrine 10 mg three times/day	3-week, four-arm, double-blind, randomized, placebo-controlled study	Standing SBP (unclear after how long) 1 h postdose Symptom improvement (yes/no) 1 h postdose	Subjects on midodrine had significantly higher standing SBP and lower burden of symptoms ($p < 0.001$) than those on placebo
Low et al. 1997 [106]	193	Placebo three/ times a day Midodrine 10 mg three times/day	3-week, double-blind, randomized, placebo-controlled study	Standing SBP (unclear after how long) 1 h postdose Lightheadedness symptoms (using a 0–10 scale) 1 h postdose	Subjects on midodrine had significantly higher standing SBP and lower burden of symptoms ($p < 0.001$) than those on placebo
Wright et al. 1998 [67]	25	Placebo Midodrine 2.5 mg Midodrine 5 mg Midodrine 10 mg	Double-blind, randomized, dose-response, placebo-controlled, crossover study	Standing SBP after 1-min standing within 3 h postdose	Midodrine increased SBP in a dose-dependent manner, peaking at 1 h postdose
Smith et al. 2016 [71]	19	Placebo Midodrine (2.5–15 mg/ day)	Double-blind, placebo-controlled, randomized, crossover, tilt-table study	Time to symptoms upon head-up tilt 1 h postdose	The time to symptoms in patients receiving midodrine was significantly longer than when receiving placebo (27-mins vs. 18-min $p = 0.0013$).

enzyme dopa-decarboxylase, a.k.a. aromatic amino acid decarboxylase [111–113]. In 1989, it was reported that in patients with familial amyloid polyneuropathy and symptomatic nOH treatment with 600 mg of droxidopa increased plasma norepinephrine levels and standing blood pressure [114]. Subsequent studies in Japan led to its approval in 1989 for the treatment of nOH in patients with PD, MSA, and familial amyloid polyneuropathy [115].

Droxidopa (L-threo-3,4-dihydroxyphenyl-serine, L-DOPS) is an oral synthetic amino acid that is converted to norepinephrine in the body [116]. Droxidopa is decarboxylated to norepinephrine by the enzyme aromatic amino-acid decarboxylase (AAAD) the same enzyme the converts L-dopa to dopamine. Extensive clinical experience shows that droxidopa is safe and well tolerated [117–125]. Peak plasma concentrations of droxidopa are reached ~3 h after oral administration. The dosage used in clinical trials was 100–600 mg three times/day although clinical experience indicates that the dosage should be tailored to each patient's needs considering the periods of time when he/she is going to be active or inactive [116, 118, 123].

9.4.1   Phase 2 Studies

Initial studies with droxidopa in patients with nOH were conducted with a racemic mixture of the D- and L-stereoisomers, and showed conflicting results [126]. In 1984, a placebo-controlled study using racemic DL-threo-DOPS failed to show improvement in six patients with nOH (due to diabetic autonomic neuropathy and PAF). Only 2% of the administered DL-DOPS was converted to biologically active L-norepinephrine [127]. A later study in patients with nOH showed that DL-DOPS did increase supine and upright BP as well as norepinephrine [128]. An early, small ($n = 6$), open-label study performed in the USA showed that droxidopa increased upright BP in patients with MSA, but less in PAF [129]. A subsequent, better-powered double-blind, randomized, placebo-controlled study showed that a single dose of droxidopa (200–2000 mg) effectively increased BP and improved symptoms in 19 patients with nOH due to MSA and PAF [89]. An open-label European study showed that droxidopa (100–300 mg three times daily) improved symptoms of nOH in patients with MSA and PAF after 6 weeks of treatment [130] (Table 2).

9.4.2   Phase 3 Studies

Because of the variable pressor response in patients with nOH, phase 3 trials had a dose optimization period with a forced upward titration from 100 to 600 mg three times daily. Dose titration lasted a maximum of 14 days with patients receiving escalating dosages in 100 mg three times daily increments until (a) becoming asymptomatic (with a score of 0 in the item 1 of the Orthostatic Hypotension Questionnaire [OHQ] [*see below*] [66]), (b) developing sustained supine hypertension >180/110 mmHg, (c) reaching the maximum dose of 600 mg three/times day; or (d) patient complains of intolerable side effects. Responders were defined as patients that had at least a one-point improvement in item 1 of the OHQ (dizziness/lightheadedness) in conjunction with a 10 mmHg or more increase in standing systolic BP.

Study NOH301 was a double-blind, randomized, placebo-controlled, parallel-group study conducted at 94 US, Canadian, and European sites. The trial enrolled 162 patients with nOH due to PD, MSA, PAF, or nondiabetic autonomic neuropathy [80]. Following an open-label dose titration phase (starting at 100 mg

Table 2
Phase 2 clinical studies with L-DOPS and DL-DOPS in patients with neurogenic OH

Study	Subjects (*N*)	Active agent	Design	Results
Hoeldtke et al., 1984 [127]	6 (2 PAF and 4 autonomic diabetic neuropathy)	DL-DOPS	Open-label crossover study using a single dose of DL-DOPS (600 or 800 mg) and single dose of placebo in separate days	No change in supine or upright BP after DL-DOPS
Kaufmann et al., 1991 [129]	6 (2 PAF; 4 MSA)	L-DOPS	Open-label dose-titration study followed by 2 days of open-label administration (700–1000 mg/day)	L-DOPS increased supine BP in MSA and PAF, and standing BP only in MSA
Freeman et al., 1999 [128]	10 (6 MSA; 4 PAF)	DL-DOPS	Randomized, double-blind, placebo-controlled, crossover study using a single dose of 1000 mg of DL-DOPS and single dose of placebo	DL-DOPS significantly increased supine and upright BP
Mathias et al., 2001 [130]	32 (26 MSA; 6 PAF)	L-DOPS	Open-label incremental study (4 weeks) followed by a 6-week maintenance study (100–300 mg twice/day)	L-DOPS significantly reduced the fall in systolic BP upon standing and the symptoms of OH
Kaufmann et al., 2003 [89]	19 (11 MSA; 8 PAF)	L-DOPS	Single-blind dose-titration study followed by a 3-day double-blind, placebo-controlled, crossover trial	L-DOPS significantly increased supine and standing BP for several hours and improved symptoms of OH. After L-DOPS, BP increases were associated with increases in plasma NE levels

and escalating to 600 mg three times/day), "responders" (162 patients, 62% of enrolled) proceeded to a washout period and were then randomized to receive placebo or droxidopa for 1 week (double-blind). Differences between groups were compared on day 7. Compared with placebo, patients randomized to droxidopa had better improvement in overall OHQ scores ($p < 0.003$) and standing systolic BP that was 7 mmHg higher ($p < 0.001$). Differences between groups were modest, partly because the placebo group failed to worsen back to their original state, suggesting a possible carry over effect of droxidopa requiring a longer washout period to disappear.

Study NOH302 was a double-blind, placebo-controlled, randomized *withdrawal* trial including 181 patients with nOH due to PD, MSA, PAF, or nondiabetic autonomic neuropathy [81]. Study 302 began with an open-label titration and 1 week treatment phase. After completing the open-label treatment phase, responders were randomized to withdraw to placebo or continue taking droxidopa. Differences between groups were assessed 2 weeks after randomization. The mean dose of droxidopa was 386 ± 178 mg (three times/day), slightly lower than in study 301 ($p > 0.33$). After withdrawal, there were no significant differences in symptoms of dizziness/lightheadedness (OHQ item 1), the primary outcome measure, between droxidopa and placebo patients. This was because placebo patients experienced continuing relief of the dizziness/lightheadedness score and standing SBP at the end of the study compared with baseline. This again raised the possibility of a carryover effect of droxidopa. Substantial carryover effects have been observed for levodopa, which like droxidopa undergoes decarboxylation to become a neurotransmitter. Alternatively, patients may have improved simply by their participation in the study, for example, by better adherence to nonpharmacologic treatment recommendations [81]. Post hoc analysis of the overall composite score for OH symptoms and activities (OHQ composite), however, did show a significant improvement with droxidopa ($p = 0.026$). Standing systolic BP was not different between the two treatment groups. Patients' self-ratings of clinical global impression (CGI) of symptom severity showed a significant difference favoring droxidopa ($p = 0.008$).

Study NOH306 was originally designed to evaluate the clinical efficacy of droxidopa over an 8-week double-blind period [84]. In a preplanned interim efficacy analysis, the initial 51 subjects (study NOH306A) [83] showed no significant difference across groups in the change in OHQ composite score, the trial primary endpoint. Exploratory analyses suggested efficacy for dizziness/lightheadedness score and led to a change in the trial's primary efficacy measure while data for subsequent subjects remained blinded. The subsequent 171 enrolled patients formed study NOH306B, multicenter, double-blind, randomized, placebo-controlled, parallel-group study restricted to patients with symptomatic nOH associated with PD and conducted exclusively in the USA [84]. Subjects were randomized initially and underwent up to 2 weeks of double-blind titration with either droxidopa or placebo. Patients taking droxidopa had a significant improvement in symptoms of dizziness/lightheadedness (OHQ item 1) compared with placebo after 1 week of treatment ($p = 0.018$) [84]. At that time, patients taking droxidopa also had significantly higher systolic BPs in the standing position compared with placebo ($p = 0.032$).

Table 3
Phase 3 clinical studies with droxidopa (L-DOPS) in patients with neurogenic OH

Study	Subjects (N)	Design	Enrichment design	L-DOPS dosage	Length	Titration	Primary efficacy endpoints	Results
NOH301 [80]	162 (PD, MSA, PAF, NDAN, DBHD)	Multicenter, multinational, double-blind, randomized controlled trial parallel group induction design	Yes	100–600 mg three times daily	Initial open-label titration, 1-week washout, 7-day randomized, placebo-controlled phase	Open label—before randomization	OHQ composite score at 1 week	Compared with placebo, patients on droxidopa had a significant improvement in overall OHQ scores ($p < 0.003$); and standing BP that was 7 mmHg higher ($p < 0.001$)
NOH302 [81]	101 (PD, MSA, PAF, NDAN, DBHD)	Multicenter, double-blind, randomized controlled trial with withdrawal design	Yes	100–600 mg three times daily	Initial open-label titration, 7-day open-label treatment, 2 weeks of double-blind, placebo-controlled phase	Open label—before randomization	OHQ item 1 (dizziness/lightheadedness) at 2 weeks	No differences in the OHQ item 1. Significant improvement of the overall OHQ composite score on droxidopa than placebo ($p = 0.013$). Standing systolic BP was not different between the two groups. Patient self-reported symptom severity showed a significant improvement with droxidopa ($p = 0.008$)

| NOH306A [83] NOH306B [142] | 51 PD (306A) 171 PD (306 B) | Multicenter, double-blind, randomized, controlled trial parallel group induction design | No | 100–600 mg three times daily | 2-week double-blind titration; 8 weeks of double-blind maintenance treatment | Double blind–after randomization | OHQ composite score at 8 weeks (306A) OHQ item 1 at 1 week (306B) | For the initial 51 subjects (Study 306A), the primary efficacy measure (OHQ composite score) did not show significant change versus placebo at week 8. For the subsequent 171 subjects (Study 306B) patients taking droxidopa had a significant improvement in symptoms of dizziness/lightheadedness (OHQ item 1) compared to placebo after 1 week ($p = 0.018$); and higher systolic BP standing ($p = 0.032$) |

(continued)

Table 3
(continued)

Study	Subjects (*N*)	Design	Enrichment design	L-DOPS dosage	Length	Titration	Primary efficacy endpoints	Results
NOH303 (unpublished)	75 (PD, MSA, PAF, NDAN, DBHD)	Multicenter, open-label extension to 301 and 302 studies	No	100–600 mg three times daily	3-month open-label treatment; 2 week placebo-controlled withdrawal phase; after the controlled phase of the trial some patients continued on open-label droxidopa for up to 2 years	No	OHQ composite score	No significant differences between treatment arms
NOH304 (unpublished)	255 (PD, PAF, NDAN, DBHD)	Long term, open-label safety extension to 301, 302 and 306 studies	No	100–600 mg three times daily	Up to 1–2 years	No	Safety measures	N/A

| NOH305 [74] | 18 (PD, PAF, NDAN, DBHD) | Multicenter, open-label extension to 301 and 304 studies | No | 100–600 mg three times daily | An off drug 24-h ambulatory BP assessment was compared to a 4–6 week on-drug 24-h ambulatory BP assessment | No | Mean 24-h systolic and diastolic BP | Overall mean 24-h SBP and DBP were higher compared to off drug (137/81 mmHg vs. 129/76 mmHg; $p = 0.017/0.002$). Mean daytime SBP was significantly higher with droxidopa (8.4 ± 3.1 mmHg; $p = 0.014$). Nocturnal BP was not significantly higher on droxidopa versus off treatment ($p = 0.122$) |

The pooled results of NOH306A and NOH306b (composite NOH306) showed that patients taking droxidopa had a significant improvement in symptoms of dizziness/lightheadedness (OHQ item 1) compared with placebo after 1 week of treatment ($p < 0.01$) [84] (Table 3).

9.4.3 Long-Term Clinical Studies

The long-term efficacy of droxidopa has not been documented. *Study NOH303* was a long-term continuation of studies NIH301 and NIH302 and featured a withdrawal design. A total of 75 patients with nOH received droxidopa for a further 3 months and entered into a 2-week randomized withdrawal phase. This study did not show any statistically significant difference between treatment arms. Potential carryover of droxidopa may have influenced the results, as patients on placebo did not return to their baseline symptom score or blood pressure standing within the 2-week withdrawal phase. Moreover, this was an exploratory study not powered to reach significance.

In *Study NOH306 overall*, differences in change in OHQ item 1 scores from baseline to maintenance weeks 2, 4, and 8 showed numerical trends favoring droxidopa that approached statistical significance ($p = 0.077$ at week 8) [84].

Eventually, these double-blind clinical trials led to the FDA approval of droxidopa in the USA. The trials showed that patients with symptomatic nOH receiving droxidopa had both symptomatic improvement and higher blood pressure when standing than those on placebo. Integrated analysis and meta-analysis have been published [131, 132]. In summary, 226 patients received droxidopa and 236 received placebo. Symptoms of nOH were measured with the OHQ [66]. Those who received droxidopa improved in virtually all nOH symptom scores compared to those receiving placebo. Droxidopa also increased upright systolic blood pressure significantly ($+11.5 \pm 20.5$ mmHg vs. placebo $+4.8 \pm 21.0$ mmHg; $p < 0.001$).

9.4.4 Does Carbidopa Inhibit Droxidopa?

Another important issue is whether patients taking dopa-decarboxylase inhibitors (DDCI, e.g., carbidopa), which are always combined with levodopa in the treatment of PD, get less benefit from droxidopa. DDCI block the conversion of droxidopa to norepinephrine and could, theoretically, block droxidopa's blood pressure raising effect. In the integrated analysis, the magnitude of improvement observed in patients on droxidopa *not* taking DDCI was more pronounced than in those taking DDCI [114]. However, because none of the studies were designed to specifically assess this, no statistical model could confirm the significance of the difference. Moreover, the dosages of DDCI were not collected and the dose-response effect could not be analyzed. The combined analysis did show, however, that patients receiving droxidopa still had an increase in their blood pressure and symptomatic improvement when taking DDCI at clinically indicated dosages.

Droxidopa has been available in the USA for almost 4 years now. For many patients with different autonomic disorders droxidopa is well tolerated and improves their symptoms of nOH and quality of life. Still, like with any drug, some patients fail to respond. Why droxidopa appears not to be effective in this subpopulation remains an important clinical and research question. The FDA-approved droxidopa under the accelerated approval program based on clinical data showing that the drug has a significant effect on an intermediate clinical measure (in this case, short-term relief of dizziness/lightheadedness) that is reasonably likely to predict the outcome of ultimate interest (relief of dizziness/lightheadedness during chronic treatment). The approval was contingent upon the company that commercializes droxidopa conducting postapproval clinical trials to verify the drug's clinical benefit. Therefore, additional clinical trials to assess the long-term efficacy of droxidopa in patients with nOH are underway (ClinicalTrials.gov: NCT02586623). It would be interesting to compare droxidopa with midodrine, the only other FDA–approved drug for the treatment of nOH.

9.5 Pyridostigmine

Pyridostigmine, a cholinesterase inhibitor, potentiates cholinergic neurotransmission in sympathetic and parasympathetic autonomic ganglia. The evidence for using pyridostigmine in patients with nOH is scarce, and limited to one double-blind, randomized, crossover study enrolled 58 patients with nOH [133]. In this study, patients were successfully administered four different treatments (placebo, pyridostigmine 60 mg, pyridostomine 60 mg + midodrine 2.5 mg, pyridostigmine 60 mg + midodrine 5 mg) in a random order. For unknown reasons, they did not include a group of midodrine alone. The primary endpoint was the ΔDBP at 1 h after drug administration. Patients receiving pyridostimine alone had a ΔDBP of −27.6 mmHg compared to −34 mmHg receiving placebo ($p = 0.04$). The combination of pyridostigmine and midodrine 5 mg was more effective than pyridostigmine alone (ΔDBP of 27.2 mmHg vs. to 34 mmHg with placebo, $p = 0.002$).

A recent randomized, open-label clinical trial enrolled 87 patients with OH who were randomized to receive one of three treatments: midodrine 2.5 mg only, pyridostigmine 30 mg only, or midodrine 2.5 mg + pyridostigmine 30 mg [134]. The primary endpoint was the ΔBP within 3 min of standing at 3 months after treatment. Secondary endpoints were improvement of the ΔBP within 3 min of standing at 1 month, and the OHQ. The authors concluded that the ΔSBP and ΔDBP were less severe after treatment regardless of the treatment. This trial was severely biased in several ways: it was not blinded; although the authors report that only patients with nOH were included, they did not perform any test to confirm this, and patients with OH of any other cause could have been enrolled. Also, most of the patients had diabetes or "idiopathic OH," so the diagnoses were quite heterogeneous. Finally, the primary outcome measure was the ΔBP, with all its associated limitations.

9.6 Norepinephrine Transporter (NET) Blockade

Atomoxetine and similar medications increase norepinephrine concentration in the sympathetic neurovascular junction by selectively blocking the norepinephrine transporter (NET). Atomoxetine is currently FDA-approved in the USA for the treatment of attention deficit and hyperactivity disorder (ADHD) in children, adolescents, and adults. Atomoxetine has been studied in three randomized, placebo-controlled trials of patients with nOH. In the first one, 17 patients with nOH were administered a single dose of either placebo or yohimbine 5.4 mg or atomoxetine 18 mg, and the combination yohimbine and atomoxetine in a single-blind, crossover study [91]. The primary outcome was seated BP and standing BP 1 h postdrug. Neither yohimbine nor atomoxetine significantly increased seated systolic blood pressure or orthostatic tolerance compared with placebo. The combination, however, significantly increased seated SBP ($p < 0.001$) in a synergistic manner. The second trial studied 21 patients with nOH who were administered a single dose of atomoxetine 18 mg or placebo in a randomized, single-blind, crossover fashion [90]. The primary outcome was SBP every 5 min for 1 h postdrug administration. Atomoxetine increased significantly seated and standing SBP in patients with MSA (i.e., central autonomic failure) compared to placebo ($p = 0.016$). However, in patients with PD or PAF (i.e., peripheral autonomic failure), there were not differences in seated or standing BP compared to placebo ($p = 0.546$), suggesting that atomoxetine might be only effective to raise BP in patients with intact postganglionic sympathetic nerves (i.e., MSA).

The third, and most recent trial, was a randomized, single-blind, crossover trial with 65 patients with nOH of a single dose of atomoxetine 18 mg vs. midodrine 5–10 mg [70]. The primary endpoint was the standing SBP after 1 min 1 h postdrug. Secondary endpoints included posttreatment seated SBP and DBP, standing DBP and HR, and OHQ and OHQ item 1 scores. Atomoxetine significantly increased standing SBP by 20 mmHg and standing DBP by 11 mmHg compared to placebo ($p < 0.001$ in both cases). Likewise, midodrine increased standing SBP by 12 mmHg and standing DBP by 7 mmHg compared to placebo ($p < 0.001$ in both cases). Atomoxetine, however, improved standing SBP to a greater extent than midodrine did (mean difference = 7.5 mmHg; 95% CI, 0.6 to 14.5; $p = 0.03$).

An FDA-sponsored phase 2, oligocenter, double-blind, placebo-controlled crossover clinical trial to study the efficacy of atomoxetine for nOH ($n = 40$) is underway (ClinicalTrials.gov: NCT02796209). Thus, atomoxetine appears a very promising alternative for the treatment of nOH. However, because atomoxetine has a short biological effect (~3–4 h), longer active NET blockers are required. In this regard, ampreloxetine (TD-9855) an investigational long-acting NET blocker and serotonin reuptake inhibitor showed preliminary results in a phase 2, multicenter, single-blind, placebo-controlled trial with an open-label extension (ClinicalTrials.gov: NCT02705755). Large phase 3 multicenter studies are currently underway to confirm if ampreloxetine is effective for patients with nOH.

9.7 Other Treatments   Domperidone is a peripheral D_2 receptor antagonist, and since presynaptic dopamine receptors on sympathetic nerve endings also modulate noradrenaline release, it has been proposed as a treatment for OH. Indeed, domperidone has been extensively used in clinical practice to blunt the fall in BP associated with levodopa and dopaminergic agonists in patients with PD. However, it has been systematically studied only in one randomized crossover trial of 13 patients with PD [99]. Patients were randomized to receive domperidone 30 mg/day or fludrocortisone 0.1 mg/day for 3 weeks and, after 1 week of washout period, they were crossed over to the other active agent. At the end of each treatment period, a tilt-table test and COMPASS were administered, however the endpoints were unclear. There were no statistically significant differences in the ΔBP either during fludrocortisone or during domperidone treatment.

Several open-label studies investigated the efficacy of octreotide, a somatostatin analog, for the treatment of nOH [135, 136]. One randomized, placebo-controlled, crossover, double-blind study analyzed the effect of otreotide in nine patients with MSA [137]. Patients had a 60° tilt-table test on 3 consecutive days: the first one without medication; the second, 30 min after subcutaneous administration of octreotide (100 µg); and the third, 30 min after subcutaneous administration of placebo. Duration of the head-up tilt position was significantly longer with octreotide compared to placebo or no intervention ($p = 0.02$) although the minimal SBP, DBP, and mean BP were not different. Plasma norepinephrine levels were not different either. The same group performed a 6-month open-label trial showing that five patients with MSA experienced a functional improvement during the treatment period.

Several small, open-label studies reported a beneficial effect of erythropoietin on patients with nOH with or without anemia [138–140], although no randomized placebo-controlled studies have been performed so far.

10   Conclusions

Experimental therapeutics for rare disorders such as nOH face many challenges. Despite these challenges, there have been several accomplishments in the field of nOH in the last decade, including the development and validation of a disease-specific symptom scale (i.e., OHQ), the approval of droxidopa by the FDA, and several ongoing clinical trials. Careful consideration of study design, end-point selection, and patient recruitment and retention strategies are key when planning a clinical trial for nOH in patients with PD and other synucleinopathies. Future priorities are to continue to improve our understanding of nOH (e.g., by developing an animal model of the disease), to define the natural history of nOH in patients with synucleinopathies, to improve collaboration with patient-advocacy groups, and to determine what causes individuals to respond differently to the same medication. In this regard, it is well described that natural human genetic variations are linked to differences to antihypertensive drugs, such as β-blockers, angiotensin receptor blockers, and angiotensin-converting enzyme (ACE) inhibitors (i.e., pharmacogenomics) [141]. Thus, it is likely that potential genetic variations in drug targets of nOH may alter therapeutic efficacy and safety of drugs. Identification of these genetic variations would result in improved enrichment strategies in clinical trials of nOH and, ultimately, enhanced personalized medicine.

References

1. Obeso JA, Stamelou M, Goetz CG, Poewe W, Lang AE, Weintraub D, Burn D, Halliday GM, Bezard E, Przedborski S, Lehericy S, Brooks DJ, Rothwell JC, Hallett M, DeLong MR, Marras C, Tanner CM, Ross GW, Langston JW, Klein C, Bonifati V, Jankovic J, Lozano AM, Deuschl G, Bergman H, Tolosa E, Rodriguez-Violante M, Fahn S, Postuma RB, Berg D, Marek K, Standaert DG, Surmeier DJ, Olanow CW, Kordower JH, Calabresi P, Schapira AHV, Stoessl AJ (2017) Past, present, and future of Parkinson's disease: a special essay on the 200th anniversary of the shaking palsy. Mov Disord 32(9):1264–1310. https://doi.org/10.1002/mds.27115

2. Calne DB, Stern GM, Spiers AS, Laurence DR (1969) L-dopa in idiopathic parkinsonism. Lancet 2(7628):973–976

3. McDowell FH, Lee JE (1970) Levodopa, Parkinson's disease, and hypotension. Ann Intern Med 72(5):751–752

4. Schatz IJ, Podolsky S, Frame B (1963) Idiopathic orthostatic hypotension. Diagn Treat JAMA 186:537–540

5. Vanderhaeghen JJ, Perier O, Sternon JE (1970) Pathological findings in idiopathic orthostatic hypotension. Its relationship with Parkinson's disease. Arch Neurol 22(3):207–214

6. Thomas JE, Schirger A (1963) Neurologic manifestations in idiopathic orthostatic hypotension. Arch Neurol 8:204–208

7. Roessmann U, Van den Noort S, McFarland DE (1971) Idiopathic orthostatic hypotension. Arch Neurol 24(6):503–510

8. Gross M, Bannister R, Godwin-Austen R (1972) Orthostatic hypotension in Parkinson's disease. Lancet 1(7743):174–176

9. Kaufmann H, Norcliffe-Kaufmann L, Palma JA, Biaggioni I, Low PA, Singer W, Goldstein DS, Peltier AC, Shibao CA, Gibbons CH, Freeman R, Robertson D, Autonomic Disorders C (2017) Natural history of pure autonomic failure: a United States prospective cohort. Ann Neurol 81(2):287–297. https://doi.org/10.1002/ana.24877

10. Berg D, Postuma RB, Adler CH, Bloem BR, Chan P, Dubois B, Gasser T, Goetz CG, Halliday G, Joseph L, Lang AE, Liepelt-Scarfone I, Litvan I, Marek K, Obeso J, Oertel W, Olanow CW, Poewe W, Stern M, Deuschl G (2015) MDS research criteria for prodromal Parkinson's disease. Mov Disord 30(12):1600–1611. https://doi.org/10.1002/mds.26431

11. Ooi WL, Hossain M, Lipsitz LA (2000) The association between orthostatic hypotension and recurrent falls in nursing home residents. Am J Med 108(2):106–111

12. Jonsson PV, Lipsitz LA, Kelley M, Koestner J (1990) Hypotensive responses to common daily activities in institutionalized elderly. A potential risk for recurrent falls. Arch Intern Med 150(7):1518–1524

13. Goldstein DS, Holmes C, Sharabi Y, Wu T (2015) Survival in synucleinopathies: a prospective cohort study. Neurology 85(18):1554–1561. https://doi.org/10.1212/WNL.0000000000002086

14. Merola A, Sawyer RP, Artusi CA, Suri R, Berndt Z, Lopez-Castellanos JR, Vaughan J, Vizcarra JA, Romagnolo A, Espay AJ (2017) Orthostatic hypotension in Parkinson disease: impact on health care utilization. Parkinsonism Relat Disord 47:45. https://doi.org/10.1016/j.parkreldis.2017.11.344

15. Freeman R, Wieling W, Axelrod FB, Benditt DG, Benarroch E, Biaggioni I, Cheshire WP, Chelimsky T, Cortelli P, Gibbons CH, Goldstein DS, Hainsworth R, Hilz MJ, Jacob G, Kaufmann H, Jordan J, Lipsitz LA, Levine BD, Low PA, Mathias C, Raj SR, Robertson D, Sandroni P, Schatz I, Schondorff R, Stewart JM, van Dijk JG (2011) Consensus statement on the definition of orthostatic hypotension, neurally mediated syncope and the postural tachycardia syndrome. Clin Auton Res 21(2):69–72. https://doi.org/10.1007/s10286-011-0119-5

16. Freeman R (2008) Clinical practice. Neurogenic orthostatic hypotension. N Engl J Med 358(6):615–624. https://doi.org/10.1056/NEJMcp074189

17. Masaki KH, Schatz IJ, Burchfiel CM, Sharp DS, Chiu D, Foley D, Curb JD (1998) Orthostatic hypotension predicts mortality in elderly men: the Honolulu heart program. Circulation 98(21):2290–2295

18. Shibao C, Grijalva CG, Raj SR, Biaggioni I, Griffin MR (2007) Orthostatic hypotension-related hospitalizations in the United States. Am J Med 120(11):975–980. https://doi.org/10.1016/j.amjmed.2007.05.009

19. Fotherby MD, Potter JF (1994) Orthostatic hypotension and anti-hypertensive therapy in the elderly. Postgrad Med J 70(830):878–881

20. Rutan GH, Hermanson B, Bild DE, Kittner SJ, LaBaw F, Tell GS (1992) Orthostatic hypotension in older adults. The cardiovascular health study. CHS Collaborative Research Group. Hypertension 19(6 Pt 1):508–519

21. Arenander E (1960) Hemodynamic effects of varicose veins and results of radical surgery. Acta Chir Scand Suppl Suppl 260:1–76

22. Sarasin FP, Louis-Simonet M, Carballo D, Slama S, Junod AF, Unger PF (2002) Prevalence of orthostatic hypotension among patients presenting with syncope in the ED. Am J Emerg Med 20(6):497–501

23. Feldstein C, Weder AB (2012) Orthostatic hypotension: a common, serious and under-recognized problem in hospitalized patients. J Am Soc Hypertens 6(1):27–39. https://doi.org/10.1016/j.jash.2011.08.008

24. Aung AK, Corcoran SJ, Nagalingam V, Paul E, Newnham HH (2012) Prevalence, associations, and risk factors for orthostatic hypotension in medical, surgical, and trauma inpatients: an observational cohort study. Ochsner J 12(1):35–41

25. Angelousi A, Girerd N, Benetos A, Frimat L, Gautier S, Weryha G, Boivin JM (2014) Association between orthostatic hypotension and cardiovascular risk, cerebrovascular risk, cognitive decline and falls as well as over-all mortality: a systematic review and meta-analysis. J Hypertens 32(8):1562–1571;. discussion 1571. https://doi.org/10.1097/HJH.0000000000000235

26. Rose KM, Eigenbrodt ML, Biga RL, Couper DJ, Light KC, Sharrett AR, Heiss G (2006) Orthostatic hypotension predicts mortality in middle-aged adults: the atherosclerosis risk in communities (ARIC) study. Circulation 114(7):630–636. https://doi.org/10.1161/CIRCULATIONAHA.105.598722

27. Xin W, Lin Z, Mi S (2014) Orthostatic hypotension and mortality risk: a meta-analysis of cohort studies. Heart 100(5):406–413. https://doi.org/10.1136/heartjnl-2013-304121

28. Kaufmann H, Biaggioni I (2003) Autonomic failure in neurodegenerative disorders. Semin Neurol 23(4):351–363. https://doi.org/10.1055/s-2004-817719

29. Kaufmann H, Norcliffe-Kaufmann L, Palma JA, Biaggioni I, Low PA, Singer W, Goldstein DS, Peltier AC, Shibao CA, Gibbons CH, Freeman R, Robertson D, Autonomic Disorders C (2017) The natural history of pure autonomic failure: a U.S. prospective cohort. Ann Neurol 81(2):287–297. https://doi.org/10.1002/ana.24877

30. Palma JA, Gomez-Esteban JC, Norcliffe-Kaufmann L, Martinez J, Tijero B, Berganzo K, Kaufmann H (2015) Orthostatic hypotension in Parkinson disease: how much you fall or how low you go? Mov Disord 30(5):639–645. https://doi.org/10.1002/mds.26079

31. Velseboer DC, de Haan RJ, Wieling W, Goldstein DS, de Bie RM (2011) Prevalence of orthostatic hypotension in Parkinson's disease: a systematic review and meta-analysis. Parkinsonism Relat Disord 17(10):724–729. https://doi.org/10.1016/j.parkreldis.2011.04.016

32. Palma JA, Kaufmann H (2017) Epidemiology, diagnosis, and management of neurogenic orthostatic hypotension. Mov Disord Clin Pract 4(3):298–308. https://doi.org/10.1002/mdc3.12478

33. Thaisetthawatkul P, Boeve BF, Benarroch EE, Sandroni P, Ferman TJ, Petersen R, Low PA (2004) Autonomic dysfunction

in dementia with Lewy bodies. Neurology 62(10):1804–1809

34. Horimoto Y, Matsumoto M, Akatsu H, Ikari H, Kojima K, Yamamoto T, Otsuka Y, Ojika K, Ueda R, Kosaka K (2003) Autonomic dysfunctions in dementia with Lewy bodies. J Neurol 250(5):530–533. https://doi.org/10.1007/s00415-003-1029-9

35. Gilman S, Low P, Quinn N, Albanese A, Ben-Shlomo Y, Fowler C, Kaufmann H, Klockgether T, Lang A, Lantos P, Litvan I, Mathias C, Oliver E, Robertson D, Schatz I, Wenning G (1998) Consensus statement on the diagnosis of multiple system atrophy. American Autonomic society and American Academy of Neurology. Clin Auton Res 8(6):359–362

36. Low PA, Reich SG, Jankovic J, Shults CW, Stern MB, Novak P, Tanner CM, Gilman S, Marshall FJ, Wooten F, Racette B, Chelimsky T, Singer W, Sletten DM, Sandroni P, Mandrekar J (2015) Natural history of multiple system atrophy in the USA: a prospective cohort study. Lancet Neurol 14(7):710–719. https://doi.org/10.1016/S1474-4422(15)00058-7

37. Roncevic D, Palma JA, Martinez J, Goulding N, Norcliffe-Kaufmann L, Kaufmann H (2014) Cerebellar and Parkinsonian phenotypes in multiple system atrophy: similarities, differences and survival. J Neural Transm (Vienna) 121(5):507–512. https://doi.org/10.1007/s00702-013-1133-7

38. Berganzo K, Diez-Arrola B, Tijero B, Somme J, Lezcano E, Llorens V, Ugarriza I, Ciordia R, Gomez-Esteban JC, Zarranz JJ (2013) Nocturnal hypertension and dysautonomia in patients with Parkinson's disease: are they related? J Neurol 260(7):1752–1756. https://doi.org/10.1007/s00415-013-6859-5

39. Fanciulli A, Gobel G, Ndayisaba JP, Granata R, Duerr S, Strano S, Colosimo C, Poewe W, Pontieri FE, Wenning GK (2016) Supine hypertension in Parkinson's disease and multiple system atrophy. Clin Auton Res 26(2):97–105. https://doi.org/10.1007/s10286-015-0336-4

40. Umehara T, Matsuno H, Toyoda C, Oka H (2016) Clinical characteristics of supine hypertension in de novo Parkinson disease. Clin Auton Res 26(1):15–21. https://doi.org/10.1007/s10286-015-0324-8

41. Kaufmann H, Goldstein DS (2013) Autonomic dysfunction in Parkinson disease. Handb Clin Neurol 117:259–278. https://doi.org/10.1016/B978-0-444-53491-0.00021-3

42. Jain S, Goldstein DS (2012) Cardiovascular dysautonomia in Parkinson disease: from pathophysiology to pathogenesis.

Neurobiol Dis 46(3):572–580. https://doi.org/10.1016/j.nbd.2011.10.025

43. Goldstein DS, Holmes CS, Dendi R, Bruce SR, Li ST (2002) Orthostatic hypotension from sympathetic denervation in Parkinson's disease. Neurology 58(8):1247–1255

44. Nagayama H, Yamazaki M, Ueda M, Nishiyama Y, Hamamoto M, Katayama Y, Mori O (2008) Low myocardial MIBG uptake in multiple system atrophy with incidental Lewy body pathology: an autopsy case report. Mov Disord 23(7):1055–1057. https://doi.org/10.1002/mds.22031

45. Nagayama H, Ueda M, Yamazaki M, Nishiyama Y, Hamamoto M, Katayama Y (2010) Abnormal cardiac [(123)I]-meta-iodobenzylguanidine uptake in multiple system atrophy. Mov Disord 25(11):1744–1747. https://doi.org/10.1002/mds.23338

46. Goldstein DS (2014) Dysautonomia in Parkinson disease. Compr Physiol 4(2):805–826. https://doi.org/10.1002/cphy.c130026

47. Orimo S, Oka T, Miura H, Tsuchiya K, Mori F, Wakabayashi K, Nagao T, Yokochi M (2002) Sympathetic cardiac denervation in Parkinson's disease and pure autonomic failure but not in multiple system atrophy. J Neurol Neurosurg Psychiatry 73(6):776–777

48. Braune S (2001) The role of cardiac metaiodobenzylguanidine uptake in the differential diagnosis of parkinsonian syndromes. Clin Auton Res 11(6):351–355

49. Kaufmann H, Oribe E, Miller M, Knott P, Wiltshire-Clement M, Yahr MD (1992) Hypotension-induced vasopressin release distinguishes between pure autonomic failure and multiple system atrophy with autonomic failure. Neurology 42(3 Pt 1):590–593

50. Shannon JR, Jordan J, Diedrich A, Pohar B, Black BK, Robertson D, Biaggioni I (2000) Sympathetically mediated hypertension in autonomic failure. Circulation 101(23):2710–2715

51. Goldstein DS, Pechnik S, Holmes C, Eldadah B, Sharabi Y (2003) Association between supine hypertension and orthostatic hypotension in autonomic failure. Hypertension 42(2):136–142. https://doi.org/10.1161/01.HYP.0000081216.11623.C3

52. Hanby MF, Panerai RB, Robinson TG, Haunton VJ (2017) Is cerebral vasomotor reactivity impaired in Parkinson disease? Clin Auton Res 27(2):107–111. https://doi.org/10.1007/s10286-017-0406-x

53. Norcliffe-Kaufmann L, Galindo-Mendez B, Garcia-Guarniz AL, Villarreal-Vitorica E, Novak V (2017) Transcranial Doppler in autonomic testing: standards and clinical applica-

tions. Clin Auton Res 28:187. https://doi.org/10.1007/s10286-017-0454-2

54. Kempf L, Goldsmith JC, Temple R (2017) Challenges of developing and conducting clinical trials in rare disorders. Am J Med Genet A 176:773. https://doi.org/10.1002/ajmg.a.38413

55. Arnold AC, Biaggioni I (2012) Management approaches to hypertension in autonomic failure. Curr Opin Nephrol Hypertens 21(5):481–485. https://doi.org/10.1097/MNH.0b013e328356c52f

56. de la Fuente-Fernandez R, Ruth TJ, Sossi V, Schulzer M, Calne DB, Stoessl AJ (2001) Expectation and dopamine release: mechanism of the placebo effect in Parkinson's disease. Science 293(5532):1164–1166. https://doi.org/10.1126/science.1060937

57. Administration FDA (2017) The voice of the patient: a series of reports from FDA's patient-focused drug development initiative. FDA, Silver Spring, MD. https://www.fda.gov/forindustry/userfees/prescriptiondruguserfee/ucm368342.htm

58. Fuente Mora C, Palma JA, Kaufmann H, Norcliffe-Kaufmann L (2016) Cerebral autoregulation and symptoms of orthostatic hypotension in familial dysautonomia. J Cereb Blood Flow Metab 37:2414. https://doi.org/10.1177/0271678X16667524

59. Horowitz DR, Kaufmann H (2001) Autoregulatory cerebral vasodilation occurs during orthostatic hypotension in patients with primary autonomic failure. Clin Auton Res 11(6):363–367

60. Palma JA, Norcliffe-Kaufmann L, Kaufmann H (2016) An orthostatic hypotension mimic: the inebriation-like syndrome in Parkinson disease. Mov Disord 31(4):598–600. https://doi.org/10.1002/mds.26516

61. Sletten DM, Suarez GA, Low PA, Mandrekar J, Singer W (2012) COMPASS 31: a refined and abbreviated composite Autonomic symptom score. Mayo Clin Proc 87(12):1196–1201. https://doi.org/10.1016/j.mayocp.2012.10.013

62. Visser M, Marinus J, Stiggelbout AM, Van Hilten JJ (2004) Assessment of autonomic dysfunction in Parkinson's disease: the SCOPA-AUT. Mov Disord 19(11):1306–1312. https://doi.org/10.1002/mds.20153

63. Robinson-Papp J, Sharma SK, George MC, Simpson DM (2017) Assessment of autonomic symptoms in a medically complex, urban patient population. Clin Auton Res 27(1):25–29. https://doi.org/10.1007/s10286-016-0384-4

64. Schrag A, Selai C, Mathias C, Low P, Hobart J, Brady N, Quinn NP (2007) Measuring health-related quality of life in MSA: the

MSA-QoL. Mov Disord 22(16):2332–2338. https://doi.org/10.1002/mds.21649

65. Schrag A, Geser F, Stampfer-Kountchev M, Seppi K, Sawires M, Kollensperger M, Scherfler C, Quinn N, Pellecchia MT, Barone P, Del Sorbo F, Albanese A, Ostergaard K, Dupont E, Cardozo A, Tolosa E, Nilsson CF, Widner H, Lindvall O, Giladi N, Gurevich T, Daniels C, Deuschl G, Coelho M, Sampaio C, Abele M, Klockgether T, Schimke N, Eggert KM, Oertel W, Djaldetti R, Colosimo C, Meco G, Poewe W, Wenning GK, European MSASG (2006) Health-related quality of life in multiple system atrophy. Mov Disord 21(6):809–815. https://doi.org/10.1002/mds.20808

66. Kaufmann H, Malamut R, Norcliffe-Kaufmann L, Rosa K, Freeman R (2012) The orthostatic hypotension questionnaire (OHQ): validation of a novel symptom assessment scale. Clin Auton Res 22(2):79–90. https://doi.org/10.1007/s10286-011-0146-2

67. Wright RA, Kaufmann HC, Perera R, Opfer-Gehrking TL, McElligott MA, Sheng KN, Low PA (1998) A double-blind, dose-response study of midodrine in neurogenic orthostatic hypotension. Neurology 51(1):120–124

68. Frith J, Newton JL (2016) Validation of a questionnaire for orthostatic hypotension for routine clinical use. Geriatr Gerontol Int 16(7):785–790. https://doi.org/10.1111/ggi.12553

69. Busner J, Targum SD (2007) The clinical global impressions scale: applying a research tool in clinical practice. Psychiatry (Edgmont) 4(7):28–37

70. Ramirez CE, Okamoto LE, Arnold AC, Gamboa A, Diedrich A, Choi L, Raj SR, Robertson D, Biaggioni I, Shibao CA (2014) Efficacy of atomoxetine versus midodrine for the treatment of orthostatic hypotension in autonomic failure. Hypertension 64(6):1235–1240. https://doi.org/10.1161/HYPERTENSIONAHA.114.04225

71. Smith W, Wan H, Much D, Robinson AG, Martin P (2016) Clinical benefit of midodrine hydrochloride in symptomatic orthostatic hypotension: a phase 4, double-blind, placebo-controlled, randomized, tilt-table study. Clin Auton Res 26(4):269–277. https://doi.org/10.1007/s10286-016-0363-9

72. Norcliffe-Kaufmann L, Kaufmann H (2014) Is ambulatory blood pressure monitoring useful in patients with chronic autonomic failure? Clin Auton Res 24(4):189–192. https://doi.org/10.1007/s10286-014-0229-y

73. Jordan J, Biaggioni I (2002) Diagnosis and treatment of supine hypertension in autonomic failure patients with orthostatic

hypotension. J Clin Hypertens (Greenwich) 4(2):139–145

74. Kaufmann H, Norcliffe-Kaufmann L, Hewitt LA, Rowse GJ, White WB (2016) Effects of the novel norepinephrine prodrug, droxidopa, on ambulatory blood pressure in patients with neurogenic orthostatic hypotension. J Am Soc Hypertens 10(10):819–826. https://doi.org/10.1016/j.jash.2016.07.009

75. Pickering RM, Grimbergen YA, Rigney U, Ashburn A, Mazibrada G, Wood B, Gray P, Kerr G, Bloem BR (2007) A meta-analysis of six prospective studies of falling in Parkinson's disease. Mov Disord 22(13):1892–1900. https://doi.org/10.1002/mds.21598

76. Rascol O, Perez-Lloret S, Damier P, Delval A, Derkinderen P, Destee A, Meissner WG, Tison F, Negre-Pages L (2015) Falls in ambulatory non-demented patients with Parkinson's disease. J Neural Transm (Vienna) 122(10):1447–1455. https://doi.org/10.1007/s00702-015-1396-2

77. van der Marck MA, Klok MP, Okun MS, Giladi N, Munneke M, Bloem BR, Force NPFFT (2014) Consensus-based clinical practice recommendations for the examination and management of falls in patients with Parkinson's disease. Parkinsonism Relat Disord 20(4):360–369. https://doi.org/10.1016/j.parkreldis.2013.10.030

78. Perez-Lloret S, Rey MV, Fabre N, Ory F, Spampinato U, Montastruc JL, Rascol O (2012) Do Parkinson's disease patients disclose their adverse events spontaneously? Eur J Clin Pharmacol 68(5):857–865. https://doi.org/10.1007/s00228-011-1198-x

79. Coote S, Sosnoff JJ, Gunn H (2014) Fall incidence as the primary outcome in multiple sclerosis falls-prevention trials: recommendation from the international MS falls prevention research network. Int J MS Care 16(4):178–184. https://doi.org/10.7224/1537-2073.2014-059

80. Kaufmann H, Freeman R, Biaggioni I, Low P, Pedder S, Hewitt LA, Mauney J, Feirtag M, Mathias CJ, Investigators NOH (2014) Droxidopa for neurogenic orthostatic hypotension: a randomized, placebo-controlled, phase 3 trial. Neurology 83(4):328–335. https://doi.org/10.1212/WNL.0000000000000615

81. Biaggioni I, Freeman R, Mathias CJ, Low P, Hewitt LA, Kaufmann H, Droxidopa I (2015) Randomized withdrawal study of patients with symptomatic neurogenic orthostatic hypotension responsive to droxidopa. Hypertension 65(1):101–107. https://doi.org/10.1161/HYPERTENSIONAHA.114.04035

82. Hauser RA, Heritier S, Rowse GJ, Hewitt LA, Isaacson SH (2016) Droxidopa and reduced falls in a trial of Parkinson disease patients with neurogenic orthostatic hypotension. Clin Neuropharmacol 39(5):220–226. https://doi.org/10.1097/WNF.0000000000000168

83. Hauser RA, Hewitt LA, Isaacson S (2014) Droxidopa in patients with neurogenic orthostatic hypotension associated with Parkinson's disease (NOH306A). J Parkinsons Dis 4(1):57–65. https://doi.org/10.3233/JPD-130259

84. Hauser RA, Isaacson S, Lisk JP, Hewitt LA, Rowse G (2015) Droxidopa for the short-term treatment of symptomatic neurogenic orthostatic hypotension in Parkinson's disease (nOH306B). Mov Disord 30(5):646–654. https://doi.org/10.1002/mds.26086

85. Espay AJ, Bonato P, Nahab FB, Maetzler W, Dean JM, Klucken J, Eskofier BM, Merola A, Horak F, Lang AE, Reilmann R, Giuffrida J, Nieuwboer A, Horne M, Little MA, Litvan I, Simuni T, Dorsey ER, Burack MA, Kubota K, Kamondi A, Godinho C, Daneault JF, Mitsi G, Krinke L, Hausdorff JM, Bloem BR, Papapetropoulos S, Movement Disorders Society Task Force on T (2016) Technology in Parkinson's disease: challenges and opportunities. Mov Disord 31(9):1272–1282. https://doi.org/10.1002/mds.26642

86. Arneric SP, Cedarbaum JM, Khozin S, Papapetropoulos S, Hill DL, Ropacki M, Rhodes J, Dacks PA, Hudson LD, Gordon MF, Kern VD, Romero K, Vradenburg G, Au R, Karlin DR, Facheris MF, Fitzer-Attas CJ, Vitolo OV, Wang J, Miller BM, Kaye JA (2017) Biometric monitoring devices for assessing end points in clinical trials: developing an ecosystem. Nat Rev Drug Discov 16(10):736. https://doi.org/10.1038/nrd.2017.153

87. El-Saifi N, Moyle W, Jones C, Tuffaha H (2017) Medication adherence in older patients with dementia: a systematic literature review. J Pharm Pract. https://doi.org/10.1177/0897190017710524

88. Dodge HH, Zhu J, Mattek NC, Austin D, Kornfeld J, Kaye JA (2015) Use of high-frequency in-home monitoring data may reduce sample sizes needed in clinical trials. PLoS One 10(9):e0138095. https://doi.org/10.1371/journal.pone.0138095

89. Kaufmann H, Saadia D, Voustianiouk A, Goldstein DS, Holmes C, Yahr MD, Nardin R, Freeman R (2003) Norepinephrine precursor therapy in neurogenic orthostatic hypotension. Circulation 108(6):724–728. https://doi.org/10.1161/01.CIR.0000083721.49847.D7

90. Shibao C, Raj SR, Gamboa A, Diedrich A, Choi L, Black BK, Robertson D,

Biaggioni I (2007) Norepinephrine transporter blockade with atomoxetine induces hypertension in patients with impaired autonomic function. Hypertension 50(1):47–53. https://doi.org/10.1161/HYPERTENSIONAHA.107.089961

91. Okamoto LE, Shibao C, Gamboa A, Choi L, Diedrich A, Raj SR, Black BK, Robertson D, Biaggioni I (2012) Synergistic effect of norepinephrine transporter blockade and alpha-2 antagonism on blood pressure in autonomic failure. Hypertension 59(3):650–656. https://doi.org/10.1161/HYPERTENSIONAHA.111.184812

92. Shibao C, Gamboa A, Diedrich A, Dossett C, Choi L, Farley G, Biaggioni I (2007) Acarbose, an alpha-glucosidase inhibitor, attenuates postprandial hypotension in autonomic failure. Hypertension 50(1):54–61. https://doi.org/10.1161/HYPERTENSIONAHA.107.091355

93. Okamoto LE, Diedrich A, Baudenbacher FJ, Harder R, Whitfield JS, Iqbal F, Gamboa A, Shibao CA, Black BK, Raj SR, Robertson D, Biaggioni I (2016) Efficacy of servo-controlled splanchnic venous compression in the treatment of orthostatic hypotension: a randomized comparison with midodrine. Hypertension 68(2):418–426. https://doi.org/10.1161/HYPERTENSIONAHA.116.07199

94. Guyatt G, Sackett D, Taylor DW, Chong J, Roberts R, Pugsley S (1986) Determining optimal therapy—randomized trials in individual patients. N Engl J Med 314(14):889–892. https://doi.org/10.1056/NEJM198604033141406

95. Lillie EO, Patay B, Diamant J, Issell B, Topol EJ, Schork NJ (2011) The n-of-1 clinical trial: the ultimate strategy for individualizing medicine? Per Med 8(2):161–173. https://doi.org/10.2217/pme.11.7

96. Close S, Smaldone A, Fennoy I, Reame N, Grey M (2013) Using information technology and social networking for recruitment of research participants: experience from an exploratory study of pediatric Klinefelter syndrome. J Med Internet Res 15(3):e48. https://doi.org/10.2196/jmir.2286

97. Jordan J, Shannon JR, Black BK, Ali Y, Farley M, Costa F, Diedrich A, Robertson RM, Biaggioni I, Robertson D (2000) The pressor response to water drinking in humans : a sympathetic reflex? Circulation 101(5):504–509

98. May M, Jordan J (2011) The osmopressor response to water drinking. Am J Physiol Regul Integr Comp Physiol 300(1):R40–R46. https://doi.org/10.1152/ajpregu.00544.2010

99. Schoffer KL, Henderson RD, O'Maley K, O'Sullivan JD (2007) Nonpharmacologic treatment, fludrocortisone, and domperidone for orthostatic hypotension in Parkinson's disease. Mov Disord 22(11):1543–1549. https://doi.org/10.1002/mds.21428

100. Fanciulli A, Goebel G, Metzler B, Sprenger F, Poewe W, Wenning GK, Seppi K (2016) Elastic abdominal binders attenuate orthostatic hypotension in Parkinson's disease. Mov Disord Clin Pract 3(2):156–160. https://doi.org/10.1002/mdc3.12270

101. Grijalva CG, Biaggioni I, Griffin MR, Shibao CA (2017) Fludrocortisone is associated with a higher risk of all-cause hospitalizations compared with midodrine in patients with orthostatic hypotension. J Am Heart Assoc 6(10). https://doi.org/10.1161/JAHA.117.006848

102. Chobanian AV, Volicer L, Tifft CP, Gavras H, Liang CS, Faxon D (1979) Mineralocorticoid-induced hypertension in patients with orthostatic hypotension. N Engl J Med 301(2):68–73. https://doi.org/10.1056/NEJM197907123010202

103. Norcliffe-Kaufmann L, Axelrod FB, Kaufmann H (2013) Developmental abnormalities, blood pressure variability and renal disease in Riley day syndrome. J Hum Hypertens 27(1):51–55. https://doi.org/10.1038/jhh.2011.107

104. Schreglmann SR, Buchele F, Sommerauer M, Epprecht L, Kagi G, Hagele-Link S, Gotze O, Zimmerli L, Waldvogel D, Baumann CR (2017) Pyridostigmine bromide versus fludrocortisone in the treatment of orthostatic hypotension in Parkinson's disease – a randomized controlled trial. Eur J Neurol 24(4):545–551. https://doi.org/10.1111/ene.13260

105. Jankovic J, Gilden JL, Hiner BC, Kaufmann H, Brown DC, Coghlan CH, Rubin M, Fouad-Tarazi FM (1993) Neurogenic orthostatic hypotension: a double-blind, placebo-controlled study with midodrine. Am J Med 95(1):38–48

106. Low PA, Gilden JL, Freeman R, Sheng KN, McElligott MA (1997) Efficacy of midodrine vs placebo in neurogenic orthostatic hypotension. A randomized, double-blind multicenter study. Midodrine Study Group. JAMA 277(13):1046–1051

107. McTavish D, Goa KL (1989) Midodrine. A review of its pharmacologic properties and therapeutic use in orthostatic hypotension and secondary hypotensive disorders. Drugs 38(5):757–777

108. Kaufmann H, Brannan T, Krakoff L, Yahr MD, Mandeli J (1988) Treatment of orthostatic hypotension due to autonomic failure with a peripheral alpha-adrenergic agonist (midodrine). Neurology 38(6):951–956

109. Schirger A, Sheps SG, Thomas JE, Fealey RD (1981) Midodrine. A new agent in the management of idiopathic orthostatic hypotension and shy-Drager syndrome. Mayo Clin Proc 56(7):429–433

110. Rosenmund KW, Dornsaft H (1919) Über oxy- und hioxyphenylserin und die muttersubstanz des adrenalins. Ber Dtsch Chem Ges 52:1734–1749

111. Beyer KH, Blaschko H, Burn JH, Langemann H (1950) Enzymic formation of noradrenaline in mammalian tissue extracts. Nature 165(4206):926

112. Blaschko H, Burn JH, Langemann H (1950) The formation of noradrenaline from dihydroxyphenylserine. Br J Pharmacol Chemother 5(3):431–437

113. Blaschko H, Holton P, Stanley GH (1948) The decarboxylation of -3: 4-dihydroxyphenylserine (noradrenaline carboxylic acid). Br J Pharmacol Chemother 3(4):315–319

114. Suzuki T, Higa S, Sakoda S, Hayashi A, Yamamura Y, Takaba Y, Nakajima A (1981) Orthostatic hypotension in familial amyloid polyneuropathy: treatment with DL-threo-3,4-dihydroxyphenylserine. Neurology 31(10):1323–1326

115. Kachi T, Iwase S, Mano T, Saito M, Kunimoto M, Sobue I (1988) Effect of L-threo-3,4-dihydroxyphenylserine on muscle sympathetic nerve activities in shy-Drager syndrome. Neurology 38(7):1091–1094

116. Kaufmann H, Norcliffe-Kaufmann L, Palma JA (2015) Droxidopa in neurogenic orthostatic hypotension. Expert Rev Cardiovasc Ther 13(8):875–891. https://doi.org/10.1586/14779072.2015.1057504

117. Kaufmann H (2017) Droxidopa for symptomatic neurogenic orthostatic hypotension: what can we learn? Clin Auton Res 27(Suppl 1):1–3. https://doi.org/10.1007/s10286-017-0426-6

118. Gupta F, Karabin B, Mehdirad A (2017) Titrating droxidopa to maximize symptomatic benefit in a patient with Parkinson disease and neurogenic orthostatic hypotension. Clin Auton Res 27(Suppl 1):15–16. https://doi.org/10.1007/s10286-017-0430-x

119. Vernino S, Claassen D (2017) Polypharmacy: droxidopa to treat neurogenic orthostatic hypotension in a patient with Parkinson disease and type 2 diabetes mellitus. Clin Auton Res 27(Suppl 1):33–34. https://doi.org/10.1007/s10286-017-0435-5

120. Kremens D, Lew M, Claassen D, Goodman BP (2017) Adding droxidopa to fludrocortisone or midodrine in a patient with neurogenic orthostatic hypotension and Parkinson disease. Clin Auton Res 27(Suppl 1):29–31. https://doi.org/10.1007/s10286-017-0434-6

121. Mehdirad A, Karabin B, Gupta F (2017) Managing neurogenic orthostatic hypotension with droxidopa in a patient with Parkinson disease, atrial fibrillation, and hypertension. Clin Auton Res 27(Suppl 1):25–27. https://doi.org/10.1007/s10286-017-0433-7

122. Claassen D, Lew M (2017) Initiating droxidopa for neurogenic orthostatic hypotension in a patient with Parkinson disease. Clin Auton Res 27(Suppl 1):13–14. https://doi.org/10.1007/s10286-017-0429-3

123. Goodman BP, Claassen D, Mehdirad A (2017) Adjusting droxidopa for neurogenic orthostatic hypotension in a patient with Parkinson disease. Clin Auton Res 27(Suppl 1):17–19. https://doi.org/10.1007/s10286-017-0431-9

124. Goodman BP, Gupta F (2017) Defining successful treatment of neurogenic orthostatic hypotension with droxidopa in a patient with multiple system atrophy. Clin Auton Res 27(Suppl 1):21–23. https://doi.org/10.1007/s10286-017-0432-8

125. Gupta F, Kremens D, Vernino S, Karabin B (2017) Managing neurogenic orthostatic hypotension in a patient presenting with pure autonomic failure who later developed Parkinson disease. Clin Auton Res 27(Suppl 1):9–11. https://doi.org/10.1007/s10286-017-0428-4

126. Bartholini J, Constantinidis J, Puig M, Tissot R, Pletscher A (1975) The stereoisomers of 3,4-dihydroxyphenylserine as precursors of norepinephrine. J Pharmacol Exp Ther 193(2):523–532

127. Hoeldtke RD, Cilmi KM, Mattis-Graves K (1984) DL-Threo-3,4-dihydroxyphenylserine does not exert a pressor effect in orthostatic hypotension. Clin Pharmacol Ther 36(3):302–306

128. Freeman R, Landsberg L, Young J (1999) The treatment of neurogenic orthostatic hypotension with 3,4-DL-threo-dihydroxyphenylserine: a randomized, placebo-controlled, crossover trial. Neurology 53(9):2151–2157

129. Kaufmann H, Oribe E, Yahr MD (1991) Differential effect of L-threo-3,4-dihydroxyphenylserine in pure autonomic failure and multiple system atrophy with autonomic failure. J Neural Transm Park Dis Dement Sect 3(2):143–148

130. Mathias CJ, Senard JM, Braune S, Watson L, Aragishi A, Keeling JE, Taylor MD (2001) L-Threo-dihydroxyphenylserine (L-threo-DOPS; droxidopa) in the management of neurogenic orthostatic hypotension: a multinational, multi-center, dose-ranging study in

multiple system atrophy and pure autonomic failure. Clin Auton Res 11(4):235–242

131. Biaggioni I, Arthur Hewitt L, Rowse GJ, Kaufmann H (2017) Integrated analysis of droxidopa trials for neurogenic orthostatic hypotension. BMC Neurol 17(1):90. https://doi.org/10.1186/s12883-017-0867-5

132. Elgebaly A, Abdelazeim B, Mattar O, Gadelkarim M, Salah R, Negida A (2016) Meta-analysis of the safety and efficacy of droxidopa for neurogenic orthostatic hypotension. Clin Auton Res 26(3):171–180. https://doi.org/10.1007/s10286-016-0349-7

133. Singer W, Sandroni P, Opfer-Gehrking TL, Suarez GA, Klein CM, Hines S, O'Brien PC, Slezak J, Low PA (2006) Pyridostigmine treatment trial in neurogenic orthostatic hypotension. Arch Neurol 63(4):513–518. https://doi.org/10.1001/archneur.63.4.noc50340

134. Byun JI, Moon J, Kim DY, Shin H, Sunwoo JS, Lim JA, Kim TJ, Lee WJ, Lee HS, Jun JS, Park KI, Lee ST, Jung KH, Jung KY, Lee SK, Chu K (2017) Efficacy of single or combined midodrine and pyridostigmine in orthostatic hypotension. Neurology 89(10):1078–1086. https://doi.org/10.1212/WNL.0000000000004340

135. Hoeldtke RD, Israel BC (1989) Treatment of orthostatic hypotension with octreotide. J Clin Endocrinol Metab 68(6):1051–1059. https://doi.org/10.1210/jcem-68-6-1051

136. Bordet R, Benhadjali J, Libersa C, Destee A (1994) Octreotide in the management of orthostatic hypotension in multiple system atrophy: pilot trial of chronic administration. Clin Neuropharmacol 17(4):380–383

137. Bordet R, Benhadjali J, Destee A, Belabbas A, Libersa C (1995) Octreotide effects on orthostatic hypotension in patients with multiple system atrophy: a controlled study of acute administration. Clin Neuropharmacol 18(1):83–89

138. Hoeldtke RD, Streeten DH (1993) Treatment of orthostatic hypotension with erythropoietin. N Engl J Med 329(9):611–615

139. Perera R, Isola L, Kaufmann H (1995) Effect of recombinant erythropoietin on anemia and orthostatic hypotension in primary autonomic failure. Clin Auton Res 5(4):211–213

140. Biaggioni I, Robertson D, Krantz S, Jones M, Haile V (1994) The anemia of primary autonomic failure and its reversal with recombinant erythropoietin. Ann Intern Med 121(3):181–186

141. Hauser AS, Chavali S, Masuho I, Jahn LJ, Martemyanov KA, Gloriam DE, Babu MM (2017) Pharmacogenomics of GPCR drug targets. Cell 172:41. https://doi.org/10.1016/j.cell.2017.11.033

142. Hauser RA, Isaacson S, Lisk JP, Hewitt LA, Rowse G (2014) Droxidopa for the short-term treatment of symptomatic neurogenic orthostatic hypotension in Parkinson's disease (nOH306B). Mov Disord 30:646. https://doi.org/10.1002/mds.26086

Chapter 14

Clinical Trials for Erectile Dysfunction in Parkinson's Disease

Shen-Yang Lim, Ai Huey Tan, and Mathis Grossmann

Abstract

Impaired sexual function is a very frequent feature of Parkinson's disease (PD). For example, erectile dysfunction (ED), defined as the inability to attain or maintain penile erection sufficient for satisfactory sexual performance, develops in the majority of men with PD. Management of sexual dysfunction often needs to be multifaceted and may sometimes need to include the partner. The first-line pharmacological treatments for ED are the selective cyclic guanosine monophosphate (cGMP) phosphodiesterase type 5 inhibitors (PDE5-I), of which sildenafil has the best evidence, being in clinical use since 1998. Dopamine agonists induce erections in some patients with PD. Results with testosterone are controversial in PD. Clinical trials for treatments of ED in PD remain scarce, and existing studies were limited by small sample size and open-label methodology. Therefore, well-designed double-blind RCTs are required to further understand the efficacy and adverse effect profiles of these treatments in PD.

Key words Parkinson's disease, Erectile dysfunction, Sexual function, Treatment, Clinical trials

1 Background

Impaired sexual function in Parkinson's disease (PD) occurs in both men and women. One study found that more male than female PD patients (68% vs. 36%) reported having moderate-to-severe sexual problems [1]. Erectile dysfunction (ED), defined as the inability to attain or maintain penile erection sufficient for satisfactory sexual performance, develops in the majority of men with PD [1–3]. One study (n = 268 male PD patients; mean age ≈61 years) reported an overall 55% rate of erection problems vs. 27% in age-matched control subjects [4]. A recent large study involving 1132 men with recent-onset PD (diagnosed within the past 3 years; mean age ≈67 years) reported ED in 56.1% of patients [5]. These studies found increasing prevalence of ED with worsening disease severity as measured by Hoehn and Yahr stage, although increasing age could have been an important confounder. Interestingly, population-based studies have demonstrated an

Santiago Perez-Lloret (ed.), *Clinical Trials In Parkinson's Disease*, Neuromethods, vol. 160,
https://doi.org/10.1007/978-1-0716-0912-5_14, © Springer Science+Business Media, LLC, part of Springer Nature 2021

increased risk (1.3- to 3.8-fold) of developing PD in men with ED [6–8], suggesting that ED can be a very early prodromal feature of PD. A clinicopathological study showed that earlier appearance of autonomic dysfunction including ED was associated with more rapid disease progression and shorter survival in PD patients; potential confounding by multiple system atrophy (MSA) was avoided since all 100 cases of PD were autopsy-confirmed; only 4% of patients were reported to have died of cardiovascular causes [9]. ED is associated with impaired quality of life [10], and frustration with this problem is sometimes further compounded when patients experience an increase in sexual urge or even frank hypersexuality due to treatment with dopamine agonists (DA) [10–12]. Recognition of ED is the first step in management, and can be probed either directly in a sensitive and nonjudgmental way, or indirectly, for example, with the use of questionnaires, leaflets, posters, and other relevant information in the clinic to invite patients to present their sexual concerns [10, 13].

2   Causative and Contributing Factors

An erection is a complex event involving vascular, neural, endocrine, and psychological inputs contributing to its initiation and maintenance [14]. Aging greatly influences the development of ED, via multiple mechanisms [15]. PD is a multisystem disease involving both the central and peripheral nervous systems, including the autonomic nervous system [16, 17]. Other causative or contributing factors include psychological issues (e.g., poor self-esteem, depression, anxiety, stress, cognitive impairment, apathy, relationship problems), impaired mobility, fatigue, and pain, which are common problems in PD [1, 2, 18, 19]. The partner's reaction to the effects of the PD (e.g., loss of physical attraction) may be a further compounding factor [1]. Excessive use of alcohol or tobacco can also contribute to ED [15].

While some patients experience hypersexuality with DA used to treat PD, bromocriptine has also been reported to reduce libido and induce ED in a dose-dependent manner [20]. In one case series, ED improved when levodopa was substituted for bromocriptine [21]. The use of bromocriptine has declined substantially over the past decade due to the association of ergot-derived DA and cardiac valvular, pleural, and peritoneal fibrosis [22], although still prescribed by physicians in some parts of the world [23]. Conversely, other reports suggest that DA can improve erectile function (discussed below). Antidepressant medications such as selective serotonin reuptake inhibitors (SSRIs) and tricyclics may also be associated with ED or premature ejaculation [20, 24], whereas mirtazapine, bupropion, and reboxetine are antidepressants with a lower propensity to cause sexual dysfunction [2, 25].

Antihypertensives such as thiazide diuretics and beta blockers may also contribute to ED [24]. Finasteride, commonly used to treat benign prostatic hyperplasia and male hair loss, is also a cause of ED [26].

A link between PD and diabetes has been increasingly recognized [9]; diabetes is a common cause of ED due to its vascular effects, and in causing autonomic neuropathy [27]. Men with diabetes are four times more likely to experience ED with onset on average 15 years earlier compared to men without diabetes [15]. In the general non-PD population, there is a strong association between cardiovascular disease and ED, with authors recommending a thorough cardiovascular evaluation in patients presenting with ED [15]. To our knowledge, this association has not been specifically studied in PD.

3 Clinical Trials and Management

Management of sexual dysfunction often needs to be multifaceted and may sometimes need to include the partner. Referral to other specialists (urologist, psychiatrist/psychologist, sex/couple therapist, gynecologist for the female partner) should be considered (e.g., sex counseling may enable pleasure-oriented "outercourse" instead of goal/orgasmic-oriented intercourse) [1, 2, 10, 18, 28].

The first-line pharmacological treatments for ED are the selective cyclic guanosine monophosphate (cGMP) phosphodiesterase type 5 inhibitors (PDE5-I), of which sildenafil has the best evidence, being in clinical use since 1998 [15, 29, 30]. Clinical trials of pharmacological treatments for ED in PD are summarized in Table 1. Novel drugs that are being tested but not yet in clinic use have recently been reviewed [14].

3.1 Selective Cyclic GMP Phosphodiesterase Type 5 Inhibitors (PDE5-I)

These agents enhance nitric oxide–mediated relaxation of corpus cavernosum smooth muscle (with increased blood flow into cavernosal spaces), and are first-line therapy for most men with ED. Sildenafil (Viagra®) is the most commonly used agent [19, 31]. A meta-analysis involving 3254 subjects from ten RCTs found that more men experienced a global improvement in erections with sildenafil than with placebo (79% vs. 21% responding affirmatively to the question, "Has the treatment you have been taking over the past four weeks improved your erections?" with a number needed to treat (NNT) of 1.7) [32]. One study of older men aged >60 years reported a 54% response rate to sildenafil, with a mean increase in International Index of Erectile Function Erectile Function domain (IIEF-EF) score of 5.7 [33].

In the PD population, one small RCT ($n = 10$ patients completing the study) demonstrated that sildenafil was efficacious in nine of the ten patients, with improvements in the ability to achieve

Table 1
Clinical trials of treatments for erectile dysfunction in Parkinson's disease

Study	Sample size and patient characteristics	Design	Result	Additional comments
Selective cyclic GMP phosphodiesterase type 5 inhibitors				
Zesiewicz et al., 2000 [35]	10 PD patients; no control group Mean age 72.8 years Mean PD duration 7.5 years Two patients with mild HTN; none with DM or other cardiovascular disease	Open-label	Sildenafil (50–100 mg; mean 70 mg) resulted in significant improvements in multiple outcomes including Sexual Health Inventory—M version (SHI-M) score, and ability to achieve and maintain erections	A primary outcome measure was not specified
Hussain et al., 2001 [34]	12 PD patients (ten completing the study) Median age 61 years (range 48–68) PD duration not reported Patients with DM or history of stroke or significant cardiac disease excluded	Double-blind, placebo-controlled crossover RCT The 1° outcome measure was the IIEF questions 3 and 4 (ability to obtain and maintain erections, respectively); and a QoL questionnaire	Sildenafil (25–100 mg; mean 92 mg) efficacious in 9/10 patients, with significant improvements in IIEF questions 3 and 4 and in the quality of sex life	This study was assigned a low quality score (67%) by the MDS EBM review [29, 51]
Bernard et al., 2017 [36]	20 PD patients participated and completed Mean age 60 years Mean PD duration 7.8 years Patients with history of stroke or significant cardiovascular disease excluded; no mention of DM	Double-blind, placebo-controlled crossover RCT The 1° outcome measure was the IIEF erectile function (IIEF-EF) domain	Sildenafil (25–100 mg; mean 96 mg) resulted in significant improvement in the IIEF-EF; however, there was no treatment effect for quality of life (PDQ-39; a 2° outcome measure)	This study was assigned a high quality score (82%) by the MDS EBM review [29, 51]
Testosterone treatment				

(continued)

Table 1
(continued)

Study	Sample size and patient characteristics	Design	Result	Additional comments
Okun et al., 2002 [48]	Ten PD patients; no control group	Open-label	50% and 30% of patients had improved libido and erectile function, respectively, at 1 month	–
Okun et al., 2006 [49]	30 PD patients (15 receiving testosterone; 15 receiving placebo)	Double-blind, placebo-controlled RCT	No significant difference in the St. Louis Testosterone Deficiency Questionnaire (the 1° outcome variable)	This study was not aimed specifically at evaluating treatment effects on ED; of the ten yes/no items in the St. Louis Testosterone Deficiency Questionnaire, only one relates to libido and one relates to erections ("Are your erections less strong?")

These are arranged by category of treatment, and chronology (publication year)

DM diabetes mellitus, *HTN* hypertension, *IIEF* International Index of Erectile Function, *MDS EBM* International Parkinson and Movement Disorder Society Evidence-Based Medicine Task Force review, *RCT* randomized clinical trial

and maintain erections and in the quality of sex life [34]. Similar improvements were reported in an open-label study by Zesiewicz et al. ($n = 10$) [35]. The best quality evidence thus far comes from a recent RCT by Bernard et al., in which patients had a mean IIEF-EF score of 23.2 during the treatment phase vs. 12.3 during the placebo phase (corresponding to "mild" vs. "moderate" degrees of ED, respectively) [36]. The impressive improvements seen in these RCTs [34, 36] could be due in part to the selection criteria excluding patients with concomitant diabetes or significant vascular disease—factors that may be associated with poorer treatment response in real-life settings (*see* Table 1).

Other PDE5-I such as tadalafil (Cialis®), vardenafil (Levitra®), and avanafil (Stendra®) are also FDA-approved and are believed to be comparably efficacious. Anecdotally, some patients prefer tadalafil because of its longer duration of action, possibly allowing couples greater freedom regarding timing of intercourse [14]. To the best of our knowledge, there have been no studies using these newer agents in PD patients [2, 3, 14].

The usual starting dose of sildenafil is 50 mg, which can be titrated up to 100 mg, or down to 25 mg, depending on efficacy and tolerability (mean doses are reported in Table 1) [34–36]. Sildenafil is usually taken half or 1 h before sexual activity, but in some PD patients the onset of action may be delayed up to 4 h, presumably due to slowed gastric motility and delayed drug absorption [10, 28]. Efficacy is optimal when taken on an empty stomach [31].

Sildenafil is usually well tolerated, with only minor and transient adverse effects, related to vasodilation, such as headache (the most frequent side effect, affecting approximately 5–10% of patients) and flushing [34–36]. Transient visual symptoms, mainly disturbances of color vision, were reported predominantly at the 100 mg dose and is due to transient inhibition of phosphodiesterase in the retina [31]. Postural hypotension is a potential concern, but PD patients in the study by Hussain et al. showed minimal change in blood pressure between sildenafil and placebo treatment (in contrast to a severe drop in postural blood pressure induced in patients with MSA) [34]. Supine and standing blood pressure measurements should be taken before prescribing sildenafil, and patients should be advised to seek medical advice if symptoms of orthostatic hypotension develop [3, 34]. Sildenafil is contraindicated if blood pressure is below 90/50 mmHg [37], and in patients receiving nitrate therapy since their coadministration can produce life-threatening hypotension [38]. Due to the potential for orthostatic hypotension, some authors recommend using only short-acting PDE5-I (e.g., sildenafil) in PD patients. Coexistent heart disease is probably not a contraindication to treatment with PDE5-I, and there is evidence to suggest that these agents are safe and exert beneficial effects in patients with concomitant coronary artery disease or heart failure, in part through direct myocardial actions independent of vascular effects [38, 39]. Nitrates can be safely administered 24 h after the last use of sildenafil or vardenafil, and 48 h after tadalafil (since this drug has a prolonged duration of action) [38].

The RCT of sildenafil for PD-related ED by Safarinejad et al. was positive, but this paper was retracted by the publisher. Curiously, in two studies from the group of Giammusso and Raffaele (reporting improved erections in 85% of PD patients), the results for the various ED and depression efficacy measures, at baseline and follow-up, were exactly the same, despite the fact that the cohorts would appear to be mutually exclusive (all patients in the report by Giammusso et al. were stated to be on pramipexole but not levodopa treatment; whereas all patients in the report by Raffaele et al. were receiving levodopa therapy but not DA) [40, 41]. Clinical characteristics of both cohorts (mean age 63.3 years, Hoehn and Yahr stage 2.1) were also identical. The findings of these studies should therefore be interpreted with extreme caution.

3.2 Dopamine Agonists (DA)

These drugs can induce erections in some patients with PD, thought to be mediated via dopamine D2-like (i.e., D2, D3, or D4) receptors in the paraventricular nucleus (PVN) of the hypothalamus, although spinal and peripheral mechanisms have also been implicated [19, 25, 42, 43]. Apomorphine (subcutaneous injections) and pergolide have been the best studied DA in terms of effect on erectile function, and apomorphine has also been used to treat ED in the general population [43]. Pergolide, like other ergot DA, is associated with fibrotic complications and for this reason it has been removed from the market. O'Sullivan and Hughes first reported that some PD patients given subcutaneous injections of apomorphine to treat motor fluctuations started using the injections specifically for ED (erections result within 10 min and last up to 60 min). However, sublingual apomorphine (2–4 mg) was found to be ineffective for a majority of patients and as such was never approved by the FDA and has since been discontinued in the UK; this agent has not been studied specifically for ED in PD patients [14, 25, 44]. Nausea, dizziness, and hypotension are relatively common adverse events [30]. There are isolated case reports of oral DA such as ropinirole regularly inducing erections in some patients [42].

3.3 Testosterone Treatment

Endogenous testosterone plays an important role in all aspects of sexual function, especially in libido. In men with organic hypogonadism due to pituitary or testicular disease, testosterone replacement therapy consistently increases spontaneous sexual interest and facilitates sexual activities. In older men, ED is commonly due to neurovascular disease, and modestly reduced testosterone may be a marker of ill health, rather than a causal contributor. In an RCT of stringently selected older men ≥65 years with serum testosterone <275 ng/dL (<27.5 pg/mL or <9.5 nmol/L) without overtly pathological hypogonadism, testosterone treatment provided a modest benefit on all measures of sexual function, including erectile function [45]. These Testosterone Trials ("T-Trials") sponsored by the National Institutes of Health also found improved bone density and hemoglobin [46].

The literature examining the effects of testosterone treatment on sexual function in PD is very limited. In one study involving 68 male patients, a lowered free testosterone (defined as <70 pg/mL) was present in 35% of men and was more common with increasing age [47]. In a small case series ($n = 5$ PD patients with free testosterone <70 pg/mL), testosterone administration resulted in improvements in libido in four patients and erectile function in two patients, in addition to improvements in other nonmotor symptoms [47]. Similar results were seen in an open-label study involving ten PD patients with free testosterone <80 pg/mL: 50% and 30% of patients had improved libido and erectile function, respectively, at 1 month [48]. These studies have to be viewed with caution given the small sample size and lack of a placebo control group. Indeed, a subsequent blinded placebo-controlled RCT by the same authors, recruiting patients with free testosterone <100 pg/mL, showed no treatment effect of intramuscular testosterone enanthate injections (200 mg/mL) fortnightly for 8 weeks, in the St. Louis Testosterone Deficiency Questionnaire (the primary outcome variable) [49].

3.4 Local Treatments   These include vacuum constriction devices (VCD) and intracavernosal injections (ICI) of vasodilators (e.g., prostaglandin E1/ alprostadil, phentolamine, vasoactive intestinal polypeptide (VIP) and papaverine, either alone or in combination) [2, 18]. VCD are considered very safe [15]. Due to the invasive route of administration with adverse effects such as penile pain, ICI have become second-line treatment but are very effective, including in patients not responding to PDE5-I [14]. Surgical implantation of a penile prosthesis is generally a last resort, although patient satisfaction is reported to be high [15]. There are no published studies of these treatments in PD patients.

4   Conclusion and Future Directions

ED is a common and probably underreported symptom in PD, and can affect patients across different stages of the disease. ED is usually recognized as part of a broader spectrum of autonomic dysfunction in PD, although the specific pathophysiology is not well studied [30, 50]. Clinical trials for treatments of ED in PD remain scarce, and existing studies were limited by small sample size and open-label methodology. PDE5-I, in particular sildenafil, appear to be efficacious and safe, while current evidence regarding testosterone treatment remains inconclusive. Well-designed double-blind RCTs are required to further understand the efficacy and adverse effect profiles of these treatments in PD. Preclinical studies aiming to understand the pathophysiology of ED in PD may unravel potential novel therapeutic targets.

References

1. Brown RG, Jahanshahi M, Quinn N, Marsden CD (1990) Sexual function in patients with Parkinson's disease and their partners. J Neurol Neurosurg Psychiatry 53(6):480–486. https://doi.org/10.1136/jnnp.53.6.480

2. Bronner G (2011) Sexual problems in Parkinson's disease: the multidimensional nature of the problem and of the intervention. J Neurol Sci 310(1–2):139–143. https://doi.org/10.1016/j.jns.2011.05.050

3. Pfeiffer RF (2012) Autonomic dysfunction in Parkinson's disease. Expert Rev Neurother 12(6):697–706. https://doi.org/10.1586/ern.12.17

4. Verbaan D, Marinus J, Visser M, van Rooden SM, Stiggelbout AM, van Hilten JJ (2007)

Patient-reported autonomic symptoms in Parkinson disease. Neurology 69(4):333–341. https://doi.org/10.1212/01.wnl.0000266593.50534.e8

5. Malek N, Lawton MA, Grosset KA, Bajaj N, Barker RA, Burn DJ, Foltynie T, Hardy J, Morris HR, Williams NM, Ben-Shlomo Y, Wood NW, Grosset DG, Consortium PRC (2016) Autonomic dysfunction in early Parkinson's disease: results from the United Kingdom tracking Parkinson's study. Mov Disord Clin Pract 4(4):509–516. https://doi.org/10.1002/mdc3.12454

6. Gao X, Chen H, Schwarzschild MA, Glasser DB, Logroscino G, Rimm EB, Ascherio A (2007) Erectile function and risk of Parkinson's disease. Am J Epidemiol 166(12):1446–1450. https://doi.org/10.1093/aje/kwm246

7. Schrag A, Horsfall L, Walters K, Noyce A, Petersen I (2015) Prediagnostic presentations of Parkinson's disease in primary care: a case-control study. Lancet Neurol 14(1):57–64. https://doi.org/10.1016/s1474-4422(14)70287-x

8. Yang Y, Liu H, Lin T, Kuo Y, Hsieh T (2017) Relationship between erectile dysfunction, comorbidity, and Parkinson's disease: evidence from a population-based longitudinal study. J Clin Neurol 13(3):250–258. https://doi.org/10.3988/jcn.2017.13.3.250

9. De Pablo-Fernandez E, Tur C, Revesz T, Lees AJ, Holton JL, Warner TT (2017) Association of autonomic dysfunction with disease progression and survival in Parkinson disease. JAMA Neurol 74(8):970–976. https://doi.org/10.1001/jamaneurol.2017.1125

10. Bronner G, Korczyn AD (2017) The role of sex therapy in the management of patients with Parkinson's disease. Mov Disord Clin Pract 5(1):6–13. https://doi.org/10.1002/mdc3.12561

11. Lim S-Y, Evans AH, Miyasaki JM (2008) Impulse control and related disorders in Parkinson's disease. Ann N Y Acad Sci 1142(1):85–107. https://doi.org/10.1196/annals.1444.006

12. Codling D, Shaw P, David AS (2015) Hypersexuality in Parkinson's disease: systematic review and report of 7 new cases. Mov Disord Clin Pract 2(2):116–126. https://doi.org/10.1002/mdc3.12155

13. Jitkritsadakul O, Jagota P, Bhidayasiri R (2015) Postural instability, the absence of sexual intercourse in the past month, and loss of libido are predictors of sexual dysfunction in Parkinson's disease. Parkinsonism Relat Disord 21(1):61–67. https://doi.org/10.1016/j.parkreldis.2014.11.003

14. Peak TC, Yafi FA, Sangkum P, Hellstrom WJG (2015) Emerging drugs for the treatment of erectile dysfunction. Expert Opin Emerg Drugs 20(2):263–275. https://doi.org/10.1517/14728214.2015.1021682

15. Mobley DF, Khera M, Baum N (2017) Recent advances in the treatment of erectile dysfunction. Postgrad Med J 93(1105):679–685. https://doi.org/10.1136/postgradmedj-2016-134073

16. Lim S-Y, Fox SH, Lang AE (2009) Overview of the extranigral aspects of Parkinson disease. Arch Neurol 66(2). https://doi.org/10.1001/archneurol.2008.561

17. Poewe W, Seppi K, Tanner CM, Halliday GM, Brundin P, Volkmann J, Schrag A-E, Lang AE (2017) Parkinson disease. Nat Rev Dis Primers 3(1). https://doi.org/10.1038/nrdp.2017.13

18. Basson R (1996) Sexuality and Parkinson's disease. Parkinsonism Relat Disord 2(4):177–185. https://doi.org/10.1016/s1353-8020(96)00020-x

19. Papatsoris AG, Deliveliotis C, Singer C, Papapetropoulos S (2006) Erectile dysfunction in Parkinson's disease. Urology 67(3):447–451. https://doi.org/10.1016/j.urology.2005.10.017

20. Bronner G, Royter V, Korczyn AD, Giladi NIR (2004) Sexual dysfunction in Parkinson's disease. J Sex Marital Ther 30(2):95–105. https://doi.org/10.1080/00926230490258893

21. Cleeves L, Findley LJ (1987) Bromocriptine induced impotence in Parkinson's disease. Br Med J (Clin Res Ed) 295(6594):367–368. https://doi.org/10.1136/bmj.295.6594.367-a

22. Fox SH, Katzenschlager R, Lim S-Y, Barton B, de Bie RMA, Seppi K, Coelho M, Sampaio C (2018) International Parkinson and movement disorder society evidence-based medicine review: update on treatments for the motor symptoms of Parkinson's disease. Mov Disord 33(8):1248–1266. https://doi.org/10.1002/mds.27372

23. Lim S-Y, Tan AH, Ahmad-Annuar A, Klein C, Tan LCS, Rosales RL, Bhidayasiri R, Wu Y-R, Shang H-F, Evans AH, Pal PK, Hattori N, Tan CT, Jeon B, Tan E-K, Lang AE (2019) Parkinson's disease in the Western Pacific region. Lancet Neurol 18(9):865–879. https://doi.org/10.1016/s1474-4422(19)30195-4

24. Francis ME, Kusek JW, Nyberg LM, Eggers PW (2007) The contribution of common medical conditions and drug exposures to erectile dysfunction in adult males. J Urol 178(2):591–596. https://doi.org/10.1016/j.juro.2007.03.127

25. Simonsen U, Comerma-Steffensen S, Andersson K-E (2016) Modulation of dopaminergic pathways to treat erectile dysfunction. Basic Clin Pharmacol Toxicol 119:63–74. https://doi.org/10.1111/bcpt.12653

26. Shin YS, Karna KK, Choi BR, Park JK (2019) Finasteride and erectile dysfunction in patients with benign prostatic hyperplasia or male androgenetic alopecia. World J Mens Health 37(2):157–165. https://doi.org/10.5534/wjmh.180029

27. Kouidrat Y, Pizzol D, Cosco T, Thompson T, Carnaghi M, Bertoldo A, Solmi M, Stubbs B, Veronese N (2017) High prevalence of erectile dysfunction in diabetes: a systematic review and meta-analysis of 145 studies. Diabet Med 34(9):1185–1192. https://doi.org/10.1111/dme.13403

28. Bronner G (2009) Practical strategies for the management of sexual problems in Parkinson's disease. Parkinsonism Relat Disord 15:S96–S100. https://doi.org/10.1016/s1353-8020(09)70791-6

29. Seppi K, Ray Chaudhuri K, Coelho M, Fox SH, Katzenschlager R, Perez Lloret S, Weintraub D, Sampaio C, The collaborators of the Parkinson's Disease Update on Non-Motor Symptoms Study Group on behalf of the Movement Disorders Society Evidence-Based Medicine C (2019) Update on treatments for nonmotor symptoms of Parkinson's disease-an evidence-based medicine review. Mov Disord 34(2):180–198. https://doi.org/10.1002/mds.27602

30. Palma J-A, Kaufmann H (2018) Treatment of autonomic dysfunction in Parkinson disease and other synucleinopathies. Mov Disord 33(3):372–390. https://doi.org/10.1002/mds.27344

31. Briganti A, Salonia A, Gallina A, Saccà A, Montorsi P, Rigatti P, Montorsi F (2005) Drug insight: oral phosphodiesterase type 5 inhibitors for erectile dysfunction. Nat Clin Pract Urol 2(5):239–247. https://doi.org/10.1038/ncpuro0186

32. Moore RA, Edwards JE, McQuay HJ (2002) Sildenafil (Viagra) for male erectile dysfunction: a meta-analysis of clinical trial reports. BMC Urol 2:6–6. https://doi.org/10.1186/1471-2490-2-6

33. Müller A, Smith L, Parker M, Mulhall JP (2007) Analysis of the efficacy and safety of sildenafil citrate in the geriatric population. BJU Int 100(1):117–121. https://doi.org/10.1111/j.1464-410x.2007.06915.x

34. Hussain IF, Brady CM, Swinn MJ, Mathias CJ, Fowler CJ (2001) Treatment of erectile dysfunction with sildenafil citrate (Viagra) in Parkinsonism due to Parkinson's disease or multiple system atrophy with observations on orthostatic hypotension. J Neurol Neurosurg Psychiatry 71(3):371–374. https://doi.org/10.1136/jnnp.71.3.371

35. Zesiewicz TA, Helal M, Hauser RA (2000) Sildenafil citrate (viagra) for the treatment of erectile dysfunction in men with Parkinson's disease. Mov Disord 15(2):305–308. https://doi.org/10.1002/1531-8257(200003)15:2<305::aid-mds1015>3.0.co;2-w

36. Bernard BA, Metman LV, Levine L, Ouyang B, Leurgans S, Goetz CG (2016) Sildenafil in the treatment of erectile dysfunction in Parkinson's disease. Mov Disord Clin Pract 4(3):412–415. https://doi.org/10.1002/mdc3.12456

37. Bronner G, Vodušek DB (2011) Management of sexual dysfunction in Parkinson's disease. Ther Adv Neurol Disord 4(6):375–383. https://doi.org/10.1177/1756285611411504

38. Chrysant SG (2013) Effectiveness and safety of Phosphodiesterase 5 inhibitors in patients with cardiovascular disease and hypertension. Curr Hypertens Rep 15(5):475–483. https://doi.org/10.1007/s11906-013-0377-9

39. Hutchings DC, Anderson SG, Caldwell JL, Trafford AW (2018) Phosphodiesterase-5 inhibitors and the heart: compound cardioprotection? Heart 104(15):1244–1250. https://doi.org/10.1136/heartjnl-2017-312865

40. Giammusso B, Raffaele R, Vecchio I, Giammona G, Ruggieri M, Nicoletti G, Malaguarnera M, Rampello L, Nicoletti F (2002) Sildenafil in the treatment of erectile dysfunction in elderly depressed patients with idiopathic Parkinson's disease. Arch Gerontol Geriatr 35:157–163. https://doi.org/10.1016/s0167-4943(02)00124-3

41. Raffaele R (2002) Efficacy and safety of fixed-dose oral sildenafil in the treatment of sexual dysfunction in depressed patients with idiopathic Parkinson's disease. Eur Urol 41(4):382–386. https://doi.org/10.1016/s0302-2838(02)00054-4

42. Fine J, Lang AE (1999) Dose-induced penile erections in response to ropinirole therapy for Parkinson's disease. Mov Disord 14(4):701–702. https://doi.org/10.1002/1531-8257(199907)14:4<701::aid-mds1026>3.0.co;2-r

43. Sakakibara R, Kishi M, Ogawa E, Tateno F, Uchiyama T, Yamamoto T, Yamanishi T (2011) Bladder, bowel, and sexual dysfunction in Parkinson's disease. Parkinsons Dis 2011:924605–924605. https://doi.org/10.4061/2011/924605

44. Perez-Lloret S, Rey MV, Pavy-Le Traon A, Rascol O (2013) Emerging drugs for autonomic dysfunction in Parkinson's disease. Expert Opin Emerg Drugs 18(1):39–53. https://doi.org/10.1517/14728214.2013.766168

45. Snyder PJ, Bhasin S, Cunningham GR, Matsumoto AM, Stephens-Shields AJ, Cauley JA, Gill TM, Barrett-Connor E, Swerdloff RS, Wang C, Ensrud KE, Lewis CE, Farrar JT, Cella D, Rosen RC, Pahor M, Crandall JP, Molitch ME, Cifelli D, Dougar D, Fluharty L, Resnick SM, Storer TW, Anton S, Basaria S, Diem SJ, Hou X, Mohler ER 3rd, Parsons JK, Wenger NK, Zeldow B, Landis JR, Ellenberg SS, Testosterone Trials I (2016) Effects of testosterone treatment in older men. The New England journal of medicine 374(7):611–624. https://doi.org/10.1056/NEJMoa1506119

46. Yeap BB, Page ST, Grossmann M (2018) Testosterone treatment in older men: clinical implications and unresolved questions from the testosterone trials. Lancet Diabet Endocrinol 6(8):659–672. https://doi.org/10.1016/s2213-8587(17)30416-3

47. Okun MS (2002) Refractory nonmotor symptoms in male patients with Parkinson disease due to testosterone deficiency. Arch Neurol 59(5):807. https://doi.org/10.1001/archneur.59.5.807

48. Okun MS, Walter BL, McDonald WM, Tenover JL, Green J, Juncos JL, DeLong MR (2002) Beneficial effects of testosterone replacement for the nonmotor symptoms of Parkinson disease. Arch Neurol 59(11):1750. https://doi.org/10.1001/archneur.59.11.1750

49. Okun MS, Fernandez HH, Rodriguez RL, Romrell J, Suelter M, Munson S, Louis ED, Mulligan T, Foster PS, Shenal BV, Armaghani SJ, Jacobson C, Wu S, Crucian G (2006) Testosterone therapy in men with Parkinson disease. Arch Neurol 63(5):729. https://doi.org/10.1001/archneur.63.5.729

50. Coon EA, Cutsforth-Gregory JK, Benarroch EE (2018) Neuropathology of autonomic dysfunction in synucleinopathies. Mov Disord 33(3):349–358. https://doi.org/10.1002/mds.27186

51. Seppi K, Weintraub D, Coelho M, Perez-Lloret S, Fox SH, Katzenschlager R, Hametner E-M, Poewe W, Rascol O, Goetz CG, Sampaio C (2011) The Movement Disorder Society evidence-based medicine review update: treatments for the non-motor symptoms of Parkinson's disease. Mov Disord 26(Suppl 3):S42–S80. https://doi.org/10.1002/mds.23884

Chapter 15

Clinical Trials for Constipation in Parkinson's Disease

Patricio Millar Vernetti

Abstract

Constipation is the most frequent gastrointestinal disturbance in Parkinson's disease, and its prevalence increases with disease duration. It is believed to be caused by a combination of direct affection of the enteric nervous system, the dorsal motor nucleus of the vagus, and pelvic floor dyssynergia. Parkinson disease subtype as well as medication, as well as the usual factors considered in the general population, may also play a role. Constipation in Parkinson's disease can be a cause of daily discomfort and may lead to serious complications.

This chapter will approach the use of different tools to assess subjective and objective outcome measures, and the pitfalls and considerations pertaining each of these methods. Clinical trials will be reviewed to identify strengths and weaknesses, and lastly, a series of recommendations will be outlined for the design of clinical trials for constipation in Parkinson's disease.

Key words Parkinson's disease, Constipation, Clinical trials, Nonmotor symptoms, Rating scales, Outcome measures

1 Introduction

Regarding nonmotor symptoms, constipation is the most frequent gastrointestinal disturbance in Parkinson's disease (PD). The prevalence of constipation in PD (CPD) has been reported to vary between 24.6% and 63% [1]. This variability may be in part accounted by the type of method utilized. The use of certain questionnaires may yield a prevalence as low as 3%. As a matter of fact, the use of questionnaires compared to objective measurements, may underestimate the true prevalence of this disorder. When relying on objective methods, up to 79% of PD patients may show a delayed colonic transit time, and 66% an increased colonic volume [2].

Moreover, the prevalence of CPD increases along with the duration of the disease. In a longitudinal study, recently diagnosed patients with PD had a prevalence of 33%, which increased to up to

Santiago Perez-Lloret (ed.), *Clinical Trials In Parkinson's Disease*, Neuromethods, vol. 160,
https://doi.org/10.1007/978-1-0716-0912-5_15, © Springer Science+Business Media, LLC, part of Springer Nature 2021

50% after 2 years of follow-up, compared to 22% prevalence in healthy controls, which remained stable over the same period of time [3].

As a matter of fact, CPD may precede the onset of overt motor symptoms for up to 10–20 years. Furthermore, in spite of being a nonspecific symptom, it may confer a relative risk for PD between 1.78 and 2.56 [4]. This may be explained by the "dual-hit" hypothesis, following Braak's staging of caudorostral progression, in which the gastrointestinal autonomic ganglia and nerves would be one of the first sites of onset of this pathology [5, 6], and evidence of alfa-synuclein deposition can be found in colonic tissue as much as 20 years prior to PD diagnosis. Thus, delayed colonic transit time is thought to be due to a combination of direct affection of the enteric nervous system within the colon, and of the dorsal motor nucleus of the vagus nerve in the brainstem, in combination with pelvic floor dyssynergia arising from supraspinal modulation dysfunction of the defecatory reflex [7]. CPD seems to increase along with duration of the disease (33–39%, for de novo patients; 41–47% for a duration <8 years; 52–72% for a diagnosis >8 years), and severity in PD, as measured by the Hoen and Yahr scale. Currently, it is debated whether this is only due to progression of the disease, or by accompanying factors, specifically dopaminergic medication (and particularly dopaminergic agonists as compared to LDOPA). As PD progresses and motor symptoms worsen, antiparkinsonian medication is increased. If indeed an association exists, type of antiparkinsonian medication and levodopa equivalent daily dose should be reported.

Moreover, constipation has been shown to be more prevalent in patients with akinetic-rigid phenotype (45–69%) than those with tremor-dominant features (21–49%) [8].

Factors influencing the presence of constipation in the general population may as well affect CPD, such as insufficient physical activity, decreased water intake, and low calorie intake [9–11].

Besides being a cause of daily discomfort, CPD may lead to serious complications, which may require surgical treatment, such as sigmoid megacolon and volvulus, ultimately progressing to intestinal necrosis and perforation [12, 13].

Diverse treatments have been developed to treat this frequent nonmotor symptom in PD, including lifestyle and dietary modifications, the use of specific medications, as well as the evaluation of the effect of previously approved interventions for motor symptoms of PD for this disturbance.

With these considerations in mind we will approach in this chapter the appropriate methodology for the design of clinical trials for constipation in PD and how to avoid pitfalls, in order to provide high quality of evidence.

2   Measurements and Outcomes

2.1 Subjective Measurements

There are a series of issues worth noting when trying to define and assess CPD. First, even though listed within the items of the NMSQuest scale, there is no specific scale designed to particularly address CPD. Secondly, constipation is usually defined by ROME III criteria, although there is currently no published information regarding the validation process of this questionnaire either. Moreover, in 2009, a Movement Disorder Society's task force was set out to review the available rating scales used to evaluate different symptoms attributed to autonomic dysfunction (constipation among them), and concluded that none of the scales or questionnaires reviewed met the criteria to suggest or recommend its use [14].

Even though constipation is usually defined by the ROME III criteria (Table 1), many different scales and scoring systems have been used in clinical trials for this symptom, most of them addressing number of bowel movements per week, and the presence of difficulty or straining with defecation (Table 2). In addition to these symptoms, ROME III criteria also delve into signs of anorectal dysfunction such as sensation of incomplete evacuation or anorectal obstruction, and the need to use digital maneuvers during defecation, which take into consideration the physiopathological aspect of pelvic floor dyssynergia in PD. Most of the previous clinical trials have used different versions of the ROME-III scale or modifications of it, usually emphasizing on the frequency of bowel movements (Tables 2 and 3).

Table 1
ROME III criteria for the diagnosis of functional constipation

1. Must include two or more of the following:
(a) Straining during at least 25% of defecations
(b) Lumpy or hard stools in at least 25% of defecations
(c) Sensation of incomplete evacuation for at least 25% of defecations
(d) Sensation of anorectal obstruction/blockage for at least 25% of defecations
(e) Manual maneuvers to facilitate at least 25% of defecations (e.g., digital evacuation, support of the pelvic floor)
(f) Fewer than three defecations per week
2. Loose stools are rarely present without the use of laxatives
3. Insufficient criteria for irritable bowel syndrome

[a]Criteria fulfilled for the last 3 months with symptom onset at least 6 months prior to diagnosis

Table 2
Comparison of variables assessed by subjective questionnaires

Reference/scale	Definition of constipation
ROME III (Drossman 2006 [37])	Frequency of BM; straining; consistency of stools; incomplete evacuation; sensation of anorectal obstruction; use of digital maneuvers; use of laxatives; infrequently loose stools in absence of laxatives
Constipation scoring system (CSS; Agachan 1996 [38])	Frequency of BM; straining; incomplete evacuation; use of digital maneuvers; use of laxatives
Sakakibara 2015 [30]	Frequency of BM; straining; patient self-report report of constipation
Sakakibara 2005 [39]	Frequency of BM; straining
Cadeddu 2005[a] [33]	Frequency of BM; straining; incomplete evacuation; use of laxatives; use of digital maneuvers
Questionnaire-based assessment of pelvic organ dysfunction in Parkinson's disease (Sakakibara 2001 [40])	Frequency of BM; straining; use of laxatives
Eichorn 2001 [26]	Frequency of BM; straining
Ashraf 1997 [23]	Frequency of BM

BM bowel movements, *CSS* constipation scoring system
[a]The definition of constipation in this study contains the use of objective measurements as well

Nevertheless, recall bias seems to be a major determinant and source of misallocation or later exclusion, as subjects may underestimate the frequency of bowel movements prior to enrollment. This problem is not rarely encountered in patients with constipation, and as many as 49% of them may overestimate the severity of their condition [15]. The inclusion of a pre-enrollment period in

Table 3
Other rating scales and their components used as outcome measures for CPD

Scale	Components evaluated
Neurogenic bowel dysfunction score (NBDS) [41]	Frequency of BM; time spent defecating; physical discomfort during defecation; use of laxatives; use of digital maneuvers; incontinence; flatus; perianal problems
Knowles–Eccersley scoring system (KESS) [42]	Frequency of BM; straining; incomplete evacuation; use of digital maneuvers; use of laxatives
Gastrointestinal symptoms rating scale (GSRS)[a] [43]	Abdominal pain Dyspeptic syndrome Indigestion (borborygmus, abdominal distention, eructation, increase flatus) Bowel dysfunction syndrome (frequency of BM, consistency of stool, incomplete evacuation)

[a]GSRS was originally designed for patients with irritable bowel syndrome and peptic ulcer disease

which constipation indicators are recorded by the patient (usually by the means of a diary) would prevent this issue.

As an example of this, in a recent clinical trial, Pfeiffer and colleagues have observed a high exclusion rate subsequent to an increase in bowel movements after a period of placebo administration, compared to what was reported by the patients prior to starting the trial. This rendered most patients ineligible for their study and resulting in failure to achieve their recruitment goals [16]. Recall bias and placebo effect may have been the cause of this problem, as well as Hawthorne effect (i.e., the change in baseline conditions associated with patient awareness of being part of a study) and the use of diaries in a pre-enrollment period may also help assessing the true extent of placebo effect in these scenarios. Another issue that the authors were able to identify was that one of the patients' main complaints were not related to a decreased frequency of bowel movements but to incomplete evacuation, causing multiple unsatisfactory bowel movements throughout the day.

When addressing this issue, in a study where patients were initially considered eligible by ROME-III criteria, were evaluated by means of a stool diary. Only 73% of patients were found to fulfil constipation criteria [17]. In any case, this was performed over a 2-week period, in comparison to the 3-month time frame required by ROME-III.

Most questionnaires used in previous trials include in their definition of constipation a predetermined number of bowel movements per week or day, where in fact, multiple, incomplete, strained bowel movements seem to be the norm. The problem of selecting outcome measures applies to constipation trials in general. Ervin et al. acknowledging this issue, and keeping in line with the FDA's guidance on the use of patient-reported outcome measures, stating that endpoint selection for clinical trials should be based on patient input, and all symptoms important to patients should be measured, set out to develop a list of symptoms as reported by patients themselves, and which of those they would most like to see improved by treatment. Symptoms most frequently reported were infrequent BM, incomplete evacuation, straining, hard or lumpy stools, abdominal pain, abdominal discomfort, and bloating. Among the top five symptoms that they would prefer to be resolved with treatment were incomplete evacuation, straining, unsuccessful BM, and bloating (Table 4) [18].

2.2 Objective Measurements

As previously mentioned, CPD is caused mainly by a delay in colonic transit time (CTT), anorectal dysfunction, or a combination of both. CTT can be evaluated by quantifying the presence of radiopaque markers (ROM) ingested throughout a period of time, present in the colon, or by total colonic volume as measured by CT.

Colonic volume can be calculated by the use of dedicated software and manual delineation of volume of interest, yielding 3D

Table 4
Symptoms most frequently reported by constipated patients and their preference of resolution with treatment

Symptoms reported	Most important symptoms to treat
Infrequent BM	Incomplete BM
Incomplete BM	Straining
Straining	Bloating
Hard or lumpy stools	Unsuccessful BM
Abdominal pain	Infrequent BM
Abdominal discomfort	Abdominal discomfort
Bloating	
Unsuccessful attempts	

BM bowel movements [18]

volumes from which gas volume is later subtracted [19]. In a study assessing colonic volume, a total volume of 1024 cc showed a 66% sensitivity, 85% specificity, and likelihood ratio of 4.3 for separating PD patients from controls [19].

We remark that the use of objective measures of colonic transit may not be readily available, be costly, and require specialized equipment or training, and the exposure to ionizing radiation.

The evaluation of anorectal function can be approached by the use of barium defecography (assessment of anorectal dyssynergia and incomplete evacuation), balloon distention, and expulsion test or manometry. Although, it must be considered that said studies usually show discordant results when comparing different methods [20].

Due to the aforementioned limitations on the use of questionnaires and rating scales for constipation, an ideal trial should either use a previously validated tool (when and if one is designed) or include a validation of the used tool. Due to the heterogeneity of the clinical definition of constipation, the authors could opt to state if the desired outcome is related to the objective term "constipation" associated with delayed transit time and increased volume, or to the discomfort or complications associated with it, such as completeness of evacuations, straining, abdominal pain, the need to use maneuvers or rescue medication, or abdominal symptoms such as bloating, pain, or discomfort, as reported and emphasized by the patients. The role and need of objective measures should be evaluated in cases where, as mentioned above, are warranted to provide a validation, or are essential to the chosen outcome. The potential risks and discomfort associated with the use of objective measures should be kept in mind when considering its use in clinical trials.

3   Previous and Ongoing Trials

3.1   Considerations on Dietary Habits and Use of Laxatives

While most studies included the use of laxatives either as part of the outcomes or part of the constipation criteria, some stated a particular conduct as part of the study methods, by either completely restricting, reducing, or instituting a protocol for its use.

In their trials of 1993 and 1997, Jost et al. [21, 22] did not allow for the use of laxatives during the study. In addition to this, Ashraf et al. [23] instructed patients to stop laxatives for at least 1 week before starting the intervention. Patients were advised to reduce laxative use as much as possible in Barichella et al. [17]. During the MOVE PD trial [16], patients were allowed to use Macrogol 3350 after 4 days with no BM, and the use of an enema after 5 or 6 days with no BM, if previous measures failed.

In the study by Tateno et al. [24] patients were specifically instructed not to change neither frequency nor quantity of laxative use.

Furthermore, as was previously mentioned, daily habits such as hydration and fiber intake, and physical activity may influence constipation in the general population as well as in PD patients. Taking this into consideration, in Barichella et al.'s study [17], patients were told to maintain or increase fluid and fiber consumption in order to achieve 1.5 L and 15–20 g daily intake, including the use of a daily food diary for reporting these measures. In contrast to this, Sullivan [25] and Ashraf [23] were advised not to make changes in their diet, while Jost [22], patients were fed hospital food.

Physical activity was not regularly accounted for, except in Barichella et al. [17], were sedentarism was defined as >4 h of weekly activity, while patients in Jost [21] received regular physiotherapeutic exercise.

3.2   Medical

3.2.1   Oral Laxatives

The use of fiber as a bulk forming agent (*Psyllium*) was evaluated in a small RCT involving three patients in the treatment group vs. four in the placebo arm. Patients were first evaluated during a 4-week baseline period, by the means of a stool diary, after which CTT and anorectal manometry was performed, as a form of validating patients report of constipation. Patients then entered an 8-week treatment period, and another 4-week washout period. After each phase, subjective and objective measurements were repeated. Constipation as defined by <3 BM per week was found in 7 out of 12, as well as decreased threshold for first distention rectal sensation, although CTT was not significatively reduced. Stool frequency and weight was found to be increased in the treatment group compared to both baseline and placebo. No change in stool consistency, straining or completeness were found [23].

There are two trials regarding the use of macrogol (polyethylene glycol), an osmotic laxative for the treatment of CPD. In a first publication of a small open-label trial assessed the efficacy of 13 g of *macrogol 3350* in eight PD patients with self-reported constipation. Even though there was not a fixed observation period (follow-up varied from 9 to 21 weeks), patients reported a relief of constipation with a latency of 2–10 days, All patients reported a marked improvement in a Global Impression of Change assessment and ease of defecation. For softening of stool, all patients reported marked improvement, except one who reported moderate improvement. At baseline, BM frequency varied from 1 per 2 weeks to 1 per week, compared to 3–7 per week. Patients were able to reduce the dose of macrogol at 2 weeks. There is no mention of any statistical methods. None of the patients required the use of further laxatives, nor reported any adverse events [26].

A more recent RCT comparing *macrogol 4000* at a dose of 7.3 g to placebo for CPD, using ROME II definition, included a total of 57 patients (29 in the treatment group vs. 28 in the placebo group), having previously calculated a statistical power of 80% with a significance level of 5% for an estimated population of 60. Patients were evaluated at baseline, 4 and 8 weeks by the use of a diary which included number of BM, stool consistency, difficulty in evacuation, degree of abdominal discomfort, use of oral and rectal laxatives, and treatment efficacy was defined as complete relief of symptoms or a marked improvement of two of the variables assessed in the diary. A total of 15 patients were excluded from the final analysis: ten in the treatment group (four did not present to further visits, two due to poor compliance, two due to adverse events, and one withdrew consent) and five from the placebo group (three did not present to further visits, one due to poor compliance, and one withdrew consent). Nausea and diarrhea were reported as adverse events in two patients. At 8 weeks, 80% of patients the treatment group were considered responders (93.8% response for BM frequency, 68.8% for stool consistency) compared to 30% in the placebo group (46.7% in BM frequency, and 20% in consistency), no statistically significative differences were found for straining [27].

| 3.2.2 | Prokinetic Agents |

Different trials evaluated 5-HT4 agonists for CPD. *Cisapride* was first assessed in two open-label trials published by the same group, which defined constipation by prolonged CTT (calculated by using a 10x6 ROM protocol). At doses of 5 mg BID, the study of 20 PD patients showed a significant reduction of CTT (53.8 SD 4.74 to 30.4 SD 6.54), and reported a marked reduction of hard stools, abdominal fullness and straining after a week [21]. Nevertheless, when assessing long term improvement with doses of 10 mg BID in 25 patients with PD, there was a marked reduction of CTT after the first week of treatment (reduction in CTT from 131 to 81 h), although this response tended to wane at 6 months (99 h) and 12 months (118 h) [22]. It is worth noting the difference between baseline CTT times in both studies, as the same CTT protocol was used, and patients appear to be clinically similar (age 64.4 SD 10.45, Webster score 16.9 SD 3.9 in Jost [22] vs. age 67.5, range 52–85 years; Webster score 17, range 12–23).

Mosapride effectiveness at a dose of 15 mg/day was also assessed in an open-label trial in 14 patients with PD defined by a questionnaire-based assessment of pelvic organ dysfunction in PD. Patients were evaluated using CTT (using a 20 × 6 ROM protocol) and rectoanal manometry. Epigastric discomfort was reported by one patient as an adverse event. Reported subjective improvement in decreased BM frequency (>3 BM/w), difficulty defecating (less time, less strain, less digitation). Similar to the previous trials, the authors found a significant reduction in CTT (110.8–73.1 h).

Also, rectoanal videomanometry evidenced a reduction in volume to first rectal distention sensation (136–94 mL), and rectal compliance and maximum rectal capacity, with a nonsignificant increase in amplitude of phasic rectal contraction [28].

Tegaserod, a 5-HT4 partial agonist, was evaluated in a small RCT involving eight PD patients in the treatment arm and seven in the placebo arm, being constipation defined by ROME III criteria. Patients were first evaluated during a 1 week period, followed by a 4 week intervention period. The authors used the Subjects Global Assessment [29], for the evaluation of abdominal discomfort/pain, symptom relief, bowel habits, satisfaction with bowel habits, bloating, stool frequency and straining; even though this scale was validated for symptoms associated with irritable bowel syndrome. There were no statistically significative findings regarding these measures, although there was a trend for decreased constipation ($p = 0.14$), improvement in SGA satisfaction ($p = 0.1$), and total SGA ($p = 0.1$) vs. placebo [25].

Though it was not reported, or explicitly looked for in any of the mentioned trials, tegaserod and cisapride have since been withdrawn due to concerns on cardiovascular function.

An H2 antagonist usually used for peptic-ulcer disease, *nizatidine*, was evaluated in an open-label trial of 20 PD patients, out of which 12 complained of constipation (defined by less than 3BM per week and straining), and evaluated with CTT (20×6 ROM protocol) at baseline and week 12. The study reports that there was a subgroup of patients who showed normal CTT, although it does not specify if those were the same who did not complain of constipation. In patients with abnormal CTT, there was a significant reduction compared to baseline (94.7–67 h) [30].

A previously mentioned trial assessed a ghrelin agonist, *relamorelin*, in an RCT in which constipation was defined by modified ROME-III criteria (it was considered mandatory to have fewer than 3 BM/week). Initially, 37 patients met inclusion criteria, but during a first period of 2 weeks placebo intervention, 19 patients were excluded (ten of which due to not meeting BM frequency); the remaining 18 patients continued through a 2 week intervention period (ten receiving treatment, eight placebo). The results showed no significant differences between the relamorelin and placebo groups with respect to bowel movement outcomes [16].

3.2.3 Secretagogues

Lubiprostone, a chloride-channel activator, which enhances fluid secretion to the intestinal lumen, was assessed in a RCT that included a total of 52 patients (25 in the treatment group vs. 27 in the placebo group), defining constipation by ROME III criteria, plus a score of 10 in the ROME II constipation assessment. Patients were first evaluated through a 2 week pretreatment period and after a 4 week intervention period, by the use of a BM diary, reported use of laxatives and a visual analog scale. Loose stools were reported as an adverse event by 48% of patients, although severity was rated as mild, the events were self-limited and caused no withdrawals. Statistically significant findings were improvement in global impression of change, BM frequency (0.75 ± 0.8 to 0.97 ± 0.88 BM per day) vs. placebo (0.84 ± 0.76–0.83 ± 0.76), visual analog scale score (51.4 ± 8.5–71.2 ± 16.6 treatment group vs. 50.7 ± 5.9–56.8 ± 13.0 in placebo group), constipation questionnaires (no further information on this outcome is reported) [31].

Domperidone, a dopaminergic antagonist that does not cross the blood-brain barrier and exerts prokinetic effects primarily on the upper GI tract was evaluated for CPD in an open-label trial of 11 patients with constipation defined as abdominal bloating and self-reported constipation (not otherwise defined). Patients were evaluated at baseline, at week 2, and during a variable period of follow-up from 2 to 6 years, by the use of a gastrointestinal symptom questionnaire that included self-reported severity of nausea, vomiting, anorexia, abdominal bloating, heartburn, dysphagia, regurgitation, and constipation. Results revealed no effect of domperidone in constipation or dysphagia, with a significant reduction in the rest of the symptoms (although no further information is offered in respect to this) [32].

3.2.4 Dopaminergic Medication

The effect of LDOPA on CPD was evaluated in an open-label study including 19 PD patients with constipation defined by the use of a questionnaire assessing BM frequency and difficulty with defecation. Recently diagnosed patients were evaluated at baseline and after 12 weeks after receiving LDOPA/carbidopa 200 mg/20 mg BID with CTT (20 × 6 ROM protocol) and anorectal manometry. Significant changes were detected in anorectal function, mainly in lessening of first sensation to rectal distention (178–121 mL), less amplitude in paradoxical sphincter contraction upon defecation (29 to −7.1 cmH$_2$O), and reduced postdefecation residuals (142–54 mL) [24].

3.2.5 Botulinum Toxin

The injection of *botulinum toxin type A* into the puborectalis muscle was evaluated in an open-label trial in 18 patients. Constipation was defined by a combination of subjective and objective measures, including (1) incomplete, prolonged and difficult evacuation with constant use of enemas, laxatives, and manual maneuvers to facilitate bowel movement, (2) number of evacuations less than three per week, (3) failure to relax perineal floor during straining at physical examination, (4) inability to achieve evacuation of barium paste during defecography, with lack of a measurable increase in the anorectal angle between rest and attempted evacuation, (5) increased activity of the puborectalis muscle at electromyography performed with needle electrode, (6) high pressure levels during straining at anorectal manometry. Initially, 100 units were injected into the puborectalis muscles guided by transrectal ultrasonography. Improvement in symptoms was reported by eight patients at 4 weeks, and ten patients at 8 weeks. The eight patients who failed to respond received a dose of 200–300 units, with further improvement in four patients. Anorectal manometry evidenced a decrease in pressure during straining, and defecography showed a reduced anorectal angle during straining. There are no clear indicators or measures regarding what was considered as "symptom improvement" [33].

3.2.6 Ongoing Trials

There are currently two RCT undergoing recruitment, one of which is a phase 1b trial to evaluate *RQ-10* for gastroparesis and CPD, as defined by the regular need of medication for constipation, a score of 2 or higher on the Gastroparesis Cardinal Symptoms Index, or a score of 5 or higher on the gastroparesis or constipation subscale of the Gastrointestinal Symptoms in Neurodegenerative Disease Scale (NCT02838797).

The RAMSET study (Evaluation of safety and tolerability of *ENT-01* or the treatment of Parkinson's Disease Related Constipation) is a phase 1/2a in PD patients with constipation for over 6 months, unresponsive to milk of magnesia, and requiring at least weekly treatment using an oral laxative, stool softener, bulking agent, and/or a suppository, and dissatisfaction with current treatment, and at least two of the ROME-IV criteria (NCT03047629). Besides safety, tolerability and dosing, secondary outcomes will assess frequency of BM through a period of 11 weeks at each dose increment.

3.3 Instrumental

In one randomized uncontrolled trial, comparing *abdominal massage plus lifestyle advice* (fluid and dietary intake, physical activity, correct defecation position) versus lifestyle advice only. Patients were instructed on abdominal massage technique and/or lifestyle modifications during a 6 week period. Sixteen patients were enrolled in each group. Data was compared between baseline, after intervention period, and 4 week after. Constipation symptoms were assessed using the GSRS, NBDS and CSS. A caveat of this study, is that at baseline, the control group was not comparable to the intervention group in that, regarding outcomes related to constipation, they had lower scores. Related to baseline, both groups had a nonsignificative reduction on constipation symptoms, with no significative differences between groups [34].

An open-label trial assessed the efficacy of twice-daily *magnetic stimulation* on T9 and L3 spinal levels in 16 patients. Objective measures such as CTT and defecography, as well as the KESS score were taken at baseline and week 3. Authors found significant decrease in scores of: frequency of BM, unsuccessful evacuation, enema/digitations, relative difficulty of evacuation, time needed for defecation, CTT, anorectal angles, changes in pelvic floor muscles and amount of barium remaining after evacuation. Nevertheless, it should be pointed that the use of sham-procedure could have been introduced as a comparator [35].

There is a registered open-label trial, completed in 2009, with no published results, for use of *biofeedback* for CPD, by the use of surface EMG, perianal sensors and biofeedback PC equipment. No information is available regarding the number of patients enrolled or the criteria used to define constipation (NCT00869830).

Another open-label trial currently undergoing recruitment is set out to evaluate the effect of *osteopathic manipulative medicine* for CPD. Constipation will be diagnosed according to ROME-III criteria and outcomes measured using the CSS and stool consistency and the PAC-SYM questionnaire, which assesses self-reported severity of abdominal discomfort, pain and bloating; abdominal cramps, painful BM, rectal discomfort related to defecation, incomplete BM, stool consistency, straining, and unsuccessful attempts (NCT02344485).

3.4 Diet

In a more recent RCT [17], a formulation of fermented milk, containing multiple *probiotic strains and prebiotic fiber* compared to placebo for the treatment of CPD was assessed. This larger study included a total of 120 patients (80 allocated to treatment vs. 40 in the placebo arm). Initially eligible PD patients fulfilling modified ROME III criteria (were fewer than 3 complete BM per week and less than 6 BM per week, for a period of 2 weeks), entered a 2 week phase in which these criteria were evaluated by the means of a stool diary. After this period, 43 out of the initially eligible patients were excluded for failing to meet constipation criteria. During the treatment period five patients discontinued the intervention: three in the treatment group (one disliked the product, one due to abdominal discomfort and one for requiring the use of antibiotics due to a medical condition) and two in the placebo group (one disliked the product, one due to abdominal discomfort). Authors found that the probiotic and fiber compound significatively increased the number of complete BM, the total number of BM, improved stool consistency and patients achieve a reduction in the use of laxatives.

3.5 Surgery

Only one reported study has evaluated the effect of surgery (specifically *subthalamic DBS*) on CPD. This was a cohort study, originally devised to assess motor and nonmotor symptoms after said intervention [36]. CPD is vaguely defined as a nominal variable extracted from clinical charts (presence or absence of constipation), with no further specifications. The authors found that, compared to baseline evaluation, constipation was significantly lower after 1 and 2 years of follow-up (66% vs. 50% vs. 40.9%), attributing this change to the reduction of dopaminergic medication, though no correlation was made between these two variables, nor other potentially influencing factors were accounted for.

4 Recommendations

Throughout this chapter we have described several pitfalls on the published trials in CPD that need to be urgently overcome in the design of future studies. (a) *Interdisciplinary approach.* Both movement disorders specialists and gastroenterologist are necessary to properly design and carry out a trial on CPD. (b) *Lack of a unified definition of constipation:* Most regrettable, a consensus on constipation definition in PD patients has not been reached. This critically influences patient inclusion and exclusion and precludes for a structured comparison of published literature. (c) *Definition of outcome measures and variables to be measured through the study.* Objective quantification of constipation by instrumental methods or validated scales and questionnaires for PD patients have not been always used and is a sine qua non requirement. (d) *Pretreatment period:* PD is an extremely fluctuating disorder, therefore an observation period, by the use of patient diaries, enables to accurately assess and validate patient reported symptoms and control for recall bias or Hawthorne effect prior to intervention. (e) *Confounding factors and Control group:* As for the evaluation of most of the non motor symptoms patient groups should be similar regarding duration of the disease, severity of PD (as measured by the Hoen and Yahr) and use of LDOPA and dopaminergic agonists (levodopa equivalent daily dose should be reported). Factors unrelated to PD, such as physical activity, hydration and fiber consumption should be accounted for (and reported in a patient diary), patients could receive specific indications to ensure homogeneity in this aspect. Preexisting medical conditions or the use of medication that could alter gastrointestinal motility should be considered before inclusion and recorded if occurring during the trial. In conclusion more work has to be done in order to clearly evaluate most of the clinical correlates and drug effects on CPD.

References

1. Stirpe P, Hoffman M, Badiali D, Colosimo C (2016) Constipation: an emerging risk factor for Parkinson's disease? Eur J Neurol 23:1606–1613. https://doi.org/10.1111/ene.13082

2. Knudsen K, Fedorova TD, Bekker AC et al (2017) Objective colonic dysfunction is far more prevalent than subjective constipation in Parkinson's disease: a colon transit and volume study. J Parkinsons Dis 7:359–367. https://doi.org/10.3233/JPD-161050

3. Simuni T, Caspell-Garcia C, Coffey CS et al (2018) Baseline prevalence and longitudinal evolution of non-motor symptoms in early Parkinson's disease: the PPMI cohort. J Neurol Neurosurg Psychiatry 89:78–88. https://doi.org/10.1136/jnnp-2017-316213

4. Adams-Carr KL, Bestwick JP, Shribman S et al (2016) Constipation preceding Parkinson's disease: a systematic review and meta-analysis. J Neurol Neurosurg Psychiatry

87:710–716.   https://doi.org/10.1136/jnnp-2015-311680

5. Braak H, Del Tredici K, Rüb U et al (2003) Staging of brain pathology related to sporadic Parkinson's disease. Neurobiol Aging 24:197–211.   https://doi.org/10.1016/S0197-4580(02)00065-9

6. Hawkes CH, Del Tredici K, Braak H (2007) Parkinson's disease: a dual-hit hypothesis. Neuropathol Appl Neurobiol 33:599–614. https://doi.org/10.1111/j.1365-2990.2007.00874.x

7. Cersosimo MG, Benarroch EE (2012) Pathological correlates of gastrointestinal dysfunction in Parkinson's disease. Neurobiol Dis 46:559–564. https://doi.org/10.1016/j.nbd.2011.10.014

8. Pont-Sunyer C, Hotter A, Gaig C et al (2015) The onset of nonmotor symptoms in parkinson's disease (the onset pd study). Mov Disord 30:229–237. https://doi.org/10.1002/mds.26077

9. Everhart JE, Go VLW, Johannes RS et al (1989) A longitudinal survey of self-reported bowel habits in the United States. Dig Dis Sci 34:1153–1162. https://doi.org/10.1007/BF01537261

10. Towers AL, Burgio KL, Locher JL et al (1994) Constipation in the elderly: influence of dietary, psychological, and physical factors. J Am Geriatr Soc 42:701–706. https://doi.org/10.1111/j.1532-5415.1994.tb06527.x

11. Ueki A, Otsuka M (2004) Life style risks of Parkinson's disease: association between decreased water intake and constipation. J Neurol 251:18–23. https://doi.org/10.1007/s00415-004-1706-3

12. Caplan LH, Jacobson HG, Rubinstein BM, Rotman MZ (1965) Megacolon and volvulus in Parkinson's disease. Radiology 85:73–79. https://doi.org/10.1148/85.1.73

13. Rosenthal M, Marshall CE (1987) Sigmoid volvulus in association with Parkinsonism. Report of four cases. J Am Geriatr Soc 35:683–684

14. Evatt ML, Chaudhuri KR, Chou KL et al (2009) Dysautonomia rating scales in Parkinson's disease: Sialorrhea, dysphagia, and constipation – critique and recommendations by movement disorders task force on rating scales for Parkinson's disease. Mov Disord 24:635–646. https://doi.org/10.1002/mds.22260

15. Ashraf W, Park F, Lof J, Quigley EMM (1996) An examination of the reliability of reported stool frequency in the diagnosis of idiopathic constipation. Am J Gastroenterol 91:26–32

16. Pfeiffer RF, Parkinson T, Group S (2017) A randomized trial of relamorelin for consti-pation in Parkinson's disease (MOVE-PD): trial results and lessons learned. Parkinsonism Relat Disord 37:101–105. https://doi.org/10.1016/j.parkreldis.2017.02.003

17. Barichella M, Pacchetti C, Bolliri C et al (2016) Probiotics and prebiotic fiber for constipation associated with Parkinson disease. Neurology 87:1274–1280. https://doi.org/10.1212/WNL.0000000000003127

18. Ervin CM, Fehnel SE, Baird MJ et al (2014) Assessment of treatment response in chronic constipation clinical trials. Clin Exp Gastroenterol 7:191–198. https://doi.org/10.2147/CEG.S58321

19. Knudsen K, Krogh K, Østergaard K, Borghammer P (2017) Constipation in parkinson's disease: subjective symptoms, objective markers, and new perspectives. Mov Disord 32:94–105. https://doi.org/10.1002/mds.26866

20. Palit S, Thin N, Knowles CH et al (2016) Diagnostic disagreement between tests of evacuatory function: a prospective study of 100 constipated patients. Neurogastroenterol Motil 28:1589–1598. https://doi.org/10.1111/nmo.12859

21. Jost WH, Schimrigk K (1993) Cisapride treatment of constipation in parkinson's disease. Mov Disord 8:339–343. https://doi.org/10.1002/mds.870080315

22. Jost WH, Schimrigk K (1997) Long-term results with cisapride in Parkinson's disease. Mov Disord 12:423–425. https://doi.org/10.1002/mds.870120324

23. Ashraf W, Pfeiffer RF, Park F et al (1997) Constipation in Parkinson's disease: objective assessment and response to Psyllium. Mov Disord 12:946–951. https://doi.org/10.1002/mds.870120617

24. Tateno F, Sakakibara R, Yokoi Y et al (2011) Levodopa ameliorated anorectal constipation in de novo Parkinson's disease: the QL-GAT study. Parkinsonism Relat Disord 17:662–666. https://doi.org/10.1016/j.parkreldis.2011.06.002

25. Sullivan KL, Staffetti JF, Hauser RA et al (2006) Tegaserod (Zelnorm) for the treatment of constipation in Parkinson's disease. Mov Disord 21:115–116. https://doi.org/10.1002/mds.20666

26. Eichorn TE, Oertel WH (2001) Macrogol 3350-electrolyte improves constipation in Parkinson's disease and multiple system atrophy. Mov Disord 16:1176–1177. https://doi.org/10.1002/mds.1204

27. Zangaglia R, Martignoni E, Glorioso M et al (2007) Macrogol for the treatment of

constipation in Parkinson's disease. A random-ized placebo-controlled study. Mov Disord 22:1239–1244. https://doi.org/10.1002/mds.21243

28. Liu Z, Sakakibara R, Odaka T et al (2005) Mosapride citrate, a novel 5-HT4 agonist and partial 5-HT3 antagonist, amerliorates consti-pation in Parkinsonian patients. Mov Disord 20:680–686. https://doi.org/10.1002/mds.20387

29. Thompson WG, Longstreth GF, Drossman DA et al (1999) Functional bowel disorders and functional abdominal pain. Gut 45:ii43. https://doi.org/10.1136/gut.45.2008.ii43

30. Sakakibara R, Doi H, Sato M et al (2015) Nizatidine ameliorates slow transit constipa-tion in Parkinson's disease. J Chem Inf Model 63:399–400. https://doi.org/10.1017/CBO9781107415324.004

31. Ondo WG, Kenney C, Sullivan K et al (2012) Placebo-controlled trial of lubiprostone for constipation associated with Parkinson dis-ease. Neurology 78:1650–1654. https://doi.org/10.1212/WNL.0b013e3182574f28

32. Soykan I, Sarosiek I, McCallum RW (1997) The effect of chronic oral domperidone ther-apy on gastrointestinal symptoms, gastric emptying, and quality of life in patients with gastroparesis. Am J Gastroenterol 92:976–980

33. Cadeddu F, Bentivoglio AR, Brandara F et al (2005) Outlet type constipation in Parkinson's disease: results of botuli-num toxin treatment. Aliment Pharmacol Ther 22:997–1003. https://doi.org/10.1111/j.1365-2036.2005.02669.x

34. McClurg D, Hagen S, Jamieson K et al (2016) Abdominal massage for the alleviation of symptoms of constipation in people with Parkinson's: a randomised controlled pilot study. Age Ageing 45:299–303. https://doi.org/10.1093/ageing/afw005

35. Chiu CM, Wang CP, Sung WH et al (2009) Functional magnetic stimulation in consti-pation associated with Parkinson's disease. J Rehabil Med 41:1085–1089. https://doi.org/10.2340/16501977-0456

36. Zibetti M, Torre E, Cinquepalmi A et al (2007) Motor and nonmotor symptom fol-low-up in Parkinsonian patients after deep brain stimulation of the subthalamic nucleus. Eur Neurol 58:218–223. https://doi.org/10.1159/000107943

37. Drossman D, Dumitrascu DL (2006) Rome III: new standard for functional gastrointestinal disorders. J Gastrointest liver Dis 15:237–241

38. Agachan F, Chen T, Pfeifer J et al (1996) A constipation scoring system to simplify evalua-tion and management of constipated patients. Dis Colon Rectum 39:681–685. https://doi.org/10.1007/BF02056950

39. Sakakibara R, Odaka T, Lui Z et al (2005) Dietary herb extract dai-kenchu-to amelio-rates constipation in parkinsonian patients (Parkinson's disease and multiple system atro-phy) [3]. Mov Disord 20:261–262. https://doi.org/10.1002/mds.20352

40. Sakakibara R, Shinotoh H, Uchiyama T et al (2001) Questionnaire-based assessment of pel-vic organ dysfunction in Parkinson's disease. Auton Neurosci Basic Clin 92:76–85. https://doi.org/10.1016/S1566-0702(01)00295-8

41. Krogh K, Christensen P, Sabroe S, Laurberg S (2006) Neurogenic bowel dysfunction score. Spinal Cord 44:625–631. https://doi.org/10.1038/sj.sc.3101887

42. Knowles CH, Eccersley J, Scott M et al (2000) Linear discriminant analysis of symp-toms in patients with chronic constipation: validation of a new scoring system (KESS). Gastroenterology 118:A1181. https://doi.org/10.1016/s0016-5085(00)80553-3

43. Svedlund J, Sjödin I, Dotevall G (1988) GSRS-A clinical rating scale for gastrointesti-nal symptoms in patients with irritable bowel syndrome and peptic ulcer disease. Dig Dis Sci 33:129–134. https://doi.org/10.1007/BF01535722

INDEX

Santiago Perez-Lloret (ed.), *Clinical Trials In Parkinson's Disease*, Neuromethods, vol. 160,
https://doi.org/10.1007/978-1-0716-0912-5, © Springer Science+Business Media, LLC, part of Springer Nature 2021

Printed in the United States
by Baker & Taylor Publisher Services